Demographic Methods across the Tree of Life

T0202348

Demographic Methods across the Tree of Life

EDITED BY

Roberto Salguero-Gómez

Associate Professor in Ecology, University of Oxford, UK
Tutorial Fellow in Ecology, Pembroke College, UK
Guest Researcher, Max Planck Institute for Demographic Research, Germany
Honorary Researcher, University of Queensland, Australia

Marlène Gamelon

Chargée de Recherche, Centre National de la Recherche Scientifique (CNRS), University of Lyon, France
Researcher, Norwegian University of Science and Technology, Norway

OXFORD
UNIVERSITY PRESS

UNIVERSITY PRESS

Great Clarendon Street, Oxford, OX2 6DP,
United Kingdom

Oxford University Press is a department of the University of Oxford.
It furthers the University's objective of excellence in research, scholarship,
and education by publishing worldwide. Oxford is a registered trade mark of
Oxford University Press in the UK and in certain other countries

Published in the United States of America by Oxford University Press
198 Madison Avenue, New York, NY 10016, United States of America

British Library Cataloguing in Publication Data
Data available

Library of Congress Control Number: 2021936265

ISBN 978-0-19-883860-9 (hbk.)
ISBN 978-0-19-883861-6 (pbk.)

DOI: 10.1093/oso/9780198838609.001.0001

Printed: Printed and bound by
CPI Group (UK) Ltd, Croydon, CR0 4YY

Cover image: d3sign/Getty Images (front main) and Cecilie Sonsteby/Getty Images (back)

Foreword

Demography is at the heart of biology. Populations change in size over time due to temporal variation in demographic rates, and spatial variation in population dynamics is due to different demographic rates being observed across populations of the same species at different locations. Community dynamics are determined by the demographic rates of constituent species. Evolution occurs when different genotypes have different demographies. A good biologist consequently needs to have a firm grasp of demography.

Despite the central role of demography in biology, it is not an easy subject to master for two reasons. First, demographic data are neither cheap nor easy to collect. The most detailed data come from repeated captures and observations of individuals that can be individually identified. Not all species are amenable to such study—try collecting such data on *Caenorhabditis elegans* or *Anopheles gambiae* in the wild—and those species that are suitable for such study are often long-lived, surviving for many years. Studying evolution in such long-lived species is unfeasible, as it can take numerous consecutive research grants, or even careers, to gain robust insight. Fortunately, there are approaches being developed to collect demographic data without marking individuals that can be used to construct powerful demographic models.

The second reason that demography is a difficult subject is because it is mathematical. All populations are structured, be it by genotypes, phenotypic traits, strategies, age, sex, life history stage, or some other quantity. Modelling structured population usually requires matrices, and the analysis of matrices requires, at the bare minimum, an understanding of linear algebra. When models become frequency-dependent, and include trait-mediated interactions, two sexes, or multiple interacting species, analytical methods become even more challenging.

These challenges should not deter you from studying demography, because the rewards from mastering the subject are myriad. Demography has become popular in biology in recent years. The first reason for this popularity is in part because it is now widely accepted as a powerful way to study eco-evolution, evolution in structured populations, life history evolution, population dynamics, species interactions, and epidemiology. The reward from mastering the maths is worth it. The second reason is that the development of integral projection models (IPMs) that allow the study of continuous phenotypic traits has solved a challenge that faced demography for many years—how to study quantitative characters without arbitrarily splitting them into wide, discrete bins. IPMs allow researchers to go seamlessly from the identification of statistical patterns in data to demographic models. The third reason for the rise of demography is the technological advances that now allow increased ease of data collection from wild populations. These methods include miniaturised tags, ruggedised lower-power wireless sensor networks, cheaper and more accurate satellite tags for animals, and even automated traps that target which individuals to live trap.

I imagine that when I am asked to write the foreword for the second edition of this book, I have no doubt that the fourth reason I will list will be this tome. It is a very well-structured book that covers not only data collection but also model construction and analysis, and the contributors are the very people who have advanced the field of demography so substantially in the last decade. The book succeeds, where so many fail, in not only providing an accessible introduction to the field for those who have not been immersed in it for years, but also containing some state-of-the-art methods and applications. The editors, who are leaders of their generation in the field of demography, have put together a very thoughtful document: the book has something for everyone, from the novice to the demographic wizard.

Part of demography is about prediction. As a demographer, it would consequently be remiss of me to fail to make a prediction. So here it is: within a year all good biologists will have this book, well-thumbed, centre-stage on their book case.

Tim Coulson
Professor of Zoology
University of Oxford, UK

Acknowledgements

The community of individuals to whom we are grateful for their impact on this book is as heterogeneous as the diversity of questions, methods, and applications of demography itself.

First and foremost, this attempt to bring together demographic methods across the Tree of Life would have been impossible without the excellent contributions of all 77 authors. They brought a depth, breadth, and clarity in explanations that we feel confident will be extremely useful to the demographic community and beyond. The diligence and resilience of these authors—even through a pandemic—kept us focused on the delivery of their individual contributions and the book. Likewise, we are indebted to the 44 anonymous reviewers who provided excellent, timely, and on-time suggestions to improve the content and stylistic presentation of each chapter.

The inception of this book started almost as a matter of unconscious intellectual development during our respective postdoctoral times at the Max Planck Institute for Demographic Research (Germany) for Roberto Salguero-Gómez and at the Centre for Biodiversity Dynamics (CBD), Norwegian University of Science and Technology (NTNU) (Norway) for Marlène Gamelon. The exposure to demographers working on a wide range of systems and with an even broader repertoire of methodologies and questions encouraged us to test whether those methods would be applicable within and beyond our specific study systems. We owe a great deal of inspiration to Ruth Archer, Oskar Burger, Brenda Casper, Hal Caswell, Fernando Colchero, Dalia Conde, David N. Koons, Steve Ellner, Steinar Engen, Jean-Michel Gaillard, Olivier Gimenez, Vidar Grøtan, Owen Jones, Eelke Jongejans, Hans de Kroon, Jean-Dominique Lebreton, Cory Merow, Sean McMahon, Jessica Metcalf, Jim Oeppen, Peter Petraitis, Brett K. Sandercock, Ralph Schaible, Alex Scheuerlein, Iain Stott, Bernt-Erik Sæther, James Vaupel, and Cyrille Violle for inspiring us to think beyond our particular one-species, one-population, one-method, one-question origins.

We warmly thank our past and present academic institutions: the University of Queensland (Australia) and the Universities of Sheffield and Oxford (UK) for Roberto, and the Centre d'Ecologie Fonctionnelle et Evolutive in Montpellier, the Laboratoire de Biométrie et Biologie Evolutive in Lyon (France), and the CBD in Trondheim (Norway) for Marlène. All these academic institutions, although in different countries, with different cultures and languages (and versions of English!), share common denominators: stimulating scientific environments, and open discussions in ecology, evolution, and conservation biology, from which we greatly benefitted while editing and writing this book. In this regard, we are extremely thankful for the intellectual, logistical, and emotional support provided by Aziz Aboobaker, Mike Bonsall, Stephanie Borrelle, Natalie Briscoe, Pol Capdevila, Dylan Childs, Tim Coulson, Shaun Coutts, Katrina Davis, John Dwyer, Jane Elith, Rob Freckleton, Ivar Herfindal, Matt Holden, Irja Ida Ratikainen, Aline M. Lee, E.J. Milner-Gulland, Maria Paniw, Henrik Pärn, Christophe Pélabon, Hugh Possingham, Mark Rees, Ben C. Sheldon, Jarle Tufto, and Pete Vesk.

The maturation of the corpus of this book was carried over numerous internet conference calls in different time zones. However, the whole thing originated from discussions between us and other colleagues at multiple international meetings. It is worth here acknowledging the most influential venue: the Evolutionary Demography Society (EvoDemoS) annual meetings. The intellectual exchange supported by those meetings allowed us to fine-tune the content of the book as well as to recruit invaluable contributions by excellent junior and established researchers in the field of demography. Interactions with EvoDemoS members, as well as members from the British Ecological Society and the Ecological Society of America have further allowed us to tailor this book for our target audience.

We also acknowledge the financial support provided by various funding agencies and schemes.

Roberto Salguero-Gómez was supported by an Australian Research Council DECRA fellowship while at the University of Queensland, and a NERC Independent Research Fellowship while at the University of Sheffield and University of Oxford. Further grants to him and his research group by the ARC, NERC, BBSRC, Royal Society, Oxford–Africa Funds, Oxford University Press John Fell Funds, and the Spanish Ministerio de Economía y Competitividad allowed for interactions with key collaborators, as well as developing and applying demographic methods to a diversity of species. Marlène Gamelon was supported by a grant of the French National Agency for Wildlife (formerly ONCFS - now OFB), the Research Council of Norway through its Centres of Excellence funding scheme to CBD, and the Centre National de la Recherche Scientifique (CNRS).

This work has greatly benefitted from the timely, supportive, and encouraging advice and guidance from the OUP team. Special thanks go to Ian Sherman, Charles Bath, and Bethany Kershaw.

Finally, we would like to thank the loved ones who have provided their input to this book while being very patient for the time that this book took away from them. Roberto Salguero-Gómez thanks Katrina Davis and Manon Salguero-Davis for their unconditional support, love, critical input, and patience. Marlène Gamelon dedicates this book to her family, who have always been supportive along the academic scientific path.

Roberto Salguero-Gómez
Marlène Gamelon

Contents

21 Demographic processes in socially structured populations   341

Maria Paniw, Gabriele Cozzi, Stefan Sommer, and Arpat Ozgul

22 Demographic methods in epidemiology   351

Petra Klepac and C. Jessica E. Metcalf

List of Contributors

Michelle E. Afkhami Assistant Professor, University of Miami, USA

Pia Backmann Post-Doctoral Researcher, Technische Universität Dresden, Germany

Uta Berger Professor, Technische Universität Dresden, Germany

Bryan A. Black Associate Professor, University of Arizona, USA

Christophe Bonenfant Chargé de Recherche, CNRS, University of Lyon, France

Carlo Giovanni Camarda Researcher, Institut National d/Études Démographiques, France

Pol Capdevila Research Associate, University of Bristol, UK

Hal Caswell Professor, University of Amsterdam, The Netherlands

Marie J. E. Charpentier Directrice de Recherche, CNRS, University of Montpellier, France

Dylan Z. Childs Senior Lecturer (Associate Professor), University of Sheffield, UK

Alan A. Cohen Associate Professor, University of Sherbrooke, Canada

Fernando Colchero Associate Professor, University of Southern Denmark, Denmark

Geoffrey S. Cook Assistant Professor, University of Central Florida, USA

Gabriele Cozzi Post-Doctoral Researcher in Movement Ecology, University of Zurich, Switzerland

Sarah Cubaynes Maître de Conférences, University of Montpellier, France

Johan Ehrlén Professor, Stockholm University, Sweden

Steinar Engen Professor, Norwegian University of Science and Technology, Norway

Margaret E. K. Evans Assistant Professor, University of Arizona, USA

Donald A. Falk Professor, University of Arizona, USA

Erola Fenollosa Post-Doctoral Researcher, University of Barcelona; Post-Doctoral Researcher, IRBio (Institut de Recerca de la Biodiversitat), Spain

Josh A. Firth Researcher, University of Oxford, UK

Jean-Michel Gaillard Directeur de Recherche, CNRS, University of Lyon, France

Simon Galas Professor, University of Montpellier, France

Marlène Gamelon Chargée de Recherche, CNRS, University of Lyon, France; Researcher, Norwegian University of Science and Technology, Norway

Courtney L. Giebink Graduate Fellow, University of Arizona, USA

Edgar J. González Associate Professor, Universidad Nacional Autónoma de México, Mexico.

Volker Grimm Professor, University of Potsdam; Senior Researcher, Helmholtz Centre for Environmental Research–UFZ, Germany

Caspar A. Hallmann Post-Doctoral Researcher, Radboud University Nijmegen, The Netherlands

David T. Iles Quantitative Wildlife Biologist, Canadian Wildlife Service, Environment and Climate Change Canada, Canada

Henrik Jensen Professor, Norwegian University of Science and Technology, Norway

Owen R. Jones Associate Professor, University of Southern Denmark, Denmark

Eelke Jongejans Assistant Professor, Radboud University Nijmegen, The Netherlands

Petra Klepac Assistant Professor, London School of Hygiene and Tropical Medicine, UK

Jonas Knape Associate Professor, Swedish University of Agricultural Sciences, Sweden

Stephanie Kramer-Schadt Professor, Technische Universität Berlin; Head of Department of Ecological Dynamics, Leibniz Institute for Zoo and Wildlife Research, Germany

David N. Koons Professor, James C. Kennedy Endowed Chair of Wetland and Waterfowl Conservation, Colorado State University, USA

Anna Kuparinen Professor, University of Jyväskylä, Finland

Christie Le Coeur Post-Doctoral Researcher, University of Oslo, Norway

Jean-François Lemaître Chargé de Recherche, CNRS, University of Lyon, France

Mathilde Le Moullec Post-Doctoral Researcher, Norwegian University of Science and Technology, Norway

Andreas Lindén Senior Scientist, Wildlife Ecology, Natural Resources Institute Finland (Luke), Finland

Suzanne T. E. Lommen Post-Doctoral Researcher, Leiden University and University of Fribourg, Germany

Maurizio Mencuccini Professor, ICREA (Catalan Institute for Research and Advances Studies), Barcelona; Professor, CREAF (Research Center in Ecology and Forestry Applications), Spain

Eric S Menges Program Director and Senior Researcher, Archbold Biological Station, USA

C. Jessica E. Metcalf Associate Professor, Evolutionary Biology and Public Affairs, Princeton University, USA

Sergi Munné-Bosch Professor, University of Barcelona, Spain

James D. Nichols Adjunct Professor, University of Florida, USA

Alina K. Niskanen Post-Doctoral Researcher, University of Oulu, Finland

Arpat Ozgul Associate Professor, University of Zurich, Switzerland

Maria Paniw Post-Doctoral Researcher, University of Zurich, Switzerland

Marie Pelé Chargée de Recherche, Lille Catholic University, France

Fanie Pelletier Professor, University of Sherbrooke, Canada

Guillaume Péron Chargé de Recherche, CNRS, University of Lyon, France

William K. Petry Assistant Professor, North Carolina State University, USA

Roger Pradel Directeur de Recherche, CNRS, University of Montpellier, France

Pedro F. Quintana-Ascencio Professor, Department of Biology, University of Central Florida, USA

Viktoriia Radchuk Senior Researcher, Leibniz Institute for Zoo and Wildlife Research, Austria

Julien P. Renoult Chargé de Recherche, CNRS, University of Montpellier, France

Myriam Richaud PhD, University of Montpellier, France

Victor Ronget Post-Doctoral Researcher, University of Paris, France

Sébastien Roques PhD, University of Montpellier, France

Roberto Salguero-Gómez Associate Professor, University of Oxford, UK; Tutorial Fellow, Pembroke College, UK; Guest Researcher, Max Planck Institute for Demographic Research, Germany; Honorary Researcher, University of Queensland, Australia

Bernt-Erik Sæther Director/Professor, Norwegian University of Science and Technology, Norway

Ana Sanz Aguilar Researcher, Ramon y Cajal Researcher, IMEDEA, Spain

Cédric Scherer Post-Doctoral Researcher, Leibniz Institute for Zoo and Wildlife Research, Austria

Emily L. Schultz Post-Doctoral Researcher, University of Arizona, USA

Richard Shefferson Associate Professor, University of Tokyo, Japan

Emily G. Simmonds Post-Doctoral Researcher, Norwegian University of Science and Technology, Norway

Steven Smith Head of Ecological Genomics, University of Veterinary, Austria

Stefan Sommer Research Assistant, University of Zurich, Switzerland

Iain Stott Senior Lecturer, University of Lincoln, UK

Cédric Sueur Maître de Conférences, University of Strasbourg, and Institut Universitaire de France, France

Giacomo Tavecchia Researcher, IMEDEA, Spain

Oldřich Tomášek Researcher, Institute of Vertebrate Biology of the Czech Academy of Sciences, Czechia; Researcher, Masaryk University, Czechia

Shripad Tuljapurkar Professor, Stanford University, USA

Yngvild Vindenes Associate Professor, University of Oslo, Norway

Marcel E. Visser Professor, Groningen University and Wageningen University, The Netherlands

Stefan J. G. Vriend PhD, Norwegian University of Science and Technology, Norway

Wenyun Zuo Researcher, Stanford University, USA

From lions, to lion's manes, to dandelions: how different types of demographic data and methods are selected and why

Roberto Salguero-Gómez and Marlène Gamelon

1 On the origins of this book

The three pillars upon which this book has been engineered are the diversity of demographic schedules found in nature, coupled with the common currencies of demography and the myriad of existing quantitative methods in demography. These pillars are present along the book and its 22 chapters in the following aspects:

1. High diversity in living forms: the vast amount of diversity across the Tree of Life is expressed at all levels of biological organisation, from genes to ecosystems. Individuals within a population, populations within species, and species within a community differ a great deal in terms of variation in their demographic schedules (i.e. rates of survival and reproduction through time). For instance, longevity varies by orders of magnitude, from days as in the case of some mayfly species (e.g. *Ephemera simulans*, Carey 2002) to millennia as in bristlecone pines (*Pinus longevata*, Lanner and Connor 2001). But why is there variation in demographic schedules across individuals, populations, and species? What are the implications of this demographic variation in ecology, evolution, and conservation/management contexts?

2. The common currencies of demography: Despite variation in demographic schedules, all individuals of all species are born, then develop, survive, reproduce, and, sooner or later, die. These *rates of life* (Latin '*vita*'), or *vital rates* of survival, development, reproduction, and recruitment, are common across all species, from bacteria to humans. Thus, similar demographic methods can accommodate all the species across the Tree of Life to tackle important questions regarding the universality of rules in ecology and evolution (Sutherland et al. 2013).

3. High diversity in quantitative methods: demography, through its very origin (see Part II), is well equipped to address all of the aforementioned questions and more (Metcalf and Pavard 2007), as it is strongly cemented in statistics and actuarial science. Demographic methods—as the reader will have an opportunity to learn from this book—allow one to accommodate a high variation in data quality and sources, to address fundamental questions in ecology, evolution, and conservation/management, for all species across the Tree of Life. From that angle, we introduce here a book that goes beyond taxonomic boundaries. The data and applications of most chapters include trans-kingdom comparisons that highlight the universality of said common currencies.

We realised that there was an empty niche in trying to bring together *Demographic Methods across the Tree of Life*. This was the initial aim of this book! Later, the beginnings of this book were *developed* via frequent conversations between us and with all of the contributing authors, as well as rigorous peer-reviewing, to

Roberto Salguero-Gómez and Marlène Gamelon, *From lions, to lion's manes, to dandelions: how different types of demographic data and methods are selected and why*. In: *Demographic Methods Across the Tree of Life*. Edited by Roberto Salguero-Gómez and Marlène Gamelon, Oxford University Press. © Oxford University Press (2021). DOI: 10.1093/oso/9780198838609.002.0001 make sure that our primary motivation was fulfilled: to provide a cohesive book that brings together field ecologists, ecological modellers, and practitioners working with animal and plant kingdoms. To reach this goal, the book is divided into three parts: demographic data collection (Part I), demographic analyses (Part II), and demographic applications (Part III). This phase of *growth/development* took some years and luckily ended with the *birth* of this book. We will do our best to slow down the *inevitable decline of performance with increasing ages* (but see Jones et al. 2014) by maintaining a repository containing the materials linked to this book to prevent it from becoming outdated. We now hope that this book's reproductive output will be high with the *recruitment* of the next generation of demographers, where questions, methods, and systems will not be siloed taxonomically or biogeographically.

2 Origins of demographic analyses: from humans to nonhuman animals, plants, fungi, and microbes

Classical demography is broadly concerned with the size, distribution, structure, and change of populations through time. *Demography* (Gr. *demo-*: people, *-graphy*: writing, description, or measurement) literally means 'description of people'. Not surprisingly, this science has long been centred on the study of human populations (Preston et al. 2000). It is worth noting that such descriptions of people and their populations have been greatly enhanced through the partnering of human demographers (self-named 'formal demographers', Hinde 1998) with statisticians and actuarial scientists (i.e. researchers examining ageing patterns). Although demographic tools were initially developed to examine human populations, these tools have been progressively applied to nonhuman species. For instance, competing risk statistical models (Crowder 2001; Heisey and Patterson 2006), used to estimate age- and cause-specific mortality probability when the fate of an individual (i.e. alive or dead) and the cause of death (e.g. cancer, car accident) are precisely known, have been used in medicine, human demography, and more broadly on captive animals (Southey et al. 2004) and on plant populations (Shefferson and Roach 2013).

Though these kinds of demographic analyses are possible in other taxonomic corners of the Tree of Life—perhaps because the data are scarcer—the applicability of the tool is very much there. To further build on our example on mortality probability, advances in radiotelemetry and GPS technologies now allow for the fate of wild animals to be known and the causes of death to be identified (Tomkiewicz et al. 2010).

However, in most wild populations, knowing the fate of an individual remains challenging. An individual that is alive is not necessarily detected by the observer (Barthold et al. 2016; Eikelboom et al. 2019). This component adds an additional layer of complexity, as identifying the cause of death is difficult because death is rarely observed. This example demonstrates why we need to develop new techniques for data acquisition, as well as new demographic methods for data analyses, across the Tree of Life to address key and timely questions in ecology, evolution, and conservation biology.

3 Why use demographic methods across the Tree of Life?

A myriad of questions can be addressed from demographic data collected *in natura* or in the lab when analysed with the appropriate demographic methods. This book provides a nonexhaustive review of the questions that can be tackled, ranging from the study of eco-evolutionary feedbacks (Chapter 20), to the assessment of the extinction risk in harvested populations and the best way to manage them (Chapters 17 and 19), through the evaluation of the effects of environmental disturbances and perturbations on short- and long-term population dynamics (Chapters 10 and 11), as well as the importance of demography in epidemic dynamics (Chapter 22). All these questions are particularly timely, given the ongoing context of global change where the Earth has entered into the sixth major extinction event of its history (Lewis and Maslin 2015) and where we are all experiencing a pandemic!

The development and use of demographic tools for the analysis of demographic data requires strong quantitative inclinations. However, the reliance of demography on quantitative methods does not mean that if the current holder of this book is not *good* at math, matrix calculus, statistics, or the like they should immediately close it. What it does mean, though, is that investing adequate time to become quantitatively trained (or collaborate with somebody who is) will undeniably result in important benefits in the understanding of, say, (1) through which evolutionary mechanisms populations of dandelions (*Taraxacum officinale*) became established and invasive in New Zealand (Vitousek et al. 1997), (2) the current ecological influences and socioecological implications of hunting by male lions (*Panthera leo*) in South Africa (Funston et al. 1998), or (3) the projected viability of the edible yet endangered lion's mane fungus (*Hericium erinaceus*) in the United Kingdom (Boddy et al. 2011).

The choice of examples described above regarding how demographic tools help biologists and managers understand the biology of three different species of *'lion'*, on three different topics, three different corners of the Tree of Life, three different locations around the world, and at three different temporal scales (past, present, and future) is not coincidental. Rather, these are part of a conscious effort—also represented in *every* chapter in this book—to highlight the fact that demographic tools need not be limited to biogeographic, temporal, and/or taxonomic silos. Tearing down those silos and improving cross-fertilisation among them has been the primary motivation for this book. Indeed, equipping oneself with robust quantitative methods can help the researcher describe a vast range of questions in ecology, evolution, and conservation biology. Their application, we argue, pertains to a broad range of creatures and ecosystems because of the universal currencies of demography: the vital rates of survival, development, reproduction, and recruitment. These universal currencies are shared across dandelions, lions, and lion's manes, as much as they are between bacteria, you, and us.

We believe that a significant number of questions, protocols, and methods can be applied across the entire spectrum of life, and that important research advances will emerge via the unification and standardisation of these questions, protocols, and analytical tools across taxa. This book, thus, aims to fulfil a need to unify both field and analytical demographic methods across all major life forms. From this unification, three historically separated disciplines (formal [human] demography, [nonhuman] animal demography, and rest of the Tree of Life [including plants and microbes] demography) will benefit in at least three foreseeable ways. First, by sharing questions and field/analytical methods, broader analyses (both globally and taxonomically) will emerge that will allow researchers to address general rules in ecology and evolution, and ultimately develop more cost-effective management plans for the eradication of invasive species and the preservation of endangered ones. Second, thanks to new techniques shared across taxa, a series of eco-evolutionary questions will be made available to a broader range of taxa. For instance, the improvement of pedigree analyses in areas that have historically not implemented them (e.g. plant and microbial biology) will allow novel eco-evolutionary questions to be tackled, questions addressed predominantly in animal populations so far (Wilson et al. 2011; Bonnet et al. 2019). Third, this book strengthens the link between data collectors and demographic modellers, thus bringing a tighter biological understanding to the statistical analyses and vice versa (Gimenez et al. 2013).

4 From data acquisition to demographic analyses

Techniques for data acquisition have been developed across the entire Tree of Life, from microbes, to fungi, plants, and animals. Several textbooks (Harper 1977; Elzinga et al. 2001) have done a wonderful job at detailing specific data collection methods. However, these volumes have focused on specific taxonomic groups (e.g. plants, Gibson 2002; nonhuman animals, Williams et al. 2002; or humans, Hinde 1998). The historical taxonomic focus of demographic textbooks is based on the rationale that certain protocols that are appropriate for animals (e.g. GPS tracking) are a priori not applicable for plants, and vice versa. In recent decades, though, the number of techniques for data acquisition has drastically increased thanks, in part, to the development of technologies such as micro radio-tracking unmanned aerial vehicles (i.e. drones, Hodgson et al. 2018) and satellites (Zhao et al. 2020; Chapter 5). These technologies hold great promise to unlock the potential of classical human demographic tools, or of large mammal demographic techniques to the whole of the Tree of Life (e.g. Pons and Pausas 2007). As a consequence of the development of techniques for data acquisition, demographic data have been collected for many taxa across the Tree of Life. This key advancement has given rise to a large number of open-access databases gathering information on vital rates and/or life history traits (Table 1). However, we note that more demographic works would be particularly welcome in microbes (Levin 1990), certain taxonomic groups in the plant kingdom (e.g. algae, Salguero-Gómez et al. 2014), and some groups with important ecological value in the animal kingdom (e.g. insects, Carey 2001), as well as even among some charismatic tetrapods (Conde et al. 2019).

The last decades have also witnessed an explosion of demographic analyses for a large range of taxa. For instance, key textbooks have elaborated on demographic tools (e.g. matrix population models, Caswell 2001; integrated population models, Kéry and Schaub 2011; Schaub and Kéry 2021; integral projection models, Easterling et al. 2000; Ellner et al. 2016; stochastic population dynamics, Lande et al. 2003; and a large diversity of analytical and modelling approaches, Williams et al. 2002; Murray and Sandercock 2020). However, these textbooks assume that the data have already been collected—or provide somewhat limited information about how to obtain them. In addition, some of them are now circa 20 years old and do not reflect how methods for data acquisition and analyses have improved rapidly in the last decades. As such,

Table 1 Some of the key data sources that contain open-access demographic data of species across the Tree of Life.

Data source	Description	Source
Amniote	Life history trait database of bird, mammal, and reptile species	http://onlinelibrary.wiley.com/doi/10.1890/15-0846R.1/abstract
AnAge	Animal ageing and longevity database	https://onlinelibrary.wiley.com/doi/full/10.1111/j.1420-9101.2009.01783.x
BIDDABA	Bird demographic database	https://link.springer.com/article/10.1007/s10336-010-0582-0
BTO survey data	British Trust for Ornithology	http://www.bto.org/research-data-services/data-services
DATLife	Life tables of animals, as well as records of longevity	https://datlife.org/
demography	Forecasting mortality, fertility, migration, and population data	https://cran.r-project.org/web/packages/demography/index.html
COMADRE	Matrix population models for over 500 animal species worldwide	https://compadre-db.org/Data/Comadre
COMPADRE	Matrix population models for ca. 1000 plant species worldwide	https://compadre-db.org/Data/Compadre
Coral Trait Database	Database containing, among other traits, life history trait information for over 1500 coral species	https://coraltraits.org
demography	Forecasting mortality, fertility, migration, and population data	https://cran.r-project.org/web/packages/demography/index.html
EURING databank	Ringing recovery data of bird species across Europe	https://euring.org/data-and-codes/euring-databank
europop	Historical populations of European cities, 1500–1800	https://cran.r-project.org/web/packages/europop/index.html
fishdata	A small collection of fish population data sets	https://cran.r-project.org/web/packages/fishdata/index.html
FishTraits	Ecological and life history traits of fishes of the United States	https://pubs.er.usgs.gov/publication/70156095
Global Assessment of Reptile Distributions	Trait and geographic data on lizards	https://onlinelibrary.wiley.com/doi/full/10.1111/geb.12773
Human Cause-of-Death Database	Continuous human data series of causes of mortality for 16 countries	https://www.causesofdeath.org
Human Fertility Database	Period and cohort fertility data for human populations from 32 countries	http://www.humanfertility.org
Human Life-Table Database	Collection of human population life tables for multiple years across 141 countries	https://www.lifetable.de/cgi-bin/index.php
Human Mortality Database	Human population size and mortality data for 41 countries	http://www.mortality.org
International Database of Longevity	Information about supercentenarians (humans of ages 110 and above) from 13 countries	https://www.supercentenarians.org
International Data Base	Demographic measures of human populations across 288 countries worldwide	https://www.census.gov/programs-surveys/international-programs/about/idb.html
LEDA	A database of life history traits of the Northwest European flora	https://besjournals.onlinelibrary.wiley.com/doi/10.1111/j.1365-2745.2008.01430.x
Life history trait database of European reptile species	A database of traits (e.g. activity, energy, movement) of European reptile species	https://datadryad.org/stash/dataset/doi:10.5061/dryad.hb4ht

Name	Description	Source
Living Planet Index	Trends of 27,232 natural populations from 4,784 species through time around the world	WWF International. 2012. Living Planet Report 2012. Pp. 160. WWF International, Gland, Switzerland
Longevity Records	Life spans for mammals, birds, amphibians, reptiles, and fish	https://www.demogr.mpg.de/longevityrecords/
MALDABBA	Age-specific vital rates and life history traits for 200 mammalian species	https://www.pnas.org/content/suppl/2020/03/18/1911999117.DCSupplemental
PADRINO	Database of IPMs for hundreds of animals and plants around the globe	https://github.com/levisc8/Padrino.github.io
Pantheria	Mammal life history database	https://ecologicaldata.org/wiki/pantheria
LTER	US Long Term Ecological Research Network, including long-term records of population size and individual records in some cases	https://lternet.edu
SCALES	Securing the conservation of biodiversity across administrative levels and spatial, temporal, and ecological scales—a database of species traits	http://scales.ckff.si/scaletool/?menu=6
Serengeti bird species occurrence, abundance and habitat	Georeferenced occurrences for 568 species from 1929 to 2017. Records contain feeding location, food source, distribution status, observation locality	https://esajournals.onlinelibrary.wiley.com/doi/full/10.1002/ecy.2919
Serengeti: survey of age structure in ungulates and ostrich	Sample counts from 1926 to 2018 of 13 ungulate species and 1 ostrich species	https://www.nature.com/articles/s41597-020-00701-0
SPI-Birds	Database of bird breeding and mark/capture information	https://nioo.knaw.nl/en/spi-birds
Traits	Freshwater biological traits database	https://www.epa.gov/risk/freshwater-biological-traits-database-traits

the present book should serve as a useful crossroads of data collection, analyses, and interpretation to a broad range of disciplines.

A key feature of any research programme is that its analyses and results should be reproducible (Goodman et al. 2016). The field of demography, due to its strong quantitative inclinations, has benefitted tremendously from the move towards textbooks with scripts and open-access packages. Caswell, for instance, in his (2001) seminal overview of matrix population models where a detailed account of the pertinent theory, construction, and interpretation of this key demographic tool is provided, also provided MatLab scripts that greatly helped readers apply this tool. Likewise, conservation biologists had a unique opportunity to apply demographic tools in the context of population viability analyses in the textbook by Morris and Doak (2002), where MatLab scripts were also made available. Even earlier than these two references, Ebert (1999) already provided detailed comments and code in BASIC for the analysis of structured population data in both animals and plants. Following that spirit of research reproducibility, and with the specific aim to provide our readers with key materials to adapt to their data sets and questions, the present textbook contains carefully annotated scripts in R (Team 2010) (a nonproprietary software platform that is widely used among ecologists, evolutionary biologists, and conservation biologists), as well as full sets of demographic data. These are available at www.oup.com/companion/SalgueroGamelonDM, and it is our intention to revisit them, update them, and add new applications, in an effort to keep up with the high speed of development of demography.

The last decades have witnessed a *Malthusian* increase in R libraries specifically targeted at helping ecologists, evolutionary biologists, and conservation biologists avail from demographic models. This emergence has been further supported by the recent appearance of journals (e.g. *Methods in Ecology and Evolution*, founded in 2010) that house a lot of software for demographic analyses (e.g. Colchero et al. 2012; Metcalf et al. 2013; Kettle et al. 2018; Shefferson et al. 2021). With the purpose of making the reader aware of the diversity of demographic tools at their disposal via R, we have compiled a not at all comprehensive and yet rather long list of R libraries for demographic analysis (Table 2). The main merit of these publications and their software is that they focus on keeping the mathematics and statistics behind these developments to a manageable (but important!) size, while often providing useful vignettes with examples regarding their execution with real-world data. However, there is still a significant gap

that researchers and practitioners must fill in, linking the acquisition of data, their analyses, and applications to their specific questions. An additional motivation of this book is to provide a broad springboard from which *any* biologist may be able to walk away with a good understanding of what demographic data to collect (Part I), how to analyse those data through a multitude of possible analytical approaches (Part II), and what kinds of applications in ecology, evolution, and conservation biology are pertinent (Part III).

5 Organisation of the book

This book will guide the readers, step by step, along 22 chapters organised into three main parts: data collection, analyses, and applications. Because we have aimed to find a common ground in demographic data collection, modelling, and questions that go beyond taxonomic boundaries, all the chapters do not specialise according to taxonomic affiliations and rather include diverse taxa.

Part I of this book naturally brings together six chapters that present how data can be collected at different levels of organisation, from genes to the environment, going through traits, individuals, and populations. Chapter 1 introduces the reader to the collection and consideration of genetic data in demographic analyses. Specifically, it provides a gentle introduction to pedigrees and phylogenies, now fundamental tools of demographic analyses. Chapter 2 details methods to collect information at the biochemical and physiological levels. Examples include the quantification of biomarkers, hormones, and proxies of oxidative stress. Chapter 3 introduces the reader to the collection of social data. It provides sampling procedures to describe fine-grained patterns of social interactions and initiate the reader to social network analyses. Chapters 4 and 5 divide the heavy load of demographic data collection into two categories. In Chapter 4, methods are introduced to collect demographic data from a single visit (or few visits) to a field site using tools that allow researchers to obtain a historical perspective, such as dendrochronological and sclerochronological methods. Chapter 5, on the other hand, overviews methods to track dynamics of individuals in populations, and whole populations, in mobile and sessile species, including the usage of new technologies such as small tracking devices and drones. Chapter 6 then serves as an excellent complement to contextualise the demography of the species of interest, as populations do not occur in isolation. This chapter overviews some of the key biotic and

Table 2 A nonexhaustive review of R packages developed in the last decades that have significantly improved the understanding and applicability of demographic tools by ecologists, evolutionary biologists, and conservation biologists.

Name of package	Brief description	Source
albopictus	Age-structured population dynamics models	https://cran.r-project.org/web/packages/albopictus/index.html
BaSTA	Age-specific survival analysis from incomplete capture-recapture/recovery data	https://cran.r-project.org/web/packages/BaSTA/index.html
BayeSPsurv	Bayesian spatial split population survival model	https://cran.r-project.org/web/packages/BayesSPsurv/index.html
befproj	Makes a local-population projection	https://cran.r-project.org/web/packages/befproj/index.html
bootstrapFP	Bootstrap algorithms for finite population inference	https://cran.r-project.org/web/packages/bootstrapFP/index.html
bssm	Bayesian inference of non-Gaussian state space models	https://cran.r-project.org/web/packages/bssm/index.html
BTSPAS	Bayesian time-stratified population analysis	https://cran.r-project.org/web/packages/BTSPAS/index.html
capm	Companion animal population management	https://cran.r-project.org/web/packages/capm/index.html
capwire	Estimates population size from noninvasive sampling	https://cran.r-project.org/web/packages/capwire/index.html
CARE1	Statistical package for population-size estimation in capture-recapture models	https://cran.r-project.org/web/packages/CARE1/index.html
colorednoise	Simulate temporally autocorrelated populations	https://cran.r-project.org/web/packages/colorednoise/index.html
conStruct	Models spatially continuous and discrete population genetic structure	https://cran.r-project.org/web/packages/conStruct/index.html
Cyclops	Cyclic coordinate descent for logistic, Poisson, and survival analysis	https://cran.r-project.org/web/packages/Cyclops/index.html
DPWeibull	Dirichlet process Weibull mixture model for survival data	https://cran.r-project.org/web/packages/DPWeibull/index.html
demogR	Analysis of age-structured demographic models	https://cran.r-project.org/web/packages/demogR/index.html
dendRoAnalyst	A tool for processing and analysing dendrometer data	https://cran.r-project.org/web/packages/dendRoAnalyst/index.html
dendrometeR	Analysing dendrometer data	https://cran.r-project.org/web/packages/dendrometeR/index.html
dendroTools	Linear and nonlinear methods for analysing daily and monthly dendroclimatological data	https://cran.r-project.org/web/packages/dendroTools/index.html
demography	Forecasting mortality, fertility, migration, and population data	https://cran.r-project.org/web/packages/demography/index.html
dynamichazard	Dynamic hazard models using state space models	https://cran.r-project.org/web/packages/dynamichazard/index.html
dynsurv	Dynamic models for survival data	https://cran.r-project.org/web/packages/dynsurv/index.html
FinePop	Fine-scale population analysis	https://cran.r-project.org/web/packages/FinePop/index.html
FREQ	Estimate population size from capture frequencies	https://cran.r-project.org/web/packages/FREQ/index.html
gauseR	Lotka-Volterra models for Gause's 'struggle for existence'	https://cran.r-project.org/web/packages/gauseR/index.html
IBMPopSim	Individual-based model population simulation	https://cran.r-project.org/web/packages/IBMPopSim/index.html
IPMpack	Analysis and interpretation of simple IPMs	https://cran.r-project.org/src/contrib/Archive/IPMpack/

continued

Table 2 *Continued*

Name of package	Brief description	Source
IPMR	Analysis and interpretation of simple and complex IPMs	https://github.com/levisc8/ipmr
jacpop	Jaccard index for population structure identification	https://cran.r-project.org/web/packages/jacpop/index.html
lefko3	Historical and ahistorical population projection matrix analysis	https://cran.r-project.org/web/packages/lefko3/index.html
lhmixr	Fit sex-specific life history models with missing classifications	https://cran.r-project.org/web/packages/lhmixr/index.html
LifeHist	Life history models of individuals	https://cran.r-project.org/web/packages/LifeHist/index.html
lmf	Functions for estimation and inference of selection in age-structured populations	https://cran.r-project.org/web/packages/lmf/index.html
LTRCforests	Ensemble methods for survival data with time-varying covariates	https://cran.r-project.org/web/packages/LTRCforests/index.html
metafolio	Metapopulation simulations for conserving salmon through portfolio optimisation	https://cran.r-project.org/web/packages/metafolio/index.html
microPop	Modelling microbial populations	https://cran.r-project.org/web/packages/microPop/index.html
missDeaths	Simulating and analysing time to event data in the presence of population mortality	https://cran.r-project.org/web/packages/missDeaths/index.html
MortalityLaws	Parametric mortality models, life tables, and HMD	https://cran.r-project.org/web/packages/MortalityLaws/index.html
MortalityTables	A framework for various types of mortality/life tables	https://cran.r-project.org/web/packages/MortalityTables/index.html
morse	Modelling tools for reproduction and survival data in ecotoxicology	https://cran.r-project.org/web/packages/morse/index.html
mptools	RAMAS metapop tools	https://cran.r-project.org/web/packages/mptools/index.html
MRsurv	A multiplicative-regression model for relative survival	https://cran.r-project.org/web/packages/MRsurv/index.html
ncappc	NCA calculations and population model diagnosis	https://cran.r-project.org/web/packages/ncappc/index.html
NEff	Calculating effective sizes based on known demographic parameters of a population	https://cran.r-project.org/web/packages/NEff/index.html
LexisPlotR	Plot Lexis diagrams for demographic purposes	https://cran.r-project.org/web/packages/LexisPlotR/index.html
npsurv	Nonparametric survival analysis	https://cran.r-project.org/web/packages/npsurv/index.html
OBMbpkg	Estimate the population size for the MB capture-recapture model	https://cran.r-project.org/web/packages/OBMbpkg/index.html
openCR	Open population capture-recapture	https://cran.r-project.org/web/packages/openCR/index.html
pec	Prediction error curves for risk prediction models in survival analysis	https://cran.r-project.org/web/packages/pec/index.html
plethem	Population life course exposure to health-effects modelling	https://cran.r-project.org/web/packages/plethem/index.html
pop	A flexible syntax for population dynamic modelling	https://cran.r-project.org/web/packages/pop/index.html
popbio	Construction and analysis of matrix population models	https://cran.r-project.org/web/packages/popbio/index.html
popdemo	Demographic modelling using projection matrices	https://cran.r-project.org/web/packages/popdemo/index.html
POPdemog	Plot population demographic history	https://cran.r-project.org/web/packages/POPdemog/index.html

PopED	Population and individual optimal experimental design	https://cran.r-project.org/web/packages/PopED/index.html
popEpi	Functions for epidemiological analysis using population data	https://cran.r-project.org/web/packages/popEpi/index.html
popkin	Estimate kinship and FST under arbitrary population structure	https://cran.r-project.org/web/packages/popkin/index.html
poppr	Genetic analysis of populations with mixed reproduction	https://cran.r-project.org/web/packages/poppr/index.html
poptrend	Estimate smooth and linear trends from population-count survey data	https://cran.r-project.org/web/packages/poptrend/index.html
PROSPER	Simulation of weed population dynamics	https://cran.r-project.org/web/packages/PROSPER/index.html
PSPManalysis	Analysis of physiologically structured population models	https://cran.r-project.org/web/packages/PSPManalysis/index.html
PCAClone	Population viability analysis with data cloning	https://cran.r-project.org/web/packages/PVAClone/index.html
pyramid	Draw population pyramid	https://cran.r-project.org/web/packages/pyramid/index.html
rangeMapper	A platform for the study of macro-ecology of life history traits	https://cran.r-project.org/web/packages/rangeMapper/index.html
Rage	Calculation of life history traits from matrix population models	https://github.com/jonesor/Rage
RCOMPADRE	Manipulation of data in the COMPADRE & COMADRE databases	https://github.com/jonesor/Rcompadre
ribd	Pedigree-based relatedness coefficients	https://cran.r-project.org/web/packages/ribd/index.html
RMark	Interface to the software package MARK that constructs input files and extracts the output (from capture-mark-recapture data)	https://cran.r-project.org/web/packages/RMark/index.html
rmetasim	An individual-based population genetic simulation environment	https://cran.r-project.org/web/packages/rmetasim/index.html
Rpoppler	PDF tools based on Poppler	https://cran.r-project.org/web/packages/Rpoppler/index.html
Rramas	Matrix population models	https://cran.r-project.org/web/packages/Rramas/index.html
rSHAPE	Simulated haploid asexual population evolution	https://cran.r-project.org/web/packages/rSHAPE/index.html
SMITIDstruct	Data structure and manipulations tool for host and viral population	https://cran.r-project.org/web/packages/SMITIDstruct/index.html
sspse	Estimating hidden population size using respondent-driven sampling data	https://cran.r-project.org/web/packages/sspse/index.html
SSsimple	State space models	https://cran.r-project.org/web/packages/SSsimple/index.html
stagePop	Modelling the population dynamics of a stage-structured species in continuous time	https://cran.r-project.org/web/packages/stagePop/index.html
statespacer	State space modelling in R	https://cran.r-project.org/web/packages/statespacer/index.html
steps	Spatially and temporally explicit population simulator	https://cran.r-project.org/web/packages/steps/index.html
survival	Survival analysis methods	https://cran.r-project.org/web/packages/survival/index.html
survivalmodels	Models for survival analysis	https://cran.r-project.org/web/packages/survivalmodels/index.html
survParamSim	Parametric survival simulation with parameter uncertainty	https://cran.r-project.org/web/packages/survParamSim/index.html
timereg	Flexible regression models for survival data	https://cran.r-project.org/web/packages/timereg/index.html

continued

Table 2 *Continued*

Name of package	Brief description	Source
trackdem	Particle tracking and demography	https://cran.r-project.org/web/packages/trackdem/index.html
treestructure	Detect population structure within phylogenetic trees	https://cran.r-project.org/web/packages/TSSS/index.html
TSSS	Time series analysis with state space models	https://cran.r-project.org/web/packages/TSSS/index.html
wiqid	Quick and dirty estimates for wildlife populations	https://cran.r-project.org/web/packages/wiqid/index.html
wpp2019	World population prospects 2019	World population prospects in 2019

abiotic factors that researchers might want to consider examining, together with the demographic data.

Part II of this book contains an overview of the most commonly used methods in demography across different species, and it contains eight chapters. These methodologies are data- and research question driven. Each chapter clearly states the type of questions that can be tackled, and the data required to address them. Time series of population counts are commonly collected in the field (see Chapter 5), and Chapter 7 provides precious methodological tools to analyse these types of data and get important insight on changes in population size over time. Chapter 8 introduces life tables, the simplest kind of structured population model, where individuals are organised and followed across their lifetime based on their ages. Chapter 9 overviews matrix population models, discrete time demographic models where individuals are classified into ages, developmental stages, and/or discrete size ranges. Chapter 10 introduces integral projection models (IPMs), discrete time structured population models where individuals are classified along continuous variables such as size. Common to life tables, matrix population models, and IPMs is the expectation that a population will go to stationary equilibrium if left under a constant environment. Of course, most populations are constantly affected by changes in a/biotic conditions. The transient (short-term) responses that result from these disturbances are the focus of Chapter 11. Chapter 12 delves into individual-based models. In it, the reader is invited to move towards mechanistic models, where demographic population trends can be explained by processes (e.g. local interactions, response to changing environments, and adaptive behaviour) occurring at the individual level. Chapter 13 deals with survival analyses. It provides a useful toolkit to estimate survival, a key vital rate in demographic studies, for a diversity of species, ranging from lab organisms to humans, birds, reptiles, and plants. Chapter 14 introduces the reader to integrated population models (another kind of IPM!), where various sources of demographic data (e.g. population counts, capture-recapture data) are amalgamated, together with biological knowledge on the life cycle of the species of interest, to estimate vital rates and population dynamics.

Part III in this book contains key applications of demographic methods to ecology, evolution, epidemiology, and conservation biology. This part contains eight chapters. Our way to engage with nonexperts on any one topic is by making explicit linkages to the questions that both managers and demographers are interested in, and by providing tangible examples in the field of population ecology, evolutionary demography, and conservation biology. Therefore, this last part is specifically dedicated to applications. Chapter 15 illustrates state-of-the-art methods in spatial demography. Specifically, the author nicely applies some of the demographic tools described above (e.g. in Chapters 9, 10, 12, 13, and 14) to show that demographic methods can inform the spatial response of individuals and populations to changing environments. Chapter 16 showcases central ideas and methods in evolutionary demography, a research field that has grown rapidly in recent years. This includes theories of postreproductive life and models of mutation and selection. Chapter 17 applies demographic theory to discuss reproductive value in a conservation biology context. Chapter 18 exploits the large amount of demographic data across the Tree of Life to compare and contrast life history traits and explore whether these are constrained by main ecological realms. Chapter 19 introduces the reader to adaptive management, a specific form of decision-making in conservation biology and wildlife management. Chapter 20 brings a more genetic perspective on demography by discussing heritability, polymorphisms, and population dynamics. It provides an eco-evolutionary framework and highlights the strong relationship between changes in phenotypic traits and population dynamics. Chapter 21 details applications of demographic tools for socially structured populations. It nicely exemplifies the use of social data (Chapter 3) and demographic methods (e.g. Chapters 9, 10, and 12) to model the dynamics of socially structured populations. Finally, Chapter 22 focuses on the interplay between demography and epidemiology. It illustrates how demography shapes infectious disease dynamics and how, in turn, infectious disease dynamics shape demography.

6 Target audience

This book is primarily targeted at researchers and practitioners. The authors have assumed that the readers of this book will be at an upper-level undergraduate level or higher. Our motivation to make this book as math-friendly as possible has guided the writing of all of the chapters under the following six principles:

1. Field/lab experience: We have made no assumptions about the knowledge of the reader regarding data collection. Instead, in Part I, we walk them step by step through field methods to collect pertinent demographic data.
2. Programming: As mentioned above, demography has quantitative inclinations, and these are best

satiated through programming (rather than point-and-click software that prevents the user from truly *understanding* what is under the hood, Farrell and Carey 2018). This book assumes that the user is familiar with basic R programming at an introductory level (e.g. how to load data, save outputs, and interpret error messages), but all pertinent R scripts are accompanied with line-by-line explanations.

3. Population ecology/formal demography: The book makes no assumptions about the background knowledge of the reader regarding demography, so basic and advanced concepts are introduced and discussed.
4. Evolution: The book contains a series of chapters on applications of demography in evolutionary biology (e.g. Chapters 16 and 20). Some assumptions are made regarding the evolutionary understanding of the reader, but these are satisfied by pointing them towards the relevant literature in each chapter.
5. Statistics: Methodological and applied chapters that rely on advanced statistics provide a gentle introduction to the concepts in use (e.g. Bayesian statistics). When pertinent, the chapters (e.g. Chapter 9) include conceptual boxes to better understand and solidify these concepts.
6. Mathematics: Demographic analyses require a good understanding of algebra. Chapters based on advanced mathematics introduce the associated topics with the relevant literature (Part II).

We see this book as a springboard that integrates primary data collection and cutting-edge demographic analyses and their applications across a broad range of topics. Naturally, one could have (and other authors have!) written a book on any of these topics (see references in this chapter). Instead, we encourage the reader to treat this book as a comprehensive introduction to the vast amount of demographic approaches in demography. They are, of course, encouraged to deepen their understanding by searching into more specialised textbooks and the pertinent peer-review literature.

7 What next? A plea for a stronger, richer, broader, and more innovative demography

And now that the reader is aware of our motivations for this book, the available online support, the structure of the parts, and the tools presented, we dare to make a plea. The vast availability of demographic methods means that one might fall into the comfort of doing what others have done, with a slight tweak, and on a different system. We challenge the reader not to treat

this book nor the discipline as a cookbook. True, we have provided some utensils (data and methods), as well as some recipes (in the shape of commented R scripts). We strongly encourage the readers to make these tools truly their own by moving towards innovative areas where none of us envisioned they could go. Readers of this book have a unique opportunity to contribute to the big picture: unless their questions and interests are taxonomically bound to a specific group of species (and we emphasise, there is nothing wrong with that), we firmly support the idea of applying these tools to areas of the planet and corners of the Tree of Life where not much is yet known, demographically speaking. After all, demography is everywhere, and demographic research goes well beyond the classical definition of a population (Griffith et al. 2016)

References

Barthold J. A., C. Packer, A. J. Loveridge, D. W. Macdonald, and F. Colchero (2016). Dead or gone? Bayesian inference on mortality for the dispersing sex. Ecology and Evolution **6**(14), 4910–4923.

Boddy L., M. E. Crockatt, and A. M. Ainsworth (2011). Ecology of *Hericium cirrhatum, H. coralloides* and *H. erinaceus* in the UK. Fungal Ecology **4**(2), 163–173.

Bonnet T., M. B. Morrissey, A. Morris, et al. (2019). The role of selection and evolution in changing parturition date in a red deer population. PLOS Biology **17**(11), 23.

Carey J. R. (2001). Insect biodemography. Annual Review of Entomology **46**, 79–110.

Carey J. R. (2002). Longevity minimalists: life table studies of two species of northern Michigan adult mayflies. Experimental Gerontology **37**(4), 567–570.

Caswell H. (2001). Matrix Population Models: Construction, Analysis, and Interpretation. Sinauer Associates.

Colchero F., O. R. Jones, and M. Rebke (2012). BaSTA: an R package for Bayesian estimation of age-specific survival from incomplete mark-recapture/recovery data with covariates. Methods in Ecology and Evolution **3**(3), 466–470.

Conde D. A., J. Staerk, F. Colchero, et al. (2019). Data gaps and opportunities for comparative and conservation biology. Proceedings of the National Academy of Sciences of the United States of America **116**(19), 9658–9664.

Crowder M. J. (2001). Classical Competing Risks. Chapman & Hall.

Easterling M. R., S. Ellner, and P. Dixon (2000). Size-specific sensitivity: applying a new structured population model. Ecology **81**(3), 694–708.

Ebert T. A. (1999). Plant and Animal Populations: Methods in Demography. Academic Press.

Eikelboom J. A. J., J. Wind, E. Van De Ven, et al. (2019). Improving the precision and accuracy of animal population estimates with aerial image object detection. Methods in Ecology and Evolution **10**(11), 1875–1887.

Ellner S. P., D. Z. Childs, and M. Rees (2016). Data-Driven Modelling of Structured Populations. Springer.

Elzinga C. L., D. W. Salzer, J. W. Willoughby, and J. P. Gibbs (2001). Monitoring Plant and Animal Populations: A Handbook for Field Biologists. Blackwell.

Farrell K. J. and C. C. Carey (2018). Power, pitfalls, and potential for integrating computational literacy into undergraduate ecology courses. Ecology and Evolution **8**(16), 7744–7751.

Funston P. J., M. G. L. Mills, H. C. Biggs, and P. R. K. Richardson (1998). Hunting by male lions: ecological influences and socioecological implications. Animal Behaviour **56**, 1333–1345.

Gibson D. (2002). Methods in Comparative Plant Population Ecology. Oxford University Press.

Gimenez O. , F. Abadi, J. Y. Barnagaud et al. (2013). How can quantitative ecology be attractive to young scientists? Balancing computer/desk work with fieldwork. Animal Conservation **16**(2), 134–136.

Goodman S. N., D. Fanelli, and J. P. A. Ioannidis (2016). What does research reproducibility mean? Science Translational Medicine **8**(341), 6.

Griffith A. B., R. Salguero-Gómez, C. Merow, and S. McMahon (2016). Demography beyond the population. Journal of Ecology **104**(2), 271–280.

Harper J. L. (1977). Population Biology of Plants. Academic Press.

Heisey D. M. and B. R. Patterson (2006). A review of methods to estimate cause-specific mortality in presence of competing risks. Journal of Wildlife Management **70**(6), 1544–1555.

Hinde A. (1998). Demographic Methods. Routledge.

Hodgson J. C., R. Mott, S. M. Baylis, et al. (2018). Drones count wildlife more accurately and precisely than humans. Methods in Ecology and Evolution **9**(5), 1160–1167.

Jones O. R., A. Scheuerlein, R. Salguero-Gómez, et al. (2014). Diversity of ageing across the tree of life. Nature **505**(7482), 169–173.

Kéry M. and M. Schaub (2011). Bayesian Population Analysis Using WinBUGS. Academic Press.

Kettle H., G. Holtrop, P. Louis, and H. J. Flint (2018). microPop: modelling microbial populations and communities in R. Methods in Ecology and Evolution **9**(2), 399–409.

Lande R., E. Steinar, and B.-E. Saether (2003). Stochastic Population Dynamics in Ecology and Conservation. Oxford University Press.

Lanner R. M. and K. F. Connor (2001). Does bristlecone pine senesce? Experimental Gerontology **36**(4–6) 675–685.

Levin B. R. (1990). Microbial ecology and population biology. In: J. A. Hansen (ed.), Environmental Concerns: An Inter-disciplinary Exercise, 177–189. Springer.

Lewis S. L. and M. A. Maslin (2015). Defining the Anthropocene. Nature **519**(7542), 171–180.

Metcalf C. J. E. and S. Pavard (2007). Why evolutionary biologists should be demographers. Trends in Ecology & Evolution **22**(4), 205–212.

Metcalf C. J. E., S. M. McMahon, R. Salguero-Gómez, and E. Jongejans (2013). IPMpack: an R package for integral projection models. Methods in Ecology and Evolution **4**(2), 195–200.

Morris W. F. and D. F. Doak (2002). Quantitative Conservation Biology: Theory and Practice of Population Viability Analysis. Sinauer Associates.

Murray L. and B. K. Sandercock (2020). Population Ecology in Practice. Wiley.

Pons J. and J. G. Pausas (2007). Acorn dispersal estimated by radio-tracking. Oecologia **153**(4), 903–911.

Preston S., P. Heuveline, and M. Guillot (2000). Demography: Measuring and Modeling Population Processes. Wiley.

Salguero-Gómez R., O. R. Jones, C. R. Archer, et al. (2014). The COMPADRE plant matrix database: an open online repository for plant demography. Journal of Ecology **103**(1), 202–218.

Schaub M. and M. Kéry (2021). Integrated Population Models: Theory and Ecological Applications with R and JAGS. Academic Press.

Shefferson R. P. and D. A. Roach (2013). Longitudinal analysis in Plantago: strength of selection and reverse age analysis reveal age-indeterminate senescence. Journal of Ecology **101**(3), 577–584.

Shefferson R. P., S. Kurokawa, and J. Ehrlén (2021). LEFKO3: analysing individual history through size-classified matrix population models. Methods in Ecology and Evolution **12**(2), 378–382.

Southey B. B., S. L. Rodriguez-Zas, and K. A. Leymaster (2004). Competing risks analysis of lamb mortality in a terminal sire composite population. Journal of Animal Science **82**(10), 2892–2899.

Sutherland W. J., R. P. Freckleton, H. C. J. Godfray, et al. (2013). Identification of 100 fundamental ecological questions. Journal of Ecology **101**(1), 58–67.

Team R. D. C. (2010). R: A Language and Environment for Statistical Computing. R Foundation for Statistical Computing.

Tomkiewicz S. M., M. R. Fuller, J. G. Kie, and K. K. Bates (2010). Global positioning system and associated technologies in animal behaviour and ecological research. Philosophical Transactions of the Royal Society B: Biological Sciences **365**(1550), 2163–2176.

Vitousek P. M., C. M. Dantonio, L. L. Loope, M. Rejmanek, and R. Westbrooks (1997). Introduced species: a significant component of human-caused global change. New Zealand Journal of Ecology **21**(1), 1–16.

Williams B. K., J. D. Nichols, and M. J. Conroy (2002). Analysis and Management of Animal Populations. Academic Press.

Wilson A. J., D. Reale, M. N. Clements, et al. (2011). An ecologist's guide to the animal model (vol 79, p. 13, 2010). Journal of Animal Ecology **80**(5), 1109–1109.

Zhao P., S. M. Liu, Y. Zhou, et al. (2021). Estimating animal population size with very high-resolution satellite imagery. Conservation Biology **35**(1), 316–324.

PART 1

Demographic Data Collection: From Genes to Environment

CHAPTER 1

Genetic data collection, pedigrees, and phylogenies

Emily G. Simmonds, Alina K. Niskanen, Henrik Jensen, and Steven Smith

1.1 Introduction

Demographic studies are highly diverse; they can focus on a single population (Ozgul et al. 2010) or span many taxa (Salguero-Gómez et al. 2016). They can be geographically restricted (Simmonds and Coulson 2015) or global (Jones et al. 2013). Different types of demographic study have different challenges. In this chapter, we cover the use of genetic data in demographic analyses. There are two primary ways that genetic data enter into demographic analyses, and each addresses a particular challenge. These challenges will be the focus of the remainder of this chapter. The first is estimating relatedness between individuals in a population, and the second is identifying drivers of cross-taxon variation in life history. These are two very different challenges, but they are linked through their connection to genetic data.

The first way genetic data can help inform demographic analyses is through genetic pedigrees. Pedigrees address the challenge of estimating relatedness, which is needed to estimate reproductive success, a core demographic rate. To accurately quantify the reproductive success of an individual, we need to know exactly which offspring they produced. For lifetime reproductive success, we require a record of parentage spanning several reproductive cycles. To achieve this, we use a model of the relatedness of individuals in the population: this is called a pedigree. Once relatedness is established it is possible to calculate not only reproductive success (Robinson et al. 2006; López-Sepulcre et al. 2013; Pujol et al. 2014; Reid et al. 2014) but also other quantities, which can help capture change in life history traits, such as heritability (Charmantier and Gienapp 2014; Childs et al. 2016).

The second way genetic data can help inform demographic analyses is through genetic phylogenies, which address the challenge of identifying drivers of cross-taxon variation in life history. A key consideration when conducting comparative studies is the independence of data points, or the lack thereof. Taxa that are more closely related show a tendency to have similar life history traits (Blomberg et al. 2003). If we aim to understand the driving forces behind life history variation across multiple taxa, we need to tease apart the variation resulting from shared evolutionary history from that caused by environmental drivers (Freckleton 2009). To control for evolutionary history, we require a model of how closely related different taxa are—this is a phylogeny. Without phylogenetic correction, the assumptions of statistical analyses typically used to assess drivers of variation are violated because data points remain nonindependent (Freckleton 2009).

Pedigrees and phylogenies are two different models which use different types of genetic data to address different problems, but they also have some commonalities: they share fundamental genetic data collection methods, they are models of relatedness (although at different scales), they support demographic analyses rather than being a demographic method themselves, and the quantities they estimate allow us to ask demographic questions in a more accurate and robust way. In addition, both pedigrees and phylogenies have a history of being based on observational characteristics (e.g. social behaviour and morphology) with a more recent transition to using genetic data (Pemberton 2008; Patwardhan et al. 2014; Pyron 2015). Table 1.1 provides some examples of the use of pedigrees and phylogenies in demographic studies.

Emily G. Simmonds et al., *Genetic data collection, pedigrees, and phylogenies*. In: *Demographic Methods Across the Tree of Life*.
Edited by Roberto Salguero-Gómez and Marlène Gamelon, Oxford University Press.
© Oxford University Press (2021). DOI: 10.1093/oso/9780198838609.003.0001

Table 1.1 Literature examples of using pedigree and phylogenetic information in demographic analyses.

A. Pedigrees

Study	Demographic question	Taxa	Genetic data?	How the pedigree was used
Robinson et al. 2006	Are there trade-offs between fitness components and sexually antagonistic selection on weaponry?	Soay sheep (*Ovis aries*)	Yes	To assign parentage and estimate breeding success, which was later used to represent fitness
López-Sepulcre et al. 2013	How does the fecundity of dead males influence population growth?	Trinidadian guppies (*Poecilia reticulata*)	Yes	To assign parentage and estimate breeding success, which was later used to estimate contribution to population growth
Pujol et al. 2014	Is there a quantitative genetic signal of senescence?	White campion (*Silene latifolia*)	Yes	To assign parentage and estimate heritability with an animal model, this was later used to investigate patterns of additive genetic variance with age

B. Phylogenies

Study	Demographic question		Genetic data?	How the phylogeny was used
Tidière et al. 2016	Do animals live longer in zoos?	Mammals	Yes	To control for nonindependence of data points
Salguero-Gómez et al. 2016	What are the environmental drivers of schedules of survival, growth and reproduction?	Plants	Yes	To control for phylogenetic nonindependence of data points and to assess the strength of phylogenetic influence on demographic rates
Jones et al. 2013	What are the patterns of mortality and reproduction across the Tree of Life?	Cross-order	Yes	To assess the strength of phylogenetic influence on demographic rates and control for phylogenetic nonindependence of data points

In recent decades, there has been a rapid increase in the availability of genetic data due to improvements in noninvasive collection methods, advances in sequencing technologies, and a reduction in processing costs (Ekblom and Wolf 2014; Patwardhan et al. 2014; Pyron 2015). Large amounts of high-quality genetic data are now more easily obtainable, from single genes to whole genomes. These recent advances provide an opportunity for genetics to be used to support analyses across a wider range of systems and taxa. Despite the potential gains, the use of genetic data is far from ubiquitous. While genetic data could be used across all taxa, the magnitude of potential gains will be unequal across the Tree of Life. Some species and populations have the potential for more dramatic advances than others, particularly those where observational information is either missing, hard to obtain, or compromised. Mode of reproduction, pace of life, habitat, and conservation status all influence how easy it is to obtain reliable morphological or observational data that allow you to estimate relatedness and calculate reproductive

success of a species or a population. Prior to the availability of vast amounts of genetic material, there was a bias in our understanding of demography across the Tree of Life, resulting in over-representation of taxa for which we could obtain observational data on reproductive events or morphological features (e.g. humans and other vertebrates) and under-representation of others (e.g. arthropods, Pyron 2015). Using genetic data in demographic analyses provides a means to begin closing this gap, extending our understanding to a wider range of taxa.

This chapter gives an overview of genetic pedigree and genetic phylogeny construction, from the collection of genetic data, through the processing of genetic material, to the construction and optimisation of the models ready to be used in demographic analyses. This chapter provides a background for understanding how pedigrees and phylogenies are created so that they can be used appropriately in other analyses. It is also a starting point for how to make your own pedigrees and phylogenies.

1.2 Genetic pedigrees

A pedigree is a record of ancestry that depicts relationships between individuals. Pedigrees are used for multiple purposes in various fields, including breeding programs, disease mapping, quantitative genetics, behavioural ecology, and wildlife management. Information provided by pedigrees is important for estimating quantities such as the relatedness between individuals, inbreeding coefficients, individual reproductive success (Pemberton 2008), the heritability of quantitative traits (Wilson et al. 2010; Charmantier and Gienapp 2014), and effective population size (Engen et al. 2007). Without pedigrees it is not possible to estimate some quantities, such as number of offspring, or to follow the inheritance of traits in families (heritability and its links to demography are discussed further in Chapter 20).

The primary purpose of pedigree building is to establish links between parents and offspring, which is called parentage analysis (Flanagan and Jones 2019). In addition, other genealogical relationships can be estimated, including sibship clustering or the assignment of grandparents and avuncular links (Huisman 2017). Pedigrees can be built using observational (social) information from the study species, genetic markers (e.g. microsatellites or single nucleotide polymorphisms [SNPs]), or a combination of both methods. Microsatellites are short stretches of tandemly repeated DNA, occurring primarily in noncoding regions, that differ in the length of these repeated elements between individuals. SNPs are simple point mutations occurring anywhere throughout the genome. Social pedigrees have traditionally been built for mammal and bird species, where parental care of either one (typically maternal) or both parents is used to infer parentage. However, social pedigrees can solve only parent–offspring and some sibling relationships, and are prone to observational errors in polygamous species (Charmantier and Réale 2005). Furthermore, they are impossible to obtain when immediate evidence of parenthood is not available, such as in plants, corals, and some fish species.

Genetic pedigrees can solve some of the problems related to social pedigrees, and have become increasingly common along with the development of genetic methods (Jones et al. 2010). One major advantage of genetic pedigrees is that they are useful in many taxonomic groups, such as plants (Ashley 2010) and fish (Yue 2014). Genetic pedigrees are usually built by identifying parents and offspring across multiple generations. However, other types of relatives can also be determined based on the expected pairwise relatedness between individuals. For example, two full siblings are expected to share 50% (pairwise relatedness $r \approx 0.5$), and two half-siblings 25% ($r \approx 0.25$) of their genomes. Building a genetic pedigree requires polymorphic genetic markers and methods to determine whether two individuals' shared alleles match expectations from Mendelian inheritance. This is often combined with information about the year of birth (and in some programs the year of death), or a parental and an offspring cohort, and sex of the individuals, which is used to define a given individual's potential mothers, fathers, or other types of relatives.

Even though genetic pedigrees are usually more correct than social pedigrees (Reid et al. 2014), they are not free from errors. Problems in pedigree construction are often caused by incomplete sampling of the population and low resolution of genetic markers; as a result, parents will not be assigned to all nonfounder individuals (a *founder* is an individual with no known ancestors in the population). Another common source of error is assigning a wrong parent, which is more frequent in building social than genetic pedigrees, but may happen in either one. In species with high extra-pair mating, such as many bird species (Brouwer and Griffith 2019), the usage of social pedigree instead of genetic pedigree may for example give incorrect estimates of individual reproductive success and inbreeding (Reid et al. 2014) and bias heritability estimates downwards (Charmantier and Réale 2005; Bourret and Garant 2017). In turn, this may for example result in biased predictions of relationships between ecological and evolutionary dynamics (see Chapter 20).

1.3 Genetic phylogenies

Phylogenies, or phylogenetic trees, are models that represent the evolutionary relatedness of taxa at various scales, from families to species, or even genes. They are an essential component of demographic analyses that look for patterns in life history variation across taxa, a form of comparative analysis. As discussed in the introduction of this chapter (section 1.1), the primary use of phylogenies in comparative demographic analyses is to control for phylogenetic pseudoreplication. Without the inclusion of phylogenetic information, assumptions of most comparative statistical analyses cannot be met. A failure to control for evolutionary relatedness can lead to incorrect conclusions (Garland et al. 1999; McKechnie et al. 2006). This could have wide-reaching consequences as phylogenetic signal has been detected in 92% of studies on greater than 20 species (Blomberg et al. 2003).

As stated above, phylogenetic trees are models. These models are based on the principle of identifying homologous (resulting from shared ancestry) characters. These characters can be based on morphological or genetic data. Phylogenetic trees are built of branches that represent evolutionary paths, with internal nodes (where branches join), and end with taxa at the tips of the branches. They are considered working hypotheses (Penny et al. 1992), i.e. best estimates based on current data. Some recent studies present a full distribution of thousands of 'best' trees rather than selecting a single optimal topology (shape) in recognition of the uncertainty in estimation (Jetz et al. 2012). Phylogenetic trees, and the statistical analyses associated with them, are underpinned by several assumptions, one of which is a model of trait evolution (Freckleton 2009). This is distinct from models of sequence evolution discussed in section 1.7.2. The most common model of trait evolution is Brownian motion, which assumes no trait optimum (Freckleton 2009). But others also exist, such as the Ornstein–Uhlenbeck process, which does have an optimum and represents a trait evolving under stabilising selection (Blomberg et al. 2003). All of these models give the expected trait covariance for a certain level of phylogenetic relatedness (Freckleton 2009). The assumption of how traits have evolved can influence the amount of phylogenetic signal detected because it influences the evolutionary distance between taxa (Blomberg et al. 2003).

Phylogenetic signal is the strength of statistical dependence of species' trait values as a result of their phylogenetic relationships (Revell et al. 2008). Beyond the necessity of controlling for nonindependence of data points, understanding the strength of phylogenetic signal in shaping life history variation can also be of scientific interest (Jones et al. 2013; Salguero-Gómez et al. 2016). Testing for, or quantifying, phylogenetic signal is most commonly done through two methods: phylogenetically independent contrasts (Felsenstein 1985b; Blomberg et al. 2003; Freckleton 2009) and phylogenetic generalised least squares (PGLSs) (Blomberg et al. 2003; Freckleton 2009) (discussed in more detail in Chapter 18).

Genetic phylogenies are the cornerstone of modern phylogenetic analyses. They provide many more characters than morphological data and are the predominant form of phylogenetic tree for extant taxa (Nadler 1995). The genetic variation that can be used in phylogenetic analyses varies from whole genomes to shorter DNA sequences, or a handful of SNPs at specific loci. Different genetic markers are present in different taxa and accumulate mutations at different rates (Nadler 1995; Patwardhan et al. 2014), and thus different genetic markers are useful for resolving evolutionary relationships over different time scales. For example, the resolution of deep nodes require different markers compared to more recent nodes.

There are several standard markers used in molecular phylogenetics, Patwardhan et al. (2014) provide a good in-depth summary of these, we provide a brief overview here. Genetic information can come from primary data collection or from secondary sources. There are a wide range of online repositories of genetic information. One of the biggest and most well-known is GenBank (Benson et al. 2013). Possible genetic markers can be sourced from multiple parts of a cell (e.g. the nucleus, mitochondria, and chloroplasts). Nuclear material, often DNA encoding rRNA, is slow evolving and useful for resolving distant relationships (Patwardhan et al. 2014). Some common examples of nuclear DNA markers are Internal Transcribed Spacer (ITS) Tippery and Les 2008), 16S (Vences et al. 2005), and 28S (Vences et al. 2005). Mitochondrial DNA is faster evolving in animals and most useful for species-level relationships. Some common mitochondrial markers are Cyclooxygenase 1 (COX1) (Vences et al. 2005), 12S (Vences et al. 2005), and, for close relationships, Cytochrome-b (Patwardhan et al. 2014). For plants only, it is also possible to use chloroplast DNA to resolve mainly species-level relationships. Some common examples of chloroplast markers are ribulose bisphosphate carboxylase large chain (rbcL) (Morgan and Soltis 1993) and noncoding ribosomal protein L16 (rpl16) (Zhang 2000). These are just a few examples: in reality, there are many more.

Despite the gains that have been achieved through the incorporation of genetic data in phylogenetic analyses, genetic material does not create limitation-free phylogenies. Some of the same limitations of morphological data are also present in genetic data: (1) homology of the characters can be hard to establish (Nadler 1995) because convergent evolution can still occur at the genetic level (Stern 2013), (2) within-species variation can lead to incorrect relationship attribution (Harris and Crandall 2000), and (3) interpretation can be challenging. The latter occurs because gene divergence often precedes species change, resulting in gene trees that do not match species trees (Nei and Saitou 1987; Nadler 1995), a result which can also result from introgression due to hybridisation. For comparative analyses in particular, a drawback of existing phylogenies can be that they are limited in size. Finding a published phylogenetic tree covering all the taxa in the comparative demographic analysis may not be possible. To overcome this, phylogenetic supertrees can be created (Sanderson et al. 1998; Davies et al. 2004; Bininda-Emonds et al. 2007). These supertrees are constructed from several smaller trees to create a larger tree that encompasses more taxa than have ever been included in a single analysis.

In recognition of some of the limitations of genetic and morphological data, there has been a more recent trend to revive ideas of combining both sources of information into total-evidence phylogenetics (Pyron 2015; Gavryushkina et al. 2017). Combining the merits of each type of data can improve congruence of results and permits incorporation of the fossil record. Inclusion of fossils can often be useful for calibrating trees (Davies et al. 2004; Bininda-Emonds et al. 2007). Demographers working on cross-species comparative questions should remain abreast of developments in phylogenetics to ensure their inference is based on the latest working hypothesis. This is particularly pertinent as phylogenies used in demographic analyses have been found to lag behind statistical advances (Freckleton 2009).

1.4 Collection of genetic samples

When genetic samples are to be collected there are two main considerations: what material to collect and how best to bring it to the lab. The first point will require multilayered decisions based upon the planned analyses but also, perhaps more importantly, what is available, physically and ethically, for collection. High-quality DNA can be sourced from many origins (see Table 1.2) but the choice varies with the requirements of the project and the target taxa. Sequencing of mitochondrial DNA typically requires the lowest starting amount of material (e.g. trace material) whereas other top-end high-throughput approaches such as next generation sequencing (NGS) will require higher quantity and quality starting material (blood/tissue). For many projects, the ability to collect the ideal sample material is restricted due to target organism behaviour and/or ethical limitations. In these situations, noninvasive sampling may be the only option (Morin and Woodruff 1996; Waits and Paetkau 2005), and this will require streamlined protocols to maximise the recovery of DNA that is in a suboptimal state (degraded and in a matrix of other DNA, and/or accompanied by analysis inhibitors).

Sample transport and storage is clearly of paramount importance. Optimal methods include: sample freezing immediately on dry ice or liquid nitrogen, storage in a buffering solution to protect the DNA (e.g. Longmire et al. 1997), or immediate processing to isolate DNA before freezing. However, the ideal method of storage and transport are not always available in many field situations. Often, lack of electrical power or appropriate facilities will preclude many preferred methods. Alternatives such as desiccation of scat samples (Roeder et al. 2004) or drying blood or trace materials on specially

Table 1.2 Summary of potential source materials and handling considerations for DNA analysis.

Sampling technique	Suitable taxa	Suitable analyses	Limitations	Sample storage
Blood	Vertebrates (especially birds, reptiles, and amphibians)	Microsatellites, SNPs, mtDNA, whole genome, Amplicon Seq	Often contains inhibitors, permission needed	−20°C, ethanol, buffer, dried
Cuttings/biopsy	Corals, plants, jellyfish, animal organs	Microsatellites, SNPs, mtDNA, Whole genome, Amplicon Seq	Often need ethics permission	4°C, ethanol, buffer
Faeces	Vertebrates/intestinal parasites	Microsatellites, SNPs, mtDNA, Amplicon Seq	Often contains inhibitors and mixed species	−20°C, ethanol
Hair	Mammals	Microsatellites, SNPs, mtDNA, Amplicon Seq	Low quantity and quality DNA	4°C, dried
Shed skin	Reptiles, whales	Microsatellites, SNPs, mtDNA, Amplicon Seq	Low quantity and quality DNA	4°C, dried
Feathers	Birds	Microsatellites, SNPs, mtDNA, Amplicon Seq	Low quantity and quality DNA	4°C, dried
Swabs	Bacteria/fungi	Microsatellites, SNPs, mtDNA, Amplicon Seq	Often contains inhibitors and mixed species	4°C, dried

designed DNA preservation cards are an option for such extreme situations (Smith and Burgoyne 2004). Typically, however, there is a trade off in these cases in terms of DNA recovery and price of collection.

1.5 Processing of samples (DNA extraction, amplification, and sequencing)

Nucleic acid extraction is the critical starting point in every molecular genetic analysis as the recovery of high-quality DNA is a key factor for the success of all downstream applications. A variety of extraction methods are available to maximise the possibility of getting high-quality DNA and minimise degradation, but also to efficiently match requirements of time, materials, and labour costs. Generally, extraction methods can be divided into two main groups: solution-based or column-based protocols. Most protocols have been adopted into commercial kits that are simple, streamlined methods for nucleic acid extraction but are usually designed for specific starting materials.

In general, all successful nucleic acid extraction methods require a variation on the same important steps (Sambrook et al. 1989). First is the effective disruption of cells and tissues, followed by the denaturation of nucleoprotein complexes. This process will differ depending on sample type but basically involves physical homogenisation of tissues and chemical cell lysis. Often, detergents are used in cell lysis buffers but also chloroform or EDTA may be suitable for certain cell types. The next step is inactivation of nucleases and elimination of all contaminants (e.g. proteins, carbohydrates, or lipids), which can be achieved by buffering with Tris-EDTA, enzymatic protein removal, and centrifugation. DNA is then isolated either by precipitation with isopropanol or binding to silica (often in a column filter) and washed with ethanol to remove salts. The final step is the elution of DNA into a storage buffer, most commonly Tris-EDTA, and then storage until further processing (Smith and Morin 2005).

Typically, for downstream processing of DNA, it is desirable to investigate a particular gene or region of interest. This requires the amplification of the region of interest so that there are enough copies to visualise via sequencing or genotyping methods. Amplification is achieved via polymerase chain reaction (PCR). The idea behind PCR is simple and

can be divided into three stages which are iteratively repeated in cycles. Each cycle consists of *denaturation* (when double stranded DNA is denatured into two single stranded template strands), *annealing* (short oligonucleotide primers targeting the region of interest, bind to the complementary single stranded template ends), and *extension* (elongation of oligonucleotide primers by the activity of polymerase and the addition of DNA building blocks called dNTPs, that result in a DNA strand which is complementary to the template).

For pedigree construction, relationships between individuals are generally determined via comparisons of genotypic data at microsatellite or SNP loci. SNPs (single nucleotide polymorphisms) are mutations, or polymorphisms, which can be compared between individuals who will possess one (homozygous individuals) or two (heterozygous individuals) of the generally two possible variants at a given locus. At a given microsatellite locus, length variations can lead to more than 10 distinct allelic variants. Microsatellite loci are normally screened via some form of gel electrophoresis to separate alleles based on their size compared to a known standard. Recently NGS techniques have also been developed to sequence the amplified loci in multiplex format and visualise length differences via an alignment (Vartia et al. 2016). There are a multitude of methods for SNP genotyping, but all revolve around interrogating which variant is present at the targeted polymorphic position. Some make use of different fluorescent dyes for the two possible nucleotides that can be incorporated at the site while others detect thermokinetic differences in the amplified alleles based on the contrasting nucleotide compositions. A recent NGS-based method for SNP genotyping known as RAD sequencing (restriction associated DNA sequencing, Baird et al. 2008) uses restriction enzymes and DNA target capture to simultaneously discover and screen large numbers of individuals at thousands of SNP loci across a reduced representation of the entire genome.

Sequencing options for phylogenetic tree reconstruction range from the traditional Sanger sequencing (Sanger et al. 1977) to more sophisticated massively parallel sequencing (Metzker 2010). For Sanger sequencing, the skill set required is within the scope of most government and academic laboratories. Throughput is scalable, but mainly in the medium range, with 96 samples easily processed within a week for a sequence length of 700–1000 base pairs. A much higher throughput can be achieved via the NGS techniques. A popular application of NGS is amplicon sequencing (Schirmer et al. 2015) that, in its simplest form, involves standard PCR amplification of a target genomic region using primers that have been modified to include an individual ID tag and an adaptor used to initiate the sequencing. Many individuals with different individual tags can be PCR amplified and then pooled in equimolar amounts for sequencing. Subsequent data filtering steps separate out the individuals according to their specific tag. Rather than sequencing the entire genome of an organism, this approach targets one or more gene regions of interest for deep sequencing. As less of the genome is being sequenced, more individuals can be pooled in each sequencing run to greatly reduce the costs per sample.

1.6 Preparation of data for analyses

Once data have been generated for a project there are many steps that need to be performed before it is suitable for analysis to construct pedigrees or phylogenies. These steps are similar for all data types but will vary in specific details according to the marker. Usually, these preparation steps can be incorporated into scripts or bioinformatics pipelines, and often there is pre-existing software that facilitates the process. The first step is data cleaning and filtering for quality (Van Oosterhout et al. 2004; Teo 2008; Bokulich et al. 2013). Particularly with sequencing data, there is a need to trim off any portions that relate to priming sequences, individual ID tags, or sequencing adaptors. Quality filtering can be achieved in most cases by taking advantage of quality scores that are provided by the data provider or instrument. Low-quality data can then be flagged and excluded from further analysis.

For microsatellite and SNP data, allelic information (or missing data values) need to be assigned at every locus across all individuals to allow comparison and inference of relatedness. Automation of this process is often essential and is provided by software packages that allow a user to assign alleles based on their conformity to certain size criteria or signal to noise ratios specific to the instrument generating the data (e.g. Genemapper, Genemarker for microsatellites; ASSIsT, Illumina GenCall for SNPs). The end result is a matrix of multilocus genotypes for each individual which will then be used as input for pedigree or phylogeny reconstruction.

For sequence data the researcher must make the data comparable across individuals and taxa. It is important to align the sequences such that the corresponding base-pair positions reflect homology and any differences relate to evolutionary events. In this way, it is clear where a base has mutated to create a polymorphism or has been lost (deletion) or gained (insertion). These are the important sequence differences that will be used when constructing the phylogeny to

Table 1.3 Multiple sequence alignment software options for nucleotide sequence data.

Name	Description	Data set size	Availability	URL
CLC Main Workbench	GUI (graphical user interface) allows integration with many add on tools for sequence analysis	Memory dependent	Commercial licence	https://www.qiagenbioinformatics.com/products/clc-main-workbench/
Clustal2	Classic progressive alignment tool available as command line or GUI	Memory dependent	Free	http://www.clustal.org/clustal2/
Clustal Omega	Uses seeded guide trees and HMM profile-profile techniques to generate alignments	>100,000 sequences	Free	http://www.clustal.org/omega/
CodonCode Aligner	Supports quality scores. Features include mutation detection and BLAST searching	Memory dependent	Commercial licence	https://www.codoncode.com/aligner/
Kalign	Fast tool focuses on local region alignment	<2,000 sequences	Free	http://msa.cgb.ki.se/cgi-bin/msa.cgi
MAFFT	Sequences clustered using Fast Fourier Transforms	<30,000 sequences	Free under BSD licence	https://mafft.cbrc.jp/alignment/software/
MUSCLE	3-stage progressive alignment	Memory dependent, recommended <10,000 sequences	Free	http://www.drive5.com/muscle/

determine evolutionary relationships. The alignment can be greatly facilitated by mapping to an existing reference sequence if one is available. If not, multiple sequence alignments will need to be made often with the help of a software package incorporating algorithms to reflect probabilities of various mutational processes (see Table 1.3). The resulting output is a series of aligned sequences with mutated positions highlighted and ready for the extent of divergence to be analysed according to an appropriate evolutionary model.

1.7 Using the data to construct pedigrees and phylogenies

1.7.1 Pedigrees

After preparing the genetic data as described in section 1.6, a pedigree can be constructed using one of three main methods: exclusion method, relatedness-based method, and likelihood-based method (Flanagan and Jones 2019). The exclusion method excludes all potential parents that do not share at least one allele per locus with an offspring, but allow for some mismatching loci due to genotyping errors (Calus et al. 2011; Hayes 2011). It is fast due to simplicity, but it may lead to assignment of incorrect parents when multiple potential parents share alleles with the offspring (Huisman 2017). Relatedness-based methods estimate pairwise relatedness (r) or kinship coefficients between individuals. The relatedness estimates are then classified as first, and second-degree relatives, or unrelated (Thompson 1975). When data contain overlapping generations, other methods are needed to tell apart parent–offspring pairs, full siblings, and other related pairs of individuals that share the same degree of relatedness. Likelihood-based methods compare the likelihood of a specific relationship (e.g. parent–offspring relationship) compared to other potential relationships given the genotypes of the individuals (Flanagan and Jones 2019).

Pedigree construction methods usually make various assumptions regarding the study population and genetic data, but all require polymorphic genetic markers, such as microsatellites or SNPs, that are in low linkage disequilibrium (e.g. Flanagan and Jones 2019). Informative, highly heterozygous loci are required; minor allele frequency (MAF) is often used to assess the selection of SNP loci. The number of markers needed depends on the species, mating system, and sampling effort (Aykanat et al. 2014). Typically, 10–20 microsatellites or a few hundred SNPs are needed to gain sufficient power to distinguish individuals and their

relationships. Increasing the number and/or polymorphism of markers increases the power to establish correct pedigree links (Harrison et al. 2013; Aykanat et al. 2014). Genotyping errors and mutations may cause non-Mendelian mismatches between parent and offspring alleles. Ideally, a genotyping error rate is estimated from the data and used as a model parameter, or if that is not possible, a predefined number of mismatches are allowed in the pedigree construction. Some likelihood-based and Bayesian software estimate the error rate simultaneously with the pedigree (see e.g. Wang 2004).

In addition to genotype data, various kinds of population-level ecological and individual phenotypic data can be used to reduce the number of model assumptions one must make, and aid pedigree construction. The type of information used depends on the software; the year of birth may be included as a cue to solve which individual is an offspring and which a parent: for example, in Sequoia by Huisman (2017) and FRANz by Riester et al. (2009). Other software, such as Cervus (Kalinowski et al. 2007), MasterBayes (Hadfield et al. 2006), and Colony (Wang 2004, 2012; Jones and Wang 2010), use cohort-specific parental lists to assign parents. The group of potential parents is limited by the year of death in some programs (FRANz), and some programs (e.g. Cervus, Colony, and FRANz) can make use of information on the proportion of the adult population which is included in the list of candidate parents. If a previous version of a pedigree exists for the study system (FRANz), or information is available on expected relatedness between potential parents or between parents and offspring (Cervus), this can be incorporated to aid parental inference. Finally, some programs can allow for focal populations having various kinds of mating systems (e.g. Colony). The correctness of a genetic pedigree can be estimated using multiple methods (Jones et al. 2010), including: (1) simulation of genotype data sets for pedigree building to determine error rate in the simulated pedigrees compared to the actual constructed pedigree (Kalinowski et al. 2007), (2) Bayesian posterior probabilities (Nielsen et al. 2001), and (3) simulation of an experiment-wise expected error rate (Jones 2001).

Genetic data and software used to build pedigrees do not provide a single answer for estimates of relatedness between individuals in a population. Pedigrees can be constructed using a range of methodological approaches and different sources of genetic information. Each of these choices alters the degree of relatedness that can be established. The variety of

Table 1.4 Results from genetic parentage analyses carried out in two house sparrow (*Passer domesticus*) populations that are part of an individual-based long-term study of an insular metapopulation at the coast of northern Norway (see e.g. Niskanen et al. 2020).

Population	Relationship link/running time	50 SNPs				605 SNPs			
		Cervus	Colony	FRANz	Sequoia	Cervus	Colony	FRANz	Sequoia
Aldra	Maternal links	12	12	12	11	12	12	12	12
	Paternal links	10	10	10	10	10	10	10	10
	Fullsib links	0	0[a]	0	0[a]	0	0[a]	0	0[a]
	Running time (s)	149.5	7.0	3.0	0.6	781.4	59.0	31.0	4.7
Hestmannøy	Maternal links	57	56	56	51	54	54	54	54
	Paternal links	57	58	58	53	59	57	57	57
	Fullsib links	16	22[b]	16	14[b]	16	23[b]	16	22[b]
	Running time (s)	1,750.1	412.2	3.0	4.6	20,409.4	1,332.0	30.0	30.7

[a] Sibship clustering option included, but no dummy parents were found.
[b] Fullsib links include link(s) due to dummy parents.

results that can be obtained is illustrated in Table 1.4 and Figure 1.1, which present a case study of parentage analyses using different amounts of genetic information and different construction methods. In this case study, parentage was determined for house sparrow (*Passer domesticus*) recruits that hatched on two islands in northern Norway in 2008 ($n = 13$ on Aldra; $n = 76$ on Hestmannøy). All sampled adult birds present on the islands that year were included as putative parents.

The two populations reflect natural populations that differ in demographic histories: Aldra was colonised in 1998 and has a relatively small population with a high level of inbreeding; the Hestmannøy population is larger and more stable and has a low level of inbreeding (Niskanen et al. 2020). The estimated adult population size in 2008 was 43 individuals on Aldra and 162 on Hestmannøy, of which 40 and 137 were sampled and genotyped, respectively. The birds were genotyped on a custom Affymetrix Axiom SNP-array for house sparrows, including 200,000 SNPs (Lundregan et al. 2018). In the parentage analyses we used either a selection of 605 independent SNPs with a high level of heterozygosity in more than 3000 birds present in the total metapopulation during the years 1998–2013 (see Niskanen et al. 2020 for details), or a random subset consisting of 50 of these 605 SNPs. Additional information used by the different software was: sex of candidate parents (all four software packages), estimated adult population sizes (i.e. proportion of candidate parents sampled: Cervus, Colony, and FRANz), information on hatch year of candidate parents (Sequoia and FRANz), estimates of relatedness between true parents and other candidate parents (Cervus), and marker genotyping error rates (0.002: all four software packages).

Pedigree information is stored as a simple three-column file, which includes a focal individual, a first parent (either dam or sire), and a second parent (either dam or sire). These parental relationships, forming the pedigree, can be visualised using for example programs Pedigree Viewer (Kinghorn 1994) or R package Pedantics (Morrissey and Wilson 2010). Pedigree statistics, including the number of parental and other family links, depth of a pedigree (number of generations), and number of founders, can be acquired using, for example, the Pedantics package. Pedigrees are usually built for the specific study system to be able to answer research questions arising from that data. Pedigree information can then be incorporated into demographic analyses to estimate individual reproductive success, inbreeding coefficients, and pairwise relatedness of individuals (e.g. López-Sepulcre et al. 2013; Aykanat et al. 2014; Pujol et al. 2014).

1.7.2 Phylogenies

Reconstruction of molecular phylogenies to use in comparative demographic analyses takes several key steps: choosing an appropriate model of sequence evolution, establishing genetic characters, calculating pairwise distances between sequences, phylogenetic tree reconstruction, and phylogenetic tree evaluation. Not all steps are used in all analyses or in the same way, but the general procedure is similar across all phylogenetic analyses. Here we give an overview of this general procedure and an introduction into some of the variety available at each step. In this section we assume that phylogenetic reconstruction is being performed at the species level, but different taxonomic resolutions are also possible.

(a)

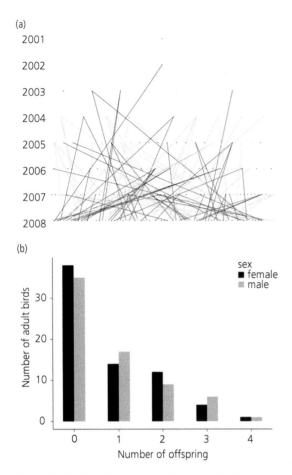

(b)

Figure 1.1 (a) The pedigree constructed by Sequoia for the insular house sparrow population on Hestmannøy using 605 SNPs as explained in the text for Table 1.4. Individuals are presented as dots, and lines connect a parent to its offspring. Black dots and lines show females and any links they have to offspring, and grey dots and links show males and any links they have to offspring. The y-axis position shows the hatch year of an individual. (b) The frequency distribution of recruits in the 2008 cohort produced by adult females (black) and males (grey) present on Hestmannøy that year.

When the data have been prepared using methods outlined in section 1.6, they can then be used for phylogenetic tree reconstruction. Alignment suggests homology, however, to reach an exact string of characters, we require a model of how the sequences have evolved. For instance, an observation of A (adenine) at a particular locus in two aligned sequences could have arisen from no evolutionary change (homology) or from multiple changes to the same locus, such as from A to G (guanine) and back to A. We call the latter homoplasy, when two characters look the same but actually result from independent evolution. To

distinguish homoplasy from homology and to estimate how much change has occurred between sequences, it is necessary to assume a model of sequence evolution (Zhang and Kumar 1997; Kalyaanamoorthy et al. 2017). Information on the amount of change between sequences forms the basis of estimating phylogenetic relatedness and is, thus, an integral part of tree reconstruction. Models of sequence evolution provide a mathematical illustration of how often nucleotide or amino acid replacements would be expected to occur (Whelan et al. 2001) and therefore influence the number of characters estimated from a particular set of sequences. Models of sequence evolution are used in all phylogenetic analyses whether they are formalised and explicit, or not (Kelchner and Thomas 2007). Many different models of sequence evolution exist, and they all include different assumptions. Two common examples are the Jukes–Cantor one-parameter model (Jukes and Cantor 1969), which assumes equal probability of all substitutions and equal frequency of all nucleotides, and the Kimura two-parameter model (Kimura 1980), which gives independent probabilities to transversions and transitions but still assumes equal frequency of all nucleotides. A model of sequence evolution can be selected using a program that runs statistical model selection for models of nucleotide substitution; two examples are jModelTest (Posada 2008) and Partition-Finder 2 (Lanfear et al. 2017), but many more exist.

Once a model of evolution has been selected, the exact number of genetic characters can be established. The next step is to use the characters to construct a phylogenetic tree. Again, there are a multitude of options available, implemented in different programs, with different optimisation principles, benefits, and drawbacks. These are summarised in Table 1.5. There is an extensive literature on the different methods (see Yang and Rannala 2012, for a nice summary), we introduce some of the most common methodologies here.

The first distinction in phylogenetic tree topology is whether the tree is rooted or unrooted. Trees that are rooted have a basal ancestor, whereas unrooted/free forming trees do not include a known root. Trees can be rooted via an outgroup (dictating the most basal relationship in the tree) or via time, using assumptions about the rate of molecular evolution (Yang and Rannala 2012).

Once you have a final tree generated with any approach, it is vital to evaluate the tree because the topology of a tree and its branch lengths are estimated with uncertainty. There are several standard methods for estimating the uncertainty associated with nodes (where branches join). The primary (non-Bayesian)

Table 1.5 Summary of phylogenetic reconstruction methods.

Name of approach:	Optimisation principle:	
Distance-matrix methods	**To minimise the tree length. The exact metric used to measure tree length and the method employed to minimise it vary.**	
Description	**Key benefits**	**Key drawbacks**	**Example programs**
These methods work in two steps: First, they take the whole sequences of genetic information and calculate pairwise differences based on a model of evolution and use these to generate a distance matrix (Yang and Rannala 2012). It does not use specific characters. Second, the distance matrix is used to construct a tree. There are several different methods available to build trees from distance matrices, for example: Clustering algorithm methods, such as UPGMA (Sneath and Sokal 1973) and Neighbour-joining (Nei and Saitou 1987). Least squares (Cavalli-Sforza and Edwards 1967). Minimum evolution (Rzhetsky and Nei 1992).	Fast computation, particularly for cluster algorithms where fewer trees are compared (Yang and Rannala 2012). Popular for big data sets (Xia 2018). Can be used with pairwise sequence alignments, rather than needing multiple sequence alignment (Thorne and Kishino 1992, Xia 2018).	Sensitive to gaps or errors in sequences (Yang and Rannala 2012; Xia 2018). Ignores information in sequences beyond the pairwise comparison. (But see Xia 2018 for more recent solutions).	RapidNJ (Simonsen et al. 2008) NINJA (Wheeler 2009) PAUP* (Swofford 2002)

Name of approach:	Optimisation principle:	
Maximum parsimony	**To minimise the number of character changes in the tree. There are various methods to minimise character changes.**	
Description	**Key benefits**	**Key drawbacks**	**Example programs**
This is a character-based method, so requires multiple sequence alignment. The principle is based on building all trees possible and finding the one that minimises the number of character changes (Farris et al. 1970). In reality, this would be too time consuming. Therefore, trees are built and searched using an algorithm developed by Fitch (1970) to find the topology that minimises the number of character changes. Maximum parsimony does not include an explicit model of sequence evolution. The implicit model is that all substitutions are equally likely and that reversion does not occur (Yang and Rannala 2012).	Simplicity of philosophy and computationally efficient (Yang and Rannala 2012). Less biased than other methods when heterogenous evolution occurs (Kolaczkowski and Thornton 2004).	Does not include an explicit model of evolution and therefore contains nonexplicit assumptions. Can suffer from long-branch attraction (Felsenstein 1978). This is a phenomenon where two taxa are incorrectly inferred to be closely related due to having long branches (i.e. having a large number of character changes since their common ancestor).	PAUP* (Swofford 2002) MEGA X (Kumar et al. 2018) TNT (Giribet 2005)

continued

Table 1.5 *Continued*

Name of approach:	Optimisation principle:	
Maximum likelihood (ML)	**Chooses the tree that maximises the likelihood given the data.**	

Description	Key benefits	Key drawbacks	Example programs
This is a character-based method, so requires multiple sequence alignment. Uses an explicit model of sequence evolution. This method is implemented in two steps. The first optimises the branches and the likelihood of the tree calculated. The trees are then searched using iterative optimisation algorithms to find the tree with the maximum likelihood value (Yang and Rannala 2012).	Heuristic searches so computationally efficient for larger data sets. No long-branch attraction if the sequence evolution model is sufficient. Robust to certain violations of assumptions and errors (Kalyaanamoorthy et al. 2017). Can include complex models of sequence evolution (Yang and Rannala 2012).	Can suffer from long-branch attraction if the model of sequence evolution is incorrect (Yang and Rannala 2012). Can be biased in some scenarios (Kolaczkowski and Thornton 2004). Can be computationally demanding.	RaxML (Stamatakis 2014) (heuristic search) GARLI (Zwickl 2006) PAUP* (Swofford 2002) IQ-TREE (Nguyen et al. 2015)

Name of approach:	Optimisation principle:	
Bayesian estimation	**Chooses the tree with best posterior support given a *priori* and the data.**	

Description	Key benefits	Key drawbacks	Example programs
This is a character-based method, so requires multiple sequence alignment. Uses an explicit model of sequence evolution. Bayesian methods also use the likelihood of the tree but estimate parameters following Bayesian principles, using priors and producing a posterior distribution for each parameter (see Rannala and Yang 1996; Yang and Rannala 1997 for more details).	No long-branch attraction if the model of sequence evolution is sufficient. Can include complex models of sequence evolution (Yang and Rannala 2012). Uncertainty has a more intuitive interpretation than other methods (Yang and Rannala 2012).	Incorrect use of priors can result in incorrect trees (Nascimento et al. 2017). Can be biased in some scenarios (Kolaczkowski and Thornton 2004). Convergence can be difficult to achieve and computationally demanding. Posterior probabilities can be overinflated (Yang and Rannala 2012).	MrBayes (Ronquist and Huelsenbeck 2003) BEAST (Bouckaert et al. 2014) (a more extensive list can be found in Nascimento et al. (2017))

approach to evaluating phylogenetic trees is the bootstrap (Felsenstein 1985a) analyses. Bootstrapping, the most popular method for determining branch support in phylogenetics (Holmes 2003), involves creating a new data set the same size as the original, by randomly sampling columns from the original data set, with replacement. This is done a set number of times and a new tree constructed each time. The bootstrap support of a node is defined as the percentage of bootstrap trees in which the clade associated with that node was recovered. For trees created from Bayesian analyses, posterior probabilities are used to represent uncertainty.

After the creation of a phylogeny using the methods outlined above, the tree is ready to be used in a comparative analysis. The phylogenetic construction methods detailed above are complex and require some level of expertise to implement them; however, reconstructed phylogenies are also widely available. Trees can be obtained from published papers (see Chaw et al. 1997 for a gymnosperm phylogeny; Bininda-Emonds et al. 1999 for a phylogeny of mammals; Qian and Jin 2016 for a vascular plant phylogeny) and from online repositories such as PHYLACINE (Faurby 2019). Phylogenies that can be taken from these sources might need to be manipulated prior to use in a comparative analysis: for example, they might need pruning (some taxa removed) or change of species names to match the demographic data or a supertree may need to be created so that the phylogeny covers all taxa in the analysis (see Sanderson et al. 1998; Bininda-Emonds et al. 2007 for examples). Lots of tools exist to manipulate phylogenetic trees to ensure they are in the right format for a demographic analysis, see https://cran.r-project.org/web/views/phylogenetics.html for a list of the options in R. More detail on comparative analyses is available in Chapter 18.

1.8 Conclusions

Over the past few decades there have been many advances in genetic analyses that we can now exploit in demographic methods. Opportunities to use genetic information exist across the Tree of Life. While some methodological practices differ by taxon, the benefits of using genetic data in demographic methods are truly cross-taxonomic. The types of models described in this chapter provide a supporting role in demographic analyses. They are used to tackle two distinct challenges, but both can improve the accuracy and robustness of our estimates of demographic quantities and patterns. Genetic data and the resulting pedigrees

and phylogenies are now among the fundamental tools of demographic analyses. These data are a complement to the rest of the demographic toolkit, and, combined with other data sources, they can help create well-informed analyses and allow us to address new biological questions across the Tree of Life.

References

Ashley M. V. (2010). Plant parentage, pollination, and dispersal: how DNA microsatellites have altered the landscape. Critical Reviews in Plant Sciences **29**(3), 148–161. doi: 10.1080/07352689.2010.481167.

Aykanat T., Johnston, S. E., Cotter, D., et al. (2014). Molecular pedigree reconstruction and estimation of evolutionary parameters in a wild Atlantic salmon river system with incomplete sampling: a power analysis. BMC Evolutionary Biology **14**(68), 1–17. doi: 10.1186/1471-2148-14-68.

Baird N. A., Etter P. D., Atwood T. S., et al. (2008). Rapid SNP discovery and genetic mapping using sequenced RAD markers. PLOS One **3**(10), e3376. doi: 10.1371/journal.pone.0003376.

Benson D. A., Cavanaugh M., Clark K., et al. (2013). GenBank. Nucleic Acids Research **41**(D1), D36–D42. doi: 10.1093/nar/gkw1070.

Bininda-Emonds O. R. P., Gittleman J. L., and Purvis A. (1999). Building large trees by combining phylogenetic information: a complete phylogeny of the extant Carnivora (Mammalia). Biological Reviews **74**(2), 143–175. doi: 10.1111/j.1469-185X.1999.tb00184.x.

Bininda-Emonds O. R. P., Cardillo M., Jones K. E., et al. (2007). The delayed rise of present-day mammals. Nature **446**(7135), 507–512. http//.doi.org.10.1038/nature05634.

Blomberg S. P., Garland T., and Ives A. R. (2003). Testing for phylogenetic signal in comparative data: behavioral traits are more labile. Evolution **57**(4), 717–745. doi: 10.1111/j.0014-3820.2003.tb00285.x.

Bokulich N. A., Subramanian S., Faith J. J., et al. (2013). Quality-filtering vastly improves diversity estimates from Illumina amplicon sequencing. Nature Method **10**, 57–59. doi: 10.1038/nmeth.2276.

Bouckaert R., Heled J., Kühnert D., et al. (2014). BEAST 2: a software platform for Bayesian evolutionary analysis. PLOS Computational Biology **10**(4), e1003537. doi: 10.1371/journal.pcbi.1003537.

Bourret A. and Garant D. (2017). An assessment of the reliability of quantitative genetics estimates in study systems with high rate of extra-pair reproduction and low recruitment. Heredity **118**, 229–238. doi: 10.1038/hdy.2016.92.

Brouwer L. and Griffith S. C. (2019). Extra-pair paternity in birds. Molecular Ecology **28**(22), 4864–4882. doi: 10.1111/mec.15259.

Calus M. P. L., Mulder H. A., and Bastiaansen J. W. M. (2011). Identification of Mendelian inconsistencies between SNP and pedigree information of sibs. Genetics Selection Evolution **43**(34), 1–12. doi: 10.1186/1297-9686-43-34.

Cavalli-Sforza L. L. and Edwards A. W. F. (1967). Phylogenetic analysis: models and estimation procedures. Society for the Study of Evolution **21**(3), 550–570.

Charmantier A. and Gienapp P. (2014). Climate change and timing of avian breeding and migration: evolutionary versus plastic changes. Evolutionary Applications **7**(1), 15–28. doi: 10.1111/eva.12126.

Charmantier A. and Réale D. (2005). How do misassigned paternities affect the estimation of heritability in the wild? Molecular Ecology **14**(9), 2839–2850. doi: 10.1111/j.1365-294X.2005.02619.x.

Chaw S. M., Zharkikh A., Sung H. M., Lau T. C., and Li W. H. (1997). Molecular phylogeny of extant gymnosperms and seed plant evolution: analysis of nuclear 18s rRNA sequences. Molecular Biology and Evolution **14**(1), 56–68. doi: 10.1093/oxfordjournals.molbev.a025702.

Childs D. Z., Sheldon B. C., and Rees M. (2016). The evolution of labile traits in sex- and age-structured populations. Journal of Animal Ecology **85**(2), 329–342. doi: 10.1111/1365-2656.12483.

Davies T. J., Barraclough T. G., Chase M. W., Soltis P. S., Soltis D. E., and Savolainen V. (2004). Darwin's abominable mystery: insights from a supertree of the angiosperms. Proceedings of the National Academy of Sciences of the United States of America **101**(7), 1904–1909. doi: 10.1073/pnas.0308127100.

Ekblom R. and Wolf J. B. W. (2014). A field guide to whole-genome sequencing, assembly and annotation. Evolutionary Applications **7**, 1026–1042. doi: 10.1111/eva.12178.

Engen S., Ringsby T. H., Sæther B. E., et al. (2007). Effective size of fluctuating populations with two sexes and overlapping generations. Evolution **61**(8), 1873–1885. doi: 10.1111/j.1558-5646.2007.00155.x.

Farris J. S., Arnold K. G., and Eckardt M. J. (1970). A numerical approach to phylogenetic systematics. Systematic Biology **19**(2), 172–189.

Faurby S. (2019). Data from PHYLACINE 1.2: the phylogenetic atlas of mammal macroecology. Journal Ecology **99**, 2626. doi: https://doi.org/10.5061/dryad.bp26v20.

Felsenstein J. (1978). Cases in which parsimony or compatibility methods will be positively misleading. Systematic Zoology **27**(4) 401–410.

Felsenstein J. (1985a). Confidence limits on phylogenies: an approach using the bootstrap. Evolution: International Journal of Organic Evolution **39**, 783–791. doi: 10.1111/j.1558-5646.1985.tb00420.x.

Felsenstein J. (1985b). Phylogenies and the comparative method. The American Naturalist **125**(1), 1–5. doi: 10.1086/284325.

Fitch W. (1970). Distinguishing homologous from analogous proteins. Systematic Biology **19**(2), 99–113.

Flanagan S. P. and Jones A. G. (2019) The future of parentage analysis: from microsatellites to SNPs and beyond. Molecular Ecology **28**(3), 544–567.

Freckleton R. P. (2009). The seven deadly sins of comparative analysis. Journal of Evolutionary Biology **22**(7), 1367–1375. doi: 10.1111/j.1420-9101.2009.01757.x.

Garland T., Midford P. E., and Ives A. R. (1999). An introduction to phylogenetically based statistical methods, with a new method for confidence intervals on ancestral values. American Zoologist **39**(2), 374–388. doi: 10.1093/icb/39.2.374.

Gavryushkina A., Heath T. A., Kspeka D. T., Stadler T., Welch D., and Drummond A. J. (2017). Bayesian total-evidence dating reveals the recent crown radiation of penguins. Systematic Biology **66**(1), 57–73. doi: 10.1093/sysbio/syw060.

Giribet G. (2005). TNT: tree analysis using new technology. Systematic Biology **54**(1), 176–178. doi: 10.1080/10635150590905830.

Hadfield J. D., Richardson D. S., and Burke T. (2006). Towards unbiased parentage assignment: combining genetic, behavioural and spatial data in a Bayesian framework. Molecular Ecology **15**(12), 3715–3730. doi: 10.1111/j.1365-294X.2006.03050.x.

Harris D. J. and Crandall K. A. (2000). Intragenomic variation within ITS1 and ITS2 of freshwater crayfishes (Decapoda: cambaridae): implications for phylogenetic and microsatellite studies. Molecular Biology and Evolution **17**(2), 284–291. doi: 10.1093/oxfordjournals.molbev.a026308.

Harrison, H. B., Saenz-Agudelo, P., Planes, S., Jones, G. P., and Berumen, M. L. (2013). Relative accuracy of three common methods of parentage analysis in natural populations. Molecular Ecology **22**(4), 1158–1170. doi: 10.1111/mec.12138.

Hayes B. J. (2011). Technical note: efficient parentage assignment and pedigree reconstruction with dense single nucleotide polymorphism data. Journal of Dairy Science **94**(4), 2114–2117. doi: 10.3168/jds.2010-3896.

Holmes S. (2003). Bootstrapping phylogenetic trees: theory and methods. Statistical Science **18**(2), 241–255. doi: 10.1214/ss/1063994979.

Huisman J. (2017). Pedigree reconstruction from SNP data: parentage assignment, sibship clustering and beyond. Molecular Ecology Resources **17**(5), 1009–1024. doi: 10.1111/1755-0998.12665.

Jetz W., Thomas G. H., Joy J. B., Hartmann K., and Mooers A. O. (2012). The global diversity of birds in space and time. Nature **491**(7424), 444–448. doi: 10.1038/nature11631.

Jones A. G. (2001). GERUD1.0: a computer program for the reconstruction of parental genotypes from progeny arrays using multilocus DNA data. Molecular Ecology Notes **1**, 215–218. doi: 10.1046/j.1471-8278.2001.00062.x.

Jones A. G., Small C. M., Paczolt K. A., and Ratterman N. L. (2010). A practical guide to methods of parentage analysis. Molecular Ecology Resources **10**(1), 6–30. doi: 10.1111/j.1755-0998.2009.02778.x.

Jones O. R. and Wang J. (2010). Molecular marker-based pedigrees for animal conservation biologists. Animal Conservation **13**(1), 26–34. doi: 10.1111/j.1469-1795.2009.00324.x.

Jones O. R., Scheuerlein A., Salguero-Gómez R., et al. (2013). Diversity of ageing across the tree of life. Nature **505**(7482), 169–173. doi: 10.1038/nature12789.

Jukes T. H. and Cantor, C. R. (1969). Evolution of protein molecules. In: H. N. Munro (ed.), Mammalian Protein Metabolism, 21–123. Academic Press.

Kalinowski S. T., Taper M. L., and Marshall T. C. (2007). Revising how the computer program CERVUS accommodates genotyping error increases success in paternity assignment. Molecular Ecology **16**(5), 1099–1106. doi: 10.1111/j.1365-294X.2007.03089.x.

Kalyaanamoorthy S., Minh B. Q., Wong T. K. F., Von Haeseler A., and Jermiin L. S. (2017). ModelFinder: fast model selection for accurate phylogenetic estimates. Nature Methods **14**(6), 587–589. doi: 10.1038/nmeth.4285.

Kelchner S. A. and Thomas M. A. (2007). Model use in phylogenetics: nine key questions. Trends in Ecology and Evolution **22**(2), 87–94. doi: 10.1016/j.tree.2006.10.004.

Kimura M. (1980). A simple method for estimating evolutionary rate of base substitution through comparative studies of nucleotide sequences. Evolution **16**, 111–120.

Kinghorn (1994). Pedigree Viewer—a graphical utility for browsing pedigreed data sets. In: C. Smith, J. S. Gavora, B. Benkel, J. Chesuais, W. Fairfull, and J. P. Gibson (eds), Proceedings of the 5th World Congress on Genetics Applied to Livestock Production, 85–86. University of Guelph.

Kolaczkowski B. and Thornton J. W. (2004). Performance of maximum parsimony and likelihood phylogenetics. Nature **431**(October), 980–984. doi: 10.1038/nature02928.1.

Kumar S., Stecher G., Li M., Knyaz C., and Tamura K. (2018). MEGA X: molecular evolutionary genetics analysis across computing platforms. Molecular Biology and Evolution **35**(6), 1547–1549. doi: 10.1093/molbev/msy096.

Lanfear R., Frandsen P. B., Wright A. M., Senfeld T., and Calcott B. (2017). Partitionfinder 2: new methods for selecting partitioned models of evolution for molecular and morphological phylogenetic analyses. Molecular Biology and Evolution **34**(3), 772–773. doi:10.1093/molbev/msw260.

Longmire J., Baker R., and Maltbie M. (1997). Use of 'lysis buffer' in DNA isolation and its implication for museum collections. Museum of Texas Tech University **163**, 1–4.

López-Sepulcre A., Gordon S. P., Paterson I. G., Bentzen P., and Reznick D. N. (2013). Beyond lifetime reproductive success: the posthumous reproductive dynamics of male Trinidadian guppies. Proceedings of the Royal Society B: Biological Sciences **280**(1763). doi:10.1098/rspb.2013.1116.

Lundregan S. L., Hagen I. J., Gohli J., et al. (2018). Inferences of genetic architecture of bill morphology in house sparrow using a high-density SNP array point to a polygenic basis. Molecular Ecology **27**(17), 3498–3514. doi:10.1111/mec.14811.

McKechnie A. E., Freckleton R. P., and Jetz W. (2006). Phenotypic plasticity in the scaling of avian basal metabolic rate. Proceedings of the Royal Society B: Biological Sciences **273**(1589), 931–937. doi:10.1098/rspb.2005.3415.

Metzker M. L. (2010). Sequencing technologies–the next generation. Nature Reviews Genetics **11**, 31–46. doi: 10.1038/nrg2626.

Morgan D. R. and Soltis D. E. (1993). Phylogenetic relationships among members of Saxifragaceae sensu lato based on rbcL sequence data. Annals of the Missouri Botanical Garden **80**(3), 631–660.

Morin P. A. and Woodruff D. S. (1996). Non-invasive sampling for vertebrate conservation. In: R. Wayne and T. Smith (eds.), Molecular Approaches in Conservation, 298–313. Oxford University Press.

Morrissey M. B. and Wilson A. J. (2010). Pedantics: an r package for pedigree-based genetic simulation and pedigree manipulation, characterization and viewing. Molecular Ecology Resources **10**(4), 711–719. doi: 10.1111/j.1755-0998.2009.02817.x.

Nadler S. A. (1995). Advantages and disadvantages of molecular phylogenetics: a case study of ascaridoid nematodes. Journal of Nematology **27**(4), 423–432.

Nascimento F. F., dos Reis M., and Yang Z. (2017). A biologist's guide to Bayesian phylogenetic analysis. Nature Ecology and Evolution **1**(10), 1446–1454. doi:10.1038/s41559-017-0280-x.

Nei M. and Saitou N. (1987) The neighbor-joining method: a new method for reconstructing phylogenetic trees. Molecular Biology and Evolution **4**(4), 406–425. doi: 10.1093/oxfordjournals.molbev.a040454.

Nguyen L. T., Schmidt H. A., Von Haeseler A., and Minh B. Q. (2015). IQ-TREE: a fast and effective stochastic algorithm for estimating maximum-likelihood phylogenies. Molecular Biology and Evolution **32**(1), 268–274. doi:10.1093/molbev/msu300.

Nielsen R., Mattila D. K., Clapham P. J., and Palsbøl, P. J. (2001). Statistical approaches to paternity analysis in natural populations and applications to the North Atlantic humpback whale. Genetics **157**(4), 1673–1682.

Niskanen, A., Billing, A., Holand, H., et al. (2020) Consistent scaling of inbreeding depression in space and time. Proceedings of the National Academy of Sciences **117**(25), 14584–14592.

Ozgul A., Childs D. Z., Oli M. K., et al. (2010). Coupled dynamics of body mass and population growth in response to environmental change. Nature **466**(7305), 482–485. doi: 10.1038/nature09210.

Patwardhan A., Ray S., and Roy A. (2014). Molecular markers in phylogenetic studies-A review. Journal of Phylogenetics & Evolutionary Biology **2**(2). doi: 10.4172/2329-9002.1000131.

Pemberton J. M. (2008). Wild pedigrees: the way forward. Proceedings of the Royal Society B: Biological Sciences **275**, 613–621. doi: 10.1098/rspb.2007.1531.

Penny D., Hendy M. D., and Steel M. A. (1992). Progress with methods for constructing evolutionary trees. Trends in Ecology and Evolution **7**(3), 73–79. doi: 10.1016/0169-5347(92)90244-6.

Posada D. (2008). jModelTest: phylogenetic model averaging. Molecular Biology and Evolution **25**(7), 1253–1256. doi: 10.1093/molbev/msn083.

Pujol B., Marrot P., and Pannell J. R. (2014). A quantitative genetic signature of senescence in a short-lived perennial plant. Current Biology **24**(7), 744–747. doi: 10.1016/j.cub.2014.02.012.

Pyron R. A. (2015). Post-molecular systematics and the future of phylogenetics. Trends in Ecology and Evolution **30**(7), 384–389. doi: 10.1016/j.tree.2015.04.016.

Qian H. and Jin Y. (2016). An updated megaphylogeny of plants, a tool for generating plant phylogenies and an analysis of phylogenetic community structure. Journal of Plant Ecology **9**(2), 233–239. doi: 10.1093/jpe/rtv047.

Rannala B. and Yang Z. (1996). Probability distribution of molecular evolutionary trees: a new method of phylogenetic inference. Journal of Molecular Evolution **43**, 304–311.

Reid J. M., Keller L. F., Marr A. B., Nietlisbach P., Sardell R. J., and Arcese P. (2014). Pedigree error due to extra-pair reproduction substantially biases estimates of inbreeding depression. Evolution **68**(3), 802–815. doi: 10.1111/evo.12305.

Revell L. J., Harmon L. J., and Collar D. C. (2008). Phylogenetic signal, evolutionary process, and rate. Systematic Biology **57**(4), 591–601. doi: 10.1080/10635150802302427.

Riester M., Stadler P. F., and Klemm K. (2009). FRANz: reconstruction of wild multi-generation pedigrees. Bioinformatics. doi: 10.1093/bioinformatics/btp064.

Robinson M. R., Pilkington J. G., Clutton-Brock T. H., Pemberton J. M., and Kruuk L. E. B. (2006). Live fast, die young: trade-offs between fitness components and sexually antagonistic selection on weaponry in Soay sheep. Evolution **60**(10), 2168. doi: 10.1554/06-128.1.

Roeder A. D., Archer F. I., Poinar H. N., and Morin P. A. (2004). A novel method for collection and preservation of faeces for genetic studies. Molecular Ecology Notes **4**, 761–764. doi: 10.1111/j.1471-8286.2004.00737.x.

Ronquist F. and Huelsenbeck J. P. (2003). MrBayes 3: Bayesian phylogenetic inference under mixed models. Bioinformatics **19**(12), 1572–1574. doi: 10.1093/bioinformatics/btg180.

Rzhetsky A. and Nei M. (1992). A simple method for estimating and testing minimum-evolution trees. Molecular Biology and Evolution **9**(5), 945–967. doi: 10.1093/oxfordjournals.molbev.a040771.

Salguero-Gómez R., Jones O. R., Jongejans E., et al. (2016). Fast–slow continuum and reproductive strategies structure plant life-history variation worldwide. Proceedings of the National Academy of Sciences **113**(1), 230–235. doi: 10.1073/pnas.1506215112.

Sambrook J., Fritsch E. F., and Maniatis T. (1989). Molecular Cloning: A Laboratory Manual. Spring Harbour Laboratory Press.

Sanderson M. J., Purvis A., and Henze C. (1998). Phylogenetic supertrees: assembling the trees of life. Trends in Ecology and Evolution **13**(3), 105–109. doi: 10.1016/S0169-5347(97)01242-1.

Sanger F., Nicklen S., and Coulson A. R. (1977). DNA sequencing with chain-terminating inhibitors. Proceedings of the National Academy of Sciences of the United States of America **74**(12), 5463–5467.

Schirmer M., Ijaz U. Z., D'Amore R., Hall N., Sloan W. T., and Quince C. (2015). Insight into biases and sequencing errors for amplicon sequencing with the Illumina MiSeq platform. Nucleic Acids Research **43**(6), e37. doi: 10.1093/nar/gku1341.

Simmonds E. G. and Coulson T. (2015). Analysis of phenotypic change in relation to climatic drivers in a population of Soay sheep *Ovis aries*. Oikos **124**(5). doi: 10.1111/oik.01727.

Simonsen M., Mailund T., and Pedersen C. N. S. (2008). Rapid neighbour-joining. In Crandall, K. A. and Lagergren, J. (eds) Algorithms in Bioinformatics, vol 5251. Springer, Berlin, pp. 113–22. doi: 10.1007/978-3-540-87361-7_10.

Smith L. M. and Burgoyne L. A. (2004). Collecting, archiving and processing DNA from wildlife samples using FTA® databasing paper. BMC Ecology **4**(4), 1–11. doi: 10.1186/1472-6785-4-4.

Smith S. and Morin P. A. (2005). Optimal storage conditions for highly dilute DNA samples: a role for trehalose as a preserving agent. Journal of Forensic Sciences **50**(5), 1101–1108. doi: 10.1520/jfs2004411.

Sneath P. H. and Sokal R. R. (1973). Numerical Taxonomy: The Principles and Practice of Numerical Classification. Freeman.

Stamatakis A. (2014). RAxML version 8: a tool for phylogenetic analysis and post-analysis of large phylogenies. Bioinformatics **30**(9), 1312–1313. doi: 10.1093/bioinformatics/btu033.

Stern D. L. (2013). The genetic causes of convergent evolution. Nature Reviews Genetics. **14**(11), 751–764. doi: 10.1038/nrg3483.

Swofford D. L. (2002). PAUP*. Phylogenetic Analysis Using Parsimony (*and Other Methods). Version 4. Sinnauer Associates. doi: 10.1007/BF02198856.

Teo Y. Y. (2008). Common statistical issues in genome-wide association studies: a review on power, data quality control, genotype calling and population structure. Current Opinion in Lipidology **19**(2), 133–143.

Thompson E. A. (1975). The estimation of pairwise relationships. Annals of Human Genetics **39**(2), 173–188. doi: 10.1111/j.1469-1809.1975.tb00120.x.

Thorne J. L. and Kishino H. (1992). Freeing phylogenies from artifacts of alignment. Molecular Biology and Evolution **9**(6), 1148–1162. doi: 10.1093/oxfordjournals.molbev.a040783.

Tidière M., Gaillard J. M., Berger V., et al. (2016). Comparative analyses of longevity and senescence reveal variable survival benefits of living in zoos across mammals. Scientific Reports **6**(June), 1–7. doi: 10.1038/srep36361.

Tippery N. P. and Les D. H. (2008). Phylogenetic analysis of the internal transcribed spacer (ITS) region in Menyanthaceae using predicted secondary structure. Molecular Phylogenetics and Evolution **49**(2), 526–537. doi: 10.1016/j.ympev.2008.07.019.

Van Oosterhout C., Hutchinson W. F., Wills D. P., and Shipley P. (2004). MICRO-CHECKER: software for identifying and correcting genotyping errors in microsatellite data. Molecular Ecology Notes **4**(3), 535–538.

Vartia S., Villanueva-Cañas, J. L., Finarelli, J., et al. (2016). A novel method of microsatellite genotyping-by-sequencing using individual combinatorial barcoding. Royal Society Open Science **3**(1), 150565.

Vences M., Thomas M., Bonett R. M., and Vieites D. R. (2005). Deciphering amphibian diversity through DNA barcoding: chances and challenges. Philosophical Transactions of the Royal Society B: Biological Sciences **360**(1462), 1859–1868. doi: 10.1098/rstb.2005.1717.

Waits L. P. and Paetkau D. (2005). Noninvasive genetic sampling tools for wildlife biologists: a review of applications and recommendations for accurate data collection. Journal of Wildlife Management **69**(4) 1419–1433. doi:10.2193/0022-541x(2005)69[1419:ngstfw]2.0.co;2.

Wang J. (2004). Sibship reconstruction from genetic data with typing errors. Genetics **166**(4), 1963–1979. doi: 10.1534/genetics.166.4.1963.

Wang J. (2012). Computationally efficient sibship and parentage assignment from multilocus marker data. Genetics **191**(1), 183–194. doi: 10.1534/genetics.111.138149.

Wheeler T. J. (2009). Large-scale neighbor-joining with NIN-JA. In Salzberg S. L. and Warnow T. (eds) Algorithms in Bioinformatics, vol 5724. Springer, Berlin, pp. 375–389. doi: 10.1007/978-3-642-04241-6_31.

Whelan S., Liò P., and Goldman N. (2001). Molecular phylogenetics: state-of-the-art methods for looking into the past. Trends in Genetics **17**(5), 262–272.

Wilson A. J., Réale D., Clements M. N., et al. (2010). An ecologist's guide to the animal model. Journal of Animal Ecology **79**(1), 13–26. doi: 10.1111/j.1365-2656.2009.01639.x.

Xia X. (2018). Distance-based phylogenetic methods. In: X. Xia, Bioinformatics and the Cell, 343–379. Springer. doi: 10.1007/978-3-319-90684-3_15.

Yang Z. and Rannala B. (1997). Bayesian phylogenetic inference using DNA sequences: a Markov chain Monte Carlo method. Molecular Biology and Evolution **14**(7), 717–724.

Yang Z. and Rannala B. (2012). Molecular phylogenetics: principles and practice. Nature Reviews Genetics **13**(5), 303–314. doi: 10.1038/nrg3186.

Yue G. H. and Xia J. H. (2014). Practical Considerations of Molecular Parentage Analysis in Fish. Journal of the World Aquaculture Society **45**, 89–103. doi: 10.1111/jwas.12107.

Zhang J. and Kumar S. (1997). Detection of convergent and parallel evolution at the amino acid sequence level. Molecular Biology and Evolution **14**(5), 527–536. doi: 10.1093/oxfordjournals.molbev.a025789.

Zhang W. (2000). Phylogeny of the grass family (Poaceae) from rpl16 intron sequence data. Molecular Phylogenetics and Evolution **15**(1), 135–146. doi: 10.1006/mpev.1999.0729.

Zwickl D. J. (2006). Genetic algorithm approaches for the phylogenetic analysis of large biological sequence data sets under the maximum likelihood criterion. University of Texas at Austin. (http://www.zo.utexas.edu/faculty/antisense/garli/garli.html).

CHAPTER 2

Biochemical and physiological data collection

Oldřich Tomášek, Alan A. Cohen, Erola Fenollosa, Maurizio Mencuccini, Sergi Munné-Bosch, and Fanie Pelletier

2.1 Introduction

The past three decades have witnessed increasing recognition that incorporation of physiological and biochemical data into demographic and population studies can provide insights into drivers of population dynamics and species distribution, as well as refine estimation of vital rates, thereby increasing predictive power of population models (Cooke et al. 2013; Merow et al. 2014; Violle et al. 2014; Salguero-Gómez et al. 2018; see Chapter 9). An emergence of an integrative field of conservation physiology is one of the most tangible results of such a recognition (Cooke et al. 2013). One of the major promises of integrating physiological data into population and conservation research is the possibility to characterise population structure and dynamics on much shorter time-scales compared to classical demographic approaches (Salguero-Gómez et al. 2018). This possibility is based on the assumption that physiological and biochemical traits or their combinations can be used as proxies (biomarkers; see Box 2.1) for individual ontogenetic stage (Cattet et al. 2018), age (Horvath 2013), sex (Cattet et al. 2017), behavioural or life history strategy (Réale et al. 2010), body condition and health state (Milot et al. 2014), fitness components (Violle et al. 2007), or environmental conditions (Dantzer et al. 2014). Apart from their use as biomarkers, physiological and biochemical traits, together with other functional traits, also emerge as direct targets of conservation science since functional diversity may be as important to protect as the rarity of species (Violle et al. 2017). In addition, demographic approaches can help answer various questions in functional ecology (Salguero-Gómez et al. 2018) or ageing and medical research (Cohen et al. 2014).

> **Box 2.1 Biomarker and functional trait definition**
>
> We use the term *biomarker* to describe any biochemical or physiological trait, or their combination measured at the individual level (from the cell to the whole-organism level) that can be used as a proxy for intrinsic or environmental states, processes, features, or conditions. These variables may be linked to individual fitness or demographic rates or may otherwise be of interest in demographic studies, but are often difficult or even impossible to measure directly (for examples, see the first paragraph of this chapter). For the term *functional trait*, we follow the definition proposed by Violle et al. (2007), who defined it as any morphological, physiological, or phenological feature measured at the individual level that impacts fitness indirectly via its effects on performance traits. Hence, functional traits may be used as biomarkers of fitness components, but other biomarkers need not fulfil the definition of functional traits. Note that, unlike Violle et al. (2007), we do not consider morphological or phenological traits among functional traits here due to the focus of this chapter on biochemical and physiological data. Nonetheless, morphological (e.g. body mass) or phenological traits can also be used as biomarkers and then the same criteria for biomarker validation apply as presented in this chapter.

Oldřich Tomášek et al., *Biochemical and physiological data collection*. In: *Demographic Methods Across the Tree of Life*.
Edited by Roberto Salguero-Gómez and Marlène Gamelon, Oxford University Press.
© Oxford University Press (2021). DOI: 10.1093/oso/9780198838609.003.0002

Despite the clear benefits of such an integration, biochemical and physiological data collection has rarely been discussed in demographic textbooks, indicating that the potential benefits still await wider recognition. Slow blending of physiological and demographic approaches may stem not only from disciplinary cloistering, but also from the difficulties of matching ecological replication numbers with biochemical replication possibilities due to field and laboratory limitations, such as laboriousness, high costs, limited sample volume, and storage difficulties. Nevertheless, overcoming those difficulties promises to bring significant advances in both ecological and medical research. Luckily, new demographic approaches, such as integral projection models (see chapter 10), are robust to low sample sizes, rendering such efforts more feasible (Salguero-Gómez et al. 2018).

In this chapter we briefly review most prominent physiological traits and biomarkers that may be interesting within demographic framework, including the methods of their collection, storage, and analysis, and the criteria to be met before the trait is validated as a biomarker. We hope that this effort will stimulate further integration of physiological and demographic approaches.

2.2 General criteria for selecting biomarkers in demographic studies

Selection of reliable biomarkers can be a difficult task, and an array of criteria should be met during biomarker selection and validation. One of the most important factors to consider is whether the biomarker is intended as a marker of stable (e.g. genetic) differences between individuals (trait biomarker) or as a marker of transient condition or environmental stress (state biomarker). Good trait biomarkers must show high within-individual repeatability over the long term, whereas state biomarkers should respond to environmental conditions. The best way to assess biomarker consistency is analysing short- and long-term within-individual repeatabilities based on intraclass correlation coefficients, which take into account the between-individual variation within the population (Hõrak and Cohen 2010). Importantly, interpretation of state markers may be context dependent and regulated and unregulated processes must be distinguished. For example, an exposure to a stressful environment can result in unregulated depletion of an antioxidant; however, long-term exposure to that

environment can lead to adaptive antioxidant upregulation (Hõrak and Cohen 2010).

Another factor to consider in biochemical biomarkers is their stability. In some cases, biomarkers may degrade or form during storage, biasing estimates, particularly if not all samples are analysed simultaneously. Biomarkers that are less stable over time need to be stored in –80°C or even in liquid nitrogen or its vapours, which poses logistic complications in the field and increases study costs. Nevertheless, dry-shipper containers are available, in which liquid nitrogen is bound in a porous material to prevent spilling. These containers can be carried safely to the field and can even be transported by air. When storage in –20°C is sufficient, non-frost-free freezers should be used because temperature rises briefly to 0°C every 24 hours in frost-free freezers (Sheriff et al. 2011). Despite these considerations, caution should be exercised when analysing/interpreting samples stored over long periods (years).

Prior to the analysis of the actual samples, the analytic methods for biomarker quantification must be validated in each species and sample matrix used (Hõrak and Cohen 2010). The criteria evaluated during the validation process include method specificity (or selectivity), sensitivity, accuracy, range of linearity, reagent and sample stability, and measurement precision

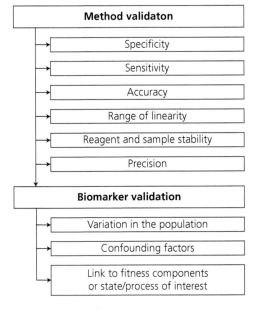

Figure 2.1 Biomarker validation process and criteria.

(Figure 2.1). Measurement precision is the ultimate criterion determining reliability and usefulness of the method. Although coefficient of variation (i.e. relative standard deviation) is usually recommended as a precision measure in the methodological literature, repeatability calculated as intraclass correlation coefficient should be preferred as it takes into account biological context (i.e. actual biomarker variation in the population; Hõrak and Cohen 2010). To this end, each sample is aliquoted, and the assay is replicated in at least two aliquots per sample, ideally including all the sample processing, not only the final measurement. In addition, repeatability among measurers should be assessed for manual methods (e.g. microscopy). As method validation is of utmost importance, we encourage readers to familiarise themselves with this process from existing reviews on this topic (e.g. Green 1996), especially if the analysis will not be done by a specialised laboratory.

Biomarker validation is the next stage of the validation process (Hõrak and Cohen 2010). This stage involves exploration of within- and between-individual variation in the population and determination of confounding factors such as sex, age, body size, reproductive state, temperature, season, or time of sample collection. Ultimately, the link between the biomarker and fitness component (or any other state/process of interest) must be established. This is probably the most challenging task of the biomarker validation process, however, as the strength and direction of correlation with fitness components may vary among contexts (e.g. Breuner et al. 2008). Hence, understanding this variation is undoubtedly one of the most important tasks of current demographic research (Shipley et al. 2016).

2.3 Designing the data collection

The physiological and biochemical data can enter demographic models either at the individual level or at the level of populations or species (Edmunds et al. 2014). The design of the data collection depends on the data type needed, and this in turn depends on the question at hand. In many cases, population- or species-level data will be sufficient. Nevertheless, both between- and within-individual variation may affect vital rates and individual-level data are needed if those effects are of interest (Edmunds et al. 2014). In such a case, the sampling design should aim to accurately quantify within-species biomarker variability in space and time, with random selection of individuals

and their repeated sampling over time, ideally over their lifetimes, being the preferred sampling methods (Hõrak and Cohen 2010; Violle et al. 2012).

In plants, repeated sample collection is usually possible even in the case of whole organs (e.g. leaves), although the extent of the damage and an herbivory-like response should be considered. In animals including humans, however, the requirement of repeated sampling largely constrains the spectrum to nondestructive and noninvasive sample types, such as bodily fluids (e.g. blood), secretions (e.g. saliva), excreta, and integumentary structures (e.g. hairs or feathers). Another limitation is posed by the need to observe both general and country- and taxon-specific ethical norms and recommendations (Lindsjö et al. 2016; Drinkwater et al. 2019). In any case, sample collection should only minimally interfere with normal life activities of the studied organism. Noninvasive samples, such as hairs or faeces, are ideal in this regard as the information can be obtained without even approaching the individual. When collecting blood, the volume should not exceed 1% of body mass per sampling event (Owen 2011). The site and means of blood collection are taxon specific and can be found in relevant literature (Sheriff et al. 2011). Blood-sucking bugs can be used to minimise stress and interference with normal activities, although applicability of this method is taxon and biomarker dependent (Bauch et al. 2013). Chelating anticoagulants (e.g. EDTA) must be avoided during blood collection for assays requiring free metal ions (Harr 2005). In arthropods, several protocols for haemolymph sampling have been developed, although in small-bodied species, samples need to be pooled (Tabunoki et al. 2019).

At the population or species level, many physiological and biochemical trait values can be obtained from published data sets and databases. We provide a (definitely not exhaustive) list of databases containing physiological and biochemical traits and related sources in Table 2.1. Additional data sets can certainly be found in public data repositories.

2.4 Physiological traits

2.4.1 Animal metabolism

Metabolic traits belong among the most promising physiological traits for demographic and large-scale comparative studies because energy is a common currency across the Tree of Life (Salguero-Gómez et al. 2018). It is tightly linked to fitness due to its importance

Table 2.1 Databases containing biochemical and physiological traits and other related sources.

Database name	Description	Link
AnAge	Database of animal life history traits including metabolic rates	http://genomics.senescence.info/species/
Biofuel Ecophysiological Traits and Yields database (BETYdb)	Open-access database of plant traits, biomass yields, and observations of ecosystem dynamics oriented on bioenergy production	www.betydb.org
Coral Trait Database	Database of coral traits	coraltraits.org
Encyclopedia of Life TraitBank	Global trait database for all organisms	eol.org/docs/what-is-eol/traitbank
GlobTherm	Global database on thermal tolerances for aquatic and terrestrial organisms; maintained on iDiv Data	datadryad.org/resource/doi:10.5061/dryad.1cv08
HormoneBase	Population-level database of steroid hormone levels from free-living, adult vertebrates	hormonebase.org
iDiv Data	Data repository of German Centre for Integrative Biodiversity Research curating multiple data sets	idata.idiv.de
Planteome	Project developing common annotation standards and reference ontologies for plant genes and traits	planteome.org
Registry of Research Data Repositories	Comprehensive registry of research data repositories	re3data.org
Thesaurus Of Plant Characteristics for Ecology and Evolution (TOP)	Project standardising terminology for plant traits and environmental associations	top-thesaurus.org
TraitNet	Collaborative network website for plant trait-based research including a list of external trait databases	traitnet.ecoinformatics.org
traits	R package enabling accessing multiple trait databases	cran.r-project.org/web/packages/traits/index.html
TRY Plant Trait Database	Global plant trait database	www.try-db.org/

for growth, reproduction, and survival (Burton et al. 2011; Pettersen et al. 2018). The field of energetics has therefore developed several tools to measure the rates at which an individual uses energy.

Minimum metabolic rates, that is, basal metabolic rate in endotherms, standard metabolic rate in ectotherms, and their less rigorous version, resting metabolic rate, are defined as the minimum rates of energy expenditure per unit time needed to maintain vital body functions (Chabot et al. 2016; Norin and Metcalfe 2019). In contrast, peak (maximum; summit) metabolic rate expresses the maximum rate of aerobic energy utilisation (Norin and Metcalfe 2019). In between lies field metabolic rate (FMR; or daily energy expenditure) that gives the free-living energy demands of the individual (Speakman 1997). In addition, routine metabolic rate has frequently been measured in fishes, involving nonexhaustive swimming in laboratory conditions (Chabot et al. 2016). Importantly, minimum, maximum, and field/routine metabolic rates

can be uncoupled and subject to different selection pressures (Norin and Metcalfe 2019). Despite pronounced plasticity, metabolic rates are repeatable within individuals and there is typically up to threefold intraspecific variation among individuals (Burton et al. 2011; Norin and Metcalfe 2019).

Except for FMR, metabolic rates are usually estimated using respirometry, which measures oxygen consumption and/or carbon dioxide production of an individual in a metabolic chamber (Lighton 2008). This requires establishing a field laboratory, which renders respirometry difficult to perform in the field and only feasible in small animals (perhaps up to the size of a rabbit). In animals that can easily be captured repeatedly, FMR estimation using doubly labelled water (DLW) to estimate carbon dioxide production may be a more convenient, but expensive, alternative (Speakman 1997; Speakman and Hambly 2016). The classical procedure involves DLW injection upon capture, collection of a first blood sample, release of an individual, and its recapture after one or more days, followed by the collection of a second blood sample. Alternatively, DLW application through a food item followed by urine or faeces collection or blood collection using blood-sucking bugs could theoretically enable DLW method to be performed without capturing the animal (Speakman and Hambly 2016). Alternative methods of FMR estimation more suitable for large animals or animals that are difficult to recapture include heart rate telemetry or accelerometry (see section 2.9).

2.4.2 Plant metabolism

Plant energy metabolism is centred on photosynthesis and respiration. In demographic studies, measurements of photosynthesis and respiration are almost universally carried out on organs, such as leaves, flowers, fruits, or photosynthetic stems, while they are still attached to the plant. Several types of equipment exist, and users need to be aware of technical characteristics and potential sources of error and bias (Long and Bernacchi 2003; cf. http://prometheuswiki.org/tiki-custom_home.php). The most common instruments are infra-red gas analysers (IRGAs). IRGAs consist of an environmentally controlled cuvette, inside which the leaf is enclosed, and of computer-controlled units transporting and measuring gas concentrations and flows. Ultimately, photosynthetic rates in units of moles of carbon dioxide exchanged with the atmosphere per unit of time and leaf area are measured (the corresponding water vapour flux being transpiration). Once organ-scale fluxes are determined, whole-organism values can be obtained using total leaf areas, sapwood volumes, or stem surface areas. Simultaneous measurement of environmental variables is essential to obtain meaningful estimates of traits, such as light-saturated photosynthesis or temperature-corrected dark respiration (Abdul-Hamid and Mencuccini 2009; Pérez-Harguindeguy et al. 2013).

Fast-response systems suitable for demographic surveys also exist to measure other metabolic quantities. Leaf stomatal conductance (related to the stomatal aperture to CO_2 and water exchange) can be obtained with relatively cheap equipment with short equilibration times (60–120 s; e.g. Vanderklein et al. 2007). The fluorescence naturally emitted by leaves upon illumination by a specific light source can be measured with modulated chlorophyll fluorometers. Modulated fluorometers consist of high-frequency light-emitting diodes and receiving transducers measuring the fluorescence re-emitted from excited chlorophyll

molecules. Depending on the protocol, the intensity and spectrum of emitted fluorescence give information on the photochemistry of photosystem II, such as F_v/F_m, the maximum efficiency of photosystem II (Murchie and Lawson 2013). Many fluorometers are suitable for rapid and large demographic surveys (Mencuccini et al. 2014; Semerci et al. 2017).

Measurement of vertical xylem sap transport in woody plant stems obtain whole-plant sap flow (closely related to photosynthesis). Implanted resistors heat the xylem, while highly sensitive thermocouples measure temperature changes as sap moves in the wood. Once installed, these sensors can monitor plant water fluxes for months to years at high time resolution (e.g. Martínez-Vilalta et al. 2007). Well-tested protocols for in-house sensor construction and calibration facilitate the task (e.g. Fuchs et al. 2017).

2.4.3 Semen physiology

Semen characteristics, especially sperm phenotypic and performance traits, represent important physiological features that may either be used as biomarkers or be of interest per se (Reinhardt et al. 2015; Lemaître and Gaillard 2017). Given their fundamental role in male reproductive success, sperm traits can serve as biomarkers of male fertility and reproductive fitness (Gomendio et al. 2006), or predict population and species differences in extrapair paternity rate (Lifjeld et al. 2010; Laskemoen et al. 2013; Šandera et al. 2013). Sperm traits further show pronounced environment-driven plasticity, implying their potential as biomarkers of environmental conditions (Reinhardt et al. 2015). Sperm traits and other semen characteristics also appear as main objects of interest in demographic studies, such as those focused on reproductive senescence (Johnson et al. 2015; Lemaître and Gaillard 2017).

The feasibility of repeated semen collection greatly varies across animal taxa. In some vertebrates, including humans, semen can be collected by massaging, applying pressure, either with or without anaesthesia, or using an artificial vagina or a dummy female (Birkhead et al. 2009; Lifjeld et al. 2010; Wasden et al. 2017; Humann-Guilleminot et al. 2018). In other vertebrate species, electroejaculation can be feasible (Lierz et al. 2013), though it can affect semen characteristics (Birkhead et al. 2009). Nonetheless, surgical or destructive semen collection is the only option in many taxa, rendering them unsuitable for longitudinal studies.

Semen characteristics commonly measured include semen volume, sperm concentration, sperm motility, sperm velocity, sperm morphometric traits, and abnormal sperm proportion (Fitzpatrick and Lüpold 2014). In animal field studies, however, sperm motility and velocity analysis may be challenging as it requires video recording of fresh samples using a microscope with a heated table, whereas, for the other analyses, samples can be stored in preservative media, such as formalin (Tomášek et al. 2017). Increasing attention has recently been devoted to more biochemical traits, including sperm ATP content or seminal plasma composition; however, those usually require immediate storage in liquid nitrogen (Poiani 2006; Fitzpatrick and Lüpold 2014). It should be emphasised that the importance of semen characteristics for male reproductive success varies greatly among taxa or even among studies on the same species (Fitzpatrick and Lüpold 2014). Hence, the link to fitness components must always be validated.

2.5 Biochemical traits

2.5.1 General biochemistry

Animal and human biochemistry

Plasma and serum biochemistry (i.e. arrays quantifying activity of various tissue enzymes and concentrations of electrolytes, minerals, proteins, lipids, and other metabolites in blood plasma or serum) are commonly used in human and veterinary medicine to assess individual health, nutritional status, and metabolic functions (Harr 2005; Cohen et al. 2014). Because clinical biochemistry is usually available in large longitudinal human studies, these tests are increasingly used in demographic studies of human ageing and age-related diseases (Cohen et al. 2014). Unlike human studies, plasma/serum biochemistry is underutilised in animal ecological research, despite its potential to provide valuable information about individual body condition and biological ageing rate (Cohen et al. 2014; Milot et al. 2014). One reason for the underutilisation is that the interpretation of individual markers can vary across species. For example, glucose levels that would be toxic in mammals are standard in birds (Tomášek et al. 2019). Similarly, bird species with carotenoid-based colouration tolerate and even seem to benefit from carotenoid levels that are known to cause pathologies in species lacking carotenoid-based displays, implying that physiological consequences of carotenoid levels are highly species specific (Cohen and McGraw 2009). Likewise, a single biomarker may indicate different things in different contexts. For example, uric acid is an antioxidant but is also related to protein catabolism during fasting (Hollmen et al. 2001), so high uric acid levels may indicate different things depending on food availability. For these reasons, while many ecophysiology studies do use specific markers in well-validated contexts (Podlesak et al. 2005), broad-scale or comparative use can be more challenging, and care should be exercised in a demographic context.

Stress and recent feeding affect concentrations of some metabolites such as glucose, uric acid, or lipids (Harr 2005). Fasting is usually unattainable in the wild, but stress effects can be minimised by taking blood samples within three minutes from the initial capture (Tomášek et al. 2019). Blood should be centrifuged immediately upon collection to prevent glucose consumption by blood cells. Plasma or serum can be stored up to 6 hours in room temperature for standard biochemistry arrays (Monneret et al. 2016). Analyses are usually done in specialised human or veterinary laboratories, although analysers are commercially available for in-house use. As always, all the analytic methods must be validated when used for nonmodel species.

Biochemistry can also provide useful information in invertebrates. In corals, for example, biochemical data, including calcification and content of chlorophyll, lipids, and proteins, have been proposed to inform demographic modelling of ecological performance of coral reefs (Edmunds et al. 2014).

Plant biochemistry and element content

Plant biochemistry deals with the structure and function of compounds such as carbohydrates, proteins, lipids, and other biomolecules. The most abundant plant molecule, however, is water. Relative water content and water potential Ψ are frequently employed to quantify abundance and chemical energy of water in plant tissues. Plant Ψ measures the unit-volume potential energy of water molecules relative to unit-volume free water under reference conditions. It quantifies the tendency of water to move from areas of high to low potential energy. Hence, the difference in Ψ between, for example, roots and leaves, drives water flow. Plant Ψ is quantified using Scholander pressure chambers (Turner 1988) or one of several psychrometric techniques. In the field, plant Ψ is measured prior to dawn (after night-time rehydration) and at midday (minimum water content). A meaningful trait for demographic studies is the seasonal absolute minimum Ψ_{min} (Ψ is negative), which measures the maximum level of water stress a population of plants can typically tolerate at a site. Ψ_{min} correlates with several other water relations and hydraulic traits, such as osmotic potential at saturation, turgor loss point, and resistance to embolism (e.g. Fu and Meinzer 2019).

Carbohydrates and lipids are important storage and communication compounds strictly related to plant metabolic status. Among the first class, starch, sucrose, glucose, fructose, and saccharose are often determined in demographic studies. Lipids (e.g. glyceride-glycerols) can be important in some species (e.g. Sala and Hoch 2009). Beside extraction, the choice of quantification method, either enzymatic, acid digestion, or ion chromatography, can affect the results (Landhäusser et al. 2018). Generalised correlations between high-throughput near-infrared spectroscopy spectra and wet-chemistry concentrations have been determined (Ramirez et al. 2015). Carbohydrate and lipid contents can vary at diurnal and seasonal time-scales and standardised sampling is required.

Pigments such as chlorophyll and carotenoids (light-absorption and photo-protection) and vitamins such as tocopherols (vitamin E) and ascorbate (vitamin C), which are both antioxidants, can provide valuable information in demographic studies (Asensi-Fabado and Munné-Bosch 2010; Mencuccini et al. 2014; Mencuccini and Munné-Bosch 2017). Similarly, macro- (N, P, K) and micro-nutrient concentrations can be strong indicators of different life history strategies (Pérez-Harguindeguy et al. 2013). Intrinsic water use efficiency (ratio of photosynthesis to stomatal conductance) integrated over the lifetime of a tissue can be estimated via $^{13}C/^{12}C$ stable isotope composition (Dawson et al. 2002). This and other stable isotope ratios ($^{1}H/^{2}H$, $^{18}O/^{16}O$, $^{15}N/^{14}N$) are determined by mass spectrometry of destructively sampled plant tissues and constitute traits useful for demographic studies (Pérez-Harguindeguy et al. 2013).

2.5.2 Hormones

Animal and human hormones

Hormones are signalling molecules that, based on information about the internal body state and environmental conditions, regulate physiological processes, phenotype, development, and other life history traits in order to promote individual fitness (Finch and Rose 1995; Bonier et al. 2009). The most studied

hormones in vertebrates are sex (e.g. testosterone or oestrogen) and stress steroids (glucocorticoids). Glucocorticoids mediate physiological stress response which evolved as a mechanism protecting internal stability (homeostasis) in the face of predictable and unpredictable environmental stimuli ('stressors'). Hence, glucocorticoid levels can provide information about environmental stress, including stress from predation risk or social conflict, as well as psychological or anticipatory stress (Dantzer et al. 2014). Their concentrations are also linked to metabolic rate (Jimeno et al. 2018). In addition, sex steroids mediate sex-specific life history trade-offs (Brooks and Garratt 2017), and a combination of multiple sex and stress hormones can be used as markers of sex, developmental stage, reproductive state, or individual ageing rate (Walther et al. 2016, Cattet et al. 2017; Cattet et al. 2018). Nonsteroid hormones that might be of interest in this context include human chorionic gonadotropin, follicle-stimulating hormone, luteinising hormone, and prolactin (Valeggia 2007).

To obtain baseline concentrations, it is necessary to collect blood samples within 2 or 3 minutes from the initial capture because glucocorticoids increase rapidly following the capture. The concentrations peak between 15 and 30 minutes from the capture, when a second sample should be collected if acute stress response is of interest (Sheriff et al. 2011). Adrenocorticotropic hormone (ACTH) may be injected to standardise the stress response (Sheriff et al. 2011). Chronic stress can also be assessed. It should be noted that blood plasma glucocorticoids are present in two states: free, and bound to corticosteroid-binding globulin (for discussion about biological importance, see Breuner et al. 2013).

Both sex and stress hormones can be analysed in plasma, serum, and dry blood spot samples, but also noninvasively in saliva, faeces, urine, feathers, or hairs (Valeggia 2007; Sheriff et al. 2011; Cattet et al. 2017; Bílková et al. 2019). While saliva steroid concentrations reflect immediate blood levels, samples of excreta and integumentary structures provide an integrated measurement reflecting endocrine status during an extended period of time from hours (excreta) to weeks (integumentary structures), rendering them suitable for analysis of chronic stress, which may be of interest in demographic studies for its fitness consequences (Dantzer et al. 2014).

Anticoagulants such as heparin or EDTA can be used for blood plasma collection. Samples can be stored up to 72 hours at 4°C or for decades at −20 or −80°C

(Valeggia 2007; Sheriff et al. 2011). Dry blood spots can even be stored up to 8 weeks at room temperature or 1 week at tropical temperature (37°C) (Valeggia 2007). Radio- or enzyme-immunoassays are available for quantification of both steroid (Sheriff et al. 2011) and nonsteroid hormones (Sturgeon and McAllister 1998; Wheeler 2006). Alternatively, LC-MS/MS methods exists that allow quantification of multiple steroid hormones in one analysis (Bílková et al. 2019).

Plant hormones

Phytohormones are molecules produced in very low concentrations that regulate all developmental processes throughout the life cycle in plants, from seed germination to plant senescence. Although no study has directly linked hormones with demographic plant projection, the role of plant hormones in relation to life history traits is well understood. Phytohormones not only include the five 'classical' hormones, auxin, gibberellins (GAs), cytokinins, abscisic acid (ABA) and ethylene, but also jasmonates, salicylates, strigolactones, brassinosteroids, and polyamines, among other newly emerging organic compounds with important regulatory function and hormone-like activity, such as melatonin and karrikins. Although some of them have been classically defined as stress hormones (e.g. ABA), this is far too simplistic, and it is now clear that all physiological processes are regulated by the so-called hormonal balance rather than by a single compound (Müller and Munné-Bosch 2011). Hormonal cross-talk is therefore essential to study hormone-related processes in plants and the understanding of plant hormonal response requires a complete hormonal profiling analysis. Moreover, hormonal action does not only depend on hormone contents (which in turn are not only influenced by biosynthesis but also conjugation and breakdown), but on hormonal transport, hormone sensitivity, and hormonal signalling.

A simultaneous quantification of multiple plant hormones is the best approximation for an accurate hormonal profiling. In that way, multiplex gas chromatography tandem mass spectrometry (GC-MS/MS) or liquid chromatography coupled to tandem mass spectrometry (LC-MS/MS) are some of the used methods to quantify phytohormones. Although plant material needed is below 0.5 g of fresh weight, samples must be strictly collected and preserved in ultracold conditions, using liquid nitrogen for sampling followed by storage at −80°C. All organs can provide information regarding plant physiological status which is potentially linkable with life history traits. Nevertheless, their

importance may depend on the research question and a completely different hormonal profiling with contrasting endogenous contents in some of the phytohormones may be obtained if either leaves, roots, fruits, flowers, or seeds are sampled.

2.5.3 Oxidative stress

Oxidative stress is an ancient stress response shared by both animals and plants (Costantini et al. 2010; Demidchik 2015). It results from excessive formation of free radicals and other reactive species (RSs) that are insufficiently eliminated by antioxidant mechanisms. Accumulating RSs induce oxidative damage to cell components causing their dysfunction (Halliwell and Gutteridge 2015). For this reason, oxidative stress has long been thought to be involved in ageing and development of numerous pathological conditions, though it has fallen out of favour as a primary ageing cause (Halliwell and Gutteridge 2015). Nonetheless, the need to maintain redox homeostasis is believed to be a major constraint in life history, possibly partially underpinning the trade-off between reproduction and longevity (Costantini et al. 2010; Isaksson et al. 2011). Elevated oxidative stress can be associated with ageing or any biotic or abiotic stressor (Costantini et al. 2010; Demidchik 2015). Redox state markers may hence serve as proxies of biological age or body condition. Moreover, their integration into demographic models may advance our mechanistic understanding of ageing process and disease development, provide insights into the roles of senescence or environmental stress in population dynamics, and shed light on mechanisms underpinning life history evolution.

Redox state analysis represents a major challenge since redox state integrates a complex system of oxidation-reduction reactions and pathways (Jones 2006; Monaghan et al. 2009). Ideally, all components determining redox state should be measured, namely RS production, antioxidant defences, oxidative damage, and repair mechanisms (Dotan et al. 2004). This is, however, rarely done in practice, and quantification of oxidative damage together with some measure of antioxidant defences is the approach that is most commonly taken (Monaghan et al. 2009). Nonetheless, the view of redox homeostasis as a single equilibrium between RSs and antioxidants is an oversimplification of reality as dysregulation in any single redox pathway can trigger tissue-specific damage (Jones 2006).

For this reason, it is highly advisable to employ multiple assays and biomarkers reflecting oxidative damage to all major biomolecule types, namely lipids (e.g. d-ROMs test, isoprostanes, and malondialde-hyde), proteins (e.g. carbonyl assay, *ortho*-tyrosine, 3-nitrotyrosine, or halogenated tyrosines) and DNA (e.g. comet assay, 8-hydroxy-2′deoxyguanosine, 8-hydroxyguanosine). Mass spectrometry-based methods are generally preferred over spectrophotometric methods due to their higher specificity and precision (Mateos and Bravo 2007; Halliwell and Gutteridge 2015; Morales and Munné-Bosch 2019).

Detection of RSs is difficult due to their very short half-lives and is rarely done in clinical and field studies, although fluorescent probes have been used successfully to detect superoxide in ecological studies (e.g. Olsson et al. 2008).

The most important antioxidant defences include antioxidant enzymes and nonenzymatic hydrophilic and lipophilic antioxidants (Valko et al. 2007; Pamplona and Costantini 2011). The hydrophilic antioxidants glutathione and ascorbic acid are active primarily in aqueous phases of cells and in the circulatory system (Yeum et al. 2004). The ratios of their reduced and oxidised forms are often used as markers of whole-organ or whole-body oxidative stress (Queval and Noctor 2007). In contrast, lipophilic antioxidants (e.g. tocopherols or carotenoids) are related to oxidative stress in cell and organelle membranes (Yeum et al. 2004). Spectrophotometric assays are available for important antioxidant enzymes, glutathione, and ascorbic acid (Queval and Noctor 2007; Halliwell and Gutteridge 2015). In addition, all nonenzymatic antioxidants and chloroplastic pigments (related to oxidative stress in chloroplasts) can be measured using chromatography-based techniques (Esteban et al. 2015; Halliwell and Gutteridge 2015).

A variety of in vitro assays have been developed intended to provide an integrated measure of total antioxidant capacity (TAC) (reviewed in Somogyi et al. 2007; Monaghan et al. 2009; Halliwell and Gutteridge 2015). Recently, a ratio of stereoisomers of hydroxy-octadecadienoic acid, an oxidation product of linoleic acid, has been proposed as a promising marker of in vivo lipophilic antioxidant capacity (Yoshida et al. 2007; Tomášek et al. 2016).

Oxidative stress can be measured in any tissue or body fluid, although nondestructive or noninvasive sample types are preferable in longitudinal studies. In plants, light conditions during sampling need to be standardised as oxidative stress may increase with the accumulation of intense light hours. Unless intact cells are required, samples must immediately be frozen in liquid nitrogen and stored at −80°C or preferably in liquid nitrogen (Barden et al. 2014). It is usually recommended to analyse samples within 1 month (Barden et al. 2014), but samples stored in −80°C for up to 60 months give reliable results for several markers and antioxidants (Piłacik et al. 2002; Jansen et al. 2017). Addition of EDTA/BHT/GSH may help prevent sample autoxidation during storage or processing (Barden et al. 2014), but one must make sure that these chemicals do not interfere with the assays.

2.6 Telomeres

Telomeres are complexes of tandem repeats of a short DNA sequence and associated proteins that cap the ends of linear chromosomes. They identify and protect chromosome ends, thus maintaining integrity of the coding sequences (Nussey et al. 2014). Importantly, in the absence of restoration processes such as telomerase activity, telomeres shorten during each cell division due to the inability of DNA polymerase to completely replicate the end of the lagging DNA strand (the 'end replication problem'). When telomeres shorten to a critical limit, cell division stops, and this state termed *cellular replicative senescence* is usually followed by cell death. This mechanism limits the number of potential cell divisions and is believed to have evolved to prevent cancer development (Young 2018).

The possible causal involvement of telomere shortening in ageing is the focus of a considerable body of research, which greatly benefits from longitudinal demographic approaches (Blackburn et al. 2015; Young 2018). In demographic studies, telomere length also emerges as a promising biomarker of ageing, survival, remaining lifespan, or stress accumulated over life. This interest stems from the presumable association between cellular replicative senescence and mortality, and from observations that stressors, including oxidative stress, increase telomere attrition during each cell division (Blackburn et al. 2015; Eastwood et al. 2018; Wilbourn et al. 2018). Moreover, telomere shortening rate, rather than telomere length itself, has been reported to predict species lifespan in birds and mammals in comparative studies (Tricola et al. 2018; Whittemore et al. 2019; Udroiu 2020).

We also note here that, while telomeres are of continued interest in the physiological ecology community, they are increasingly being questioned as a biomarker of human ageing (Sanders and Newman 2013; Ferrucci et al. 2020). Several studies have shown that telomere length correlates poorly with other measures of the ageing process (Belsky et al. 2018). It is increasingly viewed as a manifestation of psychological stress and allostatic load (Meier et al. 2019), but not of ageing per se, though there is still some debate. The extent to which these human findings translate into the physiological ecology literature is not clear, but given the history of physiological ecologists jumping naively on physiological bandwagons emerging from human or biomedical literature (Viney et al. 2005; Bonier et al. 2009; Isaksson et al. 2011; Cohen et al. 2012), we urge some caution.

Multiple methods of telomere length measurement have been developed, including terminal fragment (TRF) analysis, real-time qPCR, quantitative fluorescence in situ hybridisation (Q-FISH), single telomere length analysis (STELA), and dot blot (reviewed in Aubert et al. 2012; Montpetit et al. 2014; Nussey et al. 2014). Nonetheless, none of the methods is ideal, and the choice of an appropriate method for a given study depends on whether the desired qualities include analysis of not only the mean, but also the whole distribution of telomere lengths, high throughput, low amount of DNA needed, a lack of interference by interstitial telomeric (e.g. centromeric) sequences, or the ability to detect very short telomeres that are the most relevant to fitness (Nussey et al. 2014). Other factors that need to be considered include comparability between studies (Verhulst 2020), methods of long-term storage and extraction that ensure the highest DNA quality (Eastwood et al. 2018), or specialised equipment required. Tissue type collected is yet another important factor to consider as different cell types (even different leucocytes, for example) can vary greatly in proliferation rate or telomerase activity, and thus in telomere dynamics over lifetime (Nussey et al. 2014).

2.7 Immune markers

Immune function is a crucial aspect of physiology linking various physiological systems (endocrine, neurobiological, metabolic, etc.), psychology, stress, and various aspects of the environment (pathogen pressure, nutrition, etc.). Accordingly, it is a potent signal of organismal state, but also challenging to quantify and interpret. The immune system is immensely complex, with innate and acquired components, cellular and humoral components, and dozens if not hundreds of specific cell types, all linked in complex regulatory networks. Therefore, it is increasingly accepted that no single measure is sufficient to understand immune variation (Matson et al. 2006; Buehler et al. 2011). Moreover, differential investment in different parts of the immune system may reflect alternative strategies, either within individuals over time or across individuals (Pulendran 2004; Viney et al. 2005). For example,

changes in the balance between innate and acquired immune function that were previously considered to be 'immunosenescence' have recently been proposed to be adaptive changes in the immune system based on changing needs across ages and on ageing in other aspects of physiology (Fulop et al. 2017). While the truth may be somewhere in the middle, it is clear that we do not fully understand how to distinguish adaptive versus pathological changes in immune state, nor indications of poor condition versus alternative strategies.

In vertebrates, including humans, the most basic measures are cell counts from slides quantifying several key cell types. Beyond this, there are numerous specific assays for cytokines implicated in regulating various aspects of immune response, as well as flow cytometry approaches to get cell counts for precise cell types based on surface markers (Lu et al. 2016). Under field conditions, a series of field-adapted immune measures have been developed. In addition to PHA and slide-based cell counts, key assays include haptoglobin (an acute phase protein), immunoglobulin concentrations and globulin concentrations more generally (Cheynel et al. 2017), hemagglutination and lysis (indicators of natural antibody response in the innate system; Millet et al. 2007), and bacterial killing assays (Millet et al. 2007). Nevertheless, there are notable challenges. First, the correlation structure among these assays is unstable, such that there is no agreed-upon way to measure overall immune status, nor several key axes (Matson et al. 2006; Buehler et al. 2011; Pigeon et al. 2013). Second, a key aspect of immune function, the acquired immune response, is notably absent from field assays, yielding uncertainty whether low levels of innate immunity (the primary subject of most assays) simply reflect investments in adaptive immunity, or insufficient levels of protection. Third, while bacterial killing assays may seem to be a promising way to integrate overall function, in practice, results can be hard to interpret: scores for different strains of bacteria/yeast are often uncorrelated (Millet et al. 2007), and it is unclear how increased investment in protection against bacteria might trade off with other kinds of protections, such as against viruses or macro-parasites.

The most promising uses of immune markers in demography will thus likely rely on specific validation of particular assays/approaches in a particular context. For example, in modern humans, chronic low-grade inflammation is a major cause of chronic diseases and generally accompanies the ageing process (Franceschi et al. 2000). Individual cytokines involved in the process may also vary for other reasons (e.g. with acute infection), such that a more reliable signal of this 'inflamm-ageing' can be obtained by statistical integration of several key cytokines via principal components analysis (Morrisette-Thomas et al. 2014; Cohen et al. 2018). This analysis shows that the process involves upregulation of both pro- and anti-inflammatory cytokines, indicating activation of the system rather than pro-anti balance. Accordingly, it is possible to obtain precise measurement of a specific immune process that is of particular demographic importance in modern human populations; however, this index would likely be of little use in other species or even premodern humans. A second example is the use of antihelminth-specific antibodies in Soay sheep (*Ovis aries*), levels of which predict survival and body mass in the study population, though with relationships highly specific to age and isotype (Sparks et al. 2018). This study shows that detailed natural history can provide highly tailored approaches to specific contexts, even if they are not more broadly applicable across species and populations. In fact, we should expect such specificity to be the norm in most cases.

2.8 Epigenetic and omics data

The recent explosion in capacity to measure large quantities of biological information ('omics') cheaply has led to a revolution in our ability to measure genotype and phenotype. Roughly speaking, the main types of omics data currently available include (1) transcriptomics, quantifying messenger mRNA levels (i.e. gene expression); (2) epigenomics, usually DNA methylation data, which influences the likelihood of gene expression; (3) metabolomics, measuring large numbers of circulating small molecules, often intermediates of various metabolic pathways; and (4) proteomics, quantifying the levels of many proteins. While many of these methods are still too expensive for routine use in field contexts, that may be changing as technology improves and prices fall. Luckily, our experience with metabolomics suggests that price per sample may be reduced considerably when large numbers of samples are analysed.

The quantity of data generated has not been fully accompanied by ways to make sense of it. Multiple testing issues become daunting when assessing thousands of molecules simultaneously, particularly with the limited sample sizes often available in field studies. In some cases, algorithms have been generated that can simplify large numbers of molecules for specific purposes. Perhaps the best example is the human epigenetic clock, a set of 353 DNA methylation sites

identified using a machine learning algorithm that together can generate an index of biological age that correlates with chronological age at $r = 0.96$ across a wide array of tissues (Horvath 2013). While the stability of the epigenetic clock is a striking biological phenomenon, caution is also warranted when interpreting such indices. It is not clear to what extent the clock is measuring biological ageing per se, versus methylation sites that change with age-correlated phenomena despite having little link to the ageing process (Hertel et al. 2019). While it is clear that the omics revolution holds great promise for demography, we are of the opinion that application in most cases would still be premature, except when a precise question is asked using a well-validated method. One example could be using heat shock proteins (i.e. targeted panel of selected protein markers) as biomarkers of stress (Feder and Hofmann 1999).

2.9 Biologging

With the development of new technologies, new sets of tools are now available to measure physiological traits outside laboratory environment. The term *biologging* has been proposed to describe a process by which researchers gain information (typically position, behaviour, movement, or physiological status) from an individual remotely (Payne et al. 2014). In animals and humans, biologging units are now widely available and consist of miniaturised electronic data recorders fitted on tags, badges, or collars. These devices allow the measurement of individual vital function, including body position, activity, and physiology on prolong period (Kavanagh and Menz 2008). Biological data generally recorded on biologgers, include body temperature, heart rate and 3D body acceleration. A depth sensor and captor for dissolved oxygen levels can also be included on device used in aquatic environment (Humphries et al. 2010). Coupled with data on environmental variables, body temperature and heart rate data can be used to reconstruct energetic landscape (Steiger et al. 2009; Scharf et al. 2016). Accelerometers record whole- or partial-body acceleration in one, two or three spatial axes simultaneously (Payne et al. 2014). These data can then be used to determine individual behaviour and derive metric for energy effort (Halsey et al. 2009). More recently, it had been suggested that body-acceleration data can also be used to assess individual's internal states including health states (Wilson et al. 2014). In plants, biologging approach include (hyperspectral) cameras and eddy covariance towers used to analyse individual colour/multispectral changes and gas exchange dynamics at the community level, respectively (Jalota et al. 2018; Khanna et al. 2019).

2.10 Combining biomarkers

It is rare that a single biomarker or physiological measure can provide a clear, stable signature of demography-relevant processes, and when it is possible, it usually requires extensive validation in the study population (Sparks et al. 2018). One approach to circumvent this problem has been to measure multiple markers simultaneously and to use statistical approaches to integrate the markers. Principal components analysis and similar methods can elegantly extract a joint signal from many variables (Matson et al. 2006; Buehler et al. 2011, Renaud et al. 2019), and statistical distance techniques can quantify homeostatic dysregulation (Cohen et al. 2014; Milot et al. 2014). The challenge, however, is that both these methods rely heavily on the correlation structure among the markers, and this structure can often change markedly across time and space, depending on conditions (Matson et al. 2006; Cohen and McGraw 2009; Buehler et al. 2011; Pigeon et al. 2013). It appears that such methods may thus work better under relatively homogeneous conditions (e.g. captivity) than in highly varied conditions typical in the wild (Fowler et al. 2018). An alternative approach is to decompose the correlation structure at different ecologically relevant levels of variation (Renaud et al. 2019); while promising, this approach has yet to be widely validated. Similar challenges are likely to be present with omics data.

2.11 Conclusions and perspectives

Integration of physiological and biochemical data into demographic and population ecology research is emerging as a promising approach that may bring advances in multiple branches of ecology, conservation, and human health and ageing research (Violle et al. 2007, 2014; Cooke et al. 2013; Cohen et al. 2014; Cohen and Milot 2015; Salguero-Gómez et al. 2018). Such an approach also has its challenges, however. We must emphasise the need to stick to stringent validation procedures, including validation of the analytical methods, exploration of the extent and sources of biomarker variation in a population and ultimately validation of biomarker importance to fitness. This will always be a tedious and challenging task but, as recognised by Hõrak and Cohen (2010), understanding how to properly measure the system and what are the sources of variation yields understanding of how the system works. Hence, methodological research will often answer substantive ecological, evolutionary, or medical questions. The most challenging task will certainly be to validate the link between a biomarker and fitness components, as their relationship may often be context dependent. As a result, the potential of biomarkers to predict fitness remains largely unknown (Salguero-Gómez et al. 2018). Nevertheless, exploring this potential is of ultimate importance because functional traits, including physiological and biochemical traits, are defined as features that impact fitness (Violle et al. 2007). We also advocate efforts to standardise protocols of physiological and biochemical data collection and analysis across populations and species, and to collect these data in global, open-source repositories. Such endeavours are crucial for addressing questions on large spatial and taxonomical scales.

Acknowledgements

We thank Lumír Gvoždík for constructive comments on animal metabolic rate measurement, and Jean-François Lemaître and one anonymous reviewer for valuable comments on the early version of the manuscript. OT was supported by three grants from the Czech Science Foundation (GA17–24782S; GA19–22538S; GA21-22160S). This work was further funded by an NSERC Discovery Grant (402079–2011) to AAC. AAC is supported by a CIHR New Investigator Salary Award and is a member of the FRQ-S funded *Centre de recherche du CHUS* and *Centre de recherche sur le vieillissement*.

References

Abdul-Hamid H. and Mencuccini M. (2009). Age- and size-related changes in physiological characteristics and chemical composition of *Acer pseudoplatanus* and *Fraxinus excelsior* trees. Tree Physiology **29**, 27–38.

Asensi-Fabado M. A. and Munné-Bosch S. (2010). Vitamins in plants: occurrence, biosynthesis and antioxidant function. Trends in Plant Science **15**, 582–592.

Aubert G., Hills M., and Lansdorp P. M. (2012). Telomere length measurement—caveats and a critical assessment of the available technologies and tools. Mutation Research/Fundamental and Molecular Mechanisms of Mutagenesis **730**, 59–67.

Barden A. E., Mas E., Croft K. D., Phillips M., and Mori T. A. (2014). Minimizing artifactual elevation of lipid peroxidation products (F2-isoprostanes) in plasma during collection and storage. Analytical Biochemistry **449**, 129–131.

Bauch C., Wellbrock A. H. J., Nagel R., Rozman J., and Witte K. (2013). 'Bug-eggs' for common swifts and other small birds: minimally-invasive and stress-free blood sampling during incubation. Journal of Ornithology **154**, 581–585.

Belsky D. W., Moffitt T. E., Cohen A. A., et al. (2018). Eleven telomere, epigenetic clock, and biomarker-composite quantifications of biological aging: do they measure the same thing? American Journal of Epidemiology **187**, 1220–1230.

Bílková, Z., Adámková M., Albrecht T., and Šimek Z. (2019). Determination of testosterone and corticosterone in feathers using liquid chromatography-mass spectrometry. Journal of Chromatography A **1590**, 96–103.

Birkhead T. R., Hosken D. J., and Pitnick S. (2009). Sperm Biology: An Evolutionary Perspective. Academic Press.

Blackburn E. H., Epel E. S., and Lin J. (2015). Human telomere biology: a contributory and interactive factor in aging, disease risks, and protection. Science **350**, 1193–1198.

Bonier F., Martin P. R., Moore I. T., and Wingfield J. C. (2009). Do baseline glucocorticoids predict fitness? Trends in Ecology & Evolution **24**, 634–642.

Breuner C. W., Patterson S. H., and Hahn T. P. (2008). In search of relationships between the acute adrenocortical response and fitness. General and Comparative Endocrinology **157**, 288–295.

Breuner C. W., Delehanty B., and Boonstra R. (2013). Evaluating stress in natural populations of vertebrates: total CORT is not good enough. Functional Ecology **27**, 24–36.

Brooks R. C. and Garratt M. G. (2017). Life history evolution, reproduction, and the origins of sex-dependent aging and longevity. Annals of the New York Academy of Sciences **1389**, 92–107.

Buehler D. M., Versteegh M. A., Matson K. D., and Tieleman B. I. (2011). One problem, many solutions: simple statistical approaches help unravel the complexity of the immune system in an ecological context. PLOS One **6**, e18592.

Burton T., Killen S. S., Armstrong J. D., and Metcalfe N. B. (2011). What causes intraspecific variation in resting metabolic rate and what are its ecological consequences? Proceedings of the Royal Society B: Biological Sciences **278**, 3465–3473.

Cattet M., Stenhouse G. B., Janz D. M. et al. (2017). The quantification of reproductive hormones in the hair of captive adult brown bears and their application as indicators of sex and reproductive state. Conservation Physiology **5**, cox032.

Cattet M., Stenhouse G. B., Boulanger J. et al. (2018). Can concentrations of steroid hormones in brown bear hair reveal age class? Conservation Physiology **6**, coy001.

Chabot, D., Steffensen, J. F., and Farrell, A. P. (2016). The determination of standard metabolic rate in fishes. Journal of Fish Biology **88**, 81–121.

Cheynel L., Lemaître J.-F., Gaillard J.-M. et al. (2017). Immunosenescence patterns differ between populations but not between sexes in a long-lived mammal. Scientific Reports **7**, 13700.

Cohen A. A. and McGraw K. J. (2009). No simple measures for antioxidant status in birds: complexity in inter- and intraspecific correlations among circulating antioxidant types. Functional Ecology **23**, 310–320.

Cohen A. A., Martin L. B., Wingfield J. C., McWilliams S. R., and Dunne J. A. (2012). Physiological regulatory networks: ecological roles and evolutionary constraints. Trends in Ecology & Evolution **27**, 428–435.

Cohen A. A., Milot E., Li Q., Legault V., Fried L. P., and Ferrucci L. (2014). Cross-population validation of statistical distance as a measure of physiological dysregulation during aging. Experimental Gerontology **57**, 203–210.

Cohen A. A., Milot E., Li Q., et al. (2015). Detection of a novel, integrative aging process suggests complex physiological integration. PLOS One **10**, e0116489.

Cohen A. A., Bandeen-Roche K., Morissette-Thomas V., and Fülöp T. (2018). A robust characterization of inflamm-aging and other immune processes through multivariate analysis of cytokines from longitudinal studies. In: T. Fülöp, C., Franceschi, K. Hirokawa, and G. Pawelec (eds.), Handbook of Immunosenescence: Basic Understanding and Clinical Implications, 1–16. Springer.

Cooke S. J., Sack L., Franklin C. E. et al. (2013). What is conservation physiology? Perspectives on an increasingly integrated and essential science. Conservation Physiology **1**, cot001.

Costantini D., Rowe M., Butler M. W., and McGraw K. J. (2010). From molecules to living systems: historical and contemporary issues in oxidative stress and antioxidant ecology. Functional Ecology **24**, 950–959.

Dantzer B., Fletcher Q. E., Boonstra R., and Sheriff M. J. (2014). Measures of physiological stress: a transparent or opaque window into the status, management and conservation of species? Conservation Physiology **2**, cou023.

Dawson T. E., Mambelli S., Plamboeck A. H., Templer P. H., and Tu K. P. (2002). Stable isotopes in plant ecology. Annual Review of Ecology and Systematics **33**, 507–559.

Demidchik V. (2015). Mechanisms of oxidative stress in plants: from classical chemistry to cell biology. Environmental and Experimental Botany **109**, 212–228.

Dotan Y., Lichtenberg D., and Pinchuk I. (2004). Lipid peroxidation cannot be used as a universal criterion of oxidative stress. Progress in Lipid Research **43**, 200–227.

Drinkwater E., Robinson E. J. H., and Hart A. G. (2019). Keeping invertebrate research ethical in a landscape of shifting public opinion. Methods in Ecology and Evolution **10**, 1265–1273.

Eastwood J. R., Mulder E., Verhulst S., and Peters A. (2018). Increasing the accuracy and precision of relative telomere length estimates by RT qPCR. Molecular Ecology Resources **18**, 68–78.

Edmunds P. J., Burgess S. C., Putnam H. M. et al. (2014). Evaluating the causal basis of ecological success within the scleractinia: an integral projection model approach. Marine Biology **161**, 2719–2734.

Esteban R., Barrutia O., Artetxe U., Fernández-Marín B., Hernández A., and García-Plazaola J. I. (2015). Internal and external factors affecting photosynthetic pigment composition in plants: a meta-analytical approach. The New Phytologist **206**, 268–280.

Feder M. E. and Hofmann G. E. (1999). Heat-shock proteins, molecular chaperones, and the stress response—evolutionary and ecological physiology. Annual Review of Physiology **61**, 243–282.

Ferrucci L., Gonzalez-Freire M., Fabbri E. et al. (2020). Measuring biological aging in humans: a quest. Aging Cell **19**, e13080.

Finch C. and Rose M. (1995). Hormones and the physiological architecture of life-history evolution. Quarterly Review of Biology **70**, 1–52.

Fitzpatrick J. L. and Lüpold S. (2014). Sexual selection and the evolution of sperm quality. Molecular Human Reproduction **20**, 1180–1189.

Fowler M. A., Paquet M., Legault V., Cohen A. A., and Williams T. D. (2018). Physiological predictors of reproductive performance in the European starling (*Sturnus vulgaris*). Frontiers in Zoology **15**, 45.

Franceschi C., Bonafè M., Valensin S., et al. (2000). Inflammaging. An evolutionary perspective on immunosenescence. Annals of the New York Academy of Sciences **908**, 244–254.

Fu X. and Meinzer F. C. (2019). Metrics and proxies for stringency of regulation of plant water status (iso/anisohydry): a global data set reveals coordination and trade-offs among water transport traits. Tree Physiology **39**, 122–134.

Fuchs S., Leuschner C., Link R., Coners H., and Schuldt B. (2017). Calibration and comparison of thermal dissipation, heat ratio and heat field deformation sap flow probes for diffuse-porous trees. Agricultural and Forest Meteorology **244–245**, 151–161.

Fulop T., Larbi A., Dupuis G. et al. (2017). Immunosenescence and inflamm-aging as two sides of the same coin: friends or foes? Frontiers in Immunology **8**, 1960.

Gomendio M., Malo A. F., Soler A. J. et al. (2006). Male fertility and sex ratio at birth in red deer. Science **314**, 1445–1447.

Green J. M. (1996). A practical guide to analytical method validation. Analytical Chemistry **68**, 305A–309A.

Halliwell B. and Gutteridg, J. M. C. (2015). Free Radicals in Biology and Medicine. Oxford University Press.

Halsey L. G., Portugal S. J., Smith J. A., Murn C. P., and Wilson R. P. (2009). Recording raptor behavior on the wing via accelerometry. Journal of Field Ornithology **80**, 171–177.

Harr K. E. (2005). Diagnostic value of biochemistry. In: G. Harrison and T. Lightfoot (eds.), Clinical Avian Medicine, 611–603. Spix Publishing.

Hertel J., Frenzel S., König, J. et al. (2019). The informative error: a framework for the construction of individualized phenotypes. Statistical Methods in Medical Research **28**, 1427–1438.

Hollmen T., Franson J., Hario M., Sankari S., Kilpi M., and Lindstrom K. (2001). Use of serum biochemistry to evaluate nutritional status and health of incubating common eiders (*Somateria mollissima*) in Finland. Physiological and Biochemical Zoology **74**, 333–342.

Hõrak P. and Cohen A. A. (2010). How to measure oxidative stress in an ecological context: methodological and statistical issues. Functional Ecology **24**, 960–970.

Horvath S. (2013). DNA methylation age of human tissues and cell types. Genome Biology **14**, R115.

Humann-Guilleminot S., Blévi P., Azou-Barré A. et al. (2018). Sperm collection in black-legged kittiwakes and characterization of sperm velocity and morphology. Avian Research **9**, 24.

Humphries N. E., Queiroz N., Dyer J. R. M. et al. (2010). Environmental context explains Lévy and Brownian movement patterns of marine predators. Nature **465**, 1066–1069.

Isaksson C., Sheldon B. C., and Uller T. (2011). The challenges of integrating oxidative stress into life-history biology. BioScience **61**, 194–202.

Jalota S. K., Vashisht B. B., Sharma S., and Kaur S. (2018). Emission of greenhouse gases and their warming effect. In: S. K. Jalota, B. B. Vashisht, S. Sharma, and S. Kaur (eds.), Understanding Climate Change Impacts on Crop Productivity and Water Balance, 1–53. Academic Press.

Jansen E. H. J. M., Beekhof P. K., Viezeliene D., Muzakova V., and Skalicky J. (2017). Long-term stability of oxidative stress biomarkers in human serum. Free Radical Research **51**, 970–977.

Jimeno B., Hau M., and Verhulst S. (2018). Corticosterone levels reflect variation in metabolic rate, independent of 'stress.' Scientific Reports **8**, 13020.

Johnson S. L., Dunleavy J., Gemmell N. J., and Nakagawa S. (2015). Consistent age-dependent declines in human semen quality: a systematic review and meta-analysis. Ageing Research Reviews **19**, 22–33.

Jones D. P. (2006). Redefining oxidative stress. Antioxidants & Redox Signaling **8**, 1865–1879.

Kavanagh J. J. and Menz H. B. (2008). Accelerometry: a technique for quantifying movement patterns during walking. Gait & Posture **28**, 1–15.

Khanna R., Schmid L., Walter A., Nieto J., Siegwart R., and Liebisch F. (2019). A spatio temporal spectral framework for plant stress phenotyping. Plant Methods **15**, 13.

Landhäusser S. M., Chow P. S., Dickman L. T. et al. (2018). Standardized protocols and procedures can precisely and accurately quantify non-structural carbohydrates. Tree Physiology **38**, 1764–1778.

Laskemoen T., Albrecht T., Bonisoli-Alquati A. et al. (2013). Variation in sperm morphometry and sperm competition among barn swallow (*Hirundo rustica*) populations. Behavioral Ecology and Sociobiology **67**, 301–309.

Lemaître J.-F. and Gaillard J.-M. (2017). Reproductive senescence: new perspectives in the wild. Biological Reviews **92**, 2182–2199.

Lierz M., Reinschmidt M., Müller H., Wink M., and Neumann D. (2013). A novel method for semen collection and artificial insemination in large parrots (*Psittaciformes*). Scientific Reports **3**, 1–8.

Lifjeld J., Laskemoen T., Kleven O., Albrecht T., and Robertson R. (2010). Sperm length variation as a predictor of extrapair paternity in passerine birds. PLOS One **5**, e13456.

Lighton J. R. B. (2019). Measuring Metabolic Rates: A Manual for Scientists, 2nd edition. Oxford University Press.

Lindsjö J., Fahlman Å., and Törnqvist E. (2016). Animal welfare from mouse to moose—implementing the principles of the 3Rs in wildlife research. Journal of Wildlife Diseases **52**, S65–S77.

Long S. P. and Bernacchi C. J. (2003). Gas exchange measurements, what can they tell us about the underlying limitations to photosynthesis? Procedures and sources of error. Journal of Experimental Botany **54**, 2393–2401.

Lu Y., Tan C. T. Y., Nyunt M. S. Z. et al. (2016). Inflammatory and immune markers associated with physical frailty syndrome: findings from Singapore longitudinal aging studies. Oncotarget **7**, 28783–28795.

Martínez-Vilalta, J. Vanderklein, D. and Mencuccini M. (2007). Tree height and age-related decline in growth in Scots pine (*Pinus sylvestris* L.). Oecologia **150**, 529–544.

Mateos R. and Bravo L. (2007). Chromatographic and electrophoretic methods for the analysis of biomarkers of oxidative damage to macromolecules (DNA, lipids, and proteins). Journal of Separation Science **30**, 175–191.

Matson K. D., Cohen A. A., Klasing K. C., Ricklef R. E., and Scheuerlein A. (2006). No simple answers for ecological immunology: relationships among immune indices at the individual level break down at the species level in waterfowl. Proceedings of the Royal Society B: Biological Sciences **273**, 815–822.

Meier H. C. S., Hussein M., Needham B. et al. (2019). Cellular response to chronic psychosocial stress: ten-year longitudinal changes in telomere length in the Multi-Ethnic Study of Atherosclerosis. Psychoneuroendocrinology **107**, 70–81.

Mencuccini M. and Munné-Bosch S. (2017). Physiological and biochemical processes related to ageing and senescence in plants. In: R. P. Shefferson, O. R. Jones, and R. Salguero-Gómez (eds.), The Evolution of Senescence in the Tree of Life, 257–283. Cambridge University Press.

Mencuccini M., Oñate M., Peñuelas J., Rico L., and Munné-Bosch S. (2014). No signs of meristem senescence in old Scots pine. Journal of Ecology **102**, 555–565.

Merow C., Dahlgren J. P., Metcalf C. J. E. et al. (2014). Advancing population ecology with integral projection models: a practical guide. Methods in Ecology and Evolution **5**, 99–110.

Millet S., Bennett J., Lee K. A., Hau M., and Klasing K. C. (2007). Quantifying and comparing constitutive immunity across avian species. Developmental and Comparative Immunology **31**, 188–201.

Milot E., Cohen A. A., Vezina F., Buehler D. M., Matson K. D., and Piersma T. (2014). A novel integrative method for measuring body condition in ecological studies based on physiological dysregulation. Methods in Ecology and Evolution **5**, 146–155.

Monaghan P., Metcalfe N. B., and Torres R. (2009). Oxidative stress as a mediator of life history trade-offs: mechanisms, measurements and interpretation. Ecology Letters **12**, 75–92.

Monneret D., Godmer A., Le Guen R. et al. (2016). Stability of routine biochemical analytes in whole blood and plasma from lithium heparin gel tubes during 6-hr storage. Journal of Clinical Laboratory Analysis **30**, 602–609.

Montpetit A., Alhareeri A., Montpetit M. et al. (2014). Telomere length: a review of methods for measurement. Nursing Research **63**, 289–299.

Morales M. and Munné-Bosch S. (2019). Malondialdehyde: facts and artifacts. Plant Physiology **180**, 1246–1250.

Morrisette-Thomas V., Cohen A. A., Fülöp T. et al. (2014). Inflamm-aging does not simply reflect increases in pro-inflammatory markers. Mechanisms of Ageing and Development **139**, 49–57.

Müller M. and Munné-Bosch S. (2011). Rapid and sensitive hormonal profiling of complex plant samples by liquid chromatography coupled to electrospray ionization tandem mass spectrometry. Plant Methods **7**, 37.

Murchie E. H. and Lawson T. (2013). Chlorophyll fluorescence analysis: a guide to good practice and understanding some new applications. Journal of Experimental Botany **64**, 3983–3998.

Norin T. and Metcalfe N. B. (2019). Ecological and evolutionary consequences of metabolic rate plasticity in response to environmental change. Philosophical Transactions of the Royal Society B: Biological Sciences **374**, 20180180.

Nussey D. H., Baird D., Barrett E., et al. (2014). Measuring telomere length and telomere dynamics in evolutionary biology and ecology. Methods in Ecology and Evolution **5**, 299–310.

Olsson M., Wilson M., Uller T. et al. (2008). Free radicals run in lizard families. Biology Letters **4**, 186–188.

Owen J. C. (2011). Collecting, processing, and storing avian blood: a review. Journal of Field Ornithology **82**, 339–354.

Pamplona R. and Costantini D. (2011). Molecular and structural antioxidant defenses against oxidative stress in animals. American Journal of Physiology—Regulatory, Integrative and Comparative Physiology **301**, R843–R863.

Payne N. L., Taylor M. D., Watanabe Y. Y., and Semmens J. M. (2014). From physiology to physics: are we recognizing the flexibility of biologging tools? Journal of Experimental Biology **217**, 317–322.

Pérez-Harguindeguy N., Díaz S., Garnier E. et al. (2013). New handbook for standardised measurement of plant functional traits worldwide. Australian Journal of Botany **61**, 167–234.

Pettersen A. K., Marshall D. J., and White C. R. (2018). Understanding variation in metabolic rate. Journal of Experimental Biology **221**, jeb166876.

Pigeon G., Bélisle M., Garant D., Cohen A. A., and Pelletier F. (2013). Ecological immunology in a fluctuating environment: an integrative analysis of tree swallow nestling immune defense. Ecology and Evolution **3**, 1091–1103.

Piłacik, B., Nofer, T. W., and Wasowicz, W. (2002). F2-isoprostanes biomarkers of lipid peroxidation: their utility in evaluation of oxidative stress induced by toxic agents. International Journal of Occupational Medicine and Environmental Health **15**, 19–27.

Podlesak D. W., McWilliams S. R., and Hatch K. A. (2005). Stable isotopes in breath, blood, feces and feathers can indicate intra-individual changes in the diet of migratory songbirds. Oecologia **142**, 501–510.

Poiani A. (2006). Complexity of seminal fluid: a review. Behavioral Ecology and Sociobiology **60**, 289–310.

Pulendran B. (2004). Modulating Th1/Th2 responses with microbes, dendritic cells, and pathogen recognition receptors. Immunologic Research **29**, 187–196.

Queval G. and Noctor G. (2007). A plate reader method for the measurement of NAD, NADP, glutathione, and ascorbate in tissue extracts: application to redox profiling during Arabidopsis rosette development. Analytical Biochemistry **363**, 58–69.

Ramirez J. A., Posada J. M., Handa I. T. et al. (2015). Near-infrared spectroscopy (NIRS) predicts non-structural carbohydrate concentrations in different tissue types of a broad range of tree species. Methods in Ecology and Evolution **6**, 1018–1025.

Réale D., Garant D., Humphries M. M., Bergeron P., Careau V., and Montiglio P.-O. (2010). Personality and the emergence of the pace-of-life syndrome concept at the population level. Philosophical Transactions of the Royal Society of London B: Biological Sciences **365**, 4051–4063.

Reinhardt K., Dobler R., and Abbott J. (2015). An ecology of sperm: sperm diversification by natural selection. Annual Review of Ecology, Evolution, and Systematics **46**, 435–459.

Renaud L.-A., Blanchet F. G., Cohen A. A., and Pelletier F. (2019). Causes and short-term consequences of variation in milk composition in wild sheep. Journal of Animal Ecology, **88**, 857–869.

Sala A. and Hoch G. (2009). Height-related growth declines in ponderosa pine are not due to carbon limitation. Plant, Cell & Environment **32**, 22–30.

Salguero-Gómez R., Violle C., Gimenez O., and Childs D. (2018). Delivering the promises of trait-based approaches to the needs of demographic approaches, and vice versa. Functional Ecology **32**, 1424–1435.

Šandera M., Albrecht T., and Stopka P. (2013). Variation in apical hook length reflects the intensity of sperm competition in murine rodents. PLOS One **8**, e68427.

Sanders J. L. and Newman A. B. (2013). Telomere length in epidemiology: a biomarker of aging, age-related disease, both, or neither? Epidemiologic Reviews **35**, 112–131.

Scharf A. K., LaPoint S., Wikelski M., and Safi K. (2016). Acceleration data reveal highly individually structured energetic landscapes in free-ranging fishers (*Pekania pennanti*). PLOS One **11**, e0145732.

Semerci A., Semerci H., Çalişkan B., Çiçek N., Ekmekçi Y., and Mencuccini M. (2017). Morphological and physiological responses to drought stress of European provenances of Scots pine. European Journal of Forest Research **136**, 91–104.

Sheriff M. J., Dantzer B., Delehanty B., Palme R., and Boonstra R. (2011). Measuring stress in wildlife: techniques for quantifying glucocorticoids. Oecologia **166**, 869–887.

Shipley B., De Bello F., Cornelissen J. H. C., Laliberté E., Laughlin D. C., and Reich P. B. (2016). Reinforcing loose foundation stones in trait-based plant ecology. Oecologia **180**, 923–931.

Somogyi A., Rosta K., Puszta, P., Tulassay Z., and Nagy G. (2007). Antioxidant measurements. Physiological Measurement **28**, R41–R55.

Sparks A. M., Watt K., Sinclair R. et al. (2018). Natural selection on antihelminth antibodies in a wild mammal population. The American Naturalist **192**, 745–760.

Speakman J. (1997). Doubly Labelled Water: Theory and Practice. Springer.

Speakman J. R. and Hambly C. (2016). Using doubly-labelled water to measure free-living energy expenditure: some old things to remember and some new things to consider. Comparative Biochemistry and Physiology. Part A, Molecular & Integrative Physiology **202**, 3–9.

Steiger S. S., Kelley J. P., Cochran W. W., and Wikelski M. (2009). Low metabolism and inactive lifestyle of a tropical rain forest bird investigated via heart-rate telemetry. Physiological and Biochemical Zoology **82**, 580–589.

Sturgeon C. M. and McAllister E. J. (1998). Analysis of hCG: clinical applications and assay requirements. Annals of Clinical Biochemistry **35**, 460–491.

Tabunoki H., Dittmer N. T., Gorman M. J., and Kanost M. R. (2019). Development of a new method for collecting hemolymph and measuring phenoloxidase activity in *Tribolium castaneum*. BMC Research Notes **12**, 7.

Tomášek O., Gabrielova B., Kacer P. et al. (2016). Opposing effects of oxidative challenge and carotenoids on antioxidant status and condition-dependent sexual signalling. Scientific Reports **6**, 23546.

Tomášek O., Albrechtova J., Nemcova M., Opatova P., and Albrecht T. (2017). Trade-off between carotenoid-based sexual ornamentation and sperm resistance to oxidative challenge. Proceedings of the Royal Society B: Biological Sciences **284**, 20162444.

Tomášek O., Bobek L., Kralova T., Adamkova M., and Albrecht T. (2019). Fuel for the pace of life: baseline blood glucose concentration co-evolves with life-history traits in songbirds. Functional Ecology **33**, 239–249.

Tricola G.M., Simons M.J.P., Atema E. et al. (2018). The rate of telomere loss is related to maximum lifespan in birds. Philosophical Transactions of the Royal Society B: Biological Sciences **373**, 20160445.

Turner N. C. (1988). Measurement of plant water status by the pressure chamber technique. Irrigation Science **9**, 289–308.

Udroiu I. (2020). On the correlation between telomere shortening rate and life span. Proceedings of the National Academy of Sciences **117**, 2248–2249.

Valeggia C. R. (2007). Taking the lab to the field: monitoring reproductive hormones in population research. Population and Development Review **33**, 525–542.

Valko M., Leibfritz D., Moncol J., Cronin M. T. D., Mazur M., and Telser J. (2007). Free radicals and antioxidants in normal physiological functions and human disease. International Journal of Biochemistry & Cell Biology **39**, 44–84.

Vanderklein D., Martínez-Vilalta J., Lee, S., and Mencuccini M. (2007). Plant size, not age, regulates growth and gas exchange in grafted Scots pine trees. Tree Physiology **27**, 71–79.

Verhulst S. (2020). Improving comparability between qPCR-based telomere studies. Molecular Ecology Resources **20**, 11–13.

Viney M., Riley E., and Buchanan K. (2005). Optimal immune responses: immunocompetence revisited. Trends in Ecology & Evolution **20** 665–669.

Violle C., Navas M.-L., Vile, D. et al. (2007). Let the concept of trait be functional! Oikos **116**, 882–892.

Violle C., Enquist B. J., McGill B. J. et al. (2012). The return of the variance: intraspecific variability in community ecology. Trends in Ecology & Evolution **27**, 244–252.

Violle C., Reich P. B., Pacala S. W., Enquist B. J., and Kattge J. (2014). The emergence and promise of functional biogeography. Proceedings of the National Academy of Sciences **111**, 13690–13696.

Violle C., Thuiller W., Mouquet N. et al. (2017). Functional rarity: the ecology of outliers. Trends in Ecology & Evolution **32**, 356–367.

Walther A., Philipp M., Lozza N., and Ehlert U. (2016). The rate of change in declining steroid hormones: a new parameter of healthy aging in men? Oncotarget **7**, 60844–60857.

Wasden M. B., Roberts R. L., and DeLaurier A. (2017). Optimizing sperm collection procedures in zebrafish. Journal of the South Carolina Academy of Science **15**, 7.

Wheeler M. J. (2006). Assays for LH, FSH, and prolactin. Methods in Molecular Biology (Clifton, NJ) **324**, 109–124.

Whittemore K., Vera E., Martínez-Nevado E., Sanpera C., and Blasco M. A. (2019). Telomere shortening rate predicts species life span. Proceedings of the National Academy of Sciences of the United States of America **116**, 15122–15127.

Wilbourn R. V., Moatt J. P., Froy H., Walling C. A., Nussey D. H., and Boonekamp J. J. (2018). The relationship between telomere length and mortality risk in non-model vertebrate systems: a meta-analysis. Philosophical Transactions of the Royal Society B: Biological Sciences **373**, 20160447.

Wilson R. P., Grundy E., Massy R. et al. (2014). Wild state secrets: ultra-sensitive measurement of micro-movement can reveal internal processes in animals. Frontiers in Ecology and the Environment **12**, 582–587.

Yeum K. J., Russell R. M., Krinsky N. I., and Aldini G. (2004). Biomarkers of antioxidant capacity in the hydrophilic and lipophilic compartments of human plasma. Archives of Biochemistry and Biophysics **430**, 97–103.

Yoshida Y., Hayakawa M., Habuchi Y., Itoh N., and Niki E. (2007). Evaluation of lipophilic antioxidant efficacy in vivo by the biomarkers hydroxyoctadecadienoic acid and isoprostane. Lipids **42**, 463–472.

Young A. J. (2018). The role of telomeres in the mechanisms and evolution of life-history trade-offs and ageing. Philosophical Transactions of the Royal Society B: Biological Sciences **373**, 20160452.

CHAPTER 3

Social data collection and analyses

Marie J. E. Charpentier, Marie Pelé, Julien P. Renoult, and Cédric Sueur

3.1 Introduction

Understanding how and why animals interact non-randomly with conspecifics is at the core of evolutionary questions related to a broad range of topics (e.g. sexual selection, parental care, and kin selection). Individuals living in complex social systems with well-differentiated social bonds (i.e. social bonds differ quantitatively or qualitatively between pairs of individuals) generally benefit (i.e. have an increased fitness) from positive social interactions[1] (Silk 2007). For example, in baboons, females that are socially integrated, connected, or central into their group show improved offspring survival (Silk et al. 2003, 2009; Cheney et al. 2016) and they live longer (Archie et al. 2014). Conversely, social isolation is related to reduced life span in humans (House et al. 1988; Holt-Lunstad et al. 2010) and in nonprimate species (Silk 2007; Formica et al. 2012; Stanton and Mann 2012; Kappeler et al. 2015; Busson et al. 2019), highlighting further the strong and ubiquitous adaptive value of sociality.

Studying the functional significance of sociality requires to accurately describe social interactions and/or spatial associations, which has proven to be highly challenging. Difficulties may arise from animal behaviours themselves, for example the elusiveness of lower-ranked individuals that bias sampling towards higher-ranked, more easily observable individuals (Gimenez et al. 2019). Methods can be problematic too, as it is generally impossible to compare data collected with different approaches. Using a similar methodology to monitor different populations, or the same population in the long term, may be, however,

practically infeasible (Castles et al. 2014). Collecting social behaviours is also time-consuming and labour-intensive because of the very nature of this type of data: in several social species, individuals dedicate only a small fraction of their time socialising. Spending too little time observing animals in their social environment can be misleading (Whitehead 2008) (Figure 3.1), and a thoughtful description of animal behaviours generally requires years or decades of monitoring, especially when working with long-living animals that need years to grow. Although long-term monitoring introduces new difficulties, such as coping with interobserver variations, it is currently considered the gold standard for behavioural ecologists, ethologists, and demographists.

In section 3.2, we propose a brief history of animal social studies and an overview of the classic methods to sample social interactions and association patterns. Studying social behaviours has a long tradition with well-established methodologies; however, newly developed tools have the potential to circumvent some of the difficulties of describing social interactions and/or spatial associations. In section 3.3, we present several new technologies that are increasingly used for collecting behavioural data. These technologies can generate a large amount of data, and thus require specific analytical tools. In section 3.4, we present approaches to analyse social networks using large data sets collected when using new technologies. We also discuss how these approaches can increase our understanding of the structure of social networks, of the link between behaviour and sociality (e.g. the emergence of collective decisions) and between sociality and other parameters (e.g. disease diffusion). We further propose new analytical tools, such as the use of artificial intelligence algorithms to characterise individual behavioural patterns.

[1] 'Positive interactions' are defined here as all types of social interactions that involve cooperative or mutualistic behavior, such as grooming relationships, coalitions, and food sharing, by contrast with, for example, aggression and submission.

Marie J. E. Charpentier et al., *Social data collection and analyses*. In: *Demographic Methods Across the Tree of Life*.
Edited by Roberto Salguero-Gómez and Marlène Gamelon, Oxford University Press.
© Oxford University Press (2021). DOI: 10.1093/oso/9780198838609.003.0003

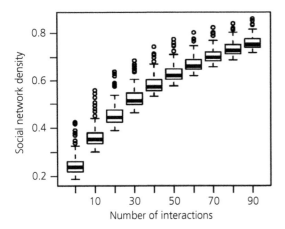

Figure 3.1 Simulated relationship between the number of interactions recorded and the density of the social network. The boxplot represents the median density (± 95% confidence intervals) obtained from 100 networks of different sizes (90–500 individuals).

3.2 Classic field-based methods for sampling social behaviours

3.2.1 A brief history of social studies, Jeanne Altmann, and the sampling of social behaviour

The first studies on animal societies arose in the nineteenth century. Emile Durkheim, Herbert Comte, and Herbert Spencer first linked natural sciences performed on animals to social science studies on humans. As well, Charles Darwin scrutinised human societies under the prism of his Theory of Evolution. In 1877, Alfred Espinas (1887) published 'Des Sociétés animales, étude de psychologie comparée' ('*Animal societies, a comparative psychology study*'), in which a parallel was drawn between human and animal social systems. Later on, Jacob Uhrich (1938) performed probably the first individually centred observations and used hierarchical social networks to understand the dominance hierarchy in mice. In the mid-century, the anthropologist Kinji Imanishi (1952) tried to understand human's aggressive and pacific behaviours by studying Japanese macaques from Koshima Island (Japan) who were provisioned with sweat potatoes. He demonstrated the first case of culture and social learning in animals. The first case of social facilitation was reported in birds from Great Britain, with tits learning from conspecifics how to open milk bottles (Fisher and Hinde 1949; Hinde and Fisher 1951), when Jane Goodall documented for the first time tool use and tool making in chimpanzees at Gombe Stream National Park (Tanzania; Goodall 1986). At this time,

the descriptive nature of behavioural studies isolated the field of ethology from other scientific domains. Since the 1960s, though, behavioural ecologists have started to systematically record social behaviours using detailed and repeatable protocols. Quantitative studies in a large range of species measuring fine-grain traits of animal social systems appeared (Carpenter 1942a; Kruuk 1972; Sade 1972; Barash 1973; Jarman 1974; Emlen and Oring 1977; Saitou 1978; Ekman 1979). This period also saw the development of theoretical works on the ultimate and proximate mechanisms of group living (Carpenter 1942b; Crook 1964, 1970; McBride 1964; Crook and Gartlan 1966; Clutton-Brock 1974; Wilson 1975). The first long-term, individually centred research on animal societies living in ecologically realistic conditions appeared in this stimulating context (e.g. in insects, Trivers and Hare 1976; birds, Stacey and Koenig 1990; ungulates, Boyd et al. 1964; Clutton-Brock 1982; cetaceans, Siebenaler and Caldwell 1956; Saayman and Tayler 1973; nonhuman primates, Seyfarth 1976; Altmann et al. 1985; and for a review see Kappeler and Watts 2012).

In 1974, Jeanne Altmann published an article in the journal *Behaviour* that would soon become a reference in behavioural ecology. In her article entitled 'Observational study of behavior: sampling methods', Altmann (1974) proposed different methods to sample behaviours in field conditions, most of the techniques described being still largely in use. For example, the *ad libitum* method allows to record every behaviour in all occurrences. It is particularly useful in conditions of poor visibility or for preliminary studies. The *ad libitum* sampling method is generally not used alone, however, because it does not allow to control for several biases, including oversampling conspicuous behaviours or individuals (Altmann 1974). Focal-animal sampling, in which the observer focuses on one particular focal individual during a predefined duration, helps to circumvent these observational biases. One at a time, the observer collects data on each individual in the group, randomly selecting focal animals. Another method consists of selecting a specific behaviour and recording it every time it occurs. This behavioural sampling is particularly indicated if the behaviour of interest is rare. In social studies, for example, grooming in primates or 'petting' in cetaceans are typically recorded through behavioural sampling. Instantaneous and scan samplings score states rather than events as in the previous methods. A *state* is a sequence of different events; for example, locomotion is a state defined by different steps (Altmann 1974). Here, the observer records at predefined time intervals different information about group members (e.g. activity, spatial distance across individuals). Instantaneous and scan samplings can simultaneously apply to a large number of individuals and thus are useful, for example, to reconstruct social networks. Altmann's sampling methods have gone far beyond the boundaries of primatology and are still widely used in a large panel of species, making her article the most quoted in the field of behavioural studies (>8,400 citations, WoS, December 2019).

3.2.2 The interplay between field and lab studies

Although long-term field sites are invaluable for behavioural ecology studies, experimental manipulations, mostly carried out in captive and controllable environments, may provide valuable and complementary inputs on animal social systems. Notably, lab studies have allowed researchers to highlight the cognitive mechanisms behind decision-making, such as delayed gratification (Dufour et al. 2007; Pelé et al. 2010, 2011) or risk calculation, potentially underlying the reciprocity of services observed between non-human primates in the wild (Pelé et al. 2014; Broihanne et al. 2018). Similarly, elaborated cognitive performances in keas or ravens have been largely evidenced in controlled environments (Heaney et al. 2017; Boucherie et al. 2019). Great apes were also shown to anticipate conspecifics' behaviour according to false beliefs (Krupenye et al. 2016) using a technology (eye tracker) not yet transposable to the field.

Initiated in primatology, the combination of field and lab experiments has increased our understanding of the social lives of many animal species (Bueno-Guerra and Amici 2018; in canids, Miklosi 2007; elephants, Garstang 2015; birds, Firth and Sheldon 2015; Emery 2016; fish, Filer 2008; and cephalopods, Godfrey-Smith 2016). Yet, the methodological approaches of lab versus field studies differ markedly because experimenters often seek to establish special relationships with their study subjects, whereas field observers try to minimise their impact on them. Observers are, however, rarely undetectable, and disturbances are sometimes unavoidable in the field. Indeed, labour-intensive habituation processes are often mandatory before starting any field studies, and, if incomplete, animals may react to the presence of observers. When habituation is not possible, indirect observations recorded from camera traps or technologies (e.g. accelerometers, radio-frequency identification (RFID) tags, and GPS) allow researchers to sample behaviours and data on spatial associa-

tions with limited disturbance. For example, RFID technologies were used to identify great tits (*Parus major*) in the wild and to monitor individuals' performance on novel feeding devices in order to better understand innovation and social diffusion (Aplin et al. 2015). Recently, artificial intelligence was also applied to individual recognition of wild primates (Schofield et al. 2019) and birds (Ferreira et al. 2019) using photographs, with possible applications in the long-term study of cognition (e.g. tool use learning), social networks, and kin selection (Charpentier et al. 2020).

3.3 New technologies to sample social behaviour in the wild

Although direct observations are sometimes mandatory, such as in species that cannot be captured, animal-attached remote sensing—necessitating the capture of animals—offers new perspectives for the study of animal behaviour. Indeed, biologging has the potential to revolutionise the sampling of social relationships because of the tremendous quantity of data sampled, the quality of which is also improving rapidly. The diversity of biologgers available allows the sampling of a large array of biological and physiological data, bringing undiscovered and sometimes unexpected outputs, and consistently nurturing new research questions (Chapter 5). Here, we propose an overview of the main devices currently used to sample social behaviours and spatial associations in animals.

3.3.1 A brief overview of animal's biologging

Starting 60 years ago, biologging technologies develop from simple, although often cumbersome devices collecting a few data per hour (e.g. Kooyman 1965) to miniaturised devices capable of continuously recording and archiving data for several years (Kays et al. 2015). Biologging studies initially flourished in marine (Hazen et al. 2012; Hussey et al. 2015) and aerial environments (Wilson et al. 1997) because of challenging visual conditions. These studies allowed researchers to decipher, for example, the large-scale migration patterns of birds (McKinnon and Love 2018) and fishes (Block et al. 2005; Bonfil et al. 2005), and the diving depths of marine mammals (Schreer and Testa 1996). Our understanding about life in the sea and in the air has therefore greatly increased thanks to various electronic devices.

In general, radio-tracking involves two devices. First, a transmitter is generally attached to an animal's

(a)

(b)

(c)

Figure 3.2 Examples of different biologgers. (a) Female vervet monkey (*Chlorocebus pygerythrus*; GPS) in South Africa (credit: Erica van de Waal, University of Lausanne, Switzerland); (b) female elephant seal (*Mirouga leonina*; GPS) on Kerguelen Island (credit: IPEV/CNRS/IPHC/DEPE); (c) Adélie penguins (*Pygoscelis adeliae*; temperature, hydrostatic pressure, and light intensity sensors) in Antarctica (credit: IPEV/CNRS/IPHC/DEPE).

body, and it sends signals at pre-set intervals. If mounted on a collar, it can be placed around the neck (Figure 3.2a), the ankle, or the leg of the study model. Alternatively, the transmitter may be embedded into a protective shell and stick on the carapace or directly in contact with the skin (e.g. on the head or on the back; Figure 3.2b,c). Transmitters may also be surgically implanted into an animal's body, generally in the abdomen. Second, a radio-receiver collects brief pulses sent by the transmitter, typically using a customisable frequency in the VHF band (30–300 MHz), with a mobile antenna that can be used manually or equipped on a vehicle such as an airplane or a helicopter. The advantages of this system include reliable and versatile ease of operation. It is also generally affordable. Disadvantages are related to a limited signal range (less than a few kilometres): animals may therefore be disturbed by human presence, and the localisation of the animal, generally obtained through triangulation, is normally coarse. Alternatively, passive receivers equipped on the animal's body may

collect positional information sent with transmitters placed on earth-orbiting satellites using, for example, GPS technology (for a review see Robertson et al. 2012). GPS systems allow remote detection and record highly accurately an animal's location; this information may be then stored in the device's memory associated to the receiver until the latter is retrieved. This necessitates recapturing the equipped animal. More conveniently, data may also be downloaded remotely via short-range radio signals and decoded using a custom receiver or using the GSM mobile phone network or the scientific satellite system ARGOS (advanced research and global observation satellite). GPS devices are, however, generally costly (Robertson et al. 2012), although prices are rapidly dropping. The energy consumption of GPS devices is also generally high, and this adds to the total weight of the device fitted on the animal and shortens the device's life span. Very recently, however, massive progress has been made through the ICARUS project (https://www.icarus.mpg.de/28874/sensor-animals-tracking) to reduce GPS dimensions, with many new applications in lightweight species such as black birds (*Turdus merula*).

Biologging is no longer restricted to localising animals. Multiple sensors may additionally record environmental and/or biological variables, helping researchers to evaluate the physiological state of their study model (Chapter 2). These sensors may include, for example, heartbeat frequency, skin humidity, breathing rates, outer and inner body temperatures, EEG, 3D body acceleration, and wing-flap rate, as well as light-level information, salinity, or conductivity of the environment (Ropert-Coudert and Wilson 2005; Chapter 6). These different options help in monitoring different characteristics of an animal's environment and biology, including its social relationships and behaviour. One of the first evidence of the use of biologging in social studies came from two female penguins (*Eudyptes chrysocome moseleyi*) that performed 286 synchronous foraging dives over 7 consecutive hours (Tremblay and Cherel 1999). In this study, the authors compared diving patterns using electronic time-depth recorders. Using the same type of recorders supplemented with cameras, others showed that Weddell seal (*Leptonychotes weddellii*) mothers modified their diving behaviour in the presence of their pups, preferring shallower dives than when diving alone, probably as a learning strategy (Sato et al. 2003). It is also now possible to decipher fine-grained motions thanks to 3D accelerometers (Wilson et al. 2008; Brown et al. 2013), with possible applications in the study of social behaviour; for example, a mutual clash/headbutt in rutting male mountain goats produces typical abrupt peaks in all three acceleration axes (Shepard 2008). When precisely calibrated, all these devices may therefore give precious indications about an animal's behaviour, including its interactions and associations with members of its social environment.

3.3.2 Biologging to determine individual social interactions and spatial associations

By providing high-resolution positions in space and time, often in fractions of seconds, lightweight GPS devices may capture fine-grain association patterns inferred from positional data (Davis et al. 2013). For example, the average spatial position of a pigeon in its flock depends on its position in the social hierarchy (Nagy et al. 2010). Other tags resulting in individual-based locations in time may provide similar information by transmitting a unique code to fixed

receivers located in relevant places in an animal's common environment, such as a sleeping tree or a water hole. When tagged animals come close to the receiver, the information is recorded through, for example, RFID (Figure 3.3a). Coded nanotags, internal or glued passive integrated transponder (PIT) tags, acoustic transmitters, or simple QR codes stuck on small animals, such as insects (Figure 3.3b), provide individual lifetime barcodes. The main advantage here is that these tags do not require much power (e.g. a PIT tag has an internal microchip that is activated when coming close to the receiver, and QR codes may be read with special cameras). Applications are multiple (Table 3.1). For example, Scardamaglia et al. (2017) studied brood parasitism in birds. They equipped 74 females of two species of cowbirds (parasites) with 1.0g coded radio-tags (transmitters) for 2.5 months and they positioned 45 fixed data loggers (receivers) just below the nests of the two parasitised species. The authors demonstrated that during prelaying visits, females from both parasite species identified nests to parasitise. Female parasites appear to form a dynamic memory library of potential laying opportunities that they use later to lay their own eggs (Scardamaglia et al. 2017).

Alternatively, transmission and reception may be conveniently combined into the same interactive talkative tag. This promising technology emerged 20 years ago and is currently bringing new perspectives in the study of animal's social networks because it allows researchers to quantify, with a tremendous amount of data, fine-scale spatial associations across individuals, in a multitude of animal species (Krause et al. 2011). These tags record contacts between individuals when they come close to each other (a few meters), using, for example, wireless technology. The information is then stored in the memory of the transmitters and downloaded with mobile or fixed receivers. This device constitutes probably the gold standard in the study of animals' spatial associations because they do not depend on contacts with fixed receivers. The number of studies using such devices has markedly increased for the past 10 years, but the studied species are still largely dominated by medium-sized mammals (Table 3.1). Studies may be classified into three main categories. A significant number are dedicated to methodological developments, with several studies involving comparisons of the efficacy of different devices in recording spatial associations or contacts (Lavelle et al. 2014; Tosa et al. 2015); others have tested different possible biases such as nonreciprocal agreements across tags or inter-logger variation in signal strength (Boyland et al. 2013; Meise et al. 2013);

(a)

(b)

Figure 3.3 Biologging on ants.
(a) *Odontomachus hastatus* equipped
with an RFID tag (credit: Cédric Sueur,
IPHC, Strasbourg); (b) *Lasius niger*
performing trophallaxis (mouth-to-mouth
food exchange) and equipped with an
Aruco tag (https://www.uco.es/investiga/
grupos/ava/node/26). The software
Usetracker is used to identify individuals
(http://www.usetracker.org/). (credit: Guy
Theraulaz, CNRS, CRCA, CBI, Toulouse.)

and a few others have proposed analytical developments (Cross et al. 2012; Duncan et al. 2012). Finally, ecological studies on natural species have increased in number and in the diversity of questions asked, going beyond the simple description of spatial associations. For example, by taking advantage of wildfires that occurred in 2009 in Victoria, Australia, Banks et al. (2011) studied the proximate effects of these environmental disturbances on mountain brushtail possums equipped with proximity data loggers. They showed that possums adapted to a decline in the number of suitable shelters by being more flexible in their selection and not by increasing shelter sharing (Banks et al. 2011). Others demonstrated that social thermoregulation was driven by kin relationships in wild raccoons (Robert et al. 2013). Using underwater acoustic loggers, Jacoby et al. (2016) recently showed also that, among

grey reef sharks from the Pacific Ocean, group leaders associate the most with conspecifics.

When compared with other techniques, proximity loggers generally outperformed tags using individual-based locations in time (Walrath et al. 2011; Lavelle et al. 2014; Tosa et al. 2015). Using proximity loggers has also proved better than using observational data (Meise et al. 2013) to estimate spatial associations across animals (see also Ryder et al. 2012). Yet, proximity data logging technology, in addition to being quite expensive, has some important pitfalls challenging both the analysis and the interpretation of results. First, proximity loggers, like GPS, do not provide any indication about the nature of the recorded contacts (e.g. aggression, affiliation, or passive spatial associations). Deciphering an animal's social structure using proximity logging technology alone

Table 3.1 A web survey of research that used proximity biologging technologies to study spatial (and social) associations across conspecifics. 'Proximity data-logger' or 'proximity logger' were the two keywords used on Web of Sciences; the survey resulted in 50 articles matching our criteria. NB: depending on the study, the sample size either indicates the total number of individuals equipped with loggers (including those that failed) or the total number of loggers that successfully recovered data ('ids' stands for individuals).

	Species	Sample size	Device (company)	Aim/main finding	Reference
Fish	Grey reef shark *Carcharhinus amblyrhynchos*	44 ids; 3 years	Acoustic transmitters associated to external receivers (VEMCO)	Leadership is predicted by the duration of co-occurrences between conspecifics	Jacoby et al. 2016
	Lemon sharks *Negaprion brevirostris*	15 ids; 17 days	Internal acoustic transmitters and receivers (Sonotronics Inc)	Methodological (loggers' testing)	Guttridge et al. 2010
	Brown trout *Salmo trutta*	9 ids; 1 week	Proximity talkative tags (Encounternet)	Males spending the most time with females mate more; temporal distribution of encounters reflects shifts in dominance status	Tentelier et al. 2016
Birds	Wire-tailed manakin *Pipra filicauda*	18 ids; 1 month	Coded nanotags with an external digitally encoded data-logger (Lotek Wireless)	Methodological (SNA)	Ryder et al. 2012
	Cowbird *Molothrus bonariensis* and *Molothrus rufoaxillaris*	74 ids; 79 days	Coded tags with an external digitally encoded data-logger (Lotek Wireless and Biotrack)	Females form a dynamic memory library of laying opportunities	Scardamaglia et al. 2017
	Barn swallow *Hirundo rustica erythrogaster*	21 ids; 13 days	Proximity talkative tags (Encounternet)	Methodological (loggers' testing)	Levin et al. 2015
	New Caledonian crows *Corvus moneduloides*	33 ids; 19 days	Proximity talkative tags (Encounternet)	Methodological (loggers' testing) Network structure responds quickly to environmental change	Bettaney et al. 2015; St Clair et al. 2015
	New Caledonian crows *Corvus moneduloides*	34 ids; 7 days	Proximity talkative tags (Encounternet)	Close-range association between nonfamily birds	Rutz et al. 2012
Reptiles	Agassiz's desert tortoise *Gopherus agassizii*	~100 ids; 6–7 months	Proximity talkative tags (Sirtrack; Encounternet)	Sex and season-biased contact interaction	Aiello et al. 2018
Mammals	Sheep *Ovis aries*	94 ids; 23 days	Proximity talkative tags (Sirtrack)	Age difference, temperature and rain, and acoustic similarity impact daily contact time	Doyle et al. 2016

continued

Table 3.1 *Continued*

Species	Sample size	Device (company)	Aim/main finding	Reference
Holstein-Friesian calf *Bos taurus*	40 ids; 1 month	Proximity talkative tags (Sirtrack)	Familiarity during rearing allows social association	Bolt et al. 2017
Cattle *Bos taurus*	20 ids; 3 weeks	Proximity talkative tags (Sirtrack)	Methodological (inter-logger variation)	Boyland et al. 2013
Belmont and Brahman Red cattle *Bos taurus*	33 ids; 29 days	Proximity talkative tags (Sirtrack)	Increased contact with the bull when cows are in oestrus	Corbet et al. 2018
Belmont Red cattle *Bos taurus*	48 ids; 48 days	Proximity talkative tags (Sirtrack)	Number and duration of bull-cow affiliations greater in oestrous cows compared to anoestrus cows	O'Neill et al. 2014
Cattle *Bos taurus*	58 ids; 11 weeks	Proximity talkative tags (Sirtrack)	More frequent association among cows of similar maternal status (pregnant or calved) and among cows with calves of similar ages	Swain et al. 2015
Angus/Limousin and Luing cattle *Bos taurus*	30 ids; 2 months	Proximity talkative tags (Sirtrack)	Methodological (disease modelisation)	Duncan et al. 2012
Cattle *Bos taurus* White-tailed deer *Odocoileus virginianus* Raccoon *Procyon lotor* Virginia opossum *Didelphis virginiana*	198 ids; 16 months	Coded tags with an external stationary data-logger (Sirtrack)	Indirect contacts between wildlife and cattle; highest contact rates between raccoons and cattle during summer and fall	Lavelle et al. 2016
Cattle *Bos taurus* Badger *Meles meles*	25 ids; 4 months	Proximity talkative tags (Sirtrack)	Specific individuals with high contact rates in both livestock and wildlife populations	Böhm et al. 2009
Cattle *Bos taurus* Badger *Meles meles*	109 ids; 17 months	Proximity talkative tags (Sirtrack)	Methodological (loggers' testing)	Drewe et al. 2012

Species	Sample	Method	Finding	Reference
Cattle *Bos taurus* Badger *Meles meles*	94 ids; 12 months	Proximity talkative tags (Sirtrack)	Heterogeneity in contact patterns between cattle and wildlife; more indirect than direct contacts	Drewe et al. 2013
Brushtail possums *Trichosurus vulpecular*	4 ids; 2 days	Proximity talkative tags (MateID, Electronics Laboratory)	Methodological (loggers' testing)	Douglas et al. 2006
Brushtail possum *Trichosurus vulpecula*	40 ids; 6 months	Proximity talkative tags (Sirtrack)	Similar female–male and male–male connection rates	Rouco et al. 2018
Brushtail possum *Trichosurus vulpecula*	22 ids; 3 years	Proximity talkative tags (Sirtrack)	Polygamous mating system, with polygyny and polyandry	Ji et al. 2005
Mountain brushtail possums *Trichosurus cunninghami*	26 ids; 3–10 months	Proximity talkative tags (Sirtrack)	Lower rate of den sharing in burnt areas compared to nonburnt areas	Banks et al. 2011
Raccoon *Procyon lotor*	42 ids; 12 months	Proximity talkative tags (Sirtrack)	Male–male pairs exhibited higher contact values than male–female or female–female pairs, especially during winter	Prange et al. 2011
Raccoon *Procyon lotor*	35 ids; 12 months	Proximity talkative tags (Sirtrack)	Methodological (loggers' testing)	Prange et al. 2006
Raccoon *Procyon lotor*	15 ids; 4–6 months	Proximity talkative tags (Sirtrack)	Correlation between home-range overlap and intraspecific contact rate (weaker in winter)	Robert et al. 2012
Raccoon *Procyon lotor*	15 ids; 17 months	Proximity talkative tags (Sirtrack)	Higher proximity rate in winter, with cold ambient temperatures (social thermoregulation) and among related individuals	Robert et al. 2013
Raccoon *Procyon lotor*	25 ids; 2 years	Proximity talkative tags (Sirtrack)	Several latrine sites visited by multiple raccoons on short time-periods	Hirsch et al. 2014
Raccoon *Procyon lotor*	42 ids; 13 months	Proximity talkative tags (Sirtrack)	Elevated likelihood of a rabies' outbreak when entering a suburban raccoon population (network modelling approach)	Reynolds et al. 2015
River otter *Lontra canadensis*	21 ids; 1–3 months	Proximity talkative tags (Encounternet)	Timing of group fissions and fusions coincides with latrine visits; spatial overlap is a good predictor of social interactions	Barocas et al. 2016

continued

Table 3.1 *Continued*

Species	Sample size	Device (company)	Aim/main finding	Reference
European badger *Meles meles*	51 ids; 1 year	Proximity talkative tags (Sirtrack)	Seasonality in daily patterns of contact frequency and duration	Silk et al. 2017
Island fox *Urocyon littoralis*	48 ids; 4–12 months	Proximity talkative tags (Sirtrack)	Mated females are in contact with other males during oestrus; mated males are in contact with other females than their mate	Ralls et al. 2013
European wild rabbit *Oryctolagus cuniculus*	126 ids; 8 months	Proximity talkative tags (Sirtrack)	Strong and highly stable intra-group associations, rare and transient inter-group associations	Marsh et al. 2011a
European wild rabbit *Oryctolagus cuniculus*	126 ids; 6 months	Proximity talkative tags (Sirtrack)	Spatial and temporal heterogeneities in contact between populations and between and within social groups in the same population	Marsh et al. 2011b
Tasmanian devil *Sarcophilus harrisii*	46 ids; 5 months	Proximity talkative tags (Sirtrack)	All individuals connected in a single large component	Hamede et al. 2009
Elk *Cervus elaphus*	24 ids; 2-3 months	Proximity talkative tags (Sirtrack)	Higher interaction rate but of lower length in males at high densities; longer interactions in females at medium densities	Vander Wal et al. 2012a
Elk *Cervus canadensis*	106 ids; 1 year	Proximity talkative tags (Sirtrack)	Higher interaction rate but of shorter duration among females than among males	Vander Wal et al. 2013
Elk *Cervus canadensis*	104 ids; 1 year	Proximity talkative tags (Sirtrack)	Close-contact interaction rate and duration are independent from genetic relatedness	Vander Wal et al. 2012b
Elk *Cervus canadensis*	about 180 ids (unclear); 6 months	Proximity talkative tags (?)	Methodological (modelling approach)	Cross et al. 2012
Elk *Cervus canadensis*	149 ids; 3 years	Proximity talkative tags (Sirtrack)	Pairwise contact rates decrease when group size increases	Cross et al. 2013
Elk *Cervus canadensis*	30 ids; 2 weeks	Proximity talkative tags (Sirtrack)	Low-density feeding led to reductions in total number of contacts and number of individuals contacting a source of brucellosis transmission	Creech et al. 2012

Species		Technology	Study focus	Reference
White-tailed deer	*Odocoileus virginianus*	22 ids; 5 months	Proximity talkative tags (Sirtrack)	Methodological (comparison of technologies)	Tosa et al. 2015
White-tailed deer	*Odocoileus virginianus*	20 ids; 8 months	Proximity talkative tags (Sirtrack)	Methodological (comparison of methods)	Walrath et al. 2011
White-tailed deer	*Odocoileus virginianus*	26 ids; 2 weeks	Proximity talkative tags (Sirtrack)	Methodological (comparison of technologies)	Lavelle et al. 2014
Koala	*Phascolarctos cinereus*	39 ids; 13 months	Proximity talkative tags (Sirtrack)	More frequent male–female, but not male–male, encounters during the breeding season	Ellis et al. 2015
Mandrill	*Mandrillus sphinx*	13 ids; 1 year	Proximity talkative tags (ELA Innovation)	No correlation between spatial associations and animal's parasitic status	Poirotte et al. 2017
Galápagos sea lion	*Zalophus wollebaeki*	23 ids; 2 months	Proximity talkative tags (Encounternet)	Methodological (comparison of methods); extended association between nonterritorial males	Meise et al. 2013

would probably not yield biologically relevant outputs, and researchers should always complement these logging data with basic information about an animal's life and ecology. Second, proximity logging devices may produce spurious data. For example, signal detection ranges may vary with the antenna angle; unique contact may be recorded as multiple events; and nonexistent 'phantom' codes are sometimes recorded (Prange et al. 2006; Böhm et al. 2009; Drewe et al. 2012; Meise et al. 2013). Signal intensity that indicates the distance between equipped individuals is further subject to environmentally related variations due to weather conditions, for example (Böhm et al. 2009). Interactions that occur at the edge of the receiving range and the nature of animal behaviours may further affect the probability to record a contact. In general, researchers tend to find higher agreement in the duration of contacts rather than in their frequency (Drewe et al. 2009; Meise et al. 2013). Preliminary testing, under various conditions, is therefore mandatory to calibrate each tag individually and to test tags' performance and reciprocity when communicating with each other. Data obtained should also be filtered using automated programs to detect spurious records. Third, and by contrast with GPS-based data, proximity loggers only provide information when two tags are within a certain distance and do not ground data points in true space, possibly limiting interpretations. Fourth, the equipment in itself may be labour-intensive and expensive, although not preventing from high probability of technological failure (Matthews et al. 2013). Capturing and fitting individuals is therefore generally limited to a fraction of the population, which inevitably generates sampling biases, especially when studying social relationships. Finally, equipping wild, often threatened animal species poses ethical issues that need to be carefully addressed and considered well in advance (Ropert-Coudert and Wilson 2005; Wilson and McMahon 2006; Forin-Wiart et al. 2019): loggers may affect animal behaviour or, worse, reproduction (Paton et al. 1991; Ackerman et al. 2004) or survival (Paton et al. 1991; Saraux et al. 2011).

3.4 Analysing social behaviour

Biologging data may be analysed at the individual level and at the group or population levels. For example, biologging allowed for measuring individual diving patterns in penguins that showed synchronisation between individuals and group foraging (Daniel et al. 2007; McInnes et al. 2017), or for studying group hunting in bats (Dechmann et al. 2009). Analysing the huge quantity of data generated by these new technologies may be complex, and new methodologies (Box 3.1) and software have been developed for this purpose. These methods are unified under the name *social network analysis* (SNA). SNA originated from mathematical graph theory and was soon applied to the social sciences (for a review, see Newman 2010). While early on SNA was employed to study social hierarchy in mice (Uhrich 1938), its development arose with studies on social relationships in macaques (Chepko and Sade 1979; Sade et al. 1988; Chepko-Sade et al. 1989; Sade 1989).

SNA is used to grasp several aspects of sociality, such as the role of individuals in their group ('node' level), the clustering of group members ('intermediary' levels) as a function of sociodemographic factors, or how these individuals also structure the network ('group' level). SNA has already proved to be an effective tool for modelling the effects of ecological factors on social relationships and association patterns such as food distribution (Foster et al. 2012), disease risk (Romano et al. 2016), and predation (Heathcote et al. 2017). SNA also provides new approaches such as multilayer, temporal, or modelling analyses to disentangle the effect of these different factors on animals' social relationships. Here, we provide a short summary of the main metrics currently used in SNA with possible applications in the study of animal social systems.

3.4.1 From data collection to social network analyses

A social network is made of nodes representing, for example, individuals, groups, populations, or species, and edges representing, for example, spatial associations or any types of relationships such as affiliative or agonistic interactions. Nodes are characterised by different sociodemographic factors, such as sex, age, personality, hierarchical rank, or body weight. Edges may take different forms depending on the research question and the sociality of the studied species. For example, grooming is interesting for studying the topology of networks in chimpanzees, macaques, or parrots in which this behaviour is frequent. However, relying on grooming in species where this behaviour is rare, like in gorillas, cows, or lizards, would produce a not well-resolved network. In such conditions, it may be preferable to analyse spatial distances, which often indicate social preferences, through instantaneous sampling or using loggers (Haddadi et al. 2011).

Box 3.1 Deep learning in behavioural ecology

With increasingly inexpensive and miniaturised sensing technologies, collecting data on individual trajectories, pose, behaviour, physiology, and other phenotypic traits in animals is more and more feasible (Krause et al. 2013). The main challenge is now to make sense of all this data to extract meaningful information. In classical data analysis, researchers select explanatory variables, manually extract these variables from the raw data and correlate them with one or several response variables. For example, to study how phenotypic variation in a bird population is associated with reproductive success, one would typically first hypothesise which trait is relevant (e.g. beak stoutness), then convert this trait into an explanatory variable (e.g. a product between beak length and depth) and eventually measure it (e.g. in pixels on photographs). Although this approach has provided major insights on animal behaviour, it presents at least five major limitations: (1) it does not guarantee that the selected trait is relevant, which limits the chance to detect an association; (2) the conversion from a trait to a variable may not be optimal; (3) reliable measurements require highly standardised procedures for data collection (e.g. head position perfectly consistent); (4) it is prone to human error; and (5) it is time-consuming and thus intractable for very large data sets. Deep learning algorithms have become explosively popular in computer science over the last 5 years, precisely because they can simultaneously overcome these limitations (LeCun et al. 2015). To understand how this is possible, we briefly present the convolutional neural network (CNN), the most commonly used deep learning algorithm.

A CNN takes raw data as inputs (e.g. images of birds), and directly predicts a response variable (e.g. individual identity) as output, thus automatically selecting those features (i.e. explanatory variables in computer science) that are the most relevant for prediction. A CNN relies on supervised learning and needs to be trained. During this training phase, the algorithm is told when it fails and when it succeeds. The training data set contains labelled data consisting of, for example, images or sound recordings of individuals with known identity, reproductive success, or longevity. The algorithm then predicts the response variable for a testing data set, for which the algorithm was blind to during training.

CNNs all present the same overall design, being composed of two main parts: the encoder that extracts features and the classifier that makes predictions (Boxed Figure 3.1). The encoder extracts features by successively repeating three elementary operations—each repetition is called a layer—with the input of a given layer being the output of the preceding one. Convolution is the central operation that extracts features. It is followed by activation that adds nonlinearity to features. Activation functions allow CNNs to solve highly nonlinear problems such as differentiating between two species at all ages, even if juveniles of either species resemble more to each other than conspecific adults do. The third operation, pooling, merges information that is spatially or temporally close and redundant. This operation makes features invariant to uninformative variations. For example, it allows to accurately estimate beak stoutness from a picture of a head even if not perfectly framed. While embedding raw data into a feature space, the encoder reduces the dimensionality of the data set, from hundred thousands of variables (typical for an image) to a few thousand features. The low-dimensional, highly informative output of the encoder eventually feeds into the classifier (which can also be a regression model for predicting continuous variables; Lathuiliére et al. 2019).

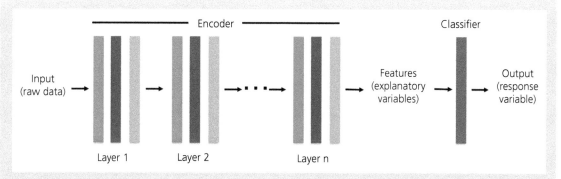

Boxed Figure 3.1 General design of a deep CNN. Each layer performs three elementary operations: convolution (blue), nonlinear activation (red), and pooling (green).

Among artificial intelligence methods, a particularity of CNNs is that the encoder and the classifier work hand-to-hand. During training, at each iteration (i.e. when a subsample of the raw training data is evaluated), the performance of the classifier is estimated and its error back-propagated into the encoder such that the encoder becomes increasingly sensitive to features relevant to the classifier over each iteration. Nevertheless, the encoder can be used in isolation if the goal is to describe the data informatively and not to predict a response variable. For example, the similarity between two phenotypes can be estimated from the Euclidean distance in the low-dimensional feature space (Schroff et al. 2015). Interestingly, one psychophysical experiment showed that such distances accurately predict the visual similarity perceived by humans (Zhang et al. 2018). In that study, the only human-specific data was the input: RGB images tuned to human colour vision. It is therefore possible that with appropriate input data, CNNs could also model the visual perceptual space of other vertebrates (Renoult et al. 2019).

The encoder can be used in isolation also when performing a classification task with a limited data set. CNNs typically need very large data sets for training, which are often unavailable in behavioural studies. For small data sets, it is usually recommended to extract features using the encoder of a trained CNN obtained from a public repository, and to feed an independent classifier with these features (e.g. a discriminant analysis or a support vector machine SVM). For example, in a study with a training data set containing only 1,550 images of 276 different wild elephants (8 images per individual on average), a SVM fed with features extracted from the ResNet50 algorithm pretrained on the ImageNet data set reached 74% accuracy in individual recognition (Körschens et al. 2018). This is remarkable given that ImageNet contains only six images of elephants out of millions. Alternatively, with small data sets one can apply transfer learning, in which a CNN pretrained with a very large data set is retrained (both the encoder and the classifier) on a smaller data set (Weiss et al. 2016). However, it is important to bear in mind that using artificial intelligence is a test and trial process and that multiple approaches should be tested to obtain the best performances. For example, in another study based on 1,550 pictures of pigs, representing 10 different individuals, training a CNN from scratch was more successful (97% in individual recognition) than transfer learning (Hansen et al. 2018).

In addition to identifying individuals based on images (Deb et al. 2018; Hansen et al. 2018) or motion patterns (Neverova et al. 2016), and to estimating similarity between phenotypes (Schroff et al. 2015), CNNs can predict the pose (Mathis et al. 2018) and behaviour (Nguyen et al. 2019) of animals. Interestingly, transfer learning allows to train algorithms with very small data sets: 200 labelled images can be enough to recognise pose (i.e. the spatial configuration of body parts) in various animal species when transferring a network pretrained with human data (Mathis et al. 2018). Behavioural ecologists will therefore certainly soon benefit from the extensive amount of research on human behaviours.

For behavioural ecologists willing to engage with CNNs, we recommend using either the Python or R programming languages. Over the last decade, Python has become a principal language of scientific programming. Most CNN tutorials and code available on public repositories are in Python. Tensorflow (developed by Google) and PyTorch (developed by Facebook) are two of the most popular Python-based frameworks, allowing the design of new models or the transfer of existing ones for all kinds of input data and applications. To those with a limited background in programming, Keras is an easy-to-use framework running on top of Tensorflow. It abstracts CNN elementary operations, which are then arranged like Lego bricks. Recently, Tensorflow and Keras have been made accessible in R.

To ensure repeatability and consistency of results obtained from SNA, the following questions should therefore be asked when collecting data in the field: (1) Are all group members observed equally and for a comprehensive period of time? (2) Are the results obtained by studying one specific group representative of the species sociality? (3) How might uncertainty be reduced? Different software or R packages are dedicated to SNA, some specifically designed to take into account observational biases (for a discussion on the different packages see Sosa et al. 2018a). While some software were specifically developed for behavioural ecologists (e.g. ANTs, Sosa et al. 2018a; Socprog, Whitehead 2009; and Asnipe, Farine 2013), others have proved useful for analysing animal societies even though they were firstly developed for other applications (e.g. Ucinet, Borgatti et al. 2002; igraph, Csardi and Nepusz 2006).

3.4.2 Individual level: the roles of nodes

Figure 3.4a, which clearly identifies two clusters, describes a typical social network generated from Table 3.2, which presents the main metrics currently in use (for comprehensive reviews of SNA metrics see Whitehead 2008; Sueur et al. 2011a; Krause et al. 2015;

Sosa et al. 2020). *Centrality* refers to the position of an individual in its group and may have different definitions (in Table 3.2, degree, strength, betweenness, eigenvector coefficient, and clustering coefficient are the main centralities used in SNA). The *degree* is the number of individuals connected to (or having a relationship with) a node. It can be interpreted as the popularity of an individual. The *strength* is the sum of interactions involving the node and it represents the social activity of an individual with group members. The *betweenness* expresses the number of times a node links to other nodes. Individuals g and j have, for example, higher betweenness because they are linked to all other nodes (Figure 3.4a). Diseases or social information should go through these two nodes (i.e. g and j) to spread rapidly into the group. By contrast, if g and j disappear, the group may split: group fission happens, for example, when an old individual linking two clusters (e.g. matrilines) dies (Lefebvre et al. 2003; Sueur and Maire 2014). Betweenness therefore matters when studying the importance of a node regarding group cohesion or social/disease diffusion. The *eigenvector* provides information about the connection between a node and its neighbours. Elevated eigenvectors indicate the high popularity of both an individual and its partners. The *clustering coefficient* represents an individual measure of cohesiveness: it estimates how partners of a node are connected. For example, a node with unconnected partners shows a clustering coefficient equal to 0, whatever the number of partners involved. Although other metrics are available to study social centrality, these five measurements (degree, strength, betweenness, eigenvector, and clustering coefficient) can all be explained by relatively simple social differences between individuals (e.g. Firth et al. 2017a) and are therefore widely used. They may all be related to sociodemographic characteristics of the nodes. For example, high-ranking individuals generally have a high centrality in primates (Sueur et al. 2011b), and old and experienced animals present a strong degree and betweenness in elephants (McComb et al. 2011).

3.4.3 Intermediary levels

Homophily (or *heterophily*) measures the probability that individuals sharing similar characteristics, such as rank, age, or nutrient requirements, have stronger (or weaker) social relationships or spatial associations (Ruckstuhl and Neuhaus 2000; Croft et al. 2005). When studying continuous traits, homophily is typically calculated using the E/I ratio of Krackhardt (Krackhardt and Stern 1988). This ratio represents a proportion between the relationships or associations observed across categories (E, for 'external', such as across sexes) with the ones observed within a given category (I, for 'internal', such as among individuals of the same sex). This metric therefore reports preferential interaction between individuals with similar (homophily) or different (heterophily) attributes. These attributes can even take the form of node metrics such as the degree (i.e. the number of links to a node) to measure assortativity. Assortativity is a special case of homophily based on centralities: nodes that share similar centralities are more associated. For example, individuals with high degrees associate more with each other. MRQAP (multiple regression quadratic assignment procedure) and matrices of correlations are particularly useful when studying homophily because a matrix of social interactions or of spatial associations may be compared to a matrix of genetic relatedness or age differences, for example. Dominant individuals are generally highly assorted because they have differentiated social relationships and share high degrees (Sueur et al. 2011b; Sosa et al. 2018b).

3.4.4 Group level and group comparison

Density is one of the primary index used to describe social networks at the group level. It represents the number of observed social relationships or spatial associations divided by the total number of all possibilities. Density is thus a measure of group cohesion and further characterises social style (Sueur et al. 2011a). *Modularity* is another index that represents the probability for a group to split into several clusters (Figure 3.4b). With this metric, the total number of clusters in a group may be compared to other variables such as the distribution of ages and relatedness within the group (but see Whitehead 2009 for limitations). For example, groups that are highly structured by kin relationships show elevated modularity indices, as observed in elephants and in several multilevel societies (e.g. hamadryas baboons, Wittemyer et al. 2005; Matsuda et al. 2012, 2015). *Network centralisation* measures how much a network is centred on some key individuals, for example because of their experience or their position in the social hierarchy (Figure 3.4c). This index is particularly useful when describing social structures and for studying diffusion processes, which typically rely heavily on these key individuals (Lusseau and Newman 2004; Carne et al. 2013). Finally, *network efficiency* measures the probability of diffusion of an entity (a knowledge or a pathogen) to the entire group and depends on the network topology. Overall, group-level indices thus

(a)

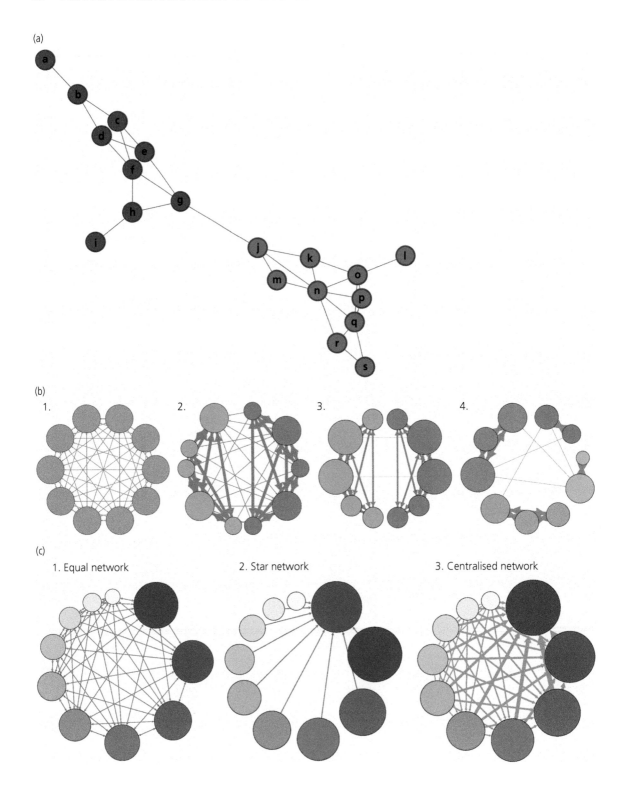

(b)

1.

2.

3.

4.

(c)

1. Equal network

2. Star network

3. Centralised network

Table 3.2 Individual network measurements of nodes illustrated in Figure 3.4a.

Node	Degree	Strength	Betweenness	Eigenvector	Clustering
a	1	1	0	0	–
b	3	3	17.5	0.02	0
c	3	3	19.667	0.04	0.33
d	2	3	10.333	0.04	0.33
e	3	4	12.833	0.06	0.5
f	5	5	41.667	0.07	0.4
g	4	4	81	0.11	0.33
h	3	3	17	0.05	0.33
i	1	1	0	0.01	–
j	4	4	81.833	0.27	0.33
k	4	3	16.5	0.29	0.67
l	2	1	0	0.1	–
m	3	2	8.333	0.19	1
n	7	7	46.333	0.51	0.33
o	5	5	6.667	0.39	0.4
p	2	3	0	0.32	1
q	4	5	8	0.4	0.5
r	4	3	9.333	0.26	0.67
s	2	2	0	0.16	1

are highly useful for comparing group or population structures and properties. However, these indices highly depend on factors such as group size or time of observations. Controlling for these factors is therefore mandatory in group-level SNA.

3.4.5 Network dynamics

While SNAs were initially applied to characterise individual social relationships, they are increasingly used with a dynamic perspective to study, for example, the stability and resilience of networks over time and generations or diffusion processes (Allen et al. 2013; Cantor and Whitehead 2013; Fisher and McAdam 2017; Sueur et al. 2017). Network dynamics can be captured in different ways. Temporality indicates, for example, how a network changes over time according to, for example, seasons, changes in group composition, or across generations (Henzi et al. 2009; Pinter-Wollman et al. 2013; Fisher et al. 2016; Wolf et al. 2018). Temporality is connected to concepts such as stability, resilience and network robustness (Naug 2009; Fushing et al. 2013; Goldenberg et al. 2016; Puga-Gonzalez et al. 2019). While the study of network efficiency provides a proxy for how an entity (i.e. a disease or an information) is transmitted, modelling dynamic diffusion processes allows a better understanding of which and when a group member is infected or informed (Hoppitt et al. 2010; Rushmore et al. 2014; Pasquaretta et al. 2016; Romano et al. 2016). Modelling is often necessary to study social network dynamics (Skyrms and Pemantle 2009; Sueur and Maire 2014; Romano et al. 2018)

Figure 3.4 Graphical representations of some theoretical network properties. (a) A node represents a group member. Nodes with the same colour code belong to the same subgroup. An edge (line) represents a relationship between two individuals. This social network corresponds to the social structure of a group of 19 individuals (labelled "a" to "s"). Individual network measures are presented in Table 3.2. (b) Four different types of networks that illustrate modularity: (1) equal network, all individuals are all equally connected (maximum modularity Q = 0); (2) two-subset network with 14 relationships between the two subsets (maximum modularity Q = 0.36); (3) two-subset network with only two relationships between subsets (maximum modularity Q = 0.48); and (4) four-subset network with five relationships between the four subsets (maximum modularity Q = 0.59). The size of the nodes represents the degree (number of relationships) and the colour represents the subset (one colour per subset). Thicker lines between nodes represent stronger relationships between individuals. (c) Three networks that illustrate the concept of centralisation: (1) equal network: all individuals are all equally connected; (2) star network: all group members are only connected to one individual; (3) centralised network: individuals only have relationships with a few (e.g. dominant) group members and the strength of relationships (line thickness) decreases with e.g. dominance rank. The larger the size of a node (individual) and the darker the colour, the higher the individual dominance rank.

because empirical data may be difficult to get or be ethically questionable (e.g. experimental removals of individuals or experimental infections, but see Flack et al. 2005; Flack et al. 2006; Aplin et al. 2015; Firth and Sheldon 2015; Firth et al. 2017b).

Finally, studying the concurrent changes and interdependencies between diffusion processes and network topology is a highly topical area of research (Farine et al. 2015; Lopes et al. 2016; Sueur et al. 2017; Fisher and Pruitt 2019) which includes both conceptual and modelling studies (Banks and Carley 1996; Puga-Gonzalez et al. 2018). Although empirical tests of these concurrent changes are rare, a recent study on ants demonstrated that workers change their self-organised behaviour following an infection to decrease network efficiency, with an influence on pathogen spread (Stroeymeyt et al. 2018). Hence, SNA can investigate how group properties influence individual properties, for example, when the number of clusters determines the diffusion of disease and/or information and thus the health and fitness of individuals. Such effects can influence in return social relationships and thus the network topology.

3.5 Conclusions

New technologies and analytical methods are revolutionising the sampling of animal behaviours, social interactions, and association patterns and are thereby opening novel research avenues (Chapter 21). The high spatiotemporal resolution of data needed when studying social processes such as disease or information diffusion (often involving rare and brief interindividual contacts) will probably be better attained using proximity logging technology rather than classic methods of observations, particularly when attempting to simultaneously track large numbers of individuals. Conveniently, new analytical tools have been developed to manage, store, and analyse large data sets. Nonetheless, despite all these benefits, biologging will probably never replace the fine-grain approach and the experience of the human observer. Whenever possible, we suggest that combining different techniques probably constitutes the best approach to catch the complexity of animal societies and the functioning of individuals that live and interact together in a myriad of ways.

References

Ackerman J. T., Adams J., Takekawa J. Y. et al. (2004). Effects of radiotransmitters on the reproductive performance of Cassin's auklets. Wildl. Soc. Bull. **32**(1973–2006), 1229–1241.

Aiello C. M., Esque T. C., Nussear K. E., Emblidge P. G., and Hudson P. J. (2018). Associating sex-biased and seasonal behaviour with contact patterns and transmission risk in *Gopherus agassizii*. Behaviour **155**, 585–619.

Allen J., Weinrich M., Hoppitt W., and Rendell L. (2013). Network-based diffusion analysis reveals cultural transmission of lobtail feeding in humpback whales. Science **340**, 485–488.

Altmann J. (1974). Observational study of behavior-sampling methods. Behaviour **49**, 227–267.

Altmann J., Hausfater G., and Altmann S. A. (1985). Demography of Amboseli baboons, 1963–1983. Am. J. Primatol. **8**, 113–125.

Aplin L., Farine D., Morand-Ferron J. et al. (2015). Experimentally induced innovations lead to persistent culture via conformity in wild birds. Nature **518**, 538–541.

Archie E. A., Tung J., Clark M., Altmann J., and Alberts S. C. (2014). Social affiliation matters: both same-sex and opposite-sex relationships predict survival in wild female baboons. Proc. R. Soc. B Biol. Sci. **281**, 20141261.

Banks D. L. and Carley K. M. (1996). Models for network evolution. J. Math. Sociol. **21**, 173–196.

Banks S. C., Knight E. J., McBurney L., Blair D., and Lindenmayer D. B. (2011). The effects of wildfire on mortality and resources for an arboreal marsupial: resilience to fire events but susceptibility to fire regime change. PLOS One **6**, e22952.

Barash D. P. (1973). Social variety in yellow bellied marmot (*Marmota flaviventris*). Anim. Behav. **21**, 579–584.

Barocas A., Golden H. N., Harrington M. W., McDonald D. B., and Ben-David M. (2016). Coastal latrine sites as social information hubs and drivers of river otter fission–fusion dynamics. Anim. Behav. **120**, 103–114.

Bettaney E. M., James R., St Clair J. J. H., and Rutz C. (2015). Processing and visualising association data from animal-borne proximity loggers. Anim. Biotelemetry **3**, 27.

Block B. A., Teo S. L. H., Walli A. et al. (2005). Electronic tagging and population structure of Atlantic bluefin tuna. Nature **434**, 1121–1127.

Böhm M., Hutchings M. R., and White P. C. L. (2009). Contact networks in a wildlife–livestock host community: Identifying high-risk individuals in the transmission of bovine TB among badgers and cattle. PLOS One **4**, e5016.

Bonfil R., Meÿer M., Scholl M. C. et al. (2005). Transoceanic migration, spatial dynamics, and population linkages of white sharks. Science **310**, 100–103.

Bolt S. L., Boyland N. K., Mlynski D. T., James R., and Croft D. P. (2017). Pair housing of dairy calves and age at pairing: effects on weaning stress, health, production and social networks. PLOS One **12**(1), e0166926. https://doi.org/10.1371/journal.pone.0166926.

Borgatti S., Everett M., and Freeman L. (2002). UCINET 6 for Windows: software for social network analysis. http://www.analytictech.com/.

Boucherie P. H., Loretto M. C., Massen J. J., and Bugnyar T. (2019). What constitutes 'social complexity' and 'social intelligence' in birds? Lessons from ravens. Behav. Ecol. Sociobiol. **73**, 12.

Boyd J. M., Mackay J. Doney R. G. G., and Jewell P. A. (1964). The Soay sheep of the island of Hirta, St. Kilda. A study of a feral population. Proc. Zool. Soc. Lond. **142**, 129–163.

Boyland N. K., James R., Mlynski D. T., Madden J. R., and Croft D. P. (2013). Spatial proximity loggers for recording animal social networks: consequences of inter-logger variation in performance. Behav. Ecol. Sociobiol. **67**, 1877–1890.

Broihanne M.-H., Romain A., Call J. et al. (2018). Monkeys (*Sapajus apella* and *Macaca tonkeana*) and great apes (*Gorilla gorilla*, *Pongo abelii*, *Pan paniscus*, and *Pan troglodytes*) play for the highest bid. J. Comp. Psychol. **133**, 301–312.

Brown D. D., Kays R., Wikelski M., Wilson R., and Klimley A. P. (2013). Observing the unwatchable through acceleration logging of animal behavior. Anim. Biotelemetry **1**, 20.

Bueno-Guerra N. and Amici F. (2018). Field and Laboratory Methods in Animal Cognition: A Comparative Guide. Cambridge University Press.

Busson M., Authier M., Barbraud C., Tixier P., Reisinger R. R., Janc A., and Guinet C. (2019). Role of sociality in the response of killer whales to an additive mortality event. Proc. Natl. Acad. Sci. **116**, 11812–11817.

Cantor M. and Whitehead H. (2013). The interplay between social networks and culture: theoretically and among whales and dolphins. Phil. Trans. R. Soc. B **368**, 20120340.

Carne C., Semple S., Morrogh-Bernard H., Zuberbühler K., and Lehmann J. (2013). Predicting the vulnerability of great apes to disease: the role of superspreaders and their potential vaccination. PLOS One **8**, e84642.

Carpenter, C. R. (1942a). Characteristics of social behavior in non-human primates. Trans. New York Acad. Sci. **4**, 248–258.

Carpenter C. R. (1942b). Societies of monkeys and apes. Biol. Symp. **8**, 177–204.

Castles M., Heinsohn R., Marshall H. H., Lee A. E., Cowlishaw G., and Carter A. J. (2014). Social networks created with different techniques are not comparable. Anim. Behav. **96**, 59–67.

Charpentier M. J. E., Harté M., Poirotte C., Meric de Bellefon J., Laubi B., Kappeler P. M., and Renoult J. P. (2020). Same father, same face: deep-learning reveals natural selection for paternally-derived signalling of kinship in a wild primate. **Sciences Advances** 6: eaba3274.

Cheney D. L., Silk J. B., and Seyfarth R. M. (2016). Network connections, dyadic bonds and fitness in wild female baboons. R. Soc. Open Sci. **3**, 160255.

Chepko-Sade B. D. and Sade D. S. (1979). Patterns of group splitting within matrilineal kinship groups. Behav. Ecol. Sociobiol. **5**, 67–86.

Chepko-Sade B. D., Reitz K. P., and Sade D. S. (1989). Sociometrics of *Macaca mulatta* IV: network analysis of social structure of a pre-fission group. Soc. Netw. **11**, 293–314.

Clutton-Brock T. (1974). Primate social organization and ecology. Nature **250**, 539–542.

Clutton-Brock T. (1982). The red deer of Rhum. Natural History **91**, 42.

Corbet N .J., Patison K. P., Menzies, D. J., and Swain D. L. (2018). Using temporal associations to determine postpartum oestrus in tropical beef cows. Anim. Prod. Sci. **58**, 1465–1469.

Creech T. G., Cross P. C., Scurlock B. M., Maichak E. J., Rogerson J. D., Henningsen J. C., and Creel S. (2012). Effects of low-density feeding on elk–fetus contact rates on Wyoming feedgrounds. J. Wildl. Manag. **76**, 877–886.

Croft D. P., James R., Ward A. J. W., Botham M. S., Mawdsley D., and Krause J. (2005). Assortative interactions and social networks in fish. Oecologia **143**, 211–219.

Crook J. H. (1964). Adaptive significance of avian social organisations. Anim. Behav. **12**, 393.

Crook J. H. (1970). Social organization and environment—aspects of contemporary social ethology. Anim. Behav. **18**, 197–209.

Crook J. H. and Gartlan J. S. (1966). Evolution of primate societies. Nature **210**, 1200–1203.

Cross P. C., Creech T. G., Ebinger M. R., Heisey D. M., Irvine K. M., and Creel S. (2012). Wildlife contact analysis: emerging methods, questions, and challenges. Behav. Ecol. Sociobiol. **66**, 1437–1447.

Cross P. C., Creech T. G., Ebinger M. R. et al. (2013). Female elk contacts are neither frequency nor density dependent. Ecology **94**, 2076–2086.

Csardi G. and Nepusz T. (2006). The Igraph software package for complex network research. InterJournal Complex Syst. **1695**, 1–9.

Daniel T. A., Chiaradia A., Logan M., Quinn G. P., and Reina R. D. (2007). Synchronized group association in little penguins, *Eudyptula minor*. Anim. Behav. **74**, 1241–1248.

Davis M. J., Thokala S., Xing X., Hobbs N. T., Miller M. W., Han R., and Mishra S. (2013). Testing the functionality and contact error of a GPS-based wildlife tracking network. Wildl. Soc. Bull. **37**, 855–861.

Deb D., Wiper S., Russo A., Gong S., Shi Y., Tymosze, C., and Jain A. (2018). Face recognition: primates in the wild. arXiv preprint arXiv:1804.08790.

Dechmann D. K. N., Heucke S. L., Giuggioli L., Safi K., Voigt C. C., and Wikelski M. (2009). Experimental evidence for group hunting via eavesdropping in echolocating bats. Proc. R. Soc. Lond. B **276**, 2721–2728.

Douglas M. E., Ji W., and Clout M. N. (2006). MateID: design and testing of a novel device for recording contacts between free-ranging animals. Wildl. Soc. Bull. **34**, 203–207.

Doyle R. E., Broster J. C., Barnes K., and Browne W. J. (2016). Temperament, age and weather predict social interaction in the sheep flock. Behav. Proc. **131**, 53–58.

Drewe J. A., Madden J. R., and Pearce G. P. (2009). The social network structure of a wild meerkat population: 1. Inter-group interactions. Behav. Ecol. Sociobiol. **63**, 1295–1306.

Drewe J. A., Weber N., Carter S. P. et al. (2012). Performance of proximity loggers in recording intra- and inter-species interactions: a laboratory and field-based validation study. PLOS One **7**, e39068.

Drewe J. A., O'Connor H. M., Weber N., McDonald R. A., and Delahay R. J. (2013). Patterns of direct and indirect contact between cattle and badgers naturally infected with tuberculosis. Epidemiol. Infect. **141**, 1467–1475.

Dufour V., Pelé M., Sterck E. H. M., and Thierry B. (2007). Chimpanzee (*Pan troglodytes*) anticipation of food return: coping with waiting time in an exchange task. J. Comp. Psychol. **121**, 145–155.

Duncan A. J., Gunn G. J., Lewis F. I., Umstatter C., and Humphry R. W. (2012). The influence of empirical contact networks on modelling diseases in cattle. Epidemics **4**, 117–123.

Ekman, J. (1979). Coherence, composition and territories of winter social groups of the willow tit *Parus montanus* and the crested tit *P. cristatus*. Ornis. Scand. **10**, 56–68.

Ellis, W., FitzGibbon, S., Pye, G. et al. (2015). The role of bioacoustic signals in koala sexual selection: insights from seasonal patterns of associations revealed with GPS-proximity units. PLOS One **10**, e0130657.

Emery N. (2016). Bird Brain: An Exploration of Avian Intelligence. Princeton University Press.

Emlen S. T. and Oring L. W. (1977). Ecology, sexual selection, and evolution of mating systems. Science **197**, 215–223.

Espinas A. (1877). Des Sociétés Animales. Etude de Psychologie Comparée. Germer Baillière.

Farine D. R. (2013). Animal social network inference and permutations for ecologists in R using Asnipe. Methods Ecol. Evol. **4**, 1187–1194.

Farine D. R., Montiglio P.-O., and Spiegel O. (2015). From individuals to groups and back: the evolutionary implications of group phenotypic composition. Trends Ecol. Evol. **30**, 609–621.

Ferreira A. C., Silva L. R., Renna F. et al. (2019). Deep learning-based methods for individual recognition in small birds. BioRxiv, 862557.

Filer J. (2008). Fish cognition and behavior. Fish and aquatic resources series 11. Fish and Fisheries **9**(2), 214–215.

Firth J. A. and Sheldon B. C. (2015). Experimental manipulation of avian social structure reveals segregation is carried over across contexts. Proc. R. Soc. Lond. B **282**, 20142350.

Firth J. A., Sheldon B. C., and Brent L. J. (2017a). Indirectly connected: simple social differences can explain the causes and apparent consequences of complex social network positions. Proc. R. Soc. Lond. B **284**, 20171939.

Firth J. A., Voelkl B., Crates R. A., Aplin L. M., Biro D., Croft D. P., and Sheldon B. C. (2017b). Wild birds respond to flockmate loss by increasing their social network associations to others. Proc. R. Soc. Lond. B **284**, 20170299.

Fisher D. N. and McAdam A. G. (2017). Social traits, social networks and evolutionary biology. J. Evol. Biol. **30**, 2088–2103.

Fisher D. N. and Pruitt J. N. (2019). Insights from the study of complex systems for the ecology and evolution of animal populations. Current Zoology **66**(1), 1–14.

Fisher D. N., Rodríguez-Muñoz R., and Tregenza T. (2016). Wild cricket social networks show stability across generations. BMC Evol. Biol. **16**, 151.

Fisher J. and Hinde R. A. (1949). The opening of milk bottles by birds. Br. Birds **42**, 347–357.

Flack J. C., Krakauer D. C., and De Waal F. B. M. (2005). Robustness mechanisms in primate societies: a perturbation study. Proc. R. Soc. Lond. B **272**, 1091–1099.

Flack J. C., Girvan M., de Waal F. B., and Krakauer D. C. (2006). Policing stabilizes construction of social niches in primates. Nature **439**, 426.

Forin-Wiart M.-A., Enstipp M. R., Le Maho Y., and Handrich Y. (2019). Why implantation of bio-loggers may improve our understanding of how animals cope within their natural environment. Integr. Zool. **14**, 48–64.

Formica V. A., Wood C. W., Larsen W. B., Butterfield R. E., Augat M. E., Hougen H. Y., and Brodie Iii E. D. (2012). Fitness consequences of social network position in a wild population of forked fungus beetles (*Bolitotherus cornutus*). J. Evol. Biol. **25**, 130–137.

Foster E. A., Franks D. W., Morrell L. J., Balcom, K. C., Parsons K. M., Van Ginneken A., and Croft D. P. (2012). Social network correlates of food availability in an endangered population of killer whales, *Orcinus orca*. Anim. Behav. **83**, 731–736.

Fushing H., Wang H., VanderWaal K., McCowan B., and Koehl P. (2013). Multi-scale clustering by building a robust and self correcting ultrametric topology on data points. PLOS One **8**, e56259.

Garstang M. (2015). Elephant Sense and Sensibility. Academic Press.

Gimenez O., Mansilla L., Klaich M. J., Coscarella M. A., Pedraza S. N., and Crespo E. A. (2019). Inferring animal social networks with imperfect detection. Ecol. Model. **401**, 69–74.

Godfrey-Smith P. (2016). Other Minds: The Octopus, the Sea, and the Deep Origins of Consciousness. Farrar, Straus and Giroux.

Goldenberg S. Z., Douglas-Hamilton I., and Wittemyer G. (2016). Vertical transmission of social roles drives resilience to poaching in elephant networks. Curr. Biol. **26**, 75–79.

Goodall J. (1986). Social rejection, exclusion, and shunning among the Gombe chimpanzees. Ethol. Sociobiol. **7**, 227–236.

Guttridge T. L., Gruber S. H., Krause J., and Sims D. W. (2010). Novel acoustic technology for studying free-ranging shark social behaviour by recording individuals' interactions. PLOS One **5**, e9324.

Haddadi H., King A. J., Wills A. P. et al. (2011). Determining association networks in social animals: choosing spatial–temporal criteria and sampling rates. Behav. Ecol. Sociobiol. **65**, 1659–1668.

Hamede R. K., Bashford J., McCallum H., and Jones M. (2009). Contact networks in a wild Tasmanian devil (*Sarcophilus harrisii*) population: using social network analysis to reveal seasonal variability in social behaviour and its implications for transmission of devil facial tumour disease. Ecol. Lett. **12**, 1147–1157.

Hansen M. F., Smith M. L., Smith L. N., Salter M. G., Baxter E. M., Farish M., and Grieve B. (2018). Towards on-farm pig face recognition using convolutional neural networks. Comp. Ind. **98**, 145–152.

Hazen E. L., Maxwell S. M., Bailey H. et al. (2012). Ontogeny in marine tagging and tracking science: technologies and data gaps. Mar. Ecol. Prog. Ser. **457**, 221–240.

Heaney M., Gray R. D., and Taylor A. H. (2017). Keas perform similarly to chimpanzees. PLoS One 12, e0169799.

Heathcote R .J. P., Darden S. K., Franks D. W., Ramnarine I. W., and Croft D. P. (2017). Fear of predation drives stable and differentiated social relationships in guppies. Sci. Rep. **7**, 41679.

Henzi S., Lusseau D., Weingrill T., Van Schaik C., and Barrett L. (2009). Cyclicity in the structure of female baboon social networks. Behav. Ecol. Sociobiol. **63**, 1015–1021.

Hinde R. A. and Fisher J. (1951). Further observations on the opening of milk bottles by birds. Br. Birds **44**, 392–396.

Hirsch B. T., Prange S., Hauver S. A., and Gehrt S. D. (2014). Patterns of latrine use by raccoons (*Procyon lotor*) and implication for Baylisascaris procyonis transmission. J. Wildl. Dis. **50**, 243–249.

Holt-Lunstad J., Smith T. B., and Layton J. B. (2010). Social relationships and mortality risk: a meta-analytic review. PLOS Medicine **7**, e1000316.

Hoppitt W., Boogert N. J., and Laland K. N. (2010). Detecting social transmission in networks. J. Theor. Biol. **263**, 544–555.

House J. S., Landis K. R., and Umberson D. (1988). Social relationships and health. Science **241**, 540.

Hussey N. E., Kessel S. T., Aarestrup K. et al. (2015). Aquatic animal telemetry: a panoramic window into the underwater world. Science **348**, 1255642.

Imanishi K. (1952). Evolution of the humanity. In: K. Imanishi (ed.) Man, 36–94. Mainichi-shinbunsha.

Jacoby D. M. P., Papastamatiou Y. P., and Freeman, R. (2016). Inferring animal social networks and leadership: applications for passive monitoring arrays. J. R. Soc. Interface **13**, 20160676.

Jarman P. J. (1974). Social organization of antelope in relation to their ecology. Behaviour **48**, 215.

Ji W., White P. C. L., and Clout M. N. (2005). Contact rates between possums revealed by proximity data loggers. J. Appl. Ecol. **42**, 595–604.

Kappeler P. M. and Watts D. P. (2012). Long-Term Field Studies of Primates. Springer Science + Business Media.

Kappeler P. M., Cremer S., and Nunn C. L. (2015). Sociality and health: impacts of sociality on disease susceptibility and transmission in animal and human societies. Philos. Trans. R. Soc. B Biol. Sci. **370**, 20140116.

Kays R., Crofoot M. C., Jetz W., and Wikelski M. (2015). Terrestrial animal tracking as an eye on life and planet. Science **348**, 2478.

Kooyman G. L. (1965). Techniques used in measuring diving capacities of Weddell seals. Polar Rec. **12**, 391–394.

Körschens M., Barz B., and Denzler J. (2018). Towards automatic identification of elephants in the wild. arXiv preprint arXiv:1812.04418.

Krackhardt D. and Stern R. N. (1988). Informal networks and organizational crises: an experimental simulation. Soc. Psychol. Q. **51**, 123.

Krause J., Krause S., Arlinghaus R., Psorakis I., Roberts S., and Rutz C. (2013). Reality mining of animal social systems. Trends Ecol. Evol. **28**, 541–551.

Krause J., Jame, R., Franks D. W., and Croft D. P. (2015). Animal Social Networks. Oxford University Press.

Krause J., Wilson A. D. M., and Croft D. P. (2011). New technology facilitates the study of social networks. Trends Ecol. Evol. **26**, 5–6.

Krupenye C., Kano F., Hirata S., Call J., and Tomasello M. (2016). Great apes anticipate that other individuals will act according to false beliefs. Science **354**, 110–114.

Kruuk H. (1972). The Spotted Hyena: A Study of Predation and Social Behaviour. University of Chicago Press.

Lathuilière S., Mesejo P., Alameda-Pineda X., and Horaud R. (2019). A comprehensive analysis of deep regression. IEEE Transactions on Pattern Analysis and Machine Intelligence **42**(9), 2065–2081.

Lavelle M. J., Fischer J. W., Phillips G. E. et al. (2014). Assessing risk of disease transmission: direct implications for an indirect science. BioScience **64**, 524–530.

Lavelle M. J., Kay S. L., Pepin K. M., Grear, D. A., Campa H., and VerCauteren K. C. (2016). Evaluating wildlife–cattle contact rates to improve the understanding of dynamics of bovine tuberculosis transmission in Michigan, USA. Prev. Vet. Med. **135**, 28–36.

LeCun Y., Bengio Y., and Hinton G. (2015). Deep learning. Nature **521**(7553), 436.

Lefebvre D., Ménard N., and Pierre, J. S. (2003). Modelling the influence of demographic parameters on group structure in social species with dispersal asymmetry and group fission. Behav. Ecol. Sociobiol. **53**, 402–

Levin I. I., Zonana D. M., Burt, J. M. and Safran, R .J. (2015). Performance of Encounternet tags: field tests of miniaturized proximity loggers for use on small birds. PLOS One **10**, e0137242.

Lopes P. C., Block P., and König B. (2016). Infection-induced behavioural changes reduce connectivity and the potential for disease spread in wild mice contact networks. Sci. Rep. **6**, 31790.

Lusseau D. and Newman M. E. J. (2004). Identifying the role that animals play in their social networks. Proc. R. Soc. Lond. B **271**, S477–S481.

Marsh M. K., McLeod S. R., Hutchings M. R., and White P. C. L. (2011a). Use of proximity loggers and network analysis to quantify social interactions in free-ranging wild rabbit populations. Wildl. Res. **38**, 1–12.

Marsh M .K., Hutchings M. R., McLeod S. R., and White P. C. L. (2011b). Spatial and temporal heterogeneities in the contact behaviour of rabbits. Behav. Ecol. Sociobiol. **65**, 183–195.

Mathis A., Mamidanna P., Cury K. M., Abe T., Murthy V. N., Mathis M. W., and Bethge M. (2018). DeepLabCut: Markerless pose estimation of user-defined body parts with deep learning. Nature Neurosci. **21**, 1281–1289.

Matsuda I., Zhang P., Swedell L., Mori U., Tuuga A., Bernard H., and Sueur C. (2012). Comparisons of intra-unit relationships in nonhuman primates living in multilevel social systems. Int. J. Primatol. **33**, 1038–1053.

Matsuda I., Fukaya K., Pasquaretta C., and Sueur C. (2015). Factors influencing grooming social networks: insights from comparisons of colobines with different dispersal patterns. In: T. Furuichi, J. Yamagiwa, Juichi, and F. Aureli (eds.), Dispersing Primate Females, 231–254. Springer.

Matthews A., Ruykys L., Ellis W. et al. (2013). The success of GPS collar deployments on mammals in Australia. Aust. Mammal. **35**, 65–83.

McBride G. (1964). A general theory of social organization and behaviour. University of Queensland Papers, Faculty of Veterinary Science 1, 75–110.

McComb K., Shannon G., Durant, S. M., Sayialel K., Slotow R., Poole J., and Moss C. (2011). Leadership in elephants: the adaptive value of age. Proc. R. Soc. B Biol. Sci. **278**, 3270–3276.

McInnes A. M., McGeorge C., Ginsberg S., Pichegru L., and Pistorius P. A. (2017). Group foraging increases foraging efficiency in a piscivorous diver, the African penguin. R. Soc. Opensci. **4**, 170918.

McKinnon E. A. and Love O. P. (2018). Ten years tracking the migrations of small landbirds: lessons learned in the golden age of bio-logging. Auk **135**, 834–856.

Meise K., Krüger O., Piedrahita P., Müller A., and Trillmich F. (2013). Proximity loggers on amphibious mammals: a new method to study social relations in their terrestrial habitat. Aquat. Biol. **18**, 81–89.

Miklosi A. 2007. Dog Behaviour, Evolution, and Cognition. Oxford University Press.

Nagy M., Ákos Z., Biro D., and Vicsek T. (2010). Hierarchical group dynamics in pigeon flocks. Nature **464**, 890–893.

Naug D. (2009). Structure and resilience of the social network in an insect colony as a function of colony size. Behav. Ecol. Sociobiol. **63**, 1023–1028.

Neverova N., Wolf C., Lacey G., Fridman L., Chandra D., Barbello B., and Taylor G. (2016). Learning human identity from motion patterns. IEEE Access **4**, 1810–1820.

Newman MEJ. (2010). Networks. An Introduction. Oxford University Press.

Nguyen N. G., Phan D., Lumbanraja F. R., et al. (2019). Applying deep learning models to mouse behavior recognition. JBiSE **12**, 183–196.

O'Neill C. J., Bishop-Hurley G. J., Williams P. J., Reid D. J., and Swain D. L. (2014). Using UHF proximity loggers to quantify male–female interactions: a scoping study of estrous activity in cattle. Anim. Reprod. Sci. **151**, 1–8.

Pasquaretta C., Klenschi E., Pansanel J., Battesti M., Mery F., and Sueur C. (2016). Understanding dynamics of information transmission in *Drosophila melanogaster* using a statistical modeling framework for longitudinal network data (the RSiena Package). Front. Psychol. **7**, 20142480.

Paton P. W. C., Zabel C. J., Neal D. L., Steger G. N., Tilghman N.G., and Noon B. R. (1991). Effects of radio tags on spotted owls. J. Wildl. Manag. **55**, 617–622.

Pelé M., Broihanne M. H., Thierry B., Call J., and Dufour V. (2014). To bet or not to bet? Decision-making under risk in non-human primates. J. Risk Uncertain. **49**, 141–166.

Pelé M., Dufour V., Micheletta J., and Thierry B. (2010). Long-tailed macaques display unexpected waiting abilities in exchange tasks. Anim. Cogn. **13**, 263–271.

Pelé M., Micheletta J., Uhlrich P., Thierry B., and Dufour V. (2011). Delay maintenance in Tonkean macaques (*Macaca tonkeana*) and brown Capuchin monkeys (*Cebus apella*). Int. J. Primatol. **32**, 149–166.

Pinter-Wollman N., Hobson E. A., Smith J. E. et al. (2013). The dynamics of animal social networks: analytical, conceptual, and theoretical advances. Behav. Ecol. **25**, 242–255.

Poirotte C., Massol F., Herbert A., Willaume E., Bomo P. M., Kappeler P. M., and Charpentier M. J. E. (2017). Mandrills use olfaction to socially avoid parasitized conspecifics. Sci. Adv. **3**, e1601721.

Prange S., Jordan T., Hunter C., Gehrt, S. D. (2006). New radiocollars for the detection of proximity among individuals. Wildl. Soc. Bull. **34**, 1333–1344.

Prange S., Gehrt S. D., and Hauver S. (2011). Frequency and duration of contacts between free-ranging raccoons: uncovering a hidden social system. J. Mammal. **92**, 1331–1342.

Puga-Gonzalez I., Ostner J., Schülke O., Sosa S., Thierry B., and Sueur C. (2018). Mechanisms of reciprocity and diversity in social networks: a modeling and comparative approach. Behav. Ecol. **29**, 745–760.

Puga-Gonzalez I., Sosa S., and Sueur C. (2019). Social style and resilience of macaques' networks, a theoretical investigation. Primates **60**, 233–246.

Ralls K., Sanchez J. N., Savage J., Coonan T. J., Hudgens B. R., and Cypher B. L. (2013). Social relationships and reproductive behavior of island foxes inferred from proximity logger data. J. Mammal. **94**, 1185–1196.

Renoult J. P., Guyl B., Mendelson T. C., Percher A., Dorignac J., Geniet F., and Molino F. (2019). Modelling the perception of colour patterns in vertebrates with HMAX. bioRxiv 552307.

Reynolds, J. J. H., Hirsch, B. T., Gehrt, S. D., and Craft, M. E. (2015). Raccoon contact networks predict seasonal susceptibility to rabies outbreaks and limitations of vaccination. J. Anim. Ecol. **84**, 1720–1731.

Robert K., Garant D., and Pelletier F. (2012). Keep in touch: does spatial overlap correlate with contact rate frequency? J. Wildl. Manag. **76**, 1670–1675.

Robert K., Garant D., Wal E. V., and Pelletier F. (2013). Context-dependent social behaviour: testing the interplay between season and kinship with raccoons. J. Zool. **290**, 199–207.

Robertson B., Holland J. D., and Minot E. (2012). Wildlife tracking technology options and cost considerations. Wildl. Res. **38**, 653–663.

Romano V., Duboscq J., Sarabian C., Thomas E., Sueur C., and MacIntosh A. J. J. (2016). Modeling infection transmission in primate networks to predict centrality-based risk. Am. J. Primatol. **78**, 767–779.

Romano V., Shen M., Pansanel J., MacIntosh A. J., and Sueur C. (2018). Social transmission in networks: global efficiency peaks with intermediate levels of modularity. Behav. Ecol. Sociobiol. **72**, 154.

Ropert-Coudert Y. and Wilso, R. P. (2005). Trends and perspectives in animal-attached remote sensing. Front. Ecol. Environ. **3**, 437–444.

Rouco C., Jewell C., Richardson K. S., French N. P., Buddle B. M., and Tompkins D. M. (2018). Brushtail possum (*Trichosurus vulpecula*) social interactions and their implications for bovine tuberculosis epidemiology. Behaviour **155**, 621–637.

Ruckstuhl K. E. and Neuhaus P. (2000). Sexual segregation in ungulates: a new approach. Behaviour **137**, 361–377.

Rushmore J., Caillaud D., Hall R. J., Stumpf R. M., Meyers L. A., and Altizer S. (2014). Network-based vaccination improves prospects for disease control in wild chimpanzees. J. R. Soc. Interface **11**, 20140349.

Rutz C., Burns Z. T., James R., Ism, S. M. H., Burt, J., Otis B., Bowen J., and St Clair J. J. H. (2012). Automated mapping of social networks in wild birds. Curr. Biol. **22**, R669–R671.

Ryder T. B., Horton B. M., van den Tillaart M., Morales De Dios J., and Moore I. T. (2012). Proximity data-loggers increase the quantity and quality of social network data. Biol. Lett. **8**, 917–920.

Saayman G. and Tayler C. (1973). Social organisation of inshore dolphins (*Tursiops aduncus* and *sousa*) in the Indian Ocean. J. Mammal. **54**, 993–996.

Sade D. S. (1972). Sociometrics of *Macaca mulatta*—linkages and cliques in grooming matrices. Folia Primatol. **18**, 196–223.

Sade D. S. (1989). Sociometrics of *Macaca mulatta* III: N-path centrality in grooming networks. Soc. Netw. **11**, 273–292.

Sade D. S., Altmann M., Loy J., Hausfater G., and Brueggeman J. A. (1988). Sociometrics of *Macaca mulatta*: II. Decoupling centrality and dominance in rhesus monkey social networks. Am. J. Phys. Anthropol. **77**, 409–425.

Saitou T. (1978). Ecological study of social organization in the great tit, *Parus major* L. I. Basic structure of the winter flocks. Jap. J. Ecol. **28**, 199–214.

Saraux C., Le Bohec C., Durant J. M. et al. (2011). Reliability of flipper-banded penguins as indicators of climate change. Nature **469**, 203–206.

Sato K., Mitani Y., Naito Y., and Kusagaya H. (2003). Synchronous shallow dives by Weddell seal mother–pup pairs during lactation. Mar. Mammal Sci. **19**, 384–395.

Scardamaglia R. C., Fiorini V. D., Kacelnik A., and Reboreda J. C. (2017). Planning host exploitation through prospecting visits by parasitic cowbirds. Behav. Ecol. Sociobiol. **71**, 23.

Schofield D., Nagrani A., Zisserman A., Hayashi M., Matsuzawa T., Biro D., and Carvalho S. (2019). Chimpanzee face recognition from videos in the wild using deep learning. Sci. Adv. **5**, eaaw0736.

Schreer J. F. and Testa J. W. (1996). Classification of Weddell seal diving behavior. Mar. Mammal Sci. **12**, 227–250.

Schroff F., Kalenichenko D., and Philbin J. (2015). Facenet: a unified embedding for face recognition and clustering. Proceedings of the IEEE Conference on Computer Vision and Pattern Recognition, 815–823.

Seyfarth R. M. (1976). Social relationships among adult female baboons. Anim. Behav. **24**, 917–938.

Shepard E. (2008). Identification of animal movement patterns using tri-axial accelerometry. Endanger. Species Res. **10**, 47.

Siebenaler J. and Caldwell D. K. (1956). Cooperation among adult dolphins. J. Mammal. **37**, 126–128.

Silk J. B. (2007). The adaptive value of sociality in mammalian groups. Philos. Trans. R. Soc. B-Biol. Sci. **362**, 539–559.

Silk J. B., Alberts S. C., and Altmann J. (2003). Social bonds of female baboons enhance infant survival. Science **302**, 1231–1234.

Silk J. B., Beehner J. C., Bergman T. J. et al. (2009). The benefits of social capital: close social bonds among female baboons enhance offspring survival. Proc. R. Soc. B-Biol. Sci. **276**, 3099–3104.

Silk M. J., Weber N., Steward L. C. et al. (2017). Seasonal variation in daily patterns of social contacts in the European badger *Meles meles*. Ecol. Evol. **7**, 9006–9015.

Skyrms B. and Pemantle R. (2009). A dynamic model of social network formation. In: T. Gross and H. Sayama (eds.), Adaptive Networks, 231–251. Springer.

Sosa S., Puga-Gonzalez I., Feng H. H., Zhang P., Xiaohua X., and Sueur C. (2018a). A multilevel statistical toolkit to study animal social networks: Animal Network Toolkit (ANT) R package. bioRxiv 347005.

Sosa S., Pele M., Debergue E. et al. (2018b). Impact of group management and transfer on individual sociality in Highland cattle (*Bos taurus*). ArXiv180511553 Q-Bio.

Sosa S., Sueur C., and Puga-Gonzalez I. (2020). Network measures in animal social network analysis: their strengths, limits, interpretations and uses. Meth. Ecol. Evol. **12**, 10–21.

St Clair J. J. H., Burns Z. T., Bettaney E. M. et al. (2015). Experimental resource pulses influence social-network dynamics and the potential for information flow in tool-using crows. Nat. Commun. **6**, 7197.

Stacey P. B. and Koenig W. D. (1990). Cooperative Breeding in Birds: Long Term Studies of Ecology and Behaviour. Cambridge University Press.

Stanton M. A. and Mann J. (2012). Early social networks predict survival in wild bottlenose dolphins. PLOS One **7**, e47508.

Stroeymeyt N., Grasse A.V., Crespi A., Mersch D. P., Cremer S., and Keller L. (2018). Social network plasticity decreases disease transmission in a eusocial insect. Science **362**, 941–945.

Sueur C. and Maire A. (2014). Modelling animal group fission using social network dynamics. PLOS One **9**, e97813.

Sueur C., Jacobs A., Amblard F., Petit O., and King A. J. (2011a). How can social network analysis improve the study of primate behavior? Am. J. Primatol. **73**, 703–719.

Sueur C., Petit O., De Marco A., Jacobs A. T., Watanabe K., and Thierry B. (2011b). A comparative network analysis of social style in macaques. Anim. Behav. **82**, 845–852.

Sueur C., Romano V., Sosa S., and Puga-Gonzalez, I. (2017). Mechanisms of network evolution: a focus on socioecological factors, intermediary mechanisms, and selection pressures. Primates **60**, 1–15.

Swain D. L., Patison K. P., Heath B. M., Bishop-Hurley G. J., and Finger A. (2015). Pregnant cattle associations and links to maternal reciprocity. Appl. Anim. Behav. Sci. **168**, 10–17.

Tentelier C., Aymes J.-C., Spitz B., and Rives J. (2016). Using proximity loggers to describe the sexual network of a freshwater fish. Environ. Biol. Fishes **99**, 621–631.

Tosa M. I., Schauber E. M., and Nielsen C. K. (2015). Familiarity breeds contempt: combining proximity loggers and GPS reveals female white-tailed deer (*Odocoileus virginianus*) avoiding close contact with neighbors. J. Wildl. Dis. **51**, 79–88.

Tremblay Y. and Cherel Y. (1999). Synchronous underwater foraging behavior in penguins. Condor **101**, 179–185.

Trivers R. L. and Hare, H. (1976). Haploidploidy and the evolution of the social insect. Science **191**, 249–263.

Uhrich J. (1938). The social hierarchy in albino mice. J. Comp. Psy. **25**, 373.

Vander Wal E., Yip H., and McLoughlin P. D. (2012a). Sex-based differences in density-dependent sociality: an experiment with a gregarious ungulate. Ecology **93**, 206–212.

Vander Wal E., Paquet P. C., and Andrés J. A. (2012b). Influence of landscape and social interactions on transmission of disease in a social cervid. Mol. Ecol. **21**, 1271–1282.

Vander Wal E., Paquet P. C., Messier F., and McLoughlin P. D. (2013). Effects of phenology and sex on social proximity in a gregarious ungulate. Can. J. Zool. **91**, 601–609.

Walrath R., Deelen T. R. V., and VerCauteren K. C. (2011). Efficacy of proximity loggers for detection of contacts between maternal pairs of white-tailed deer. Wildl. Soc. Bull. **35**, 452–460.

Weiss K., Khoshgoftaar T. M., and Wang D. (2016). A survey of transfer learning. J. Big Data 3(1), 9.

Whitehead H. (2008). Analyzing Animal Societies: Quantitative Methods for Vertebrate Social Analysis. University of Chicago Press.

Whitehead H. (2009). SOCPROG programs: analysing animal social structures. Behav. Ecol. Sociobiol. **63**, 765–778.

Wilson E. O. (1975). Sociobiology: The New Synthesis. Harvard University Press.

Wilson R. P. and McMahon C. R. (2006). Measuring devices on wild animals: what constitutes acceptable practice? Front. Ecol. Environ. **4**, 147–154.

Wilson R. P., Putz K., Peters G., Culik B., Scolaro J. A., Charrassin J. B., and Ropert-Coudert Y. (1997). Long-term attachment of transmitting and recording devices to penguins and other seabirds. Wildl. Soc. Bull. **25**, 101–106.

Wilson R. P., Shepard E. L. C., and Liebsch N. (2008). Prying into the intimate details of animal lives: use of a daily diary on animals. Endanger. Species Res. **4**, 123–137.

Wittemyer G., Douglas-Hamilton I., and Getz W. M. (2005). The socioecology of elephants: analysis of the processes creating multitiered social structures. Anim. Behav. **69**, 1357–1371.

Wolf T., Ngonga Ngomo A., Bennett N., Burroughs R., and Ganswindt A. (2018). Seasonal changes in social networks of giraffes. J. Zool. **305**, 82–87.

Zhang R., Isola P., Efros A. A., Shechtman E., and Wang O. (2018). The unreasonable effectiveness of deep features as a perceptual metric. Proceedings of the IEEE Conference on Computer Vision and Pattern Recognition, 586–595.

CHAPTER 4

Growth rings across the Tree of Life: demographic insights from biogenic time series data

Margaret E. K. Evans, Bryan A. Black, Donald A. Falk, Courtney L. Giebink, and Emily L. Schultz

4.1 Introduction

The collection of individual-level demographic data has traditionally involved repeat censuses, either through repeat visits to sessile organisms or capture-recapture of mobile organisms. An appealing and overlooked source of demographic information is biogenic time series data that can be generated in a single sampling effort. These records—found on the hard body parts of many plants and animals—offer information on variation in demographic processes over time, in many cases providing information at time scales far exceeding what is typically generated through a repeat census approach, while maintaining annual resolution. In this chapter, our goals are twofold: first, we seek to make these single-sample time series data more widely known to demographers, both from a methodological perspective and in terms of the important knowledge gaps they can fill. Second, we encourage demographers to take on the challenge and opportunity to use these data to improve demographic models and address the impacts of global change.

To make biogenic time series more widely appreciated by demographers, we begin by describing the data (i.e. how they are generated and how they have been analysed) and explaining some of the tools and practices of the 'chronology' sciences to demographers. Tree-ring science has historically been aligned with Earth sciences and geography rather than life sciences; the discipline of dendrochronology was founded by an astronomer (Andrew Ellicott Douglass). To this day, a primary use of tree-ring time series data is the study of climate system dynamics, even though dendroecology has a decades-deep literature (Fritts and Swetnam 1989). Dendrochronology has a set of principles, practices, and language surrounding these that give rise to a need for interdisciplinary translation. What are the sampling protocols of this discipline? What are 'cross-dating' and 'detrending'? What is a chronology? When borrowing methods across disciplines, it is important to understand the why behind how things are done and think carefully about their application in a different context.

In keeping with the theme of the book, we then highlight the diversity of taxa across the Tree of Life from which annually resolved demographic time series data can be generated in a single sampling effort. Further, we review other kinds of biogenic time series data, beyond growth rings, and other kinds of information to be gained from growth rings, beyond just their widths. We then survey the great variety of demographic processes into which biochronologies have offered insight. They provide a somatic record of all the individual-level processes that demographers care about. Clearly, growth increment time series make it possible to investigate and parse the many drivers of growth variation, such as individual size, climate conditions, competitive (or other) interactions, and natural or human-caused disturbances. Growth rings also act as an internal clock, recording the passing of time, so that singular demographic events in the life cycle, such as birth and death, can be dated from these time series, offering insight into their drivers. The experience of an individual with

Margaret E. K. Evans et al., *Growth rings across the Tree of Life: demographic insights from biogenic time series data.*
In: *Demographic Methods Across the Tree of Life.* Edited by Roberto Salguero-Gómez and Marlène Gamelon, Oxford University Press.

respect to dispersal or migration, which can be inferred from ring microchemistry, is also recorded in growth ring time series. Questions of evolutionary demography (i.e. life history variation among individuals and across species, individual reaction norms, and local adaptation) can also be addressed by analysing variation within and among growth time series. Hard body parts such as bones and teeth are all that remain of most extinct vertebrates (e.g. dinosaurs and hominins). Thus growth ring samples (and other biogenic time series data) can help to fill important knowledge gaps about growth, birth, death, movement, and evolution across a wide range of organisms.

While we have structured this chapter to highlight cross-taxa similarities and differences, we note that dendrochronology (i.e. the study of tree-ring time series) is the oldest and most well-developed scientific discipline to use growth rings or any other somatic records of demographic processes. The beginnings of dendrochronology date to the early twentieth century (Douglass 1914, 1920); like ecology, it is a century-old science. Hence, many examples throughout the chapter draw from the world of tree-ring science. We hope to see the kinds of demographic questions asked of and methods applied to tree-ring time series applied to other biogenic time series data.

Our second goal is for demographers to come away with new ideas about sources of demographic information that can complement more traditional data streams, new ideas about the temporal and spatial scales at which to think about the demography of their study organism, and a renewed focus on the drivers and mechanisms behind demographic outcomes. We end by suggesting how information from biochronologies may be incorporated into population models, bringing to these models a greater understanding of the drivers of vital rate variation and a new capacity to address issues of global change.

4.2 Tree and otolith growth rings—the basics

Here we lay out two prime examples of growth ring time series data: one from sessile, terrestrial organisms in the plant kingdom (trees), and the other from mobile, aquatic organisms in the animal kingdom (fish). We focus first on the nature of the rings (how they are formed) and how they are sampled and developed into time series data, along with standard practices for their analysis (i.e. a methodological focus). The subsequent two sections of this chapter then briefly survey growth rings in other organisms, informative measures other

than ring width that can be derived from growth rings, and other biogenic time series data.

Tree rings are perhaps the most well-known example of annually resolved biogenic time series data (Douglass 1920; Fritts 1976; Speer 2010). Annual increments from the ear bones of fish (i.e. otoliths) are also relatively well known, although their study is recent compared to the study of tree rings (Campana and Neilson 1985; Campana 1990; Black et al. 2019). Both are curated in open-access international repositories: in the case of tree rings, a dedicated repository known as the International Tree-Ring Data Bank (ITRDB; https://www.ncdc.noaa.gov/data-access/paleoclimatology-data/datasets/tree-ring). Both are distributed extensively, covering much of the world's temperate forest ecosystems (more than 4,000 sites on six continents), coastal marine ecosystems, and increasing coverage in freshwater systems.

In the formation of rings, the material in both trees and otoliths is not resorbed by the organism and is thus a stable archive. Tree rings represent secondary tree growth (increase in bole diameter) through the formation of xylem; the 'ring' is simply a radial cross-section of the annual growth layer. In seasonal climates (those with a predictable period each year when growth ceases), these growth rings are formed yearly. Many (though not all) conifers reliably form recognisable annual rings, with clear ring boundaries formed by darkly coloured latewood (Figure 4.1a), which are xylem cells formed late in the growth season, with thicker cell walls, smaller lumen area, and a higher concentration of lignin. Light-coloured earlywood has opposite cell characteristics. Many temperate angiosperm trees also reliably make recognisable annual growth rings, of varying wood anatomy. Tropical trees are one of the current frontiers of dendrochronology, with most species requiring additional study to identify reliable anatomical criteria and the timing of their formation (Silva et al. 2019). In the otoliths of bony fish, ratios of calcium carbonate and protein vary seasonally, forming growth increments. Yet as in trees, there are environments, such as the deep ocean, that lack the seasonality to induce annual otolith increment formation. Verification of annual periodicity is often necessary for each new species with such techniques as radiometric dating or mark and recapture (Campana and Neilson 1985; Campana 1990; Figure 4.1b).

An important distinction between these two is that sampling of otoliths from fish populations is frequently conducted to gather fish age data for stock assessment (i.e. demographic modelling) and is therefore aimed at an unbiased sample of population structure,

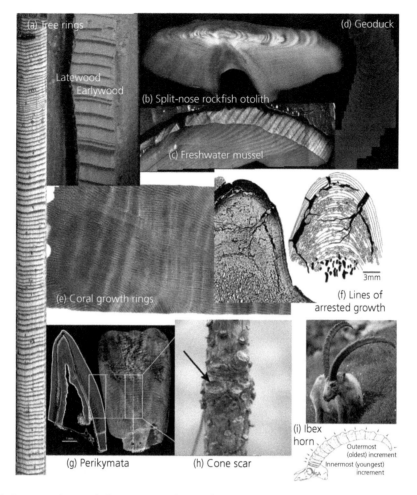

Figure 4.1 Growth rings across the Tree of Life. (a) Tree rings of the conifers Douglas-fir (*Pseudotsuga menziesii*) (left; photo by B. A. Black) and ponderosa pine (*Pinus ponderosa*) (right; photo by R. J. DeRose), with light-coloured earlywood and dark-coloured latewood. (b) Cross-section of the otolith (ear bone) of a split-nose rockfish (*Sebastes diploproa*). (c) Cross-section of a freshwater mussel shell (*Margaritifera falcata*). (d) Etched and stained cross-section from the hinge plate region of a Pacific geoduck shell (*Panopea abrupta*). (Photos in b–d by B. A. Black). (e). Growth rings of a coral (*Porites* sp.) (X-ray image by D. Thompson). (f) Growth lines, or lines of arrested growth, in the dorsal rib of *Draconyx loureiroi*, an ornithopod of the Late Jurassic (photo by K. Waskow). (g) Perikymata, or growth lines, on the teeth of Plio-Pleistocene hominins (from Le Cabec et al. 2015). (h) Cone scar on the conifer whitebark pine (*Pinus albicaulis*) (photo by J. Rapp). (i) Horn growth increments of the ibex (*Capra ibex*), an alpine ungulate (from Buntgen et al. 2014).

whereas the sampling of tree rings has often conventionally been oriented towards climate reconstruction. As a result, samples in the ITRDB are biased towards old, climate-sensitive, isolated individuals (Klesse et al. 2018; the 'site and tree selection' principles of dendrochronology). This creates a biased sample that is not appropriate for population-level inference. A sampling bias towards old individuals (desirable for their long time series) will lead to time-biased estimates of growth rates if the risk of mortality is greater for faster-growing individuals in a population, known as 'slow-grower survivorship bias' by

dendrochronologists (Brienen and Zuidema 2006) and the 'Rosa Lee phenomenon' by sclerochronologists (Morrongiello et al. 2012). However, there are relatively unbiased tree-ring collections developed in a forest inventory context (DeRose et al. 2017; Canham et al. 2018), as well as many collections created specifically to answer demographic or ecological questions that have the multiple cohort sampling needed for population-level inference (Wykoff and Clark 2002; Youngblood et al. 2004; Guiterman et al. 2018).

A key step in developing a biogenic time series, from tree rings, otoliths, or other somatic records, is the process of assigning a time of formation to each ring (or other structure). If there were no possibility of missing or falsely identified growth rings, year assignment would simply be a matter of counting backwards from the year of sampling a live organism. Given these sources of uncertainty and error, along with samples from dead individuals where the year of formation of the final ring is unknown, dendrochronologists developed the process of 'cross-dating' (Douglass 1941; Black et al. 2016), which relies upon pattern-matching across samples, analogous to the alignment of DNA sequences, but based on ring-width variation rather than nucleotide sequence variation. Whereas DNA sequence alignment is based on similarities induced through ancestor-descendant relationships (relatedness), cross-dating relies upon synchronous patterns of increment widths among individuals induced by a common (environmental) limiting factor. With enough replication, it becomes possible to identify missing or false rings (analogous to deletions and insertions) in any one sample. Unlike DNA sequence alignment, automated cross-dating of growth rings (i.e. by an algorithm) does not yet exist; a suite of tools and metrics (Holmes 1983; Wigley et al. 1984; Cook et al. 1990) are used to verify year assignments made by human judgement and experience.

Accurate year assignment is essential to answer questions about the drivers of ring-width variability. Statistical relationships between ring widths and the climatic or ecological conditions under which rings were formed can be seriously compromised even at modest error rates (Black et al. 2016; i.e. when false rings are identified as an annual growth ring by the observer, or missing rings are undetected). But there are cases where cross-dating to traditional standards is difficult, for example, because environmental conditions are not strongly limiting to growth and hence interannual variability and synchrony between samples is weak, or when lifespans and hence time series are short. This highlights the need for formal statistical methods to quantify uncertainty in year assignments, which are currently lacking (but see Ricker et al. 2020). The development of such methods would greatly support the 'chronology' sciences, since the information content of a time series is inversely proportional to the uncertainty about year assignments, and interpretations derived from time series data should be made in light of (quantified) uncertainty.

Following the assignment of a calendar year to each growth ring, it has long been standard practice in dendrochronology to *detrend* ring-width time series, with the primary goal of removing the decline in absolute ring widths that arises from measuring the growth of a 3D organism in a linear dimension (evident in Figure 4.1b,d). That is, even if the growth of a tree in area (stem basal area) is constant, with increasing stem radius, radial increments decline. Depending on the goal of the study, and the sampling that was executed towards that goal, detrending may also aim to eliminate or preserve the signal of other drivers of growth variability (e.g. climate, competition, and disturbances.). Detrending involves fitting a mathematical function to the ring-width time series, after which observed ring widths are divided by the expected ring widths described by the mathematical function (Cook et al. 1990; Speer 2010). This transforms the data from absolute ring widths to a unitless *ring width index* with (ideally) a constant mean and variance (Figure 4.2). Individual-level detrended time series are then averaged to form a *chronology*, a site-level time series of growth anomalies (Speer 2010).

The current state of the art of detrending are methods that disentangle size-related versus other causes of low-frequency ring-width variation, including regional curve standardisation (RCS; Briffa et al. 1992) and 'signal-free' detrending (Melvin and Briffa 2008). Under RCS, all ring-width time series of a given species and sampling region are aligned by tree age (rather than calendar year) and detrended with a single negative exponential curve, or other relatively stiff function (Briffa et al. 1992). Ring-width indices are then averaged with respect to calendar year to form a chronology. However, the data requirements for RCS are relatively steep: large sample sizes, individuals of a variety of ages, and ideally dead-collected material combined with live-collected samples (Helama et al. 2017). Where these requirements are not met, each measurement time series may be fit to a negative exponential (Figure 4.2c) or regression function, but at the expense of discarding low-frequency variability at wavelengths longer than the mean measurement time series (Cook et al. 1995). If the goal is to remove interdecadal variability in ring widths driven by disturbances (e.g. gap-phase dynamics) that tend to occur in closed-canopy forests, more flexible cubic smoothing splines (Figure 4.2d) may be used. The resulting chronology must then be interpreted with the knowledge that the majority of low-frequency variability—regardless of its cause—has been removed.

In the sclerochronology literature, in addition to detrending otolith ring-width time series data and forming a site-level chronology, it has also been common to use mixed effects models to analyse individual-level variability in absolute growth. These models

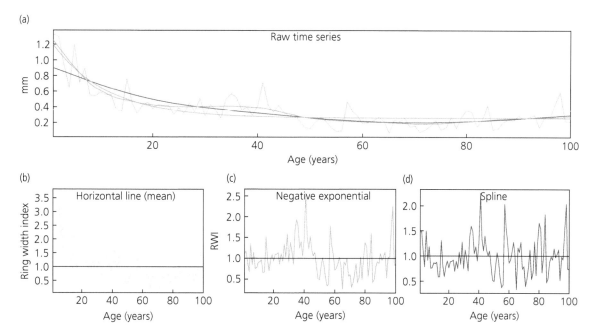

Figure 4.2 (a) Raw growth increments (grey line) of the shell of a Pacific geoduck (*Panopea abrupta*) sampled in the Tree Knob Islands of the northern British Columbia coast, Canada. Overplotted in (a) are four mathematical functions fitted to the data to 'detrend' the raw measurements. The residuals (ring width index) from three of those detrending functions are shown in panels (b) as a horizontal line, in (c) as a negative exponential, and (d) as a cubic smoothing spline, with a 50% frequency cut-off set to two-thirds of the time series length. In (a), a fourth, yet more flexible mathematical function fit to the raw measurements is Friedman's super smoother (blue line). All panels were created using the detrend.series function in the R package dplR (Bunn 2008).

are particularly effective in partitioning and attributing variance amongst a suite of predictors (i.e. parse the many factors influencing growth). For example, Morrongiello and Thresher (2015) used MEMs to attribute variation in growth of the tiger flathead (*Platycephalus richardsoni*) to effects of age, sex, cohort, year, individual, and geographic area. They found growth in poleward populations to be driven weakly by ocean temperature, whereas equatorward populations were influenced more strongly and negatively by ocean temperature. The use of hierarchical models allowed Morrongiello and Thresher (2015) to attribute growth variation across space, time, and individuals to causes ranging from individual reaction norms to environmental gradients covering the geographic distribution of a species. There are many other examples of a MEM approach to analysing fish growth from otolith time series (Weisberg et al. 2010; Helser et al. 2012; Rountrey et al. 2014; Von Biela et al. 2015; Von Biela 2016; Smolinski and Mirny 2017; van der Sleen et al. 2018; Martino et al. 2019). The MEM approach to analysing individual-level absolute growth variability has emerged in the dendroecology literature as

well (Martin-Benito et al. 2011; Fernandez-de-Una et al. 2016; Foster et al. 2016; Dorado-Linan 2018; Wright et al. 2018). Ring-width time series data may also be fit to nonlinear growth models (see Fernandez-de Una et al. 2016; Rollinson et al. 2016; Buechling et al. 2017; Canham et al. 2018). It is worth noting that the ecological literature includes arguments against the analysis of residuals over a single-stage multiple regression analysis (see Garcia-Berthou 2001; Green 2001; Freckleton 2002; Morrongiello et al. 2012), but also that a MEM approach is inherently data-hungry, requiring data on multiple covariates, which are not always available.

4.3 Growth rings in other organisms

Although trees and fish have been the most extensively studied to date, other organisms also create a record of growth (or other life events) on hard parts. The techniques of dendrochronology have been applied to the growth rings of other plant growth forms, including shrubs in the Arctic and the Himalayas (Myers-Smith et al. 2015a; Weijers et al. 2017; Gamm et al. 2018;

Dolezal et al. 2019; Le Moullec et al. 2019) and perennial herbs with a taproot ('herb-chronology'; Dee and Palmer 2016; Dolezal et al. 2018). Indeed, Buntgen (2019) has argued for doing away with plant growth form categories (e.g. of Raunkiaer) that separate trees from shrubs from herbs, making the case for fundamental anatomical and physiological similarities between them.

Sclerochronology includes, in addition to the otoliths of fish, the analysis of growth rings of bivalve shells (Figure 4.1c,d), corals (Figure 4.1e), fish scales, and turtle scutes (Morrongiello et al. 2012). Some of these organisms have surprisingly long lifespans, including the ocean quahog (*Arctica islandica*), with a maximum reported age of 507 years, making it the world's oldest known, reliably aged animal; the Pacific geoduck (*Panopea generosa*; ages of 150 years and more); and freshwater mussels (genus *Margaritifera*, with ages of 100 years and more).

Bones, teeth, and horns are other examples of hard body parts that can yield demographic information. Skeletochronology, the estimation of age by counting growth marks found in the cross-section of bones, has been used across a wide variety of vertebrates. Most (but not all) species of dinosaurs formed annual growth lines (Figure 4.1f). Extant vertebrates that are known to form annual growth lines include *Actinopterygia, Amphibia, Lepidosauria*, and *Crocodylia*. It was once thought that mammals do not form annual growth marks in bones, but recent work has shown they exist in large-bodied and slower-growing species, including cervids and primates (Woodward et al. 2013). A cautionary note is warranted: resorption or secondary reworking of bone tissue can obscure growth marks. The most reliable bones for ageing are generally long bones, and the largest of them, with methods to retrocalculate missing years (Woodward et al. 2013).

Teeth have been used to study life history variation and evolution, particularly of hominins and other primates (Smith and Tompkins 1995; Kelly and Smith 2003). Perikymata are layers of enamel that build up incrementally on the surface of a tooth as it grows (Figure 4.1g); each takes about 6–12 days to form in modern humans, and thus they can be used to estimate how quickly teeth develop. Visualisation techniques that bring a new level of detail (synchrotron microtomography-based virtual histology; Le Cabec et al. 2015) and the combination of skeletochronology and dental development data (Seselj 2017) promise to advance our understanding of human evolution. While the primary focus has been on the teeth of humans and their relatives, their study has contributed to understanding of the demography of wildebeest in the Serengeti (Sinclair and Arcese 1995) and marine mammals (Hamilton et al. 2017), as well estimates of longevity for diverse organisms (e.g. flying foxes, whale sharks; DivljanPerry-Jones et al. 2006; Ong et al. 2020).

4.4 Beyond growth rings and ring widths

Other forms of growth or developmental marks or scars can inform demography, and metrics other than the width of a growth ring provide insight as well. Myers-Smith et al. (2015b) beautifully illustrate that annual axial (stem) increments of certain tundra shrub species can be measured from bud scars or longitudinal sectioning (i.e. using winter mark septa). Cone scars make it possible to quantify and assign to specific years the reproductive effort of some pines (see Figure 4.1h; Morgan and Bunting 1992; Redmond et al. 2016) The spines of cacti and Euphorbia thorns grow in a clearly organised series, recording physiological variation over time (English et al. 2007, 2010). Further, growth rates are but one kind of information to be gained from hard body part time series. Additional characteristics of growth rings and other structures, such as isotopic composition and other forms of microchemistry variation, as well as anatomical and other forms of microstructure variation, offer insight into demographic processes or the drivers and adaptations behind demographic outcomes. Hypoplasia (i.e. deficiencies in tooth enamel thickness associated with undernutrition or disease) represent indelible, chronological records of acute stress. Stressful events recorded in xylem (wood) include frost damage, fire, defoliation, flooding, and geomorphic events. False and missing rings too are indicators of stress. While we have attempted to be comprehensive in our examples of growth rings and other somatic records of demographic processes across the Tree of Life, surely many more opportunities remain to be discovered.

4.5 Insights into demography

In this section, we survey the demographic processes that can be informed by growth ring time series, highlighting the variety of knowledge gaps filled using examples from the literature. We begin with individual growth, birth, and death, and then we consider movement and evolution.

Growth—Across a great diversity of organisms, growth time series provide long records of growth variability that are used to evaluate specific drivers, and even parse multiple drivers of growth. Based on horns (see Figure 4.1i) from more than 8,000 individuals, Buntgen et al. (2014) showed that elevated March to May temperatures, causing earlier snowmelt and increased food resources, lead to synchronised, increased growth of ibex (an alpine ungulate) across a broad area of the eastern Swiss Alps. Growth marks in the bones of loggerhead sea turtles (*Caretta caretta*) have been used to partition age, year, and size as drivers of growth variation, demonstrating compensatory growth in response to stochastic environmental variation (Bjorndal et al. 2003). Ong et al. (2015) used otolith time series data to show an ontogentic shift in the environmental drivers of growth variability of the mangrove jack (*Lutjanus argentimaculatus*), from precipitation and other factors influencing river runoff in the estuarine habitat of juveniles to atmospheric and ocean-basin conditions influencing the coastal reef habitat of adults. Analysis of annual growth rings of the coral *Montastraea faveoloata* put the severe coral bleaching event of 1998 into a long-term context, revealing that water temperatures and solar irradiance conditions were as high or higher in the past, pointing at the impact of human-caused stressors (e.g. nutrient pollution, overfishing) on top of climatic conditions (Carilli et al. 2009). Several recent analyses of tree-ring time series have focused on the joint effects of climate and competition on tree growth, including their interactions, and how they vary across forest canopy classes or species (Martin-Benito et al. 2011; Fernandez-de-Una et al. 2016; Wright et al. 2018).

Long growth time series are particularly helpful in answering certain questions about growth that inherently span time, such as ecological resilience or ecological memory. Based on growth time series data from tree rings, Lloret et al. (2011) defined a set of metrics to quantify the response to a stressful event (e.g. drought), including resistance, recovery, resilience, and relative resilience. These metrics have been used, for example, to detect increased vulnerability to drought after successive drought events (Serra-Maluquer et al. 2018), fire (Van Mantgem et al. 2018, 2020), or investigate the effect of forest density reduction treatments on resilience to drought stress (Navarro-Cerrillo et al. 2019). Given the multiple meanings of the term *resilience* in the ecological literature, let alone in the policy world of climate adaptation, it is useful to have metrics by which progress towards this goal can be measured. Surprisingly, similar questions about the consequences of stressful events have been asked about human or hominin populations

using hypoplasia in teeth. Temple (2014) found an increased risk of subsequent mortality following such events evidenced by hypoplasia. Ogle et al. (2015) and Peltier et al. (2018) used a Bayesian stochastic antecedent modelling framework to quantify ecological memory in tree-ring time series data (i.e. lagged climatic effects on growth). Indeed, one of the great strengths of long growth time series is the ability to analyse variation across temporal scales, from intrannual to centennial or more.

Improved *understanding* of the drivers of growth variation, across a great range of time scales, can then lead to improved *models* of the process of growth. The literature on climate effects on tree growth is particularly rich (Fritts 1976) because tree rings have often been used as a climate proxy. This accumulated knowledge has been the foundation for mechanistic, climate-sensitive models of tree growth such as the Vaganov-Shashkin model (Vaganov et al. 2006; Tolwinski-Ward et al. 2011) and more recent whole-tree growth process models (Holtta et al. 2010; Hayat et al. 2017; reviewed in Babst et al. 2018). Carbon and oxygen isotopes in tree rings are used to better understand the process of growth at a physiological level (McCarroll and Loader 2004; Gessler et al. 2014). Ratios of the stable isotopes $\delta^{18}O$ and $\delta^{13}C$ are used to evaluate water use efficiency, the ratio of carbon gained to water lost, which is controlled by leaf stomatal behaviour. Fractionation of both occurs when leaf stomata are open for gas exchange; a central question is whether increased atmospheric concentration of carbon dioxide leads to increased water use efficiency and enhanced tree growth, mitigating climate change, or not (Frank et al. 2015; Van der Sleen et al. 2015). Hence these isotope-based studies of tree-ring time series are contributing to a better understanding of whether tree growth is controlled by the availability of carbon dioxide versus other limiting factors (is a 'source-limited' vs. 'sink-limited' process; Fatichi et al. 2014; Zuidema et al. 2018; Ulrich et al. 2019), which will be used to improve the representation of the feedback between forests and climate in coupled atmosphere-biosphere models.

Birth and death—In contrast to growth, birth and death happen only once for each individual in a population, making them 'rare' events, particularly in long-lived organisms. Tree-ring, otolith, and other time series data can provide unique insight into these key demographic processes. For otoliths, the year the first ring was formed can be considered the year of birth, whereas in tree rings, the year the first ring was formed (i.e. the pith date) represents the year of birth (i.e.

germination) only if samples are collected near the soil surface (cores collected at other stem heights represent the year the tree reached that height). Empirically derived relationships can be used to estimate true year of birth from increment cores not sampled at ground level, although most attempts to correlate tree height with age find highly variable relationships with much less than annual precision (Wong and Lertzman 2001). Once an estimate of the year of birth is made, it becomes possible to evaluate the conditions under which recruitment occurred, or detect strong cohort structure that provides evidence of episodic recruitment. Swetnam and Betancourt (1998) suggested, from pith dates at 143 sites across the US Southwest, that drought in the late 1500s led to regional-scale mortality followed by a pulse of recruitment during a relatively cool-wet period in the early 1600s. This represents demographic inference on temporal (centennial) and spatial (10^4–10^6 km^2) scales well beyond those at which most demographers operate. Similarly, careful cross-dating of Pacific geoduck growth rings revealed episodic recruitment—two strong recruitment pulses separated by 60 years (Black et al. 2008). Pith dates have further been used to examine the interactive effects of fire, climate, and human land uses (e.g. grazing, fire suppression) on tree recruitment in certain forest types across the interior western United States, leading to the conclusion that it was lack of fire, more so than favourable climatic conditions, that drove twentieth-century patterns of recruitment (Brown and Wu 2005; Munier et al. 2014; O'Connor et al. 2014; Taylor et al. 2016). Most recently, based on pith dates of nearly 3,000 trees at 90 sites, Davis et al. (2019) inferred that climate change is pushing conditions for post-fire tree recruitment below critical climate thresholds. Further, cone scars provide unique information on fertility, making it possible to study masting (in *Pinus albicaulis*; Crone et al. 2011) and how mast years affect subsequent recruitment and age structure (in *Pinus pumila*; Kajimoto et al. 1998).

Similar to birth dates, the date of the last ring formed can offer insight into the process of mortality. This particular path towards inference on mortality is limited to tree rings, since the collection of an increment core does not harm a tree, whereas an otolith is obtained through (mortally) destructive sampling. Villalba and Veblen (1998), for example, attributed episodic mortality of the South American conifer *Austrocedrus chilensis* to synoptic variation in the El Niño-Southern Oscillation ocean-atmosphere pattern (ENSO). An alternative approach to investigating mortality is to contrast the growth time series of individuals that died versus did not, borrowing strength from repeated measures of growth to gain insight into the process of decline (and its causes) that can precede death (Ogle et al. 2000; Macalady and Bugmann 2014; Cailleret et al. 2016a, 2016b). Finally, age distribution data, along with other lines of evidence, can be used to evaluate mortality and its drivers. Indeed, Zhao et al. (2018) make the argument that, compared to mark-recapture studies, estimates of mortality rates derived from age distributions (from otoliths and skeletochronology) are an accurate and useful tool to rapidly assess mortality rates.

Movement—Otolith microchemistry is a well-established method to determine provenance and movement of fishes (Thresher 1999; Kennedy et al. 2000). Strontium isotope ratios (^{87}Sr/^{86}Sr) and other chemical fingerprints are used to infer anadromy, characterise diadromous migrations, estimate the proportion of fish populations made up by wild versus hatchery-raised individuals, and identify habitat use for conservation of wild populations (Hegg et al. 2015; Willmes et al. 2018; Brennan et al. 2019). Further, otoliths can be marked chemically en masse, to track movements and abundance (Warren-Myers et al. 2018).

Evolution—Biogenic time series data offer insight into life history variation and evolution, evidence for selection on vital rates and functional traits, as well as a means to quantify the genetic versus plastic basis of vital rate and trait variability (topics in Chapter 20). Growth lines in teeth and bones have been particularly important in studying life history variation and evolution of extinct animals such as dinosaurs and hominins (Smith and Tompkins 1995; Kelly and Smith 2003). For example, perikymata (in teeth) have been used to evaluate the role of heterochrony and neoteny in human evolution (Guatelli-Steinberg et al. 2005). Insights into basic dinosaur life history parameters can be derived from age estimates from the growth lines of bones, including growth rates (from age vs. mass curves), time to maturity, and longevity. These have been used to address some of the most eye-catching questions about dinosaurs, such as the evolution of gigantism (Erickson et al. 2004; Erickson 2005).

Tree-ring time series have been used to assess the consequences of individual-level growth heterogeneity for population-level growth rate, with the conclusion that fast-growing juveniles contribute twice as much to population growth as slow-growing juveniles (Zuidema et al. 2009). Tree rings further provided evidence for the juvenile selection effect, the preferential survival of fast-growing juveniles to the forest canopy (Rozendaal et al. 2010; Zuidema et al

2011). Microstructural variation in tree-ring time series (i.e. variability in xylem cell anatomy; Fonti et al. 2010) provides an archive of both adaptive strategy and the plastic response of an individual to environmental variation over time. The xylem is the water-transport system of a plant; hence, anatomical characteristics such as the number, length, diameter, wall thickness and pit characteristics of conduits are under selective pressure with respect to both drought, cold, and other functional roles (i.e. support, storage). The world-wide phenomenon of drought-induced tree mortality, driven by global warming through increased evaporative demand (Allen et al. 2010), has made hydraulic failure and resistance to cavitation a focus of a great deal of research (Rowland et al. 2015; Anderegg et al. 2016; Greenwood et al. 2017), with quantitative wood anatomy from tree rings being an emerging line of evidence used to improve understanding of the process of mortality (Powell et al. 2017; von Aryx et al. 2017; Zuidema et al. 2018). For example, Roskilly et al. (2019) used variation in pit membrane structure to understand how drought-tolerance, growth rate, and longevity are related to one another in *Pinus ponderosa*: individuals of greatest longevity showed selection towards pit membrane structure associated with drought-tolerance (and slower rates of growth) compared to greater variation in pit membrane structure (and growth rates) amongst younger trees. Ultimately, investigation of xylem anatomical variation unpacks the significance of so-called functional traits like wood density, which have been predictive of some aspects of tree demography and life history variation (Visser et al. 2016), but less so of others (mortality; O'Brien et al. 2017; Yang et al. 2018), into the underlying components that are truly functional (rather than emergent).

Further, biogenic time series data offer a golden opportunity to quantify the response of a single genotype (e.g. an individual tree) to varying environmental conditions, and hence the genetic (G), plastic (E), and G*E (heritable plastic) basis for adaptation to climate variation (Evans et al. 2018). In the context of a diallel mating design, Marchal et al. (2019) applied mixed effect models of the kind used in the cattle breeding literature to quantify growth reaction norms in response to soil water availability in hybrid larch compared to its parents. Morrongiello and Thresher (2015) used mixed effect modelling of otolith time series to estimate individual-level plasticity, with the conclusion that it is currently sufficient across much of the tiger flathead's range such that warming waters do not negatively impact fitness, while suggesting concern about the fitness consequences of future warming at

the equatorward edge of its distribution. Other recent examples have used genotype-phenotype association studies of tree growth ring time series data (Heer et al 2018; Housset et al. 2018). Housset et al. (2018) associated growth metrics on three timescales derived from tree rings collected in a forestry provenance trial with single nucleotide polymorphisms at loci that had demonstrated links to local climate adaptation. Their data showed opposing latitudinal clines of cold and drought sensitivity, suggesting that assisted migration of southern, drought-tolerant provenances northward might not be successful because of their sensitivity to cold stress. We expect that genome-wide association studies (i.e. pedigree-free approaches) will soon be applied to tree-ring and other biogenic time series data to better quantify the G, E, and G*E components of growth responses to climate, and further, to identify the many loci responsible for variation in the (polygenic) traits necessary to tolerate environmental extremes anticipated in the twenty-first century.

Disturbance—Forest ecologists use growth ring time series to better understand the influence of disturbances on tree demography. These range from single-tree gap dynamics (Thompson et al. 2007) to landscape-scale disturbances caused by insect outbreaks, fire, wind or ice storms, or human activities (Pederson et al. 2008; Axelson et al. 2009; Lynch 2012). One line of evidence is the detection of a 'release event', that is, an individual in a population is released from competitive pressure when disturbance kills its neighbours, leading to a change in the (detrended) width of growth rings, such as a 25% increase in the 10-year running mean (Nowacki and Abrams 1997). Other release criteria and methods (Black and Abrams 2003; Druckenbrod 2005) address variation among trees in release response because of forest canopy position at the time of disturbance (i.e. individuals in competitively suppressed vs. dominant positions) or evolutionary strategy with respect to shade tolerance. Another approach is to compare growth increment variability of individuals that are versus are not affected by a particular type of disturbance. For example, Swetnam and Lynch (1993) reconstructed regional-scale insect outbreak cycles by comparing 300-year growth time series of host versus nonhost tree species, using a spectral analysis to recover pseudo-periodic insect population dynamics.

Fire scars have been sampled and developed into extensive data networks, yielding multicentury, spatially explicit reconstruction of forest wildfires. Analyses across scales of these networks of fire-scarred trees have informed our understanding of fire regimes and their controls, revealing the interplay between bottom-up (topography, vegetation) and top-down (climate) controls of fire (Falk et al. 2007; Falk et al. 2011). Fire-scar data have further been combined with birth (pith) dates as described above (Brown and Wu 2005; Margolis et al. 2011; O'Connor et al. 2014), to better understand how the fundamental Earth system process of fire influences the demography of forest ecosystems, including the creation of spatially heterogeneous forest age structure from gap-phase and cohort responses (Heyerdahl et al. 2014, 2019). Hence multiple lines of evidence derived from tree rings (e.g. changes in growth increment time series, records of acute stress, and age structure) all have informed forest disturbance ecology; in some cases, these approaches could be applied to other species and ecosystems to understand the role of disturbance in population, range, and ecosystem dynamics.

4.6 Ecosystem and macrosystem demography

Demography has often been conducted at a small spatial scale (i.e. 1 m^2 plots), with a single-species focus, and relatively short (5–30 year) time spans, a natural consequence of the laborious process of building data sets one year at a time. Biogenic annually resolved time series data, especially when developed into a spatial network, offer insights at larger spatial and deeper temporal scales, and the opportunity to study several species in an ecosystem, even across ecosystems, to understand the drivers of demography, particularly how several organisms are affected by the changing climate system. In this final section, we highlight two examples of this from the Northeastern Pacific Ocean and Australia.

Across the ecosystems of the northeast Pacific Ocean and western North America, the winter (January–March) North Pacific High (NPH, defined as mean sea level pressure at 25°N–35°N by 125°W–145°W) exerts strong and synchronous control over productivity and hence demography of a diverse assemblage of species across marine, freshwater, and terrestrial ecosystems (Figure 4.3). In the California Current Ecosystem, the strength of winter coastal upwelling, which charges the photic zone with cold, nutrient-rich water, is associated with variability in the winter NPH, which is in turn teleconnected to ENSO (Schwing et al. 2002). This drives bottom-up trophic effects, evidenced in rockfish otolith growth increments, seabird reproductive phenology and success, copepod community, and Sacramento and Russian River chinook salmon

abundance (Black et al. 2009; Black et al. 2011, 2014; Garcia-Reyes et al. 2013). Across the adjacent terrestrial ecosystems of western North America, the NPH is further associated with storm track position, and in turn winter precipitation, as well as water year river discharge (Black et al. 2009, 2015). Years of strong NPH are associated with anomalously high levels of upwelling, but anomalously low levels of precipitation, such that productivity is antiphase across the marine versus terrestrial systems (Black et al. 2014). Centuries-long tree-ring time series indicate that the winter NPH and associated marine and terrestrial indicators are becoming increasingly variable and extreme, and that the spatial footprint of the NPH is increasing, entraining synchrony of greater spatial extent. This is expected to reduce bet-hedging effects and resilience across species, increasing the risk of extirpation; (Black et al. 2017); for example, the extinction of checkerspot butterfly populations in the San Francisco region (McLaughlin et al. 2002). Given the importance of the NPH to the key drivers of productivity in marine (upwelling) and terrestrial (precipitation) ecosystems of the region, its increasing volatility is likely to influence a wide range of demographic processes and species, from the continental slope in the sea to high-elevation forests on land.

A similar story emerges in Western Australia: coupled ocean-atmosphere patterns synchronise demography across aquatic and terrestrial ecosystems at large spatial scales. Here also, ENSO is an important driver of conditions in the marine environment, via sea surface temperatures and salinity in the tropical North Coast region and the strength of the Leeuwin current along the subtropical and temperate West and South Coasts, as well as in the terrestrial environment, via precipitation. Ong et al. (2016) showed increased growth of corals (*Porites* spp.), two species of marine fishes, and a tree in response to strong La Niña years, when North Coast sea surface temperatures and northwest Australian precipitation are higher than average, and sea surface salinity is lower than average. Ong et al. (2018) further showed synchronous ENSO-driven interannual growth variation among five species of tropical and temperate fishes spanning 23° of latitude. El Niño conditions, associated with anomalously warm sea surface temperatures, have caused widespread mortality for both fish and corals (Feng et al. 2013); indeed, coral reefs are widely considered to be the ecosystem most vulnerable to climate change–driven demographic collapse.

These multispecies, ecosystem- or macrosystem-scale studies illustrate how demographers in the

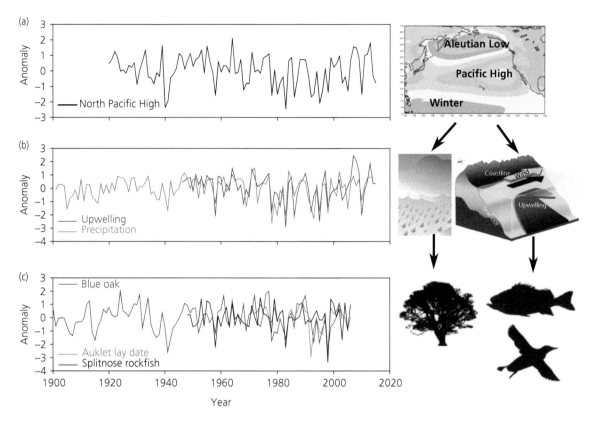

Figure 4.3 Cascading effects of (a) the strength of the winter North Pacific High on (b) precipitation in western North America and winter coastal upwelling in the northeast Pacific Ocean, and on demographic metrics of organisms of these terrestrial and aquatic ecosystems, including (c) growth increments of blue oak (*Quercus douglasii*) and the split-nose rockfish (*Sebastes diploproa*), as well as auklet (*Ptychoramphus aleuticus*) lay dates. The precipitation and blue oak time series are inverted to make synchrony easier to identify.

twenty-first century—the age of global change—could profit from a more intimate and mechanistic understanding of the atmospheric and ocean current systems that are changing and profoundly affecting the organisms they study. Indeed, biogenic time series are uniquely positioned to offer insight into both biological and physical systems, since historically their use has been for reconstructing physical system variation, and, we argue, their growing future lays in also understanding biological system (demographic) variation.

4.7 A research agenda

Across this remarkable range of insights into growth, birth, death, movement, evolution, disturbances, trophic cascades, and macrosystems gained from single-sample time series data, surprisingly little has been incorporated into population models. At the

same time, demography is sometimes criticised for being phenomenological. Here we set forth a research agenda for demographers to act upon: it is time to incorporate demographic insights from biogenic time series data into population models (Figure 4.4), capture an improved understanding of what drives population dynamics, and better anticipate the impacts of global change.

To clarify, otolith-derived age data have been used for many years to parameterise stock assessment (age-length) models for fish (e.g. a von Bertalanffy model) similar to the age-size models fit to data from dinosaur bones. However, most of these stock assessment models assume constant values for growth parameters, meaning they only use otoliths for age information (Lee and Punt 2018); they do not use the growth increment time series data contained in otoliths to quantify drivers of variation in growth or incorporate those drivers into population models. The same is true for

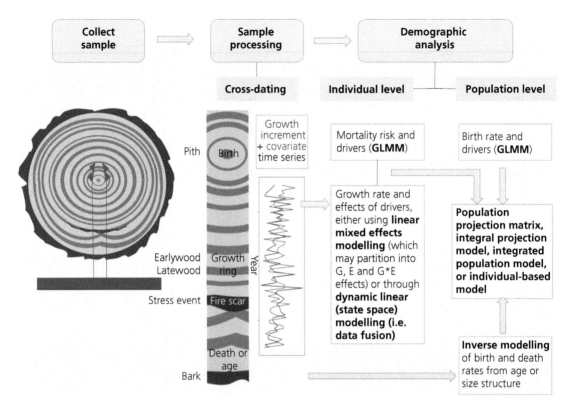

Figure 4.4 An overview of how biochronologies can inform population models. We take the example of tree rings, starting with the collection of a sample, an increment core, using an increment borer. A year of formation is then assigned to each ring using the process of cross-dating. Additional demographic information derived from the cross-dated growth increment time series may include pith date, stressful events, and age at the time of sampling or year of death. Individual-level vital rates (and drivers of their variation) that may then be inferred are growth rate (using linear modelling or dynamic linear modelling) and mortality risk. Birth rate, a population-level parameter, may also be inferred. Forward estimates of vital rates and their drivers could then be combined in a number of different kinds of demographic models. Alternatively, inverse estimation of population model parameters may be derived from time series of age or size structure data. Some of these approaches have been used in the literature (e.g. linear mixed effects modelling of growth increments) and others have not (e.g. Poisson regression of birth dates to estimate birth rates).

tree rings, which is even more surprising since the increment borer, the tool used to sample a tree for its growth rings, was invented by foresters (in the late-nineteenth century; Somerville 1891) specifically to quantify tree growth. In more than 100 years, tree-ring data have not been used to parameterise forestry growth and yield models; instead, like fisheries biologists, foresters have primarily used growth rings to age individuals.

How might estimates from biogenic time series data of vital rates and the drivers of their variation be incorporated into population models (Figure 4.4)? One possibility is to use estimates of baseline growth rate, along with any effects on growth (of individual size, climate, social status or population density), in a population projection matrix (Chapter 9), integral projection

model (IPM) (see Chapter 10), or other demographic model (Chapters 12 and 14). This would apply to any of the above cases of growth modelling—from ibex to loggerhead sea turtles to fish to coral to trees. For example, Mina et al. (2016) used climate sensitivities derived from a tree-ring-based model of growth (VS-Lite) in an individual-based forest succession model (ForClim), which improved representation of the effect of drought. Schultz and Miller (in prep.) used tree-ring time series from both live and dead individuals of *Pinus albicaulis* to parameterise multiple regression (MEM) models that form the survival-growth subkernel of an IPM. That is, ring-width time series from live individuals were used to estimate growth rates and outer-bark dates from dead individuals were used to estimate size-dependent mortality rates. A MEM of ring-width time series data can similarly be used to parameterise the growth component of a stand-level forestry growth and yield model (Giebink et al. (in prep.)), which is fundamentally similar to an IPM. When estimates of parameters are passed from one model or analysis to another in this sort of sequential manner, it is critical to quantify parameter uncertainty and include it at the stage of population projection, so that projections reflect the uncertainty associated with underlying data ('uncertainty propagation'; Dietze 2017).

Another level of complexity is to combine different sources of information on growth (i.e. data fusion) to estimate the growth parameters in a population model. One example of this used a hierarchical Bayesian framework to model tree bole diameter increments (from diameter tape measurements), radial increments (from increment cores), and their common influences at once—with two multiple regression models connected to one another via the simple fact that the diameter increment must equal twice the radial increment (Evans et al. 2017). Another example is data fusion via state space modelling. Clark et al. (2007) treated tree bole diameter at breast height (DBH) as a latent state variable; its time evolution is governed by a simple difference equation, and two data models link the available observations of growth (DBH remeasurements plus tree rings) to the (latent) true state (diameter at breast height) of the tree. This model included random effects of individual tree and year, but no fixed effects; the next step is to include fixed effects that are known to influence tree growth (e.g. tree size, climate; Heilman et al. (in prep.)).

Information contained in biogenic time series data can also be incorporated into population models through inverse modelling. Age structure data derived from vertebrate bones (including otoliths) can be used to constrain population model parameters that are difficult to estimate, such as mortality rate, which otherwise require mark-recapture data over an extended period (Zhao et al. 2018). Gonzalez and Martorell (2013) demonstrated that time series of population structure can be used to accurately estimate individual-level vital rates for an IPM. Nonidentifiability of population model parameters (i.e. the fact that more than one combination of birth and death rates can give rise to the same age structure) can be greatly reduced by constraining model structure or parameters with prior knowledge (Gonzalez and Martorell 2013). Indeed, an emerging trend in the study of demography is the use of multiple sources of data (e.g. through integrated population modelling; see Chapter 14), data fusion or data assimilation, or inverse modelling (Hartig et al.

2012; Evans et al. 2016; Gonzales et al. 2016; White et al. 2016; Vanderwel et al. 2017; Zipkin et al. 2019). These approaches should be applied to single-sample time series data to incorporate them into population models.

By incorporating biogenic time series data (or parameters derived from them), demographic models could then reflect greater understanding of the causes of variation in vital rates, and better anticipate the impacts of global change. Drawing from some of the insights described in the chapter, demographic models could then explicitly represent different environmental drivers of juvenile versus adult vital rates, how climate stress affects individuals differently across size or social status, or the lingering (lagged) effects of climate stress on performance. Models of migratory fish could explicitly represent habitat use, influences in different basins, or what proportion of the population are wild versus hatchery raised. Biogenic time series data will likely play a key role in estimating and representing in models how local adaptation and plasticity constrain responses to shifting climate. That is, demographic forecasts should reflect intraspecific variation in vital rates (see Chapter 20); they should follow reactions norms estimated from data, and biogenic time series data are an excellent source of data to do so. The demographic consequences of changing disturbance regimes are an important aspect of global change that also can be modelled with the help of biochronologies. Going one step further would be to assimilate time series data from multiple species that are linked via trophic relationships, or through common physical drivers, to trace biotically driven and abiotically driven demographic cascades through complex ecological systems. And it may be at a planetary scale that growth ring time series offer their most consequential demographic forecasts (i.e. of the impact of climate change on tree growth, the abundance and distribution of forests, and hence the expected behaviour of the planetary-scale feedback between forest ecosystems and climate, based on data from tree rings).

4.8 Conclusions

Time series data generated from natural archives, which can be obtained in a single sampling effort, are an underappreciated source of demographic information. Historically, their use has been for reconstructing physical system variation; we argue that their growing future lays in understanding biological system (demographic) variation, and ultimately, relationships between the two. They typically extend the temporal scale of demographic studies profoundly, and when developed into spatial networks, provide the opportunity to think more broadly and mechanistically about the influence of ecosystem and Earth system processes across a great range of temporal and spatial scales (e.g. disturbance processes, atmospheric and ocean circulation patterns) on demography. Biochronologies further complement other sources of demographic information in critical and unique ways, offering traction on vital rates that are otherwise difficult to estimate—stages of the life history for which information is poor because the events are rare (mortality), organisms are mobile or extinct, or for other reasons. The challenge now is to combine the information contained in biochronologies on vital rates and the causes of vital rate variation with traditional demographic data streams to better address the consequences of global change.

References

Allen C. D., Macalady A. K., Chenchouanic H., et al. (2010). A global overview of drought and heat-induced tree mortality reveals emerging climate change risks for forests. Forest Ecology and Management **259**, 660–684.

Anderegg W. R. L., Klein T., Bartlett M., et al. (2016). Meta-analysis reveals that hydraulic traits explain cross-species patterns of drought-induced tree mortality across the globe. Proceedings of the National Academy of Sciences of the United States of America **113**(49), 5024–5029.

Axelson J. N., Alfaro, R. I., and Hawkes B. C. (2009). Influence of fire and mountain pine beetle on the dynamics of lodgepole pine stands in British Columbia. Forest Ecology and Management **257**, 1874–1882.

Babst F., Bodesheim P., Charney N., et al. (2018). When tree rings go global: challenges and opportunities for retro- and prospective insight. Quarternary Science Reviews **197**, 1–20.

Bjorndal K. A., Bolten A. B., Dellinger T., Delgado C., and Rost H. (2003). Compensatory growth in oceanic loggerhead sea turtles: response to a stochastic environment. Ecology **84**, 1237–1249.

Black B. A. and Abrams M. D. (2003). Use of boundary-line growth patterns as a basis for dendroecological release criteria. Ecological Applications **13**, 1733–1749.

Black B. A., Gillespie D., Maclellan S. E., and Hand C. M. (2008). Establishing highly accurate production-age data using the tree-ring technique of crossdating: a case study for Pacific geoduck (*Panopea abrupta*). Canadian Journal of Fisheries and Aquatic Sciences **65**, 2572–2578.

Black B. A., Abrams M. D., Rentch J. S., and Gould J. S. (2009). Properties of boundary-line release criteria in North American tree species. Annals of Forest Science **66**, 205.

Black B. A., Schroeder I. D., Sydeman W. J., Bograd S. J., Wells B. K., and Schwing F. B. (2011). Winter and summer upwelling modes and their biological importance in the California Current Ecosystem. Global Change Biology **17**, 2536–2545.

Black B. A., Sydeman W. J., Frank D. C., et al. (2014). Six centuries of variability and extremes in a coupled marine–terrestrial ecosystem. Science **345**, 1498–1502.

Black B. A., Dunham J. B., Blundon B. W., Brim-Box J., and Tepley A. J. (2015) Long-term growth-increment chronologies reveal diverse influences of climate forcing on freshwater and forest biota in the Pacific Northwest. Global Change Biology **21**, 594–604.

Black B. A., Griffin D., Van Der Sleen P., et al. (2016). The value of crossdating to retain high-frequency variability, climate signals, and extreme events in environmental proxies. Global Change Biology **22**, 2582–2595.

Black B. A., Van Der Sleen P., Di Lorenzo E., et al. (2017). Rising synchrony controls western North American ecosystems. Global Change Biology **24**(6), 2305–2314.

Black B. A., Andersson C., Butler P. G., et al. (2019). The revolution of crossdating in marine palaeoecology and palaeoclimatology. Biology Letters **15**, 20180665. http://dx.doi.org/10.1098/rsbl.2018.0665.

Breinen R. J. W. and Zuidema P. A. (2006). Lifetime growth patterns and ages of Bolivian rain forest trees obtained by tree ring analysis. Journal of Ecology **94**, 481–493.

Brennan S. R., Schindler D. E.,Cline T. J. Walsworth T. E., Buck G., and Fernandez D. P. (2019). Shifting habitat mosaics and fish production across river basins. Science **364**, 783–786.

Brown P. M. and Wu R. 2005. Climate and disturbance forcing of episodic tree recruitment in a southwestern Ponderosa pine landscape. Ecology **86**, 3030–3038.

Briffa K. R., Jones P. D., Bartholin T. S., et al. 1992. Fennoscandian summers from AD 500: temperature changes on short and long timescales Climate Dynamics 7, 111–7119.

Buechling A., Matin P. H., and Canham C. D. (2017). Climate and competition effects on tree growth in Rocky Mountain forests. Journal of Ecology **105**, 1636–1647.

Bunn A. G. 2008. A dendrochronology program library in R (dplR). Dendrochronologia **26**, 115–124.

Buntgen U. (2019). Re-thinking the boundaries of dendrochronology. Dendrochronologia **53**, 1–4.

Buntgen U., Jenny H., Liebhold A., et al. (2014). European springtime temperature synchronizes ibex horn growth across the eastern Swiss Alps. Ecology Letters **17**, 303–313.

Cailleret C., Jansen S., Robert E. M. S., et al. (2016a). A synthesis of radial growth patterns preceding tree mortality. Global Change Biology **23**(4), 1675–1690.

Cailleret M., Bigler, C., Bugmann H., et al. (2016b). Towards a common methodology for developing tree logistic mortality models based on ring-width data. Ecological Applications **26**, 1827–1841.

Campana S. E. (1990). How reliable are growth back-calculations based on otoliths? Canadian Journal of Fisheries and Aquatic Science **47**, 2219–2227.

Campana S. E. and Neilson J. D. (1985). Microstructure of fish otoliths. Canadian Journal of Fisheries and Aquatic Science **42**, 1014–1032.

Canham C. D., Murphy L., Riemann R., McCullough R., and Burrill E. (2018). Local differentiation in tree growth responses to climate. Ecosphere **9**, e02368.

Carilli J. E., Norris R. D., Black B., Walsh S. M., and McField M. (2009). Century-scale records of coral growth rates indicate that local stressors reduce coral thermal tolerance threshold. Global Change Biology **16**, 1247–1257.

Clark J. S., Wolosin M., Dietze M. et al. (2007). Tree growth inference and prediction from diameter censuses and ring widths. Ecological Applications **17**, 1942–1953.

Cook E., Briffa K., Shiyatov S., and Mazepa V. (1990). Tree-ring standardization and growth-trend estimation. In: E. Cook and L. Kairiukstis (eds.), Methods of Dendrochronology: Applications in the Environmental Sciences, 104–123. Kluwer.

Cook E. R., Briffa K. R., Meko D. M., Graybill D. A., and Funkhouser, G. (1995). The 'segment length curse' in long tree-ring chronology development for paleoclimatic studies. The Holocene **5**, 229–237.

Crone, E. E., McIntire E. J. B., and Brodie J. (2011). What defines mast seeding? Spatiotemporal patterns of cone production by whitebark pine. Journal of Ecology **99**, 438–444.

Davis K. T., Dobrowski S. Z., Higuera P. E., et al. (2019). Wildfires and climate change push low-elevation forests across a critical climate threshold for tree regeneration. Proceedings of the National Academy of Sciences of the United States of America **116**(13), 6193–6198.

Dee J. R. and Palmer M. W. (2016). Application of herbochronology: annual fertilization and climate reveal annual ring signatures within the roots of U. S. tallgrass prairie plants. Botany **94**, 1–12.

DeRose R. J., Shaw J. D., and Long J. N. (2017). Building the forest inventory and analysis tree-ring data set. Journal of Forestry **115**, 283–291.

Dietze M. C. (2017). Ecological Forecasting. Princeton University Press.

Divlan A., Perry-Jones K., and Wardle G. A. (2006). Age determination in the grey-headed flying fox. Journal of Wildlife Management **70**, 607–611

Dolezal, J., Dvorsky M., Borner A., Wild J., and Schweingruber F. H. (2018). Anatomy, Age, and Ecology of High Mountain Plants in Ladakh, the Western Himalaya. Springer.

Dolezal, J., Kopecky M., Dvorsky M., et al. (2019). Sink limitation of plant growth determines tree line in the arid Himalayas. Functional Ecology **33**, 553–565.

Dorado-Linan I., Piovesan G., Martínez-Sancho E., et al. (2018). Geographic adaptation prevails over species-specific determinism in trees' vulnerability to climate change at Mediterranean read-edge forests. Global Change Biology **25**, 1296–1314.

Douglass A. E. (1914). A method of estimating rainfall by the growth of trees. Carnegie Institute of Washington Publication **192**, 101–121.

Douglass A. E. (1920). Evidence of climatic effects in the annual rings of trees. Ecology **1**, 24–32.

Douglass A. E. (1941). Crossdating in dendrochronology. Journal of Forestry **39**, 825–831.

Druckenbrod D. L. (2005). Dendroecological reconstructions of forest disturbance history using time-series analysis with intervention detection. Canadian Journal of Forest Research **35**, 868–876.

English, N. B., Dettman D. L., Sandquist D. R., and Williams D. G. (2007). Past climate changes and ecophysiological responses recorded in the isotope ratios of saguaro cactus spines. Oecologia 154(2), 247–258.

English, N. B., Dettman D. L., Sandquist D. R., and Williams D. G. (2010). Daily to decadal patterns of precipitation, humidity, and photosynthetic physiology recorded in the spines of the columnar cactus, *Carnegiea gigantea*. Journal of Geophysical Research: Biogeosciences 115. doi:10.1029/2009JG001008. hdl:20.500.11919/768.

Erickson G. M. (2005). Assessing dinosaur growth patterns: a microscopic revolution. Trends in Ecology and Evolution **20**, 677–684.

Erickson G. M. Makovicky P. J., Currie P. J., Norell M. A., Yerby S. A., and Brochu C. A. (2004). Gigantism and comparative life-history patterns of tyrannosaurid dinosaurs. Nature **412**, 405–412.

Evans M. E. K., Merow C., Record S., McMahon S. M., and Enquist B. J. (2016). Towards process-based range modeling of many species. Trends in Ecology and Evolution **31**, 860–871.

Evans M. E. K., Falk D. A., Arizpe A., Swetnam T., Babst F., and Holsinger K. E. (2017). Fusing tree-ring and forest inventory data to infer influences on tree growth. Ecosphere **8**, e01889.

Evans M. E. K., Gugger P. F., Lynch A. M., et al. (2018). Dendroecology meets genomics in the common garden: new insights into climate adaptation. New Phytologist **218**, 401–403.

Falk D. A., Miller C., McKenzie D., and Black A. E.(2007). Cross-scale analysis of fire regimes. Ecosystems **10**, 809–823.

Falk D. A., Heyerdahl E. K., Brown P. M., et al. (2011). Multiscale controls of historical forest-fire regimes: new insights from fire-scar networks. Frontiers in Ecology & Evolution **9**, 446–454.

Fatichi S., Leuzinger S., and Körner C. (2014). Moving beyond photosynthesis: from carbon source to sink-driven vegetation modeling. New Phytologist **201**(13), 1086–1095.

Feng M., McPhaden M. J., Xie S.-P., and Hafner J. (2013). La Nina forces unprecedented Leeuwin Current warming in 2011. Scientific Reports **3**, 1277.

Fernandez-de-una L., McDowell N. G., Canellas I., and Gea-Izquierdo G. (2016). Disentangling the effect of competition, CO_2, and climate on intrinsic water-use efficiency and tree growth. Journal of Ecology **104**, 678–690.

Fonti P., von Arx G., García-González I., et al. (2010). Studying global change through investigation of the plastic responses of xylem anatomy in tree rings. New Phytologist **185**, 42–53.

Foster J. R., Finley A. O., D'Amato A. W., Bradford J. B., and Banerjee S. (2016). Predicting tree biomass growth in the temperate-boreal ecotone: is tree size, age, competition, or climate response most important? Global Change Biology **22**(6), 2138–2151.

Frank D. C., Poulter B., Saurer M., et al. (2015). Water-use efficiency and transpiration across European forests during the Anthropocene. Nature Climate Change **5**(23), 579–583.

Freckleton R. P. (2002). On the misuse of residuals in ecology: regression of residuals vs. multiple regression. Journal of Animal Ecology **71**, 542–545.

Fritts H. C. (1976). Tree Rings and Climate. Academic Press.

Fritts H. C. and Swetnam T. W. (1989). Dendroecology: a tool for evaluating variations in past and present forest environments. Advances in Ecological Research **19**, 111–188.

Gamm C. M., Sullivan P. F., Buchwal A., et al. (2018). Declining growth of deciduous shrubs in the warming climate of continental western Greenland. Journal of Ecology **106**, 640–654.

Garcia-Berthou E. (2001). On the misuse of residuals in ecology: testing regression residuals vs. the analysis of covariance. Journal of Ecology **70**, 708–711.

Garcia-Reyes M., Sydeman W. J., Thompson S. A., et al. (2013). Integrated assessment of wind effects on Central California's pelagic ecosystem. Ecosystems **16**(5), 722–735.

Gessler A., Ferrio J. P., Hommel R., Treydte K., Werner R. A., and Monson R. K.(2014). Stable isotopes in tree rings: towards a mechanistic understanding of isotope fractionation and mixing processes from the leaves to the wood. Tree Physiology **34**, 796–818.

Giebink C. L., DeRose R. J., Castle M., Shaw J. D., and Evans M. E. K. (In prep.). Climatic sensitivies derived from tree rings improve predictions of the Forest Vegetation Simulator growth and yield model.

Gonzalez E. J. and Martorell C. (2013). Reconstructing shifts in vital rates driven by long-term environmental change: a new demographic method based on readily available data. Ecology & Evolution **3**, 2273–2284

Gonzalez E. J., Matorell C., and Bolker B. J. (2016). Inverse estimation of population projection parameters using time series of population-level data. Methods in Ecology and Evolution **7**, 147–156.

Green A. J. (2001). Mass/length residuals: measures of body condition or generators of spurious results? Ecology **82**, 1473–1483.

Greenwood S., Ruiz-Benito P., Martínez-Vilalta J., et al. (2017). Tree mortality across biomes is promoted by drought intensity, lower wood density and higher specific leaf area. Ecology Letters **20**, 539–553.

Guatelli-Steinberg D., Reid D. J., Bishop T. A., and Larsen C. S. (2005).Anterior tooth growth periods in Neanderthals were comparable to those of modern humans. Proceedings of the National Academy of Sciences of the United States of America **102**, 14197 e14202.

Guiterman C. H., Margolis E. Q., Allen C. D., Falk D. A., and Swetnam T. W. (2018). Long-term persistence and fire resilience of oak shrubfields in dry conifer forests of northern New Mexico. Ecosystems **21**, 943–959.

Hamilton V., Evans K., and Hindell M. A. (2017). From the forests to teeth: visual crossdating to refine age estimates in marine mammals. Marine Mammal Science **33**, 880–888.

Hartig F., Dyke J., Hickler T., et al. (2012). Connecting dynamic vegetation models to data—an inverse perspective. Journal of Biogeography **39**, 2240–2252.

Hayat A., Hacket-Pain, A. J., Pretzsch, H., Rademacher, T. T., and Friend, A. D. (2017). Modeling tree growth taking into account carbon source and sink limitations. Frontiers in Plant Science **8**, http://dx.doi.org/10.3389/fpls.2017.00182.

Heer K., Behringer D., Piermattei A., et al. (2018). Linking dendroecology and association genetics in natural populations: stress responses archived in tree rings associate with SNP genotypes in silver fir (*Abies alba* Mil.). Molecular Ecology **27**, 1428–1438.

Hegg J. C., Giarrizzo T., and Kennedy B. P. (2015). Diverse early life-history strategies in migratory Amazonian catfish: implications for conservation and management. PLOS One **10**(7), e0129697.

Heilman, K. A., M. C. Dietze., A. Arizpe., J. Aragon., A. Grey., J. D. Shaw., A. O. Finley., S. Klesse., R. J. DeRose., and M. E. K. Evans. *In Review*. Fusing tree-ring and forest inventory data to forecast the influences of climate, tree size, and stand density on tree growth.

Helama S., Melvin T. M., and Briffa K. R. (2017). Regional curve standardization: state of the art. The Holocene **27**, 172–177.

Helser T. E., Lai H., and Black B. A. (2012). Bayesian hierarchical modeling of Pacific geoduck growth increment data and climate indices. Ecological Modelling **247**, 210–220.

Heyerdahl E. K., Loehman R. A., and Falk D. A. (2014). Lodgepole pine-dominated forest in central Oregon's Pumice Plateau: historical mixed-severity fires are resistant to future climate change. Canadian Journal of Forest Research **44**, 593–603.

Heyerdahl E. K., Loehman R. A., and Falk D. A. (2019). A multi-century history of fire regimes along a transect of mixed-conifer forests in central Oregon, USA. Canadian Journal of Forest Research **49**, 76–86.

Holmes R. L. (1983). Computer-assisted quality control in tree-ring dating and measurement. Tree-Ring Bulletin **43**, 69–78.

Holtta T., Makinen H., Nojd P., Makela A., and Nikinmaa E. (2010). A physiological model of softwood cambial growth. Tree Physiology **30**, 1235–1252.

Housset J. M., Nadeau S., Isabel N., et al. (2018). Tree rings provide a new class of phenotypes for genetic associations that foster insights into adaptation of conifers to climate change. New Phytologist **218**, 630–645.

Kajimoto T., Onodera H., Ikeda S., Daimaru H., and Seki T. (1998). Seedling establishment of subalpine stone pine (*Pinus pumila*) by nutcracker (*Nucifraga*) seed dispersal on Mt. Yumori, northern Japan. Arctic and Alpine Research **30**(4), 408–417.

Kelley J. and Smith T.M. (2003). Age at first molar emergence in early Miocene *Afropithecus turkanensis* and life-history evolution in the Hominoidea. Journal of. Human Evolution **44**, 307e329.

Kennedy B. P., Blum J. D., Folt C. L., and Nislow K. H. (2000). Using natural strontium isotopic signatures as fish markers: methodology and application. Canadian Journal of Fisheries and Aquatic Science **57**, 2280–2292.

Klesse S., DeRose R. J., Guiterman C. H., et al. (2018). Sampling bias overestimates climate change impacts on forest growth in the Southwestern United States. Nature Communications **9**, 5336.

Le Cabec A., Tang N., and Tafforeau P. (2015). Accessing developmental information of fossil hominin teeth using new synchotron microtomography-based visualization techniques of dental surfaces and interfaces. PLOS One **10**(4), e0123019.

Le Moullec M., Buchwal A., Van Der Wal R., Sandal L., and Bremset Hansen B. (2019). Annual ring growth of a widespread high arctic shrub reflects past fluctuations in community-level plant biomass. Journal of Ecology **107**, 436–451.

Lee Q. and Punt A. E. (2018). Extracting a time-varying climate-driven growth index from otoliths for use in stock assessment models. Fisheries Research **200**, 93–103.

Lloret F., Keeling E. G., and Sala A. (2011). Components of tree resilience: effects of successive low-growth episodes in old ponderosa pine forests. Oikos **120**, 1909–1920.

Lynch A. M. (2012). What tree-ring reconstruction tells us about conifer defoliator outbreaks. In: P. Barbosa, D. K. Letorneau, and A. A. Agrawal (eds.), Insect Outbreaks—Revisited, 125–154. Wiley-Blackwell.

Macalady A. K. and Bugmann H. (2014). Growth-mortality relationships in Pinon Pine (*Pinus edulis*) during severe droughts of the past century: shifting processes in space and time. PLOS One **9**, e92770.

Marchal A., Schlichting C. D., Gobin R., et al., (2019). Deciphering hybrid larch reaction norms using random regression. Genes, Genomes, Genetics **9**, 21–32.

Margolis E. Q., Swetnam, T. W., and Allen, C. D. (2011). Historical stand-replacing fire in upper montane forests of the Madrean sky islands and Mogollon plateau, southwestern USA. Fire Ecology **7**, 88–107.

Martin-Benito D., Kint V., Del Rio M., Muys B., and Canellas I. (2011). Growth responses of West-Mediterranean *Pinus nigra* to climate change are modulated by competition and productivity: past trends and future perspectives. Forest Ecology and Management **262**, 1030–1040.

Martino J. C., Fowler A. J., Doubleday, Z. A., Grammar G. L., and Gillanders B. M. (2019). Using otolith chronologies to understand long-term trends and extrinsic drivers of growth in fisheries. Ecosphere **10**, e02553.

McCarroll D. and Loader N. J. (2004). Stable isotopes in tree rings. Quaternary Science Reviews **23**, 771–801.

McLaughlin J. F., Hellmann J. J., Boggs C. L., and Ehrlich P. R. (2002). Climate change hastens population extinctions. Proceedings of the National Academy of Sciences of the United States of America **99**, 6070–6074.

Melvin T. M. and Briffa K. R. (2008). A 'signal-free' approach to dendroclimatic standardisation. Dendrochronologia **26**, 71–86.

Mina M., Martin-Benito D., Bugmann H., and Cailleret M. (2016). Forward modeling of tree-ring width improves simulation of forest growth response to drought. Agricultural and Forest Meteorology 221, 13–33.

Morgan P. and Bunting S. C. (1992). Using cone scars to estimate past cone crops of whitebark pine. Western Journal of Applied Forestry (USA) **7**(3), 71–71.

Morrongiello J. R. and Thresher R. E. (2015). A statistical framework to explore ontogenetic growth variation among individuals and populations: a marine fish example. Ecological Monographs **85**, 93–115.

Morrongiello J. R., Thresher R. E., and Smith D. C. (2012). Aquatic biochronologies and climate change. Nature Climate Change **2**, 849–857.

Munier J., Brown P. M., and Romme W. H. (2014). Tree recruitment in relation to climate and fire in northern Mexico. Ecology **95**, 197–209.

Myers-Smith I. H., Elmendorf S. C., and Vellend M., et al. (2015a). Climate sensitivity of shrub growth across the tundra biome. Nature Climate Change **5**, 887–891.

Myers-Smith I. H., Hallinger M., Blok D., et al. (2015b). Methods for measuring arctic and alpine shrub growth: a review. Earth-Science Review **140**, 1–13.

Navarro-Cerrillo R. M., Sanchez-Salguero R., Rodriguez C., et al. (2019). Is thinning an alternative when trees could die in response to drought? The case of planted *Pinus nigra* and *P. sylvestris* stands in southern Spain. Forest Ecology and Management **433**, 313–324.

Nowacki G. J. and Abrams M. D. (1997). Radial-growth averaging criteria for reconstructing disturbance histories from presettlement-origin oaks. Ecological Monographs **67**, 225–249.

O'Brien M. J., Engelbrecht B. M. J., Joswig J., et al. (2017). A synthesis of tree functional traits related to drought-induced mortality in forests across climatic zones. Journal of Applied Ecology **54**, 1669–1686.

O'Connor C. D., Falk D. A., Lynch A. M., and Swetnam T. W. (2014). Fire severity, size, and climate associations diverge from historical precedent along an ecological gradient in the Pinaleno Mountains, Arizona, USA. Forest Ecology and Management **329**, 264–278.

Ogle K., Whitham T. G., and Cobb N. S. (2000). Tree-ring variation in pinon predicts likelihood of death following severe drought. Ecology **81**, 3237–3243.

Ogle K., Barber J. J., Barron-Gafford, G. A., et al. (2015). Quantifying ecological memory in plant and ecosystem processes. Ecology Letters **18**, 221–235.

Ong J. J. L., Rountrey A. N., Meeuwig J., et al. (2015). Contrasting environmental drivers of adult and juvenile growth in a marine fish: implications for the effects of climate change. Scientific Reports **5**, 10859.

Ong J. J. L., Rountrey A. N., Zinke J., et al. (2016). Evidence for climate-driven synchrony of marine and terrestrial ecosystems in northwest Australia. Global Change Biology **22**, 2776–2786.

Ong J. J. L., Rountrey A. N., Black B. A., et al. (2018). A boundary current drives synchronous growth of marine fishes across tropical and temperate latitudes. Global Change Biology **24**, 1894–1903.

Ong J. L., Meekan M. G., Hsu H. H., Fanning L.P., and Campana S. E. (2020). Annual bands in vertebrae validated by bomb radiocarbon assays provide estimates of age and growth of whale sharks. Frontiers in Marine Science **7**, 188.

Pederson N., Varner J. M., and Palik B. J. (2008). Canopy disturbance and tree recruitment over two centuries in a managed longleaf pine landscape. Forest Ecology and Management **254**, 85–95.

Peltier D. M. P.,Barber J. J., and Ogle K. (2018). Quantifying antecedent climatic drivers of tree growth in the southwestern U.S. Journal of Ecology **106**, 613–624.

Powell T. L. Wheeler J. K., de Oliveira A. A. R., et al. (2017). Differences in xylem and leaf hydraulic traits explain differences in drought tolerance among mature Amazon rainforest trees. Global Change Biology **23**, 4280–4293.

Redmond M. D., Weisberg P. J., Cobb N. S., Gehring C. A., Whipple A. V., and Whitham T. G. (2016). A robust method to determine historical cone production among slow-growing conifers. Forest Ecology and Management **368**, 1–6.

Ricker M., Gutiérrez-García G., Juarez-Guerrero D., and Evans M. E. K. (2020). Statistical age determination of tree rings. PLOS One **15**(9), e0239052.

Rollinson C. R., Kaye M. W., and Canham C. D. (2016). Interspecific variation in growth responses to climate and competition of five eastern tree species. Ecology **97**, 1003–1011.

Roskilly B., Keeling E., Hood S., Giuggiola A., and Sala A. (2019). Conflicting functional effects of xylem pit structure relate to the growth-longevity trade-off in a conifer species. Proceedings of the National Academy of Sciences of the United States of America **116**(3), 15282–15287.

Rountrey A. N., Coulson P. G., Meeuwig J. J., and Meekan M. (2014). Water temperature and fish growth: otoliths predict growth patterns of a marine fish in a changing climate. Global Change Biology **20**, 2450–2458.

Rowland L., da Costa A. C. L., Galbraith D. R., et al. (2015). Death from drought in tropical forests is triggered by hydraulics not carbon starvation. Nature **528**, 119–122.

Rozendaal D. M. A., Brienen R. J. W., Soliz-Gamboa C. C., and Zuidema P. A. (2010). Tropical tree rings reveal preferential survival of fast-growing juveniles and increased juvenile growth rates over time. New Phytologist **185**, 759–769.

Schultz E. L. and Miller T. E. X. (in prep.). Incomplete niche-filling buffers high elevation trees against loss of suitable habitat under climate change.

Schwing F. B., Murphree T., and Green P. M. (2002). The Northern Oscillation Index (NOI): a new climate index for the northeast Pacific. Progress in Oceanography **53**, 115–139.

Serra-Maluquer X., Mencuccini M., and Martinez-Vilalta J. (2018). Changes in tree resistance, recovery and resilience across three successive droughts in the northeast Iberian peninsula. Oecologia **187**, 343–354.

Sesel M. (2017). An analysis of dental development in Pleistocene *Homo* using skeletal growth and chronological age. American Journal of Physical Anthropology **163**, 531–541.

Silva M. S., Funch L. S., and Da Silva L. B. (2019). The growth ring concept: seeking a broader and unambiguous approach covering tropical species. Biological Reviews **94**, 1161–1178.

Sinclair A. R. E. and Arcese P. (1995). Population consequences of predation-sensitive foraging: the Serengeti wildebeest. Ecology **76**, 882–891.

Smith B. H. and Tompkins R. L. (1995). Toward a life history of the Hominidae. Annual Review of Anthropology **24**, 257e279.

Smolinski S. and Mirny Z. (2017). Otolith biochronology as an indicator of marine fish responses to hydroclimatic conditions and ecosystem regime shifts. Ecological Indicators **79**, 286–294.

Somerville W. (1891). An account of Pressler's growth borer. Transactions and Proceedings of the Botanical Society of Edinburgh **19**, 90–93.

Speer J. H. (2010). The Fundamentals of Tree-Ring Research. University of Arizona Press.

Swetnam T. W. and Betancourt J. L. (1998). Mesoscale disturbance and ecological response to decadal climate variability in the American Southwest. Journal of Climate **11**, 3128–3147.

Swetnam T. W. and Lynch A. M. (1993). Multicentury, regional-scale patterns of western spruce budworm outbreaks. Ecological Monographs **63**, 399–424.

Taylor A. H., Trouet V., Skinner C. N., and Stephens S. (2016). Socioecological transitions trigger fire regime shifts and modulate fire–climate interactions in the Sierra Nevadas, USA, 1600–2015 CE. Proceedings of the National Academy of Sciences of the United States of America **113**, 13684–13689.

Temple D. H. (2014). Plasticity and constraint in response to early-life stressors among late/final Jomon period foragers from Japan: evidence for life history trade-offs from incremental microstructures of enamel. American Journal of Physical Anthropology **155**, 537–545.

Thompson R. D., Daniels L. D., and Lewis K. J. (2007). A new dendroecological method to differentiate growth responses to fine-scale disturbances from regional-scale environmental variation. Canadian Journal of Forest Research **37**, 1034–1043.

Thresher R. E. 1999. Elemental composition of otoliths as a stock delineator in fishes. Fisheries Research **43**, 165–204.

Tolwinski-Ward S. E., Evans M. N., Hughes M. K., and Anchkaitis K. J. (2011). An efficient forward model of the climate controls on interannual variation in ring width. Climate Dynamics **36**, 2419–2439.

Ulrich D. E. M., Still C., Brooks J. R., Kim Y., and Meinzer F. C. (2019). Investigating old-growth Ponderosa pine physiology using tree-rings, δ^{13}C, δ^{18}O, and a process-based model. Ecology **100**, e02656.

Vaganov E. A., Hughes M. K., and Shashkin A. V. (2006). Growth Dynamics of Conifer Tree Rings: Images of Past and Future Environments. Springer.

Van Der Sleen P. Groenendijk P., Vlam M., et al. (2015). No growth stimulation of tropical trees by 150 years of CO_2 fertilization but water-use efficiency increased. Nature Geoscience **8**, 24–28

Van Der Sleen P., Stransky C., Morrongiello J. R., Haslob H., Peharda M., and Black B. A. (2018). Otolith increments in European plaice (*Pleuronectes platessa*) reveal temperature and density-dependent effects on growth (vol 75, pg 1151, 2018). ICES Journal of Marine Science **75**, 1151.

Van Mantgem P. J., Falk D. A., Williams E. C., Das A. J., and Stephenson N. L. (2018). Pre-fire drought and competition mediate post-fire conifer mortality in western U.S. National Parks. Ecological Applications **28**(7), 1730–1739.

Van Mantgem P.J., Falk D. A., Williams E. C., Das A. J., and Stephenson N. L. (2020). The influence of pre-fire growth patterns on post-fire tree mortality for common conifers in western US parks. International Journal of Wildland Fire **29**(6), 513–518.

Vanderwel, M. C., Rozendaal D. M. A., and Evans M. E. K. (2017). Predicting the abundance of forest types across the eastern United States through inverse modelling of tree demography. Ecological Applications **27**, 2128–2141.

Villalba R. and Veblen T. T. (1998). Influences of large-scale climatic variability on episodic tree mortality in northern Patagonia. Ecology **79**, 2624–2640.

Visser M. D., Bruijning M., Wright S. J., et al. (2016). Functional traits as predictors of vital rates across the life cycle of tropical trees. Functional Ecology **30**, 168–180.

Von Arx G., Arzac A., Fonti P., et al. (2017). Responses of sapwood ray parenchyma and non-structural carbohydrates of *Pinus sylvestris* to drought and long-term irrigation. Functional Ecology **31**, 1371–1382.

Von Biela V. R., Kruse G. H., Mueter F. J., et al. (2015). Evidence of bottom-up limitations in nearshore marine systems based on otolith proxies of fish growth. Marine Biology **162**, 1019–1031.

Von Biela V. R., Zimmerman C. E., Kruse G. H., et al. (2016). Influence of basin- and local-scale environmental conditions on nearshore production in the northeast Pacific Ocean. Marine and Coastal Fisheries: Dynamics, Management, and Ecosystem Science **8**, 502–521.

Warren-Myers F. W., Dempster T., and Swearer S. E. (2018). Otolith mass marking techniques for aquaculture and restocking: benefits and limitations. Reviews in Fish Biology and Fisheries **28**, 485–501.

Weijers S., Buchwal A., Blok D., Loffler J., and Elberling B. (2017). High Arctic summer warming tracked by increased *Cassiope tetragona* growth in the world's northernmost polar desert. Global Change Biology **23**, 5006–5020.

Weisberg S., Spangler G., and Richmond L. S. (2010). Mixed effects models for fish growth. Canadian Journal of Fisheries and Aquatic Sciences **67**, 269–277.

White J. W., Nickols K. J., Malone D., et al. Fitting state-space integral projection models to size-structured time series data to estimate unknown parameters. Ecological Applications **26**, 2677–2694.

Wigley T. M. L., Briffa K. R., and Jones P. D. (1984). On the average value of correlated timeseries, with applications in dendroclimatology and hydrometeorology. Journal of Applied Meteorology and Climatology **23**, 201–213.

Willmes M., Hobbs J. A., Sturrock A. M., et al. (2018). Fishery collapse, recovery, and the cryptic decline of wild salmon on a major California river. Canadian Journal of Fisheries and Aquatic Science **75**, 1836–1848.

Wong C. M. and Lertzman K. P. (2001). Errors in estimating tree age: implications for studies of stand dynamics. Canadian Journal of Forest Research **31**, 1262–1271.

Woodward H. N., Padian K., and Lee A. H. (2013). Skeletochronology. In: K. Padian and E.-T. Lamm (eds.), Bone Histology of Fossil Tetrapods: Advancing Methods, Analysis, and Interpretation, Ch. 7. University of California Press.

Wright M., Sherriff R. L., Miller A. E., and Wilson T. (2018). Stand basal area and temperature interact to influence growth in white spruce in southwest Alaska. Ecosphere **9**, e02462

Wykoff P. H. and Clark J. S. (2002). The relationship between growth and mortality for seven co-occurring tree species in the southern Appalachian Mountains. Journal of Ecology **90**, 604–615.

Yang J., Cao M., and Swenson N. G. (2018). Why functional traits do not predict tree demographic rates. Trends in Ecology & Evolution **33**, 326–336.

Youngblood A., Max T., and Coe K. (2004). Stand structure in eastside old-growth ponderosa pine forests of Oregon and northern California. Forest Ecology & Management **199**, 191–217.

Zhao M. Klaassen C. A. J., Lisovski S., and Klaassen M. (2018). The adequacy of aging techniques in vertebrates for rapid estimation of population mortality rates from age distributions. Ecology and Evolution **9**, 1394–1402.

Zipkin E. F., Inouye B. D., and Beissinger S. R. (2019). Innovations in data integration for modeling populations. Ecology **100**, e02713.

Zuidema P. A., Brienen R. J. W., During H. J., and Guneralp B. (2009). Do persistently fast-growing juveniles contribute disproportionately to population growth? A new analysis tool for matrix models and its application to rainforest trees. The American Naturalist **174**, 709–719.

Zuidema P. A., Vlam M., and Chien P. D. (2011). Ages and long-term growth patterns of four threatened Vietnamese tree species. Trees **25**, 29–38.

Zuidema P. A., Poulter B., and Frank D. C. (2018). A wood biology agenda to support global vegetation modeling. Trends in Plant Science **23**(11), 1006–1015.

CHAPTER 5

Longitudinal demographic data collection

Marlène Gamelon, Josh A. Firth, Mathilde Le Moullec,
William K. Petry, and Roberto Salguero-Gómez

5.1 Introduction: long-term field studies

Demographic data can be collected during a single visit, or multiple visits, to the population. For instance, a single visit to the field might be in principle enough to collect data on tree rings or fish otoliths, which can then be used to estimate key demographic parameters retrospectively (see Chapter 4). Alternatively, longitudinal studies involve numerous visits to the study population with repeated observations/measurements; this kind of approach typically occurs over multiple weeks, months, years, or even decades, depending on the generation time of the study species, research aims, and available support for the research programme. The collected longitudinal data are then used to estimate demographic parameters such as annual population abundances (see Chapter 7) and/or survival, growth, and reproductive rates (see Chapter 13).

Many long-term studies are running worldwide. In Antarctica, South America, and Central America, the monitoring of avian and mammalian populations has already reached over 50 years in some areas (see Taig-Johnston et al. 2017 for a review). In the northern hemisphere, some of the longest studies even reach 70 years (Clutton-Brock and Sheldon 2010). Even if there has been a strong focus on birds and mammals (Festa-Bianchet 2017; Kappeler et al. 2017 for some reviews on mammals; Smith et al. 2017; see e.g. Marshall et al. 2018), long-term studies also exist for other taxa within the animal kingdom (e.g. corals, see Connell et al. 1997; amphibians, see Cayuela et al. 2020). Long-term field studies are not restricted to animals: long-term plant studies span over 20 years in regions of North America (e.g. Ellis et al. 2012),

Central America (e.g. Condit et al. 2017), and Europe (e.g. Hutchings 2010). Longitudinal data have been, and continue to, be collected on many taxa across the Tree of Life and have provided significant insights in ecology, evolution, and demography (Clutton-Brock and Sheldon 2010; Reinke et al. 2019). Importantly, demographic data can be collected at the population level (e.g. time series of population counts) or at the individual level (e.g. monitoring of marked and/or geo-referenced individuals throughout their life).

From demographic data collected at the population level (i.e. time series of population counts) and with appropriate methodological tools to analyse them (see the next chapters), several questions can be addressed. For instance, 'What is the population size trend?', 'What is the population spatial distribution?', 'What are the effects of changes in climate on population size/distribution?', and 'What are the effects of human activities on population size/distribution?'. Finer-scale demographic data can also be recorded. For instance, population size time series can be collected with disaggregation into stage, sex, or morph (e.g. counts of adult females). This information is not recorded at the individual level, but still, it is more useful than total population counts. This better resolution can help in tackling some of the questions outlined below with individual-level data.

While long-term data collected at the population level provide significant insights into the population's dynamics, they usually do not allow a full understanding of the underlying demographic mechanisms causing changes in population distribution and abundance (Clutton-Brock and Sheldon 2010). Indeed,

Marlène Gamelon et al., *Longitudinal demographic data collection*. In: *Demographic Methods Across the Tree of Life*.
Edited by Roberto Salguero-Gómez and Marlène Gamelon, Oxford University Press.
© Oxford University Press (2021). DOI: 10.1093/oso/9780198838609.003.0005

changes in population abundance over time can result from changes in rates of births (reproduction), deaths (survival), emigration, and/or immigration. That lack of mechanistic resolution of population-level data is precisely the major limitation of longitudinal studies at the population rather than individual level.

For instance, climate-induced changes in population abundance can result from the sensitivity of a particular stage class (e.g. a particular age) to climate conditions that will exhibit lower survival or fecundity. In the well-studied Soay sheep (*Ovis aries*) population, Coulson et al. (2001) showed that juvenile survival was particularly sensitive to the North Atlantic Oscillation, a weather phenomenon associated with temperature and rainfall. Similarly, within conifer forests, recruitment and growth were associated with fire (Tepley et al. 2017). Interestingly, going beyond population abundance and exploring the effect of climate and human activities on all vital rates and at all stages can also allow one to detect reduced survival and/or fecundity even if population abundance apparently remains stable. For instance, in an eagle owl (*Bubo bubo*) population, adult survival has dramatically decreased because of electrocution, despite constant population abundance, the decline in survival being balanced through massive immigration (Schaub et al. 2010), which could only be detected because the data were collected at the individual level.

From demographic data collected at the individual level, it is possible to accurately identify the proximate causes of changes in population size/distribution and the underlying demographic mechanisms. Thus, questions such as 'What are the effects of changes in climate on survival/fecundity/immigration/emigration rates?', 'What are the effects of human activities on survival/fecundity/immigration/emigrationrates?', 'What are the stage classes the most influenced by changes in climate?', and 'Which stage classes are most influenced by human activities?' can be tackled. Identifying the stages and their vital rates (e.g. survival, fecundity) the most affected by climate changes or human activities is crucial in conservation and management to develop appropriate targeted strategies (see Frederiksen et al. 2014 for a review).

In that respect, longitudinal studies at the individual level are powerful. But as with population-level studies, they suffer from limitations. Indeed, implementing individual-based monitoring is generally costly in terms of money and human resources, oftentimes requiring expensive materials (e.g. GPS collars) and experienced fieldworkers during long periods of time. Moreover, data collected might be scarce due to low sample sizes and because researchers are reluctant to individually monitor individuals in vulnerable populations. In those cases, it might be advisable to take advantage of both population-level and individual-level data, through, for example, an integrated modelling approach (see Chapter 14).

In this chapter, first we introduce the reader to procedures that can be implemented in the field to collect specific demographic data on mobile species (e.g. birds, mammals), at both the population and individual levels. Second, we present the procedures and the type of demographic data that can be collected on sessile species (e.g. corals, plants) at both levels.

5.2 Collection of longitudinal demographic data on mobile species

5.2.1 Procedures and type of data collected at the population level

The problem of imperfect detection while monitoring populations

Within a mobile species' range, the population dynamics can be rather complex. A fragmented landscape means that individuals from the same population can occupy diverse habitat types in space and time. Determining population range size or abundance therein comes with challenges related to two key aspects: (1) spatial variability within the studied area and (2) individual detectability (Yoccoz et al. 2001). Important errors in quantification of population size and emergent dynamics can be made if individuals are missed when they are present at the site, or misidentified, or when the same individual is counted multiple times (Miller et al. 2011). Imperfect detection depends on species characteristics (e.g. camouflage), spatial and temporal variation (e.g. migration), and survey characteristics (Cressie et al. 2009; Guillera-Arroita 2017) (Figure 5.1). Uncertainties in the survey characteristics depend on the type of monitoring, the study design, and the analysis (Figure 5.1). If the aim of the monitoring is to estimate population abundance, the higher the number of repeated measurements and sample size, the higher the precision will be. Accurate estimates (i.e. precise and unbiased; Williams et al. 2002), repeated over time, enable the inference of the drivers of population dynamics, such as climate and harvest regimes (Thompson et al. 1998; Miller et al. 2011; Guillera-Arroita 2017) (Figure 5.1). Under- or

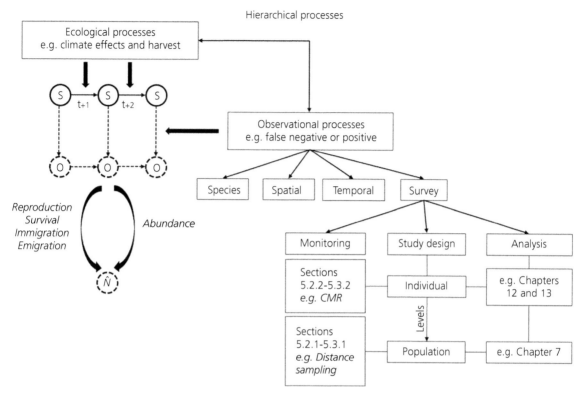

Figure 5.1 Hierarchical organisation of imperfect detection in wildlife systems, inspired from Royle and Dorazio (2008), Zuur et al. (2009), Kéry and Schaub (2012), and Guillera-Arroita (2017). The true abundance is a latent state (*S*) changing at each time-step, for example year to year (from year *t*+1 to *t*+2), due to changes in vital rates (survival, reproduction, immigration, emigration) affected by ecological processes. At each time-step, monitoring of vital rates or population abundance results in observations (*O*), such as the observed abundance ($\hat{N}$). Each of these observations is subject to observational processes, which can be decomposed into distinct components. Population survey is one of the components and consists in applying the required monitoring in the field, using the appropriate study design (at the population or individual level) and then using appropriate analyses to make inferences (modified from Le Moullec (2019)).

over-estimation of abundances or range sizes can have important consequences; for instance, they can lead to ineffective management and incorrect conservation decisions (Sinclair et al. 2006).

A large diversity of methods for monitoring populations

Monitoring populations over years to determine range size and/or abundance has been central in the field of ecology and wildlife management since the 1930s (Krebs 1998; Williams et al. 2002; Sutherland et al. 2013), and a large diversity of methods has emerged since (Seber 1992; Thompson et al. 1998; Sutherland 2006; Morellet et al. 2011; ENETWILD Consortium et al. 2020). The monitoring method chosen by a manager/scientist, the 'tools' (e.g. camera traps, permanent plot) used to increase the overall detectability of

species, as well as the spatial coverage vary according to the species characteristics, including its rarity and detectability in the environment (Figure 5.2). Thereafter, data can be collected with multiple monitoring methods and sampling designs (see Figure 5.2): for instance, camera traps located at random or stratified designs can be used with capture–recapture or distance sampling. Several of the listed monitoring approaches here are also applicable to sessile species (see section 5.3.1).

Estimating population abundance can be performed on an absolute or relative scale with direct or indirect population counts. Absolute abundance methodologies aim to estimate the true population size, that is, the state variable, while relative abundance methodologies (e.g. capture rate, hunting records, sign detection rate, activity indices) estimate the population size relative

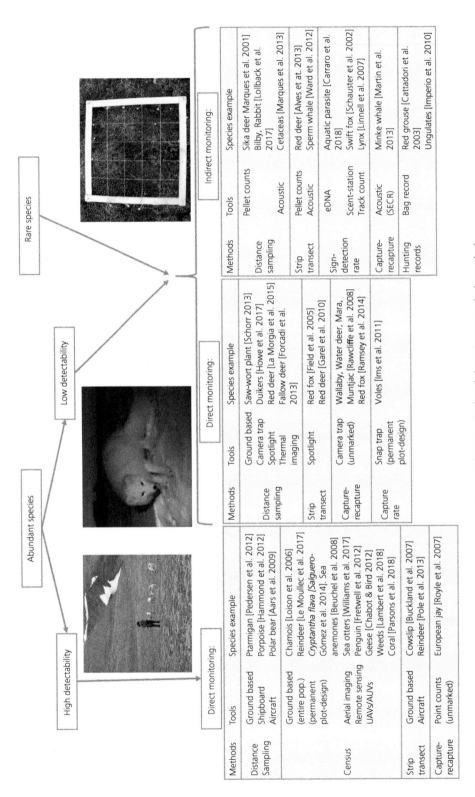

Figure 5.2 Examples of monitoring methods for the estimation of population abundance in animals and plants. A same monitoring method can be conducted with the help of different tools to increase species detectability and spatial coverage. UAVs = unmanned aerial vehicles; AUVs = autonomous underwater vehicles; SECR = spatially explicit capture–recapture (unmarked individuals); eDNA = environmental DNA (credit: A. Tholfsen / M. Le Moullec).

to a previous survey (Pollock et al. 2002; Hopkins and Kennedy 2004; O'Brien 2011; Amos et al. 2014). Importantly, the latter approach assumes a constant detection probability and thus requires constant survey design, personnel training, species behaviour, habitat use, and range size between survey events. Under such conditions, the trends in abundance can inform managers' decisions. However, relative estimates of population size cannot be compared across systems (Amos et al. 2014).

Abundance estimates from direct individual counts within a population require visual detection of the organism (i.e. total counts, camera surveys). These methodologies are often preferred (Morellet et al. 2011) and are adapted for abundant, tame, and easily detectable species, since they are 'data hungry' for robust modelling and often logistically costly (Yoccoz et al. 2001; Guillera-Arroita 2017) (Figure 5.2). Consequently, these approaches are often restricted to small-scale studies (but see Yuan et al. 2017). Monitoring methods for rare or hard-to-detect species, such as cryptic, nocturnal, forest, or aquatic species, require the use of specific tools to increase detections and spatial coverage (e.g. spotlight, thermal imaging, acoustic recorders) (Marques et al. 2013). For instance, to monitor marine species, unmanned underwater vehicles (UUVs) can be coupled with postproduction computing algorithms to obtain the relevant demographic information (Smale et al. 2012). Notably, the large diversity of existing methods to monitor populations in terrestrial systems is still valid in marine environments (see Katsanevakis et al. 2012 for a review). In some cases, indirect population counts based on signs detection (e.g. fecal pellet counts, track counts, or environmental DNA) can be better suited to the task (Pollock et al. 2002; Thompson 2004; MacKenzie et al. 2005; Jones 2011).

The methodology chosen to study a population should preferably be associated with a measure of detection probability that is inherent to certain methodologies, such as distance sampling (Buckland et al. 2007a), or capture–mark–recapture (Yoccoz et al. 2001). Capture–mark–recapture is further developed in the next section on individual-based methods, yet inferences on population abundance can be drawn from this methodology too. More generally, information on organism resightings (marked or unmarked) (Royle et al. 2013) is often combined with monitoring methodologies to access information on detection probability, that is total counts (Le Moullec et al. 2017), and camera traps (Karanth and Nichols 1998).

Methods for spatially referenced observations

Spatial referencing of individual observations (i.e. coordinates, spatial unit) within a population enables researchers to relate the frequency of detections of individuals to the surrounding conditions, that is spatially referenced environmental covariates (Aarts et al. 2012; Guillera-Arroita 2017). Thereafter, within the modelled spatial distribution of the species, population density can be predicted in areas or periods of time not surveyed, as long as extrapolations are done within the range of data monitored (Sillett et al. 2012). Random sampling of individuals across the study region is fundamental to unbiased design-based studies. However, model-based studies grant access to a large diversity of study designs and opportunistic count data collection. Aarts et al. (2012) demonstrated the similarities in spatial density estimates between count data collected in discrete space (i.e. number of observations per spatial unit) or in continuous space (e.g. use-availability, where each individual is treated as one observation) and presence–absence data (if the spatial unit corresponds to single observations). Hence, data to model the spatial density can be diverse (Baddeley et al. 2006; Zuur et al. 2009; Royle et al. 2013; Ramsey et al. 2015; Williams et al. 2017; Antún et al. 2018; Le Moullec et al. 2019). Spatiotemporal variations in detection probability are often accounted for, prior to analysing the spatial density function (i.e. two-stage approach), but in some cases the detection and density function are estimated simultaneously (i.e. one-stage approach) (Royle et al. 2013).

5.2.2 Individual-level long-term field studies: concepts, methods, and data

Methods that allow individuals within a population to be uniquely recognised have given rise to a plethora of long-term studies of wild animal systems that have been formative in our understanding of their ecology, evolution, and conservation biology. The ability to individually recognise animals has not only provided mechanistic insights into the drivers of the patterns often considered by long-term population-level studies (such as the movements between, and age structures within, populations, and the causes of changes in fitness or selection), but also proven foundational in understanding social structure, individual-level fitness, and the links between different life history stages and generations (Clutton-Brock and Sheldon 2010). This section explores current and developing procedures used to collect data at the individual level in these long-term studies, the types of data these approaches provide, and the new implications that recent advances are allowing.

Individual-level monitoring using physical captures

The fundamental premise of long-term individual-based field studies in animal populations of mobile species has long been, and continues to be, the individual-level 'capture, mark, recapture' (CMR) procedure (see Figure 5.1); an individual is 'captured' (physically or simply recorded), 'marked' in a uniquely identifiable way (through active intervention or through documenting individually recognisable characteristics, such as marks on the tails of whales; Pomilla et al. 2014), and 'recaptured' (or 'resighted') at a later time (i.e. re-recording the unique identity of that individual). The particular protocols employed for each of these stages of the individual-level CMR procedure are specific to the species under consideration and the practicalities of fieldwork. However, following this standard's underlying logic, these procedures are constantly refined as new techniques and technologies become available and applicable to these valuable long-term study systems.

One of the best examples of the progress of specific protocols comes from the earliest of the long-term individual-based field studies: wild birds, particularly tits (great tits, *Parus major*; and blue tits, *Cyanistes caeruleus*), which started as long-term study populations across Europe in the 1940–1950s (Kluijver 1951; Lack 1966). These species readily breed in nest boxes during the spring, so individual CMR protocols can quickly be developed. They are based on physically capturing individuals as breeding adults or nestling chicks, marking them with a unique identifier (a metal leg ring with a unique code), and recapturing and identifying them in following breeding seasons. This standard procedure was widely and rapidly adopted across a range of systems (not limited to animals—see below) and remains a common method across various long-term population bird studies (Grant 1986; Nisbet 1989; O'Connor 1991; Perrins et al. 1991).

Advancing longitudinal methods: progress in individual-level monitoring

Although effective, the nest box ringing procedure meant that these early individual CMR studies within these systems were limited to monitoring individuals at specific locations (e.g. nest boxes) and restricted to particular time-frames (i.e. during breeding). As such, further techniques were soon developed to expand

beyond these restrictions, either using mist-netting to capture, mark, or recapture these birds outside of the breeding period (such as during winter foraging; Perrins et al. 1991) or using colour bands (displaying unique combinations of rings of colours) to allow non-physical 'recapture' via resightings (Ekman 1989).

The aforementioned initial developments allowed for individual monitoring to be less spatiotemporally restricted than previous protocols. However, the advances with the largest potential for these particular systems (and most of the individual-based long-term study populations generally) have come from recent technological developments in automated animal tracking systems (Bridge et al. 2011; Kays et al. 2015; Jønsson et al. 2016). For instance, various long-term study populations of tits now tag captured individuals with passive integrated transponders (or PIT tags) that are either contained within plastic leg rings or injected subcutaneously. These PIT tags contain a microchip with a unique identification code that can be read, and automatically recorded (i.e. allowing 'resightings'), by radio-frequency identification (RFID) stations. The stations can be placed at nest boxes during the breeding period (Firth and Sheldon 2015; Schlicht et al. 2015; Firth et al. 2018) or at feeding stations to allow large-scale resighting during the nonbreeding season (Firth and Sheldon 2016).

More generally, RFID technology is emerging as a particularly good example of applying tracking technologies within long-term individual-based study populations. This is because RFID is well suited to these systems due to the size of the tags (often < 1 g), their relative affordability, and the lifelong readability of the microchips which do not require an internal battery (as they are passive). As such, marking individuals with unique PIT microchips is now a prevalent and staple method of individual CMR systems, not just for long-term bird studies (Bonter and Bridge 2011) but also across various populations, ranging from insects to fish to mammals (Gibbons and Andrews 2004; Rehmeier et al. 2006; Silcox et al. 2011). Remarkably, these new technologies tend to be less invasive. While toe clipping for small mammals has been the rule for a long time in CMR protocols, these new technologies (e.g. camera trapping and noninvasive DNA; see also Chapter 1), now favour noninvasive censusing.

The development of a diverse array of exemplary methods for individual-based monitoring of animals comes from long-term studies of mammalian systems (Clutton-Brock and Sheldon 2010; Hayes and Schradin 2017; Schradin and Hayes 2017). Indeed, some of the longest and most substantial individual-based study populations are mammals, partly due to the ease of 'resighting' these animals, especially for island populations of large mammals where immigration/emigration do not exist, and habituation is rapidly achievable due to the lack of predation. For example, long-term studies such as those on red-deer (*Cervus elaphus*), which began in 1971 on the Isle of Rum, Scotland (Clutton-Brock et al. 1982), and Soay sheep, which began in 1985 on St Kilda, Scotland (Clutton-Brock and Pemberton 2004), have been successful in consistently employing standardised protocols based on capturing and uniquely marking individuals shortly after birth, and then resighting these individuals via regular censuses of the study populations thereafter. Mammalian systems in less-convenient settings have benefitted substantially from recent advances in monitoring procedures (Noonan et al. 2015; Hays et al. 2016; Nowacek et al. 2016). For example, bat species hold many of the same conveniences that long-term bird systems allow (i.e. occupying researcher-made boxes, as well as ease of capture in mist nets). As such, these systems were also originally based on monitoring using individually coded rings but are now converging on the same technological approaches found to be useful for monitoring bird systems, for example using PIT tags and RFID technology (Fleischmann et al. 2013; Law 2018) that allow automated resightings instead of relying on physical captures. Similarly, many mammal systems have benefitted from the miniaturisation of GPS technologies providing high precision and constant monitoring (Tomkiewicz et al. 2010; McMahon et al. 2017).

Long-term studies based on species that spend large proportions of their time underground (therefore restricting the use of GPS) have had to consider other approaches. For example, the UK European badger *Meles meles* project, which started over 30 years ago (Macdonald and Newman 2002), was traditionally restricted to overground monitoring or capture/resighting procedures. Now, magneto-inductive tracking techniques are available which allow automated, continuous, fine-scale monitoring of individuals whilst underground (Noonan et al. 2015). These techniques hold much potential for other long-term studies for other ground-dwelling species (Schradin and Hayes 2017).

One of the most important potential applications of technology for longitudinal studies of individual animals in relatively inconvenient settings may well be for studies of marine mammals (Hazen et al. 2012; Mann and Karniski 2017). Here, long-term monitoring has long been based on identification of individuals in photos gathered from opportunistic sightings or transect surveys (Eguchi 2014; Urian et al. 2015), which makes individual monitoring often very difficult (Mann and Karniski 2017). Advances in a range of tracking technologies, from drone surveys to newly developed animal-borne tags (Hussey et al. 2015; Kays et al. 2015; Nowacek et al. 2016; Hays et al. 2019), will provide ripe opportunity to rapidly advance the monitoring of these systems, by allowing individuals to be monitored over their entire lifetimes.

The new technological monitoring methods also provide the potential for continuously tracking movements at fine scales (e.g. within resident territorial species) as well as over larger distances for migratory species, and even nomadic species moving in irregular manners (Teitelbaum and Mueller 2019). Further, it is becoming possible to integrate technology providing detailed physiological monitoring of individual states, activity, and metabolism into tracking devices (see Chapter 2). Such information is particularly valuable within long-term study systems for which the basic underlying ecology is already well researched, and this now provides the potential to allow vast advances in knowledge surrounding the causes and consequences of individual states (Hayes and Schradin 2017; Schradin and Hayes 2017). Yet, one of the most important additional advantages of large-scale automated methods is that these new approaches allow for simultaneous monitoring of all individuals over large spatiotemporal scales, thus providing unprecedented insight into the social structure of natural populations (Krause et al. 2013; Firth and Sheldon 2016) (see Chapter 3).

5.3 Collection of longitudinal demographic data on sessile species

Sessile species are those that lack the ability of (self-)locomotion. From a life cycle perspective, sessile species typically correspond to species where established life cycle stages are anchored onto a substrate. Naturally, plants adhere to this definition, though some remarkable exceptions in the Planta kingdom do exist, such as the polytomic group of tumbleweeds (e.g. genus *Kali, Amarnathus albus, Salsola* spp.) and the Rose of Jericho (*Anastatica hierochuntica*). However, sessility is not unique to plants. Indeed, entire taxonomic groups of the animal kingdom, such as corals and sponges, as well as other species of important economic value (e.g. barnacles), are sessile. In addition, fungi and a significant percentage of bacteria (e.g. Geesey et al. 1978) have sessile modes of life.

The monitoring of sessile species has some advantages over that of mobile species, in that mortality in sessile species cannot be confounded with the individual not being at its previous location due to mobility. As such, the classical equation modelling the changes in population size (N) over time as a function of the rates of births, deaths, emigration, and immigration can be reduced to just two terms, births and deaths, assuming that dispersal (as well as mobility) does not occur. However, modelling the demography of sessile species has its own specific challenges. Indeed, sessile species, because they are 'stuck' in place for most of their life cycles, have evolved strategies to cope with the local environmental conditions (Huey et al. 2002; Svensson and Marshall 2015; Žádníková et al. 2015). It is precisely the strategies of some of those sessile species that causes new challenges for the monitoring of natural populations in sessile species. These strategies include clonality, dispersal, propagule dormancy, vegetative dormancy, creeping, and mimicry, among others, and are discussed below.

5.3.1 Procedures and type of data collected at the population level

Many of the field approaches described above to quantify and estimate the population size, distribution, and structure of mobile species are also relevant to sessile species populations (Figure 5.2). However, field techniques to monitor sessile populations tend to capitalise more on their immobility. Consequently, permanent plots, quadrats, and transects are the standard methods used in sessile longitudinal demographic studies. Nonetheless, technological advances are also reshaping how natural populations of sessile species (e.g. plants and otherwise) are monitored.

The increase in satellite data resolution now means that counts of individuals are achievable in ecosystems where density is low, individuals are distributed at random or quasi-random, and where there is little three-dimensional layering (Bai et al. 2011). Desert and Mediterranean ecosystems, for instance, are ideal candidate systems where these technologies show more promise (Peters and Eve 1995). In contrast, monitoring the population dynamics of sessile species in tropical and marine ecosystems remains more challenging. This is because, in these cases, populations of sessile species tend to have complex elevational structures, where only adults who have broken through the tight canopy (or dense marine reef community) are readily observable with satellites, and thus for which little information is directly available for other stages.

Novel light detection and ranging (LIDAR) technologies can complement data acquisition through their ability to operate in high-density stands (Malhi et al. 2018). In this regard, the application of the approaches discussed regarding the detectability of different life cycle stages in mobile species is just as equally useful here (Figure 5.1). In addition, the development of low-elevation aerial technology holds great promise, though the challenge remains in how to navigate this technology in landscapes full of handicaps, such as rocks, wind/water currents, tree trunks, lianas/algae, and small stems/debris that may not be adequately identified by navigating devices equipped with smart technology (but see https://www.ox.ac.uk/research/research-impact/poetry-motion). Some key applications of drone (unmanned aerial vehicle, UAV) technologies have been targeted specifically for agricultural use (Saha et al. 2018; Jiménez López and Mulero-Pázmány 2019). These approaches, coupled with artificial intelligence to automatically distinguish and track individuals (Ampatzidis et al. 2019), represent a unique opportunity to quantify population number and even structure in sessile organisms (Figure 5.3). It is worth mentioning here that the power of these novel technologies cannot be harnessed efficiently without taking into account the pertinent advances in image analysis (e.g. Maillard et al. 2010; Bruijning et al. 2018). The monitoring of sessile populations below water is faced with important challenges due to obvious logistical considerations, such as lack of oxygen, strong currents, and high viscosity. But novel developments from physics now also allow for the treatment of seawater waves as magnifying glasses to evaluate coral reef properties at high resolution using UAVs (Chirayath and Earle 2016).

5.3.2 Procedures and type of data collected at the individual level

While age estimation methods do exist for sessile species (Chapter 4), age estimation in species such as plants, corals, or sponges in a way that is not destructive—and thus compatible with long-term censuses—is challenging. For that reason, among others related to convenience (Ebert 1998), the tracking of individuals in sessile species' populations, as well as their modelling (e.g. Chapters 9 and 10), tend to include information on not only survival and reproduction but also changes in individual size (rather than age) through time. In fact, size is the most widely used predictor of fitness components in the demography of sessile species (Caswell 2001; Caswell and Salguero-Gómez 2013).

Size in the field can be measured according to different morphometric variables, including but not limited to the diameter at breast height (e.g. in trees), basal diameter (e.g. in shrubs), area (e.g. in corals and sponges), degree of modularity/architecture, and number of stems, and so on (Salguero-Gómez et al. 2015). When deciding which morphometric variable to collect data for, it is key to consider various factors, such as (1) *a priori* knowledge on the biology of the species and its life cycle and (2) the traits that most

inform on the vitality of the individual (e.g. height may be an important trait in light-limiting terrestrial habitats, while area/volume ratio is an important one when considering exposure to the environment). The state predictor can be based on ontogeny, rather than different dimensions of size. Classical categories of ontogeny have been developed for plants (Gatsuk et al. 1980), though these ignore non-progressive life cycles. It is recommended that, if *a priori* knowledge regarding the life cycle of the species does not exist, researchers collect multiple axes of information regarding the structure and ontogeny of the individuals in the first few field seasons and quickly construct and compare vital

(a)
(b)
(c)
(d)

Figure 5.3 Examples of monitoring methods for the estimation of population abundance, density, structure, and vital rates in sessile organisms. (a) Permanent plots can be used to quantify and track population dynamics of organisms fixed on a substrate (credit R. Salguero-Gómez). (b) A pantograph allows for the spatially explicit depiction of population (and community) structures in a compact, reliable, and fast way (credit P. Adler). (c) Quadrats and transects to track populations/individuals can be used on both terrestrial and aquatic systems (credit M. Beger). (d) UAVs, coupled with artificial intelligence algorithms, are starting to allow for the accurate, fast, and convenient measurement of population dynamics of sessile organisms. Photos show the visible range (left) vs. hyperspectral range (right) of a photo containing a parasite (blue square), a mistletoe (credit R. Salguero-Gómez).

rate models to evaluate the best predictors. This information can then help guide a more efficient, fine-tuned (and less-laborious) field monitoring protocol.

The kind of data that need to be collected to examine the population dynamics of sessile species often requires marking or mapping individuals in a way that allows researchers to relocate them in the next site visit. In terrestrial systems, marking/mapping can be done with sticks, coffee stirrers, pin flags, nails, or tags attached to each individual. Each marker needs an unequivocally distinct ID—this can be done through marking each device with pen or by producing tags that have a unique ID system. Another key attribute of these IDs is that they need to 'survive' themselves to the next census. Often, and depending on the climate at the study location and the kind of material (and presence of herbivores and people!), tags can go missing, and the researcher must update missing IDs with new tags based on a 'best guess' system supported by the GPS coordinates of the surrounding individuals whenever possible. It is recommended that, when analysing the field data (e.g. Chapters 9 and 10), the models should be updated with perfect and retagged individuals to evaluate potential sensitivities of outputs to field assumptions on ID assignment. Nonetheless, since markers can go missing frequently, photography of the study plots for every census is strongly recommended. This photographic evidence can become invaluable when trying to figure out the location of not only individuals but also entire permanent plots located in dynamic ecosystems, such as dunes and marine environments.

There are pros and cons of using tags of different materials: plastic and wooden ones can rapidly degrade, whereas aluminium tags tend to attract the attention of herbivores and vandalism. An additional approach consists of marking the corners of quadrats with metal bars that can be pounded in the soil/rock and whose positions can be relocated every visit with the help of metal detectors. Once the positions of the quadrat (or transect) have been located, the Cartesian coordinates of the individuals should allow researchers to relocate established individuals and identify new recruits. However, it must be noted that sessile individuals, contrary to common belief, do move, if only a little bit. Another important consideration is that metal markers can leak materials into the soil and also affect the temperature of its microclimate (Nassar et al. 2018). Individuals of species' populations found in high-density, overlapping statures and/or small sizes can prove particularly challenging to mark and relocate. In these cases, we suggest marking/mapping the

locations of individuals in a subset of permanent plots and trying to mark them again a week later using different methods, and then comparing the accuracy and feasibility of each. Certain GPS systems that offer resolution within centimetres are of particular interest here too (Lee and Ge 2006).

Certain sessile organisms with cryptic life stages can pose interesting challenges to monitor the dynamics of their populations. A feasible approach here involves the careful exploration of such stages on the permanent plot in a nondestructive way. For instance, some species undergo dormancy: the aboveground biomass is completely lacking, but organisms are alive belowground. Dormancy comes in the shape of propagule (e.g. seed) dormancy, vegetative (e.g. established individual) dormancy, and roots. In this realm, nondestructive approaches involve carefully excavating roots or bulbs to determine if the individual is alive (Bierzychudek 1982). An alternative approach here is to make the assumption that if a number of years have elapsed since the last time that the individual was observed aboveground, this individual is declared dead. In this context, prior knowledge about age-based and environmentally driven mortality schedules can be crucial. However, recent Bayesian statistical frameworks inspired in capture–recapture methods can be applied to cope with this uncertainty (Colchero et al. 2012; Paniw et al. 2017).

Clonality is an important challenge in demographic studies. Indeed, the incorporation (or not) of explicit clonal architecture in demographic studies has been a point of contention since the very inception of modern plant population ecology (Harper 1981). The considerations here are important because, although different segments (i.e. ramets) of a clonal individual (i.e. genet) can have a great deal of independence (Hutchings and Price 1993; Vuorisalo and Hutchings 1996), and can even compete against each other (Price et al. 1996), considering them all as separate individuals or parts of the same one will affect how the individual is tagged, how the models are defined and constructed, and ultimately the outputs from such demographic analyses (Janovský et al. 2017; Salguero-Gómez 2018). For instance, data from the same permanent plots where the functional unit *individual* has been established ignoring clonal links (i.e. the researcher tags seemingly independent units that are in fact part of the same genet as separate individuals) will inevitably result in the estimation of higher generation times and more variable population growth rates than if done on the basis of genetics (Janovský et al. 2017).

5.4 The future of long-term studies: new aspects, current biases, and arising challenges

Long-term studies clearly hold a large range of benefits compared to a single visit to the field or compared to short-term field studies. They play a crucial role in our understanding of the drivers of population dynamics and in the assessment of their demographic status (e.g. decline, remain stable, increase in size over time). As such, they are important to inform policy (see Chapter 19) and to answer societally relevant questions (Hughes et al. 2017). It is also needed to consider the potential biases, the aspects still requiring advances, and the potential arising challenges of these systems. For instance, in regard to biases in the systems currently under long-term individual-based monitoring, it is notable that species that are relatively easy to study using traditional CMR techniques are still heavily favoured over ones that it may be difficult to apply these techniques to. Nevertheless, in the dawn of new technologies providing novel avenues of monitoring, it could now well be the case that other species are particularly well suited to new methods despite been previously difficult to monitor with standard approaches. As such, diversifying long-term individual-based studies across the Tree of Life may shed new light on ecology from the viewpoint of currently understudied species and allow previously unrealised potential to be recognised.

One of the most lasting and widely acknowledged challenges of long-term study systems is their 'mismatch' with the modern scientific funding structures that often work in short-term research grants designated to proposals aimed at pursuing set hypotheses over relatively short periods of time (often 2–5 years) (Clutton-Brock and Sheldon 2010). This fact, combined with the growing 'publish-or-perish' climate across biology and most other scientific disciplines, means that short-term output is often largely favoured over longer-term goals—which can, nonetheless, be better representative of the ecology and evolution of the studied population. As such, the success of individual-based long-term study systems often relies on the continuation of short-term studies taking place within their larger framework (Schradin and Hayes 2017). Thus, it is now important that technological advances be employed to address this challenge, rather than magnify it.

Continuously applying new methods to long-term study systems may allow for their value to be constantly recognised. There are already numerous examples of successfully applying new monitoring technologies for short substudies within long-term systems (rather than integrating within the whole standard protocol) to address specific hypotheses in new ways, for example using GPS to examine foraging in the long-term Kalahari meerkat (*Suricata suricatta*) project (Gall and Manser 2018). Further, the development of technologies that allow automated manipulation (rather than just monitoring) can provide a platform to carry out individual-level experiments within long-term study systems, for instance applying automated experimental treatments to individuals based on their unique RFID codes in the Wytham tit project (Firth and Sheldon 2015). Clearly, updating long-running systems with the ever-developing new approaches provides many possibilities to acquire continued research funding for these crucially important study systems.

Finally, a more contemporary challenge arising in the new age of monitoring long-term studies is how to most efficiently deal with the vast amounts of data arising from these systems. Automatically recording information for many individuals within these populations over large periods of time at fine-scale resolutions is the dream of any population ecologist, resulting in the production of millions (Firth and Sheldon 2016) or even billions (Noonan et al. 2015) of data points. The analytical techniques developed to handle this scale of data, ranging from simply dealing with the raw records (e.g. using artificial intelligence and machine-learning algorithms) to drawing biologically relevant conclusions from the patterns within them (e.g. through comparisons to simulated null models), are arguably just as important advances as those in the monitoring technologies themselves (Krause et al. 2013; Kays et al. 2015).

Vast amounts of data are also produced by citizen science. Thanks to the development of the Internet, computational techniques and statistical tools, computers, and accessible interfaces now allow people passionate about wildlife to contribute to long-term field studies. Thus, beyond biologists, research projects are engaging millions of individuals worldwide in collecting data (see e.g. https://www.ukbms.org for the UK Butterfly Monitoring Scheme). The number of citizen science projects has exploded since 2000, and volunteers gathering data have already contributed to deliver significant insights into the ecological effects of climate change (e.g. Bonney et al. 2014; Isaac et al. 2014). Even more recent, modern computational science techniques are opening up the opportunity for 'iEcology' which aims to monitor populations through passively generated internet data, such as using millions of Google search users' activity to infer bird species' occurrences or analysing Wikipedia users' locations to assess fish migration patterns (Jaric et al. 2020).

Thus, it is essential that biologists/ecologists continue to 'keep pace' with these data analytical techniques at the same rate that they adopt new data collection methods and new approaches to provide usable data storage on large scales, which have been developed for various long-term studies (Marshall et al. 2018). There is important incentive for ecologists to work more closely with engineers, computer scientists, and data scientists. Cross- and interdisciplinary work offers promising avenues for future developments in algorithms to process big data. Once the data have been gathered, processed, and stored, the question of access to these data is a final and important challenge to consider (Mills et al. 2015; Whitlock et al. 2016). Whether such databases should be open-access and available for usage by all or whether certain restrictions should be applied in an attempt to favour the continuation of these systems is currently under much debate. Although such consideration is obviously very useful, the rapidly growing trend towards unrestricted and immediate access to data across biology (Farnham et al. 2017; Culina et al. 2018; Sarabipour et al. 2019) suggests that open access to such data is not only valuable but inevitable, and emphasis should now be placed on establishing approaches to recognise and maximise the benefits of this for individual-based long-term studies.

Acknowledgements

We acknowledge financial support from the Research Council of Norway through its Centres of Excellence funding scheme (project number 223257).

References

Aars, J., T. A. Marques, S. T. Buckland, et al. 2009. Estimating the Barents Sea polar bear subpopulation size. Mar. Mammal. Sci. **25**(1): 35–52.

Aarts, G., J. Fieberg, and J. Matthiopoulos. 2012. Comparative interpretation of count, presence–absence and point methods for species distribution models. Methods Ecol. Evol. **3**: 177–187.

Alves J., A. Alves Da Silva, A. M. V. M. Soares, and C. Fonseca. 2013. Pellet group count methods to estimate red deer densities: precision, potential accuracy and efficiency. Mamm. Biol. **78**: 134–141.

Amos, M., G. Baxter, N. Finch, A. Lisle, and P. Murray. 2014. I just want to count them! Considerations when choosing a deer population monitoring method. Wildl. Biol. **20**: 362–370.

Ampatzidis, Y., V. Partel, B. Meyering, and U. Albrecht. 2019. Citrus rootstock evaluation utilizing UAV-based remote sensing and artificial intelligence. Comput. Electron. Agric. **164**: 104900.

Antún, M., R. Baldi, L. M. Bandieri, and R. L. D. Agostino. 2018. Analysis of the spatial variation in the abundance of lesser rheas using density surface models. Wildl. Res. **45**: 47–54.

Baddeley, A., P. Gregori, J. Mateu, R. Stoica, and D. Stoyan (eds). 2006. Case Studies in Spatial Point Process Modeling. Springer, New York.

Bai, J., J. Li, and S. Li. 2011. Monitoring the plant density of cotton with remotely sensed data. In: D. Li, Y. Liu, and Y. Chen (eds), Computer and Computing Technologies in Agriculture IV (pp. 90–101). Springer, Berlin.

Beuchel, F. and B. Gulliksen. 2008. Temporal patterns of benthic community development in an Arctic fjord (Kongsfjorden, Svalbard): results of a 24-year manipulation study. Polar Biol. **31**(8): 913–924.

Bierzychudek, P. 1982. Life histories and demography of shade-tolerant temperate forest herbs: a review. New Phytol. **90**: 757–776.

Bonney, R., J. L. Shirk, T. B. Phillips, et al. 2014. Next steps for citizen science. Science **343**: 1436–1437.

Bonter, D. N. and E. S. Bridge. 2011. Applications of radio frequency identification (RFID) in ornithological research: a review. J. Field Ornithol. **82**: 1–10.

Bridge, E. S., K. Thorup, M. S. Bowlin, et al. 2011. Technology on the move: recent and forthcoming innovations for tracking migratory birds. BioScience **61**: 689–698.

Bruijning, M., M. D. Visser, C. A. Hallmann, and E. Jongejans. 2018. trackdem: automated particle tracking to obtain population counts and size distributions from videos in r. Methods Ecol. Evol. **9**: 965–973.

Buckland, S. T., D. R. Anderson, K. P. Burnham, J. L. Laake, D. L. Borchers, and L. Thomas (eds). 2007a. Advanced Distance Sampling: Estimating Abundance of Biological Populations. Oxford University Press, Oxford.

Buckland, S. T., D. L. Borchers, A. Johnston, P. A. Henrys, and T. A. Marques. 2007b. Line transect methods for plant surveys. Biometrics **63**(4): 989–998.

Carraro, L., H. Hartikainen, J. Jokela, E. Bertuzzo, and A. Rinaldo. 2018. Estimating species distribution and abundance in river networks using environmental DNA. PNAS **115**: 11724–11729.

Caswell, H. 2001. Matrix Population Models: Construction, Analysis, and Interpretation. Sinauer Associates, Sunderland, MA.

Caswell, H. and R. Salguero-Gómez. 2013. Age, stage and senescence in plants. J. Ecol. **101**: 585–595.

Cattadori, I. M., D. T. Haydon, S. J. Thirgood, and P. J. Hudson. 2003. Are indirect measures of abundance a useful index of population density? The case of red grouse harvesting. Oikos **100**: 439–446.

Cayuela, H., R. A. Griffiths, N. Zakaria, et al. 2020. Drivers of amphibian population dynamics and asynchrony at local and regional scales. J. Anim. Ecol. **89**: 1350–1364.

Chabot D. and D. M. Bird. 2012. Evaluation of an off-the-shelf unmanned aircraft system for surveying flocks of geese. Waterbirds **35**: 170–174.

Chirayath, V. and S. A. Earle. 2016. Drones that see through waves—preliminary results from airborne fluid lensing for centimetre-scale aquatic conservation. Aquat. Conserv. Mar. Freshw. Ecosyst. **26**: 237–250.

Clutton-Brock, T. H. and J. M. Pemberton (eds). 2004. Soay Sheep: Dynamics and Selection in an Island Population. Cambridge University Press, Cambridge.

Clutton-Brock, T. and B. C. Sheldon. 2010. Individuals and populations: the role of long-term, individual-based studies of animals in ecology and evolutionary biology. Trends Ecol. Evol. **25**: 562–573.

Clutton-Brock, T. H., F. E. Guinness, and S. D. Albon. 1982. Red Deer: Behavior and Ecology of Two Sexes. University of Chicago Press, Chicago.

Colchero, F., O. R. Jones, and M. Rebke. 2012. BaSTA: an R package for Bayesian estimation of age-specific survival from incomplete mark–recapture/recovery data with covariates. Methods Ecol. Evol. **3**: 466–470.

Condit, R., R. Pérez, S. Lao, S. Aguilar, and S. P. Hubbell. 2017. Demographic trends and climate over 35 years in the Barro Colorado 50 ha plot. For. Ecosyst. **4**: 17.

Connell, J. H., T. P. Hughes, and C. C. Wallace. 1997. A 30-year study of coral abundance, recruitment, and disturbance at several scales in space and time. Ecol. Monogr. **67**: 461–488.

Coulson, T., E. A. Catchpole, S. D. Albon, et al. 2001. Age, sex, density, winter weather, and population crashes in Soay sheep. Science **292**: 1528–1531.

Cressie, N., C. A. Calder, J. S. Clark, J. M. V. Hoef, and C. K. Wikle. 2009. Accounting for uncertainty in ecological analysis: the strengths and limitations of hierarchical statistical modeling. Ecol. Appl. **19**: 553–570.

Culina, A., T. W. Crowther, J. J. C. Ramakers, P. Gienapp, and M. E. Visser. 2018. How to do meta-analysis of open datasets. Nat. Ecol. Evol. **2**: 1053–1056.

Ebert, T. A. 1998. Plant and Animal Populations: Methods in Demography. Academic Press, San Diego.

Eguchi, T. 2014. Estimating the proportion of identifiable individuals and group sizes in photographic identification studies. Mar. Mammal Sci. **30**: 1122–1139.

Ekman, J. 1989. Ecology of non-breeding social systems of *Parus*. Wilson Bull. **101**: 263–288.

Ellis, M. M., J. L. Williams, P. Lesica, et al. 2012. Matrix population models from 20 studies of perennial plant populations. Ecology **93**: 951–951.

ENETWILD Consortium, S. Grignolio, M. Apollonio, et al. 2020. Guidance on estimation of abundance and density data of wild ruminant population: methods, challenges, possibilities. EFSA Support. Publ. **17**: 1876E.

Farnham, A., C. Kurz, M. A. Öztürk, et al. 2017. Early career researchers want Open Science. Genome Biol. **18**: 221.

Festa-Bianchet, M. 2017. When does selective hunting select, how can we tell, and what should we do about it? Mammal Rev. **47**: 76–81.

Field, S. A., A. J. Tyre, K. H., Thorn, P. J. O'Connor, and H. P. Possingham. 2005. Improving the efficiency of wildlife monitoring by estimating detectability: a case study of foxes (*Vulpes vulpes*) on the Eyre Peninsula, South Australia. Wild. Res. **32**: 253–258.

Firth, J. A. and B. C. Sheldon. 2015. Experimental manipulation of avian social structure reveals segregation is carried over across contexts. Proc. R. Soc. B Biol. Sci. **282**: 20142350.

Firth, J. A. and B. C. Sheldon. 2016. Social carry-over effects underpin trans-seasonally linked structure in a wild bird population. Ecol. Lett. **19**: 1324–1332.

Firth, J. A., B. L. Verhelst, R. A. Crates, C. J. Garroway, and B. C. Sheldon. 2018. Spatial, temporal and individual-based differences in nest-site visits and subsequent reproductive success in wild great tits. J. Avian Biol. **49**: e01740.

Fleischmann, D., I. O. Baumgartner, M. Erasmy, et al. 2013. Female Bechstein's bats adjust their group decisions about communal roosts to the level of conflict of interests. Curr. Biol. **23**: 1658–1662.

Focardi S., B. Franzetti, and F. Ronchi. 2013. Nocturnal distance sampling of a Mediterranean population of fallow deer is consistent with population projections. Wild. Res. **40**: 437–446.

Frederiksen, M., J.-D. Lebreton, R. Pradel, R. Choquet, and O. Gimenez. 2014. REVIEW: identifying links between vital rates and environment: a toolbox for the applied ecologist. J. Appl. Ecol. **51**: 71–81.

Fretwell, P. T., M. A. LaRue, P. Morin, et al. 2012. An emperor penguin population estimate: the first global, synoptic survey of a species from space. PLOS One **7**: e33751.

Gall, G. E. C. and M. B. Manser. 2018. Spatial structure of foraging meerkat groups is affected by both social and ecological factors. Behav. Ecol. Sociobiol. **72**: 77.

Garel, M., C. Bonenfant, J. -L. Hamann, F. Klein, and J. -M. Gaillard. 2010. Are abundance indices derived from spotlight counts reliable to monitor red deer *Cervus elaphus* populations? Wild. Biol. **16**: 77–84.

Gatsuk, L. E., O. V. Smirnova, L. I. Vorontzova, L. B. Zaugolnova, and L. A. Zhukova. 1980. Age states of plants of various growth forms: a review. J. Ecol. **68**: 675–696.

Geesey, G. G., R. Mutch, J. W. Costerton, and R. B. Green. 1978. Sessile bacteria: an important component of the microbial population in small mountain streams 1. Limnol. Oceanogr. **23**: 1214–1223.

Gibbons, W. J. and K. M. Andrews. 2004. PIT tagging: simple technology at its best. BioScience **54**: 447–454.

Grant, P. R. 1986. Ecology and evolution of Darwin´s finches. Princeton University Press, Princeton, NJ.

Guillera-Arroita, G. 2017. Modelling of species distributions, range dynamics and communities under imperfect detection: advances, challenges and opportunities. Ecography **40**: 281–295.

Hammond, P. S., P. Berggren, H. Benke, et al. 2002. Abundance of the harbor porpoise and other cetaceans in the North Sea and adjacent waters. J. Appl. Ecol. **39**: 361–376.

Harper, J. L. 1981. Population Biology of Plants. Academic Press, London.

Hayes, L. D. and C. Schradin. 2017. Long-term field studies of mammals: what the short-term study cannot tell us. J. Mammal. **98**: 600–602.

Hays, G. C., L. C. Ferreira, A. M. M. Sequeira, et al. 2016. Key questions in marine megafauna movement ecology. Trends Ecol. Evol. **31**: 463–475.

Hays, G. C., H. Bailey, S. J. Bograd, et al. 2019. Translating marine animal tracking data into conservation policy and management. Trends Ecol. Evol. **34**: 459–473.

Hazen, E. L., S. M. Maxwell, H. Bailey, et al. 2012. Ontogeny in marine tagging and tracking science: technologies and data gaps. Mar. Ecol. Prog. Ser. **457**: 221–240.

Hopkins, H. L. and M. L. Kennedy. 2004. An assessment of indices of relative and absolute abundance for monitoring populations of small mammals. Wildl. Soc. Bull. 1973–2006 **32**: 1289–1296.

Howe, E. J., S. T. Buckland, M. L. Després-Einspenner, and H. S. Kühl. 2017. Distance sampling with camera traps. Methods Ecol. Evol. **8**: 1558–1565.

Huey, R. B., M. Carlson, L. Crozier, M. Frazier, H. Hamilton, C. Harley, A. Hoang, and J. G. Kingsolver. 2002. Plants versus animals: do they deal with stress in different ways? Integr. Comp. Biol. **42**: 415–423.

Hughes, B. B., R. Beas-Luna, A. K. Barner, et al. 2017. Long-term studies contribute disproportionately to ecology and policy. BioScience **67**: 271–281.

Hussey, N. E., S. T. Kessel, K. Aarestrup, et al. 2015. Aquatic animal telemetry: a panoramic window into the underwater world. Science **348**: 1255642.

Hutchings, M. J. 2010. The population biology of the early spider orchid Ophrys sphegodes Mill. III. Demography over three decades. J. Ecol. **98**: 867–878.

Hutchings, M. J. and E. A. C. Price. 1993. Does physiological integration enable clonal herbs to integrate the effects of environmental heterogeneity? Plant Species Biol. **8**: 95–105.

Imperio, S., M. Ferrante, A. Grignetti, G. Santini, and S. Focardi. 2010. Investigating population dynamics in ungulates: do hunting statistics make up a good index of population abundance? Wild. Biol. **16**: 205–214.

Ims, R. A., N. G. Yoccoz, and S. T. Killengreen. (2011). Determinants of lemming outbreaks. PNAS **108**(5): 1970–1974.

Isaac, N. J. B., A. J. Van Strien, T. A. August, M. P. De Zeeuw, and D. B. Roy. 2014. Statistics for citizen science: extracting signals of change from noisy ecological data. Methods Ecol. Evol. **5**: 1052–1060.

Janovský, Z., T. Herben, and J. Klimešová. 2017. Accounting for clonality in comparative plant demography—growth or reproduction? Folia Geobot. **52**: 433–442.

Jarić, I., R. A. Correia, B. W. Brook, et al. 2020. iEcology: harnessing large online resources to generate ecological insights. Trends Ecol. Evol. **35**: 630–639.

Jiménez López, J. and M. Mulero-Pázmány. 2019. Drones for conservation in protected areas: present and future. Drones **3**: 10.

Jønsson, K. A., A. P. Tøttrup, M. K. Borregaard, S. A. Keith, C. Rahbek, and K. Thorup. 2016. Tracking animal dispersal: from individual movement to community assembly and global range dynamics. Trends Ecol. Evol. **31**: 204–214.

Jones, J. P. G. 2011. Monitoring species abundance and distribution at the landscape scale. J. Appl. Ecol. **48**: 9–13.

Kappeler, P. M., F. P. Cuozzo, C. Fichtel, et al. 2017. Long-term field studies of lemurs, lorises, and tarsiers. J. Mammal. **98**: 661–669.

Karanth, K. U. and J. D. Nichols. 1998. Estimation of tiger densities in India using photographic captures and recaptures. Ecology **79**: 2852–2862.

Katsanevakis, S., A. Weber, C. Pipitone, et al. 2012. Monitoring marine populations and communities: methods dealing with imperfect detectability. Aquat. Biol. **16**: 31–52.

Kays, R., M. C. Crofoot, W. Jetz, and M. Wikelski. 2015. Terrestrial animal tracking as an eye on life and planet. Science **348**: aaa2478.

Kéry, M. and M. Schaub. 2012. Bayesian Population Analysis Using WinBUGS: A Hierarchical Perspective. Academic Press, Boston.

Kluijver, H. N. 1951. The population ecology of the great tit, *Parus m. major* L. Ardea **39**: 1–135.

Krause, J., S. Krause, R. Arlinghaus, I. Psorakis, S. Roberts, and C. Rutz. 2013. Reality mining of animal social systems. Trends Ecol. Evol. **28**: 541–551.

Krebs, C. J. 1998. Ecological Methodology, 2nd edition. Benjamin Cummings, Menlo Park, CA.

La Morgia, V., R. Calmanti, A. Calabrese, and S. Focardi. (2015). Cost-effective nocturnal distance sampling for landscape monitoring of ungulate populations. Eur J Wildl. Res, **61**: 285–298.

Lack, D. 1966. Population studies of birds. Clarendon Press, Oxford.

Lambert, J. P. T., H. L. Hicks, D. Z. Childs, and R. P. Freckleton. (2018). Evaluating the potential of unmanned aerial systems for mapping weeds at field scales: a case study with *Alopecurus myosuroides*. Weed Res. **58**(1): 35–45.

Law, B. S. 2018. Long-term research on forest bats: we have the technology. Aust. Zool. **39**: 658–668.

Le Moullec, M. (2019) Spatiotemporal variation in abundance of key tundra species: from local heterogeneity to large-scale synchrony. Trondheim: NTNU (ISBN 978-82-326-3653-2) Doctoral theses at NTNU, 2019:20.

Le Moullec, M., Å. Ø. Pedersen, N. G. Yoccoz, R. Aanes, J. Tufto, and B. B. Hansen. 2017. Ungulate population monitoring in an open tundra landscape: distance sampling versus total counts. Wild. Biol. 2017.

Le Moullec, M., Å. Ø. Pedersen, A. Stien, J. Rosvold, and B. B. Hansen. 2019. A century of conservation: the ongoing recovery of Svalbard reindeer. J. Wildl. Manag. **83**: 1676–1686.

Lee, I.-S. and L. Ge. 2006. The performance of RTK-GPS for surveying under challenging environmental conditions. Earth Planets Space **58**: 515–522.

Linnell, J. D. C., P. Fiske, J. Odden, H. Brseth, I. Herfindal, and R. Andersen. 2007. An evaluation of structured snow-track surveys to monitor Eurasian lynx Lynx lynx populations. Wild. Biol. **13**: 456–466.

Loison, A., J. Appolinaire, J. -M. Jullien, and D. Dubray. (2006). How reliable are total counts to detect trends in population size of chamois Rupicapra rupicapra and R. pyrenaica? Wild. Biol., **12**: 77–88.

Lollback,G. W., J. E. Dunwoody, R. Mebberson, et al. 2018. Spatial modelling of bilby (*Macrotis lagotis*) and rabbit (*Oryctolagus cuniculus*) pellets within a predator-proof enclosure. Austr. Mamm. **40**: 93–102.

Macdonald, D. W. and C. Newman. 2002. Population dynamics of badgers (*Meles meles*) in Oxfordshire, U.K.: numbers, density and cohort life histories, and a possible role of climate change in population growth. J. Zool. **256**: 121–138.

MacKenzie, D. I., J. D. Nichols, N. Sutton, K. Kawanishi, and L. L. Bailey. 2005. Improving inferences in population studies of rare species that are detected imperfectly. Ecology **86**: 1101–1113.

Maillard, D., J.-M. Gaillard, M. Hewison, et al. 2010. Ungulates and their management in France. In: M. Apollonio, R. Andersen, and R. Putman (eds), European Ungulates and their Management in the 21st Century (pp. 441–474). Cambridge University Press, Cambridge.

Malhi, Y., T. Jackson, L. Patrick Bentley, et al. 2018. New perspectives on the ecology of tree structure and tree communities through terrestrial laser scanning. Interface Focus **8**: 20170052.

Mann, J. and C. Karniski. 2017. Diving beneath the surface: long-term studies of dolphins and whales. J. Mammal. **98**: 621–630.

Marques, F. F. C., S. T. Buckland, D. Goffin, et al. (2001). Estimating deer abundance from line transect surveys of dung: sika deer in southern Scotland. J. Appl. Ecol. **38** (2): 349–363.

Marques, T. A., L. Thomas, S. W. Martin, et al. 2013. Estimating animal population density using passive acoustics. Biol. Rev. Camb. Philos. Soc. **88**: 287–309.

Marshall, H. H., D. J. Griffiths, F. Mwanguhya, et al. 2018. Data collection and storage in long-term ecological and evolutionary studies: the Mongoose 2000 system. PLOS One **13**: e0190740.

Martin, S. W., T. A. Marques, L. Thomas, et al. 2013. Estimating minke whale (*Balaenoptera acutorostrata*) boing sound density using passive acoustic sensors. Mar. Mammal Sci. **29**: 142–158.

McMahon, L. A., J. L. Rachlow, L. A. Shipley, J. S. Forbey, T. R. Johnson, and P. J. Olsoy. 2017. Evaluation of micro-GPS receivers for tracking small-bodied mammals. PLOS One **12**: e0173185.

Miller, D. A., J. D. Nichols, B. T. McClintock, E. H. C. Grant, L. L. Bailey, and L. A. Weir. 2011. Improving occupancy estimation when two types of observational error occur: non-detection and species misidentification. Ecology **92**: 1422–1428.

Mills, J. A., C. Teplitsky, B. Arroyo, et al. 2015. Archiving primary data: solutions for long-term studies. Trends Ecol. Evol. **30**: 581–589.

Morellet, N., F. Klein, and R. Andersen. 2011. The census and management of populations of ungulates in Europe. In: R. Putman, M. Apollonio, and R. Andersen (eds), Ungulate Management in Europe (pp. 106–143). Cambridge University Press, Cambridge.

Nassar, J. M., S. M. Khan, D. R. Villalva, M. M. Nour, A. S. Almuslem, and M. M. Hussain. 2018. Compliant plant wearables for localized microclimate and plant growth monitoring. Npj. Flex. Electron. **2**: 1–12. Nature Publishing Group.

Nisbet, I. C. T. 1989. Long-term ecological studies of seabirds. Colon. Waterbirds **12**: 143–147.

Noonan, M. J., A. Markham, C. Newman, et al. 2015. A new Magneto-Inductive tracking technique to uncover subterranean activity: what do animals do underground? Methods Ecol Evol. **6**: 510–520.

Nowacek, D. P., F. Christiansen, L. Bejder, J. A. Goldbogen, and A. S. Friedlaender. 2016. Studying cetacean behaviour: new technological approaches and conservation applications. Anim. Behav. **120**: 235–244.

O'Brien, T. G. 2011. Abundance, density and relative abundance: a conceptual framework. In: A. F. O'Connell, J. D. Nichols, and K. Ullas Karanth (eds), Camera Traps in Animal Ecology (pp. 71–96). Springer, New York.

O'Connor, R. J. 1991. Long-term bird population studies in the United States. Ibis **133**: 36–48.

Paniw, M., P. F. Quintana-Ascencio, F. Ojeda, and R. Salguero-Gómez. 2017. Accounting for uncertainty in dormant life stages in stochastic demographic models. Oikos **126**: 900–909.

Parsons, M., D. Bratanov, K. J. Gaston, F. Gonzalez. 2018. UAVs, hyperspectral remote sensing, and machine learning revolutionizing reef monitoring. Sensors 2018 (18): 2026.

Pedersen, Å. Ø., B.-J. Bårdsen, N. G. Yoccoz, N. Lecomte, and E. Fuglei. 2012. Monitoring Svalbard rock ptarmigan: distance sampling and occupancy modeling. J. Wild. Manag. **76**(2): 308–316.

Perrins, C. M., J. D. Lebreton, and G. J. M. Hirons. 1991. Bird Population Studies: Relevance to Conservation and Management. Oxford University Press, Oxford.

Peters, A. J. and M. D. Eve. 1995. Satellite monitoring of desert plant community response to moisture availability. Environ. Monit. Assess. 37: 273–287.

Pollock, K. H., J. D. Nichols, T. R. Simons, G. L. Farnsworth, L. L. Bailey, and J. R. Sauer. 2002. Large scale wildlife monitoring studies: statistical methods for design and analysis. Environmetrics 13: 105–119.

Pomilla, C., A. R. Amaral, T. Collins, et al. 2014. The world's most isolated and distinct whale population? Humpback whales of the Arabian Sea. PLOS One 9: e114162. Public Library of Science.

Poole, K. G., C. Cuyler, and J. Nymand. (2013). Evaluation of caribou *Rangifer tarandus groenlandicus* survey methodology in West Greenland. Wild. Biol. 19(3): 225–239.

Price, E. A. C., M. J. Hutchings, and C. Marshall. 1996. Causes and consequences of sectoriality in the clonal herb Glechoma hederacea. Vegetatio 127: 41–54.

Ramsey, D. S. L., P. A. Caley, and A. Robley. 2015. Estimating population density from presence–absence data using a spatially explicit model. J. Wildl. Manag. 79: 491–499.

Rehmeier, R. L., G. A. Kaufman, and D. W. Kaufman. 2006. An automatic activity-monitoring system for small mammals under natural conditions. J. Mammal. 87: 628–634.

Reinke, B. A., D. A. Miller, and F. J. Janzen. 2019. What have long-term field studies taught us about population dynamics? Annu. Rev. Ecol. Evol. Syst. 50: 303–333.

Rowcliffe, J. M., J. Field, S. T. Turvey, and C. Carbone. 2008. Estimating animal density using camera traps without the need for individual recognition. J. Appl. Ecol. 45: 1228–1236.

Royle, J. A. and R. M. Dorazio. 2008. Hierarchical Modeling and Inference in Ecology: the Analysis of Data from Populations, Metapopulations and Communities. Academic Press, Amsterdam.

Royle, J. A., M. Kery, R. Gautier, and H. Schmid. (2007). Hierarchical spatial models of abundance and occurrence from imperfect survey data. Ecol. Monogr. 77(3): 465–481.

Royle, J. A., R. B. Chandler, R. Sollmann, and B. Gardner. 2013. Spatial Capture–Recapture. Academic Press, Amsterdam.

Saha, A. K., J. Saha, R. Ray, S. Sircar, S. Dutta, S. P. Chattopadhyay, and H. N. Saha. 2018. IOT-based drone for improvement of crop quality in agricultural field. In: Proceedings of the 2018 IEEE 8th Annual Computing and Communication Workshop and Conference (CCWC) (pp. 612–615). IEEE, New York.

Salguero-Gómez, R. 2018. Implications of clonality for ageing research. Evol. Ecol. 32: 9–28.

Salguero-Gómez, R., H. Kempenich, I. N. Forseth, and B. B. Casper. (2014). Long-term individual-level population dynamics of a native desert chamaephyte. Ecology 95(2): 577.

Salguero-Gómez, R., O. R. Jones, C. R. Archer, et al. 2015. The compadre Plant Matrix Database: an open online repository for plant demography. J. Ecol. 103: 202–218.

Sarabipour, S., H. J. Debat, E. Emmott, S. J. Burgess, B. Schwessinger, and Z. Hensel. 2019. On the value of preprints: an early career researcher perspective. PLOS Biol. 17: e3000151.

Schaub, M., A. Aebischer, O. Gimenez, S. Berger, and R. Arlettaz. 2010. Massive immigration balances high anthropogenic mortality in a stable eagle owl population: lessons for conservation. Biol. Conserv. 143: 1911–1918.

Schauster, E. R., Gese, E. M., and A. M. Kitchen. 2002. An evaluation of survey methods for monitoring swift fox abundance. Wildl. Soc. Bull. 30(2): 464–477.

Schlicht, L., M. Valcu, and B. Kempenaers. 2015. Male extraterritorial behavior predicts extrapair paternity pattern in blue tits, *Cyanistes caeruleus*. Behav. Ecol. 26:1404–1413.

Schorr, R. A. (2013). Using distance sampling to estimate density and abundance of Saussurea weberi hultén (Weber's saw-wort). Southwest. Naturalist 58(3): 378–383.

Schradin, C. and L. D. Hayes. 2017. A synopsis of long-term field studies of mammals: achievements, future directions, and some advice. J. Mammal. 98: 670–677.

Seber, G. A. F. 1992. A review of estimating animal abundance II. Int. Stat. Rev. Rev. Int. Stat. 60:129–166.

Silcox, D. E., J. P. Doskocil, C. E. Sorenson, and R. L. Brandenburg. 2011. Radio Frequency Identification Tagging: a Novel Approach to Monitoring Surface and Subterranean Insects. Am. Entomol. 57:86–93.

Sillett, T. S., R. B. Chandler, J. A. Royle, M. Kéry, and S. A. Morrison. 2012. Hierarchical distance-sampling models to estimate population size and habitat-specific abundance of an island endemic. Ecol. Appl. 22: 1997–2006.

Sinclair, A. R. E., J. M. Fryxell, and G. Caughley. 2006. Wildlife Ecology, Conservation and Management, 2nd edition. Wiley-Blackwell, Malden, MA.

Smale, D. A., G. A. Kendrick, E. S. Harvey, et al. 2012. Regional-scale benthic monitoring for ecosystem-based fisheries management (EBFM) using an autonomous underwater vehicle (AUV). ICES J. Mar. Sci. 69: 1108–1118.

Smith, J. E., K. D. S. Lehmann, T. M. Montgomery, E. D. Strauss, and K. E. Holekamp. 2017. Insights from long-term field studies of mammalian carnivores. J. Mammal. 98: 631–641.

Sutherland, W. J. 2006. Ecological Census Techniques Handbook, 2nd edition. Cambridge University Press, Cambridge.

Sutherland, W. J., R. P. Freckleton, H. C. J. Godfray, et al. 2013. Identification of 100 fundamental ecological questions. J. Ecol. 101: 58–67.

Svensson, J. R. and D. J. Marshall. 2015. Limiting resources in sessile systems: food enhances diversity and growth of suspension feeders despite available space. Ecology **96**: 819–827.

Taig-Johnston, M., M. K. Strom, K. Calhoun, K. Nowak, L. A. Ebensperger, and L. Hayes. 2017. The ecological value of long-term studies of birds and mammals in Central America, South America and Antarctica. Rev. Chil. Hist. Nat. **90**: 7.

Teitelbaum, C. S. and T. Mueller. 2019. Beyond migration: causes and consequences of nomadic animal movements. Trends Ecol. Evol. **34**: 569–581.

Tepley, A. J., J. R. Thompson, H. E. Epstein, and K. J. Anderson-Teixeira. 2017. Vulnerability to forest loss through altered postfire recovery dynamics in a warming climate in the Klamath Mountains. Glob. Change Biol. **23**: 4117–4132.

Thompson, W. L. 2004. Sampling Rare or Elusive Species: Concepts, Designs, and Techniques for Estimating Population Parameters, 2nd edition. Island Press, Washington.

Thompson, W. L., G. C. White, and C. Gowan. 1998. Monitoring Vertebrate Populations. Academic Press, San Diego.

Tomkiewicz, S. M., M. R. Fuller, J. G. Kie, and K. K. Bates. 2010. Global positioning system and associated technologies in animal behaviour and ecological research. Philos. Trans. R. Soc. Lond. B Biol. Sci. **365**: 2163–2176.

Urian, K., A. Gorgone, A. Read, et al. 2015. Recommendations for photo-identification methods used in capture-recapture models with cetaceans. Mar. Mammal Sci. **31**: 298–321.

Vuorisalo, T. and M. J. Hutchings. 1996. On plant sectoriality, or how to combine the benefits of autonomy and integration. Vegetatio **127**: 3–8.

Ward, J. A., L. Thomas, S. Jarvis, et al. 2012. Passive acoustic density estimation of sperm whales in the Tongue of the Ocean, Bahamas. Mar. Mammal Sci. **28**: E444–E455.

Whitlock, M. C., J. L. Bronstein, E. M. Bruna, et al. 2016. A Balanced Data Archiving Policy for Long-Term Studies. Trends Ecol. Evol. **31**: 84–85.

Williams, B. K., J. D. Nichols, and M. J. Conroy. 2002. Analysis and Management of Animal Populations. Academic Press, San Diego.

Williams, P. J., M. B. Hooten, J. N. Womble, and M. R. Bower. 2017. Estimating occupancy and abundance using aerial images with imperfect detection. Methods Ecol. Evol. **8**: 1679–1689.

Yoccoz, N. G., J. D. Nichols, and T. Boulinier. 2001. Monitoring of biological diversity in space and time. Trends Ecol. Evol. **16**: 446–453.

Yuan, Y., F. E. Bachl, F. Lindgren, et al. 2017. Point process models for spatio-temporal distance sampling data from a large-scale survey of blue whales. Ann. Appl. Stat. **11**: 2270–2297.

Žádníková, P., D. Smet, Q. Zhu, D. V. D. Straeten, and E. Benková. 2015. Strategies of seedlings to overcome their sessile nature: auxin in mobility control. Front. Plant Sci. **6**: 218.

Zuur, A. F., E. N. Ieno, N. Walker, A. A. Saveliev, and G. M. Smith. 2009. Mixed Effects Models and Extensions in Ecology with R. Springer, New York.

CHAPTER 6

Drivers of demography: past challenges and a promise for a changed future

Pedro F. Quintana-Ascencio, Eric S. Menges, Geoffrey S. Cook, Johan Ehrlén, and Michelle E. Afkhami

6.1 Population dynamics and biotic and abiotic drivers

Contemporary demographic research is a daunting but significant and stimulating endeavour. Global change due to unsustainable human resource acquisition and consumption is responsible for increasing habitat degradation and fragmentation, alteration of disturbance regimes, and shifting climatic conditions. Even abundant species suffer population declines, and many others are at risk of extinction (Ceballos et al. 2017). There is a pressing need to understand how populations and metapopulations respond to changes in the environment to mitigate the consequences of these trends.

Although many demographic models assume deterministic or stationary stochastic variation (as independent and identically distributed variables) in vital rates, these rates and thus population dynamics are nearly always driven by environmental factors. Identifying the drivers of stochastic population dynamics will often be very helpful in projections and predictions. If environmental conditions are changing directionally or cyclically (i.e. are nonstationary), then we need to identify drivers and changes in drivers to model population dynamics. In this chapter, we emphasise structured population models, but many of these considerations are relevant for other types of population models. We present a nonexhaustive examination on how environmental drivers are incorporated in demographic modelling across many types of organisms. We

critically summarise examples of the main approaches and identify major accomplishments, challenges, and limitations. We point to promising approaches and possible future developments. In the initial sections, we consider models in closed systems without migration among populations. Later we focus on metapopulation models, emphasising the importance of understanding drivers affecting migration and differential extinction among populations. Finally, we conclude with a discussion of some important and general problems associated with assessing relationships between environmental drivers and population dynamics.

6.2 Challenges with incorporating drivers into structured population models

Demographic studies often require long-term and intensive data collection, but, for most species, data collection has been limited to few sampling years and few populations (Menges 2000; Crone et al. 2011, 2013; Salguero-Gómez et al. 2015; Gurevitch et al. 2016) and often collected without proper sampling (e.g. not representative of the focal systems, subjectively and/or inconsistently sampled). Demographic studies require the collection of enough information allowing identifying patterns and associations among variables to contribute to the understanding of population dynamics. Sampling from only a few populations

Pedro F. Quintana-Ascencio et al., *Drivers of demography: past challenges and a promise for a changed future.*
In: *Demographic Methods Across the Tree of Life.* Edited by Roberto Salguero-Gómez and Marlène Gamelon,
Oxford University Press. © Oxford University Press (2021). DOI: 10.1093/oso/9780198838609.003.0006

or years may not capture critical but uncommon environmental conditions (e.g. less-optimal sites, recently disturbed sites, smaller populations) that produce different trajectories of population change. Insufficient spatial and temporal sampling limits the inferences that can be drawn for the effects of drivers on populations. For example, species with sporadic recruitment, source–sink dynamics, or populations with disproportionate reproductive contributions can result in biased population growth estimates.

Concomitantly to the evaluation of the effect of drivers, there are basic statistical and modelling considerations. Estimates of critical vital rates should be assembled with sufficiently large sample sizes. Some life history stages may be infrequent but important to demography (e.g. large individuals may contribute disproportionately to fecundity). Some life history stages, such as seed banks and dormant stages in plants and relatively ephemeral planktonic stages in marine organisms, may be very hard to study without complex experiments or sophisticated detection methods. Nonetheless, these stages may be key to population dynamics (Shefferson et al. 2018; Che-Castaldo et al. 2020). For example, persistent seed banks and dormant stages in vascular plants may make major contributions to population growth and may be particularly important in buffering populations against extinction during unfavourable conditions (Doak et al. 2002). Small sample sizes may provide poor estimates of vital rate means and variances. Covariances of different life stage transitions will require much larger sample sizes than are generally collected, but these covariances may buffer variation within the life cycle (if covariances are negative) or create more extreme dynamics than would be predicted from one vital rate alone (if positive). Careful attention to the correct specification of the focal species' life history during model construction is fundamental for reliable inference and useful projections (Kendall et al. 2019).

Until recently, a significant portion of population modelling summarised vital rates from data collected without regard to temporal and spatial environmental variation (Ehrlén and Morris 2015). This means changes in vital rates and population sizes cannot be linked to environmental changes or management regimes, thus limiting our ability to understand their influence on populations. Population projections of deterministic modelling using single matrices and the asymptotic parameters derived from these models cannot summarise population change in realistic varying environments. Few environments will have such little variation that they produce consistently changing

populations. In fact, most populations never reach equilibria (Crone et al. 2013), and short-term transitory (nonequilibrium) dynamics in these species may have a strong influence on population dynamics and persistence (Kendall and Fox 2002). This has many practical ramifications, including challenging our ability to understand and evaluate future population changes in rapidly changing environments.

Incorporating drivers into population modelling is challenging. It requires identifying the most relevant drivers, estimating their relative effects on vital rates and population dynamics, and recognising their most informative modelling frameworks (e.g. data likelihood distributions, variable interactions, number of bins in matrices, number of stages). Each driver may affect multiple demographic transitions in different ways. Drivers can also interact with each other in their effects on vital rates and population dynamics (e.g. Tye et al. 2016). Although in some systems key drivers of population dynamics are easily identified, the number of environmental factors that potentially drive the population dynamics of a focal species in many systems is large. For logistical reasons, it is often impossible to examine the effects of all potential drivers. Fitting a large number of environmental predictors in statistical models of a limited number of replicates poses other problems. As a consequence, how we select which drivers to study will, of course, affect our conclusions about the study system. Attentive consideration to major regional disturbances, limiting factors, and the main human impacts should aid in the identification of relevant drivers. Understanding the demographic effects of major drivers, particularly those that can be manipulated or mitigated, can lead to promising conservation and management strategies.

One challenge associated with assessing the effects of climatic variables is that they are multidimensional. For example, effects of temperature might differ among different periods of the year, and the minimum, mean, maximum, or range might be most relevant for different species. Another difficulty is that all these aspects of temperature might be correlated to each other and to other climatic variables, such as precipitation and snow cover, making strong inferences about exact causality difficult. One approach suggested to avoid arbitrary selection of variables when there are many potential drivers of population dynamics is to use functional smoothing methods (Teller et al. 2016). These methods can, for example, explore how climate during different time windows before an event influence the response. Lastly, a major challenge with assessing the effects of climatic variables is the temporal scale; years are usually the replicates, and sufficiently long time series are required to disentangle the different effects. Dagleish et al. (2011) used data from 19 years to examine the effects of climate on three species in a sagebrush steppe community. They found that the effects of precipitation, late-winter snowfall, and summer temperatures were important to different species. In all species, the responses of population growth rates to climate lagged one year or more, suggesting that models examining the effects of climatic drivers, as well as of other types of drivers, need to consider time-lagged effects.

Assessments of the effects of drivers are usually based on correlations between levels of drivers and vital rates and population growth rates across sites or years. Collecting information about variation in different types of environmental drivers is often relatively easy, and in most cases it represents considerably less work than collecting the demographic data. From that perspective, it is surprising that many demographic studies collect data on only a single driver or no drivers at all. The main limitations of assessing the effects of environmental drivers using correlations between driver levels and population growth rates lie in identifying causality and in replicating labour-intensive demographic studies across a sufficient number of sites or over a sufficient number of years.

6.3 Quantifying drivers by observation and by experiments

There are two principal ways to quantify the effects of environmental drivers on vital rates: either through observations or with experiments. Most studies investigating the effects of environmental drivers on plant and animal population dynamics use only observational data (Heppell et al. 2000; Ehrlén et al. 2016; Cayuela et al. 2018). Observational data have the advantage of being relevant in the sense that they examine performance over the actual range of environmental conditions as they are experienced by the focal organisms. However, causal relationships between environmental drivers and vital rates and population growth rates are difficult to detect and establish with purely observational studies because different drivers are correlated with each other and to intraspecific population density. This is because we expect population density to be positively correlated with carrying capacity, which in turn is correlated with the level of the environmental drivers. Although in principle it is possible to address these problems by including all relevant drivers and intraspecific densities in the analyses, this

is challenging in practice, and most studies include only one or a few drivers. Of 136 examined plant demographic studies examining the effects of environmental drivers published from 1995 to 2015, the mean number of examined drivers was 1.3, more than 70% examined only one driver, and only 14% of the studies included effects of intraspecific density (Ehrlén et al. 2016).

Given the problems associated with observational demography studies, one important complementary way to identify and quantify drivers is by carrying out experiments. Experimental approaches allow the detection of causal relationships that are hidden in observational data, corroborate causal relationships suggested by correlations, and break up correlations between different environmental drivers. There are two main types of experimental approaches to examine the effects of environmental drivers. First, we can manipulate the level of the focal driver at a given site. Second, we may carry out transplantation experiments and move species into new environments, representing gradients of environmental drivers.

Experimental manipulations of drivers of plant species in the field have been done relatively frequently. Examples of experimental approaches include the effects of land use (e.g. Sletvold et al. 2013), herbivory (e.g. Eckberg et al. 2014), mutualists (e.g. Yule et al. 2013), pathogens (e.g. Roy et al. 2011), competition (e.g. Prevéy et al. 2010), climatic factors (e.g. Williams et al. 2007), and other abiotic drivers (e.g. Scanga 2014). To be able to quantitatively assess the sensitivity of vital rates and population growth rate to changes in a driver from the results of an experimental manipulation, it is necessary to quantify the level of the driver in the different treatments. However, only 17% of studies that used experimental manipulations quantified the level of the drivers (Ehrlén et al. 2016). One study (da Silveira Pontes et al. 2012) quantified browsing treatments in terms of total edible stem biomass removed, finding that only heavy browsing was able to reverse positive population growth rate in the invasive shrub species *Cytisus scoparius*. Still, most studies examining the effects of grazing have compared performance in plots with ambient levels of grazing against plots where grazers are excluded, without quantifying grazing intensity in the treated plot. The lack of information about the levels of the driver means that most experimental studies carried out only have provided qualitative information about whether the driver has an effect or not. Experimental manipulation of population drivers of animal species has been less frequent, and examples include effects of population density (Aars and Ims 2002), food supplementation (Dobson and Oli 2001), disturbance (Perez-Jorge et al. 2016), and population bycatch by fisheries (Genovart et al. 2017).

A limitation in experimental approaches is that parameter combinations used are unlikely ever to occur naturally, meaning that the relevance of the results to natural conditions is questionable. However, the potential strength of experimental manipulations of drivers in the field is that they enable us to examine the effects of environmental conditions that do not currently exist but are likely to occur in the future (e.g. due to climate change) or at other places. For example, Scanga (2014) examined the importance of light regime for population growth rates of the gap-dependent plant *Trollius laxus* and extended the gradient of light availability present in natural subpopulations by including experimentally created canopy gaps. Along this extended gradient, she found a unimodal relationship, with the highest population growth rates at intermediate levels of diffuse light.

In field experiments, the level of one or several focal drivers is manipulated while other factors are allowed to vary naturally. We might grow plants under controlled conditions, for example in a common garden and keep the level of all nonfocal drivers at similar levels. Researchers have conducted common garden/lab-based experiments with corals to assess resilience of different genotypes of symbiodinium to thermal stress and outplanting from coral 'nurseries' to various habitats to try to restore/reintroduce coral populations. However, controlling the level of nonfocal drivers often means the effects of many interactions that are relevant in the field are excluded and that conditions thus might be somewhat artificial.

The effects of some drivers, such as soil nutrients, soil moisture, and light availability, are often explored using spatial variation over different scales (e.g. in terms of variation among populations or patches within populations). The effects of climatic factors, such as temperature and precipitation, are mostly examined using variation among years. Sometimes an integrative approach where both spatial and temporal variation are combined and evaluated with mixed-model regressions may be used to assess the effects of drivers such as climatic factors and herbivory on population growth rates (e.g. Doak and Morris 2010; Eckhart et al. 2011). A critical assumption in such studies is that spatial and temporal variations in drivers have the same effects on demography. However, this might not always be true. For example, spatial variation might be correlated both with other environmental factors and with intraspecific density in ways that temporal variation in the same factor is not. It is also true that differences in environmental conditions might have different effects depending on whether they appear every year (as for spatial variation) or in single years (as for temporal variation). Another method is to use information about the effects of the driver on individual performance and then estimate the effect of the driver at the population level by simulating the proportion of individuals experiencing different levels of the driver (e.g. Garcia et al. 2010).

Levels of drivers have been altered in situ, but transplantation experiments moving organisms into new environments also allow the exposure of organisms to contrasting conditions. Most transplantation experiments have focused on one or a few stages of the life cycle (Hargreaves et al. 2014). However, if these transplantations involve all life cycle stages and are designed to cover gradients of putative environmental drivers, then they can provide us with useful information about the effects of drivers. Transplantation experiments also have the advantage that they, if properly randomised, break up correlations between the history of an individual and effects of exposure to current environmental conditions. Finally, transplantation experiments allow us to break up correlations between environmental conditions and density that might bias estimates of driver effects in observational studies. One limitation with transplantation experiments is that they do not necessarily break up correlations among different environmental drivers, thus suffering partially from the same limitations as observational studies. In spite of their potential usefulness, few demographic studies have transplanted organisms along environmental gradients. One example is Stevens and Latimer (2015), who used a field transplant experiment to 14 sites to examine the importance of snowpack and disturbance in limiting montane invasions of two nonnative lowland shrubs, *Cytisus scoparius* and *Spartium junceum*. Based on demographic data analysed by matrix population models, they concluded that future predictions of reductions in winter snowpack and increases in forest disturbance are likely to increase the risk of invasion.

Given that it is difficult to know how relevant the parameter combinations used in experiments are for natural conditions, experimental manipulations should ideally be used in conjunction with observational approaches. However, only a small proportion of studies examining the effects of environmental drivers on plant population dynamics have used a combination of natural variation and experiments (Ehrlén et al. 2016). In the study by Scanga (2014) mentioned above, effects of diffuse light were examined using a combination of natural variation in light availability and experimentally created gaps. In another study, Dahlgren and Ehrlén (2011) examined the effect of soil potassium concentration on the performance of the forest herb *Actaea spicata*, using both natural variation and experimental potassium additions. Both observational data and experimental treatments indicated that higher concentrations of potassium increased individual growth and population growth rates.

6.4 Continuous drivers ('stress')

Environmental drivers constitute a gradient from those that change in a continuous fashion to those involving relatively discrete events that disrupt vital rates. This difference makes them amenable to distinct analytical and modelling approaches. In this section, we discuss drivers that vary continuously, and in section 6.5 we discuss those involving discrete events. We give an overview of the continuous drivers that have been investigated in demographic studies, how they have been studied, and some potential problems associated with specific drivers.

The effects of a range of climatic factors, including temperature, precipitation, rainfall anomaly, and snowfall, on population growth rate have been examined in many demographic studies. Ehrlén et al. (2016) found that climatic factors constituted 15.5% of all drivers examined for plants. Given the many abiotic factors other than climate that are expected to be drivers of population dynamics, it is somewhat surprising that only ~10% of the studies examining the effects of drivers on plant population growth rate included such effects (Ehrlén et al. 2016). Still, a wide range of nonclimatic abiotic factors have been investigated. These include soil water content, dissolved oxygen level, soil nutrients, soil type, pH, CO_2 levels, altitude, flooding, inclination, aspect, rock type, and light availability. Like climatic variables, other abiotic variables might often be strongly intercorrelated, and including all relevant variables might be difficult. Dahlgren and Ehrlén (2009) examined the effect of eight abiotic drivers on *Actaea spicata*, a perennial herb, finding that only soil potassium concentrations had a clear effect on population growth rates.

Effects of land use have frequently been examined in terms of intensity of grazing and different mowing regimes. Typically, these studies compare areas exposed to different grazing or cutting regimes. Winter et al. (2014) compared the effects of five different cutting treatments on the performance of the toxic grassland plant *Colchicum autumnale*. They found that population growth rate was significantly lower in plots cut earlier in the season.

Given that competition is assumed to be one of the most important factors structuring ecological communities, surprisingly few demographic studies have examined the effects of interspecific and intraspecific competition. Studies having examined the role of interspecific competition differ considerably regarding how competition has been quantified; measures used include vegetation height, neighbour density, presence of a specific neighbour species, or more detailed measures of the neighbouring community. It is worth noting that studies examining the effects of canopy cover or shading also might be regarded as having examined the effects of competition, albeit effects mediated by a particular abiotic factor. One example of a study with a detailed measure of interspecific density is Adler et al. (2012), which used 22 years of demographic data to explicitly quantify both direct effects of climate and indirect effects mediated by interspecific competition with other species for four species in a sagebrush steppe community. They found strong effects of climate acting via the abundance of neighbouring species, showing that the relative importance of interspecific competition decreased with increased differences in niches between species.

Other types of antagonistic species interactions have also been examined, including herbivory, predation, and pathogens. For studies of plant demography, herbivory is one of the most-often investigated drivers (e.g. Maron and Crone 2006). The way in which intensity of herbivory has been quantified varies considerably among studies. Many experimental studies have compared ambient levels of herbivory with some type of herbivore exclusion (Eckberg et al. 2014), while correlational studies usually estimate intensity as the proportion of individuals attacked or average level of damage, for example, in terms of proportion of biomass removed. For studies of predispersal seed predation, the proportion of seeds damaged at the population level is often used as an estimate. There are few studies of the effects of pathogens, but Davelos and Jarosz (2004) used matrix models to examine how chestnut blight infection affects population growth rates of American chestnuts (*Castanea dentata*).

Incorporation of positive interactions (e.g. mutualism and facilitation) can have important consequences for demography of the participating species, although relatively few studies have examined these effects in population models. Ehrlén et al. (2016) found that studies considering mutualism made up only 5.5% of all drivers examined in environmentally explicit population models of plants. These studies of mutualistic effects on population dynamics were focused on interactions between plants and animals (pollinators, ants, and dispersal partners). For example, Parker (1997) simulated the consequences of increased pollination on the growth of new populations of an invasive plant, *Cytisus scoparius*, by iterating transition matrices over short time scales. Her work predicted minimal effects of pollen limitation for the slow-growing urban populations of this invasive plant in contrast to the large, positive effects of increased pollinator visitation predicted for the rapidly growing prairie populations, thereby considering effects of species interactions and land use on population demography. In a paper investigating a defensive mutualism, Palmer et al. (2010) incorporated the effects of a whole community of ant species on population growth simulations, finding that *Acacia drepanolobium*, interacting with all four members of the ant community, had a higher 50-year population growth rate compared with individuals associated with the putative best mutualist. Surprisingly, this was despite two of the other species of ants being putative parasites.

More recently there has been an interest in the role of beneficial interactions with microbes in population demography. All organisms interact with a complex microbiome that can include beneficial symbionts, and much work has established the importance of these associations for individual level effects in both animals (e.g. critical for proper digestion in animals ranging from termites to cows to humans; Backhed et al. 2005) and plants (e.g. rhizobia, mycorrhizal fungi, and fungal endophytes confer stress amelioration through resource provisioning and even defence against herbivores; Rodriguez et al. 2009; Larimer et al. 2010). However, the consequences of these interactions for population dynamics have not been deeply investigated in most cases. For example, to our knowledge, only one study has examined the effects of soil microbiomes in the context of a plant matrix or integral projection model, despite the substantial literature on the importance of these interactions. This recent study demonstrated that the microbiome increased population growth rates of an endangered plant, especially in stressful habitats (David et al. 2019). A few studies have examined the consequences of beneficial symbionts with vertically transmitted fungal endophytes

that live in the aboveground tissue of grasses, finding that these resource provisioning and defensive mutualists generally increase population growth rates (Yule et al. 2013; Chung et al. 2015).

6.5 Episodic drivers ('disturbance')

Drivers that have been characterised by alternative environmental states include ecological disturbances and site differences. In this section, we explore different approaches to evaluating the effects of episodic drivers on demography, using fire, hurricanes, harvesting, drought, species interactions, and multiple disturbances as instructive examples.

Fire is a dominant ecological disturbance throughout the world (Bond and Keeley 2005) and has strong effects on ecosystems, communities, and populations (Bond et al. 2005; Koltz et al. 2018). It is also a strong evolutionary force (Bond and Keeley 2005) and is frequently manipulated by humans, through either fire suppression or prescribed fire. Fire severity and frequency are changing in many parts of the world due to climate change (Bowman et al. 2009). Given all this, it is not surprising that fire effects (and fire management effects) as drivers of demography have received a fair amount of study.

The simplest approach considers the effects of individual fires, often comparing populations affected by fire with those remaining unburned using projection matrices of fire and no-fire. A study of siamang (*Symphalangus syndactylus*), a frugivorous, Southeast Asian rainforest primate, showed impacts of El Niño–Southern Oscillation (ENSO)-related fires on the demography and persistence by comparing burned and unburned populations (O'Brien et al. 2003). Alternatively, if prefire data are also available, then a before–after/control–impact design can be used to analyse fire as a driver, as in the case of bird assemblages showing increases in some species' abundances after severe fire in Montana (Smucker et al. 2005). Finally, other aspects of the fire regime can be contrasted for their effects on demography. A comparison of burn season and fire frequency on populations of the invasive spotted knapweed (*Centaurea maculosa*) showed that summer burning was the only treatment that reduced population growth rates, and a life table response experiment showed that this effect happened because of reduced reproduction (Emery and Gross 2005).

A more involved approach uses Markovian sequences of postfire matrices in simulations to characterise postfire demographic changes. These simulations can characterise observed fire sequences, but they can also be used in stochastic analyses to predict optimal fire return intervals and estimate time postfire to quasi-extinction. For example, Menges and Quintana-Ascencio (2004) used matrix selection to draw from 54 matrices representing different times-since-fire in Florida scrub. Simulations of *Eryngium cuneifolium* population dynamics showed that populations were growing only in the first decade postfire and that populations would disappear 30 or more years after fire—patterns that were reflected in observed dynamics (Menges and Quintana-Ascencio 2004). Prescribed fire was therefore recommended at intervals of 15 years or less. Whether fires are single or repeated can have important consequences for the dynamics of species that recover slowly after fire. Most functional groups of arthropods were more resilient to single fire events than to multiple fires, although resilience varied depending on the functional group (Moretti et al. 2006).

Demographic modelling can also incorporate disturbances and postdisturbance patterns into projections of population dynamics. Demographic models can be used to predict species distributions (e.g. Diez et al. 2014). Merow et al. (2014) modelled Proteaceae shrubs in South Africa in relation to fire and weather and their interaction. These approaches can be combined with spatially explicit models. For example, the landscape dynamics of the Florida scrub plant *Hypericum cumulicola* were affected by patch size and patch isolation, in addition to fire and demographic factors in integral projection models (Quintana-Ascencio et al. 2018). Population growth was highest under frequent fire return intervals, at higher elevations, and in larger and less-isolated populations, and the models did a fair job of predicting occupancies and densities across the landscape. Fire patchiness is another key spatial factor that can drive populations. For example, low-intensity fires will tend to be patchy, which may reduce mortality of obligately seeding shrubs and promote their persistence in a pyrogenic landscape (Ooi et al. 2006).

While less well studied than fire, disturbances from hurricanes and other wind/storm events can be strong drivers of demography in the tropics and subtropics, particularly for plants and benthic coral communities (Pascarella and Horvitz 1998; Vardi et al. 2012; Edmunds 2015). Vardi et al (2012) used *Acropora palmata* (Elkhorn coral) demographic data to simulate the effect of hurricane recurrence on the population viability of this endangered species, suggesting that hurricane frequency less than 20 years would result in population extirpation. Pascarella and Horvitz (1998) assembled demographic data for the understory shrub *Ardisia escallonioides* into a megamatrix (a set of matrices nested together in a larger matrix), where each demographic matrix was part of a patch-transition matrix summarising openness with hurricanes and vegetation recovery. The use of the megamatrix revealed some patterns not seen in analyses of individual matrices, such as the importance of small juveniles to elasticity. Hurricane frequency also interacted with seed predation, with seed predator effects being greatest at intermediate hurricane frequencies (Horvitz et al. 2005).

Similar approaches have been used in characterising the effect of human harvesting on population dynamics. Matrices representing harvested and unharvested populations and varying the extent of harvest have been usefully applied to various organisms (e.g. Gaoue and Ticktin 2010). Crowder et al. (1994) used matrix population models to assess the effect of requiring turtle exclusion devices (TEDs) on shrimp harvesting trawls in the Gulf of Mexico. In this simulation they showed that TEDs would have a benefit on the loggerhead sea turtle (*Caretta caretta*) population, by reducing unintentional bycatch and drowning of turtles. These modelling exercises can define sustainable harvest levels.

Another abiotic factor that can strongly influence demography of populations is episodic drought events. For example, stochastic matrix models predicted that higher drought frequency negatively influences population growth rate of the endangered yellow-bellied toad (*Bombina variegata*), potentially threatening population persistence and requiring management practices that reduce exposure to prolonged desiccation stress (Cayuela et al. 2016). The importance of drought as an episodic driver of population dynamics may be growing if climate change increases the frequency of severe droughts. For instance, a recent IPM examined how the increased frequency of drought events predicted by Intergovernmental Panel on Climate Change (IPCC) models for Patagonia would impact the demography of a dominant evergreen tree (Molowny-Horas et al. 2017). The simulations varying the frequency of drought events projected a reduction in density of these tree populations in the mid-term under an increased drought scenario.

While species interactions can be ongoing throughout an organism's life (see section 6.4), both positive and negative interactions may instead be discrete. For example, rare events like mass emergence of periodical cicadas and pathogen outbreaks generate discrete species interactions that may strongly impact demography by enhancing predator growth or reducing host survival (respectively). Further, many interactions are discrete on shorter time scales. In facultative mutualisms, organisms may or may not associate with mutualists, and this can vary throughout the organism's lifetime. For example, the cactus *Ferocactus wislizeni* produces extrafloral nectaries intended to attract an ant defender mutualist that reduces herbivory and increases fruit production; however, over three years of monitoring, roughly half the plants were left unguarded (Ford et al. 2015). Because fruit production had only weak effects on population growth, there were no differences in population growth rates between ant and ant-free cacti.

6.6 Population dynamics in the landscape: influence of spatial habitat structure and heterogeneity

The challenges associated with incorporating continuous and episodic drivers into structured population models (see sections 6.4 and 6.5) must be considered when assessing how spatial habitat structure and landscape heterogeneity influence metapopulation dynamics. By necessity this extension to the metapopulation requires an understanding of not only the shifting mosaic of environmental stressors and disturbances influencing dynamics at the population or local scale but also the critical importance of factors influencing dispersal of individuals among the local habitat patches within the broader metapopulation.

The importance of dispersal to metapopulation dynamics has been acknowledged explicitly for over a century (e.g. Reid 1899; Tansley 1923; Elton 1925; Nicholson 1933). This long-held interest of ecologists in dispersal, the immigration and emigration of individuals, and its influence on the dynamics of spatially structured populations should not be a surprise, however, because, as Elton (1927) eloquently stated, 'When we are studying any particular animal or community of animals, we are brought up, sooner or later, against questions connected with dispersal'.

While dispersal was studied by ecologists in the early 1900s, it was not, with few exceptions, formalised analytically until almost half a century later (Skellam 1951, 1952). Research on dispersal continued through the twentieth century, with numerous theoretical advances. MacArthur and Wilson (1963, 1967) invoked the role of dispersal capacity as an implicit mechanism of colonisation in their theory of island biogeography. Levins (1969, 1970) adapted this

framework to apply spatial population dynamics to the control of agricultural pests. The balance between extinction rates and colonisation rates governed the rise and fall of spatially separated but interacting populations, termed metapopulations. Since Levins' seminal work, the field of metapopulation ecology has flourished, and the spatially implicit Levins' metapopulation model has been refined to increase realism. For example, variability in environmental conditions can impact the vital rates of local populations. Some subpopulations can 'rescue' declining subpopulations in habitat patches through dispersal. Disturbances can synchronise these dynamics at relatively large spatial and temporal scales, influencing the long-term persistence of the broader metapopulation (e.g. Brown and Kodric 1977; Ranta et al. 1995; Hanski and Ovaskainen 2000; Benton et al. 2001; Brook et al. 2009; Desharnais et al. 2018).

A growing number of studies have started to shed light on the frequency and magnitude with which spatially separated populations interact with one another (e.g. Paradis et al. 1998; Ricketts 2001; Hanski and Ovaskainen 2003; Smith and Green 2005; Stevens et al. 2010; Duncan et al. 2015). As more empirical dispersal data are generated, it has become clear that metapopulations may function differently than the original patch occupancy model conceptualised by Levins (1969). Thus, the definition of a metapopulation has evolved to include regular (rather than sporadic) dispersal among populations, relatively infrequent extinction, and both local-scale drivers that independently impact single subpopulations and regional-scale drivers that influence all subpopulations comprising the broader metapopulation.

At the local scale, habitat heterogeneity is ubiquitous in nature, and the influence of continuous and episodic drivers on vital rates of populations inhabiting this heterogeneous environment is somewhat dependent on population size and density. At small population sizes and low densities, the vagaries of sexual selection (e.g. finding a mate) and environmental disturbances (e.g. a hurricane, tsunami, or wildfire) can result in stochastic extinction of small populations (Courchamp et al. 1999) (see section 6.5). In addition to these potential stochastic effects, vital rates can be impacted by habitat geometry (patch area and shape). Hokit and Branch (2003) combined a stage-based matrix model with stochastic numerical simulation to explore the effect of habitat patch area on *Sceloperus woodi* vital rates. They found fecundity and survivorship were positively correlated with habitat patch size. In spatially heterogeneous landscapes, populations can be categorised as sources if they have positive growth rates and sinks if they have negative growth rates (Pulliam 1988). Gundersen et al. (2001) experimentally tested this source–sink framework using root voles (*Microtus oeconomus*) to assess the effect of demography in sink populations on source populations connected via vole dispersal. Dispersal rates tended to vary with population density; density-dependent dispersal of voles from source populations to sink habitat patches inhibited the growth rates within the larger source populations (Gunderson et al. 2001). In plants, regional source–sink dynamics are closely related to life history strategy and environmental stability. Clonal plants or those with extensive seed banks tend to support numerous local populations that can persist through relatively unfavourable environmental conditions, while plants with relatively high dispersal rates (e.g. *Jacobaea vulgaris*) tend to generate metapopulation patterns in the landscape, with occupied and unoccupied patches intermittently connected via dispersal (Eriksson 1996). However, environmental and climatic factors vary temporally and spatially. If these broad-scale environmental gradients generate variability in habitat quality, the loss of small populations in lower-quality habitat patches may result in decreased dispersal among patches and lower metapopulation-level fitness (Grear and Burns 2007).

For metapopulations, an assessment of the relative importance of environmental variation and dispersal on population dynamics is warranted. Density-independent spatial autocorrelation of population fluctuations controlled by environmental stochasticity has been called the Moran effect (cf. Moran 1953). Experimental and empirical support for the role of broad-scale environmental conditions synchronising dynamics in spatially separated populations is known from multiple taxa (e.g. Ranta et al. 1995 [fish], Grenfell et al. 1998 [sheep], Benton et al. 2001 [mites], Gouhier et al. 2010 [mussels], Massie et al. 2015 [phytoplankton], Kahilainen et al. 2018 [butterflies]). However, in weakly regulated populations, theoretical models suggest dispersal can modify the magnitude and time-scale of population synchrony (Lande et al. 1999; Duncan et al. 2015; Desharnais et al 2018).

Empirical studies generally suggest that greater dispersal increases synchrony. Paradis et al. (1998) were one of the first to empirically investigate the effects of dispersal and the feedbacks between environmental and demographic factors on population synchrony in 53 species of birds. While evidence for population synchrony was evident across species, the scale of synchrony varied with population size, natal and

breeding dispersal distance, and habitat type. Trenham et al. (2001) assessed spatial synchrony of demographic patterns in tiger salamanders (*Ambystoma californiense*) due to dispersal among a network of breeding ponds. Relatively high rates of salamander dispersal resulted in positively correlated breeding populations among nearby ponds. However, as distance between ponds increased, the correlation among mass and age of individuals decreased. Quintana-Ascencio et al. (2019) assessed the relative importance of seed bank and seed dispersal on the metapopulation persistence of an endangered and rare plant species. They found metapopulation viability of this species depended on a strong seed bank coupled with limited dispersal.

At the metapopulation scale, disturbances can result in greater extinction rates, but these effects are landscape dependent and tend to be aggregated in certain regions. If dispersal is nonrandom and biased toward regions with greater probability of disturbance, it will result in higher extinction rates (Elkin and Possingham 2008). Conversely, if dispersal is asymmetric and skewed toward regions of relatively low disturbance, the metapopulation persistence will be greater over time. Gouhier et al. (2010) assessed spatial synchrony for *Mytilus californianus* along the Pacific Coast of the United States, assessing the relative importance of demographic processes and dispersal with environmental conditions. In a nearshore rocky intertidal system with a relatively persistent gradient in environmental conditions, there was a weak and inconsistent spatial correlation among environmental conditions and population size at the local scale (Gouhier et al 2010). Yet the correlation in abundance of *M. californianus* switched from synchrony to asynchrony with distance from a given location, suggesting dispersal may control regional-scale patterns of spatial synchrony. Carson et al. (2011) found two species of congeneric mussels (*Mytilus californianus* and *M. galloprovincialis*) spawn and release larvae in different seasons. This results in divergent dispersal patterns and demographic structure due to seasonal shifts in the broad-scale oceanographic currents. Fletcher et al. (2018) explored the effect of random vs. aggregated habitat loss on the dynamics of an insect herbivore, *Chelinidea vittiger*, comparing the demographic consequences of local-scale effects of habitat fragmentation (e.g. edge effects) with the landscape-scale effects related to the scale of dispersal. They found the scale of herbivore dispersal and the insect's ability to move among habitat patches were the primary determinants of whether habitat fragmentation had a negative or positive effect on population growth. If the scale of fragmentation was large relative to the dispersal capacity of the herbivore, movement and reproduction decreased, resulting in a negative impact on population size.

To characterise the variability inherent in the environment, dispersal, and ensuing connectivity patterns, useful tools have been developed, particularly in the fields of landscape and conservation ecology. For example, graph and network approaches have been adopted for their ease of comprehension and ability to simplify large amounts of connectivity data (Urban et al. 2009). From a conservation standpoint, these analytical frameworks have proven useful in examining actual and hypothetical systems of protected areas (Rayfield et al. 2011). In part, the growing acceptance of the use of graphs and networks can be attributed to the ease with which patches of habitat separated by an uninhabitable matrix can (at least visually) be viewed as a graph (i.e. a set of nodes connected by edges) or as a network when empirical data are available describing the magnitude and directionality of individuals dispersing among habitat patches (Dale and Fortin 2010; Rayfield et al. 2011). As connectivity matrices continue to be generated for additional taxa, along with complementary large-scale environmental data sets, additional metrics (e.g. modularity; Fletcher et al. 2018) can be used to assess how dispersal connects demography among habitat patches. This can increase our understanding of how these are impacted by episodic and continuous environmental drivers.

6.7 New and promising approaches

To develop a predictive understanding of how current and future environmental conditions may influence demography at relatively large spatial and temporal scales will require novel approaches. Lee (2017) and Iles et al. (2019) developed modelling approaches allowing for the assessment of increasing and variable environmental drivers in 'traditional' elasticity analyses. These analytical approaches make it possible to separate the influence of environmental driver intensity as well as variability on population growth. While these approaches were developed for single populations, they lay the groundwork for extending these methods to metapopulations. Similarly, these emerging large-scale data sets and dynamic approaches, coupled strategically with regional-scale vital-rate data, can generate fundamental knowledge regarding the role of large-scale environmental variability on population and metapopulation dynamics.

By combining information about temporal changes in drivers with data on how drivers affect vital rates, we can model not only stationary environments but also how population dynamics change over time in environments that undergo directional change. This is particularly relevant for examining the effects of changes in climate and environment. Gotelli and Ellison (2006) used such an approach and combined data from whole-plot fertilisation experiments on the effect of nitrogen availability on the performance of the pitcher plant, *Sarracenia purpurea*, with predicted long-term trends in nitrogen deposition. They found that an increase in nitrogen deposition rate of only 1% substantially increased the local extinction risk. Similar models have been used to explore the effects of secondary succession mediated by changes in different abiotic and biotic factors (e.g. Dahlgren and Ehrlén 2011; Lehtilä et al. 2016).

Some more general aspects of assessing the effects of drivers on population growth rates are important to consider. One problem is that effects on vital rates and population growth rate might be nonlinear. Dahlgren et al. (2011) compared models that accounted vs. not accounted for nonlinear relationships between vital rates and state variables, finding that parameterisation method had a large effect on predicted population growth rates. They suggested that because even weak nonlinearity in relationships can have large effects on model predictions, restricted cubic regression splines should be considered for parameterising models whenever linearity cannot be assumed.

Another complication is that the effect of a focal driver might depend on the level of another driver— that is, that effects of the two drivers are interactive. However, such interaction effects have rarely been incorporated in environmentally explicit demographic models. One obvious reason for this is that detecting interaction effects requires considerably larger sample sizes. Nicolè et al. (2011) investigated how local habitat quality and climatic variation influenced population dynamics of the endangered alpine plant *Dracocephalum austriacum*. They found that populations on steeper slopes had lower population growth rates than populations on more gentle slopes and that population growth rates were lower in years with high summer temperatures. Interestingly, there was also an effect of the interaction between these two factors, the negative effects of high summer temperatures being more pronounced in populations on steeper slopes. In another example, the mountain date palm (*Phoenix loureirin*) in south India is affected by fire, grazing, and human leaf harvest (Mandle and Ticktin

2012). Vital rates were influenced by interactions among these drivers; harvesting was better tolerated in areas that had been recently burned. Other studies have considered how additional drivers might affect outcomes by combining demographic models with simulations.

To achieve a good understanding of the overall effect of a given environmental driver, it is important to include possible indirect effects. For example, effects of climatic variables may act not only directly on performance but also via changes in biotic interactions, such as interspecific competition or facilitation. The study by Adler et al. (2012), mentioned above, did indeed find not only direct effects of several climate variables but also strong indirect effects via the abundance of neighbouring species.

Neglecting intraspecific density might result in biased estimates of environmental drivers. As a consequence, estimates of environmental factors when not considering density effects may be incorrect if these correlations are strong. Understanding density effects is necessary to predict the long-term impact of environmental drivers and equilibrium abundances. Although effects of intraspecific density are likely to be important in many systems, few demographic studies with plants have explored the simultaneous effects of environmental drivers and density, but there are several with animals (Aars and Ims 2002). It is important to note that negative feedback effects of population density on population growth may occur even at low densities (e.g. Rodenhouse et al. 1997). Effects of intraspecific density can explicitly be incorporated into IPMs by modelling variation in vital rates as functions of both environmental variation and intraspecific density. Dahlgren et al. (2016) examined the simultaneous impact of climate and intraspecific density on vital rates of the dwarf shrub *Fumana procumbens* over 20 years, using generalised additive mixed models and IPMs. The analyses suggested that the population was regulated by density but with annual fluctuations in response to variation in weather. Importantly, omission of intraspecific density in these models resulted in overestimations of the effects of climatic variation and of extinction risk.

Understanding demographic change as a function of drivers' variation will be critical for better sustainable resource use, management, and conservation. This provides a more mechanistic and potentially predictive approach to understanding population dynamics. It allows scientists to consider a range of scenarios that can reasonably bound likely outcomes. In this way, studying the demographic effect of drivers can help

to project population changes due to novel conditions caused by human actions.

References

Aars, J. and Ims, R. A. (2002). Intrinsic and climatic determinants of population demography: the winter dynamics of tundra voles. *Ecology*, **83**, 3449–3456.

Adler, P.B., Dalgleish, H.J., and Ellner, S.P. (2012). Forecasting plant community impacts of climate variability and change: when do competitive interactions matter? *Journal of Ecology*, **100**, 478–487.

Backhed, F., Ley, R. E., Sonnenburg, J. L., Peterson, D. A., and Gordon, J. I. (2005). Host–bacterial mutualism in the human intestine. *Science*, **307**, 1915–1920.

Benton, T. G., Lapsley, C. T., and Beckman, A. P. (2001). Population synchrony and environmental variation: an experimental demonstration. *Ecology Letters*, **4**, 236–243.

Bond, W. J. and Keeley, J. E. (2005). Fire as a global 'herbivore': the ecology and evolution of flammable ecosystems. *Trends in Ecology and Evolution*, **20**, 387–394.

Bond, W. J., Woodward, F. I., and Midgley, G. F. (2005). The global distribution of ecosystems in a world without fire. *New Phytologist*, **156**, 525–538.

Bowman, D. M., Balch, J. K., Artaxo, P., et al. (2009). Fire in the earth system. *Science*, **324**, 481–484.

Brook, B. W., Akcakaya, H. R., Keith, D. A., Mace, G. M., Pearson, R. G., and Araujo, M. B. (2009). Integrating bioclimate with population models to improve forecasts of species extinctions under climate change. *Biology Letters*, **5**, 723–725.

Brown, J. H. and Kodric-Brown, A. (1977). Turnover rates in insular biogeography: effect of immigration on extinction. *Ecology*, **58**, 445–449.

Carson, H. S., Cook, G. S., Lopez-Duarte, P.C., and Levin, L. A. (2011). Evaluating the importance of demographic connectivity in a marine metapopulation. *Ecology*, **92**, 1972–1984.

Cayuela, H., Arsovski, D., Thirion, J. -M., et al. (2016). Demographic responses to weather fluctuations are context dependent in a long-lived amphibian. *Global Change Biology*, **22**, 2676–2687.

Cayuela, H., Rougemont, Q., Prunier, J. G., et al. (2018). Demographic and genetic approaches to study dispersal in wild animal populations: a methodological review. *Molecular Ecology*, **27**, 3976–4010.

Ceballos, G., Ehrlich, P.R., and Dirzo, R. (2017). Biological annihilation via the ongoing sixth mass extinction signaled by vertebrate population losses and declines. *Proceedings of the National Academy of Sciences*, **114**, 6089–6096.

Che-Castaldo, J., Jones, O. R., Kendall, B. E., et al. (2020). Comments to 'Persistent problems in the construction of matrix population models'. *Ecological Modelling* **416**, 108913

Chung, Y. Y., Miller, T. E. X., and Rudgers, J. A. (2015). Fungal symbionts maintain a rare host plant population but demographic advantage drives the dominance of a common host. *Journal of Ecology*, **103**, 967–977.

Courchamp, F., Clutton-Brock, T., and Grenfell, B. (1999). Inverse density dependence and the Allee effect. *Trends in Ecology and Evolution*, **14**, 405–410.

Crone, E. E., Menges, E. S., Ellis, M. M., et al. (2011). How do plant ecologist use matrix population models? *Ecology Letters*, **14**, 1–8.

Crone, E. E., Ellis, M. M., Morris, W. F., et al. (2013). Ability of matrix models to explain the past and predict the future of plant populations. *Conservation Biology*, **27**, 968–978.

Crowder, L. B., Crouse, D. T., Heppell, S. S., and Martin T. H. (1994). Predicting the impact of turtle excluder devices on loggerhead sea turtle populations. *Ecological Applications* **4**, 437–445.

Da Silveira Pontes, L., Magda, D., Jarry, M., Gleizes, B., and L Agreil, C. (2012). Shrub encroachment control by browsing: targeting the right demographic process. *Acta Oecologica*, **45**, 25–30.

Dahlgren, J. P. and Ehrlén, J. (2009) Linking environmental variation to population dynamics of a forest herb. *Journal of Ecology*, **97**, 666–674.

Dahlgren, J. P. and Ehrlén, J. (2011). Incorporating environmental change over succession in an integral projection model of population dynamics of a forest herb. *Oikos*, **120**, 1183–1190.

Dahlgren, J. P., García, M. B., and Ehrlén, J. (2011). Nonlinear relationships between vital rates and state variables in demographic models. *Ecology*, **92**, 1181–1187.

Dahlgren, J. P., Bengtsson, K., and Ehrlén, J. (2016). The demography of climate-driven and density-regulated population dynamics in a perennial plant. *Ecology*, **97**, 899–907.

Dale, M. R. T. and Fortin, M. -J. (2010). From graphs to spatial graphs. *Annual Review of Ecology, Evolution, and Systematics*, **41**, 21–38.

Dalgleish, H. J., Koons, D. N., Hooten, M. B., Moffet, C. A., and Adler, P. B. (2011). Climate influences the demography of three dominant sagebrush steppe plants. *Ecology*, **92**, 75–85.

Davelos, A. L. and Jarosz, A. M. (2004). Demography of American chestnut populations: effects of a pathogen and a hyperparasite. *Journal of Ecology*, **92**, 675–685.

David, A. S., Quintana-Ascencio, P. F., Menges, E., Thapa-Magar, E., Afkhami, M. E. A., and Searcy, C. A. S. (2019). Living in a microbial world: endangered plant requires soil microbiome for population persistence. *The American Naturalist*, **194**, 488–494.

Desharnais, R. A., Reuman, D. C., Costantino, R. F., and Cohen. J. E. (2018). Temporal scale of environmental correlations affects ecological synchrony. *Ecology Letters*, **21**, 1800–1811.

Diez, J. M., Giladi, I., Warren, R., and Pulliam, H. R. (2014). Probabilistic and spatially variable niches inferred from demography. Journal of Ecology **102**, 544–554.

Doak, D. F. and Morris, W. F. (2010). Demographic compensation and tipping points in climate-induced range shifts. *Nature*, **467**, 959–962.

Doak, D. F., Thomson, D., and Jules, E. S. (2002). Population viability analysis for plants: understanding the demographic consequences of seed banks for population health. In: Beissinger, S. R., McCullough, D. R. (eds), Population Viability Analysis, Chapter 15. University of Chicago Press, Chicago, IL.

Dobson, F. S. and Oli, M. K. (2001). The demographic basis of population regulation in Columbian ground squirrels. *The American Naturalist*, **158**, 236–247.

Duncan, A. B., Gonzalez, A., and Kaltz, O. (2015). Dispersal, environmental forcing, and parasites combine to affect metapopulation synchrony and stability. *Ecology*, **96**, 284–290.

Eckberg, J.O., Tenhumberg, B., and Louda, S. M. (2014). Native insect herbivory limits population growth rate of a non-native thistle. *Oecologia*, **175**, 129–138.

Eckhart, V. M., Geber, M. A., Morris, W. F., Fabio, E. S., Tiffin, P., and Moeller, D. A. (2011). The geography of demography: long-term demographic studies and species distribution models reveal a species border limited by adaptation. *The American Naturalist*, **178**, S26–S43.

Edmunds, P. J. (2015). A quarter-century demographic analysis of the Caribben coral, *Orbicella annularis*, and projections of population size over the next century. *Limnology and Oceanography*, **60**, 840–855.

Ehrlén, J. and Morris, W. F. (2015). Prediction changes in the distribution and abundance of species under environmental change. *Ecology Letters*, **18**, 303–314.

Ehrlén, J., Morris, W. F., Von Euler, T., and Dahlgren, J. P. (2016). Advancing environmentally explicit structured population models of plants. *Journal of Ecology*, **104**, 292–305.

Elkin, C. M. and Possingham, H. (2008). The role of landscape-dependent disturbance and dispersal in metapopulaiton persistence. *The American Naturalist*, **172**, 563–575.

Elton, C. S. (1925). The dispersal of insects to Spitsbergen. *Transactions of the Entomological Society of London*, **73**, 289–299.

Elton, C. S. (1927). Animal Ecology. Sidwick & Jackson, London.

Emery, S. M. and Gross, K. L. (2005). Effects of timing of prescribed fire on the demography of an invasive plant, spotted knapweed Centaurea maculosa. *Journal of Applied Ecology*, **42**, 60–69.

Eriksson, O. (1996). Regional dynamics of plants: a review of evidence for remnant, source–sink, and metapopulations. *Oikos*, **77**, 248–258.

Fletcher, R. J., Reichert, B. E., and Holmes, K. 2018. The negative effects of habitat fragmentation operate at the scale of dispersal. *Ecology*, **99**, 2176–2186.

Ford, K. R., Ness, J. H., Bronstein, J. L., and Morris, W. F. (2015). The demographic consequences of mutualism:

ants increase host-plant fruit production but not population growth. *Oecologia*, **179**, 435–446.

Gaoue, O. G. and Ticktin, T. (2010). Effects of harvest of non-timber forest products and ecological differences between sites on the demography of African mahogany. *Conservation Biology*, **24**, 605–614.

García, M. B., Goni, D., and Guzman, D. (2010). Living at the edge: local versus positional factors in the long-term population dynamics of an endangered orchid. *Conservation Biology*, **24**, 1219–1229.

Genovart, M., Doak, D. F., Igual, J. M., Sponza, S., Kralj, J., and Oro, D. (2017). Varying demographic impacts of different fisheries on three Mediterranean seabird species. *Global Change Biology*, **23**, 3012–3029.

Gotelli, N. J. and Ellison, A. M. (2006). Forecasting extinction risk with non-stationary matrix models. *Ecological Applications*, **16**, 51–61.

Gouhier, T., Guichard, F., and Menge, B. A. (2010). Ecological processes can synchronize marine population dynamics over continental scales. *Proceedings of the National Academy of Sciences*, **107**, 8281–8286.

Grear, J. S., and Burns, C. E. (2007). Evaluating effects of low-quality habitats on regional population growth in *Peromyscus leucopus*: insights from field parameterized spatial matric model. *Landscape Ecology*, **22**, 45–60.

Grenfell, B. T., Wilson, K., Finkenstadt, B. F., et al. (1998). Noise and determinism in synchronized sheep dynamics. *Nature*, **394**, 674–677.

Gunderson, G., Johannesen, E., Andreassen, H. P., and Ims, R. A. (2001). Source–sink dynamics: how sinks affect demography of sources. *Ecology Letters*, **4**, 14–21.

Gurevitch, J., Fox, G. A., Fowler, N. L., and Graham, C. H. (2016). Landscape demography: population change and its drivers across spatial scales. *Quarterly Review of Biology*, **91**, 459–485.

Hanski I. and Ovaskainen, O. (2000). The metapopulaiton capacity of a fragmented landscape. *Nature*, **404**, 755–758.

Hanski I. and Ovaskainen. O. (2003). Metapopulation theory for fragmented landscapes. *Theoretical Population Biology*, **64**, 119–127.

Hargreaves, A. L., Samis, K. E., and Eckert, C. G. (2014) Are species' range limits simply niche limits writ large? A review of transplant experiments beyond the range. *The American Naturalist*, **183**, 157–173.

Heppell, S. S., Caswell, H., and Crowder, L. B. (2000). Life histories and elasticity patterns: perturbation analysis for species with minimal demographic data. *Ecology*, **81**, 654–665.

Hokit, D. G. and Branch, L. C. (2003). Associations between patch area and vital rates: consequences for local and regional populations. *Ecological Applications*, **13**, 1060–1068.

Horvitz, C. C., Tuljapurkar, S., and Pascarella, J. B. (2005). Plant–animal interactions in random environments: habitat-stage elasticity, seed predators, and hurricanes. *Ecology*, **86**, 3312–3322.

Iles, D. T., Rockwell, R. F., and Koons. D. N. (2019). Shifting vital rate correlations alter predicted population responses to increasingly variable environments. *The American Naturalist*, **193**, E57–E64.

Kahilainen, A., Van Nouhuys, S., Schulz, T., and Saastamoinen, M. (2018). Metapopulation dynamics in a changing climate: increasing spatial synchrony in weather conditions drives metapopulation synchrony of a butterfly inhabiting a fragmented landscape. *Global Change Biology*, **24**, 4316–4329.

Kendall, B. E. and Fox, G. A. (2002). Variation among individuals and reduced demographic stochasticity. *Conservation Biology*, **16**, 109–116.

Kendall, B. E., Fujiwara, M., Diaz-Lopez, J., Schneider, S., Voigt, J., and Wiesner, S. (2019). Persistent problems in the construction of matrix population models. *Ecological Modeling* **406**, 33–43.

Koltz, A. M., Burkle, L. A., Pressler, Y., et al. (2018). Global change and the importance of fire for the ecology and evolution of insects. *Current Opinion in Insect Science*, **29**, 110–116.

Lande, R., Engen, S., and Saether, B. -E. (1999). Spatial scale of population synchrony: environmental correlation versus dispersal and density regulation. *The American Naturalist*, **154**, 271–281.

Larimer, A. L., Bever, J. D., and Clay, K. (2010). The interactive effects of plant microbial symbionts: a review and meta-analysis. *Symbiosis*, **51**, 139–148.

Lee, C. T. (2017). Elasticity of population growth with respect to the intensity of biotic or abiotic driving factors. *Ecology*, **98**, 1016–1025.

Lehtilä, K., Dahlgren, J. P., Garcia, M. B., Leimu, R., Syrjänen, K., and Ehrlén, J. (2016) Forest succession and population viability of grassland plants: long repayment of extinction debt in *Primula veris*. *Oecologia*, **181**, 125–135.

Levins, R. (1969). Some demographic and genetic consequences of environmental heterogeneity for biological control. *Bulletin of the Entomological Society of America*, **15**, 237–240.

Levins, R. (1970). Extinction. In: M. Gesternhaber (ed.), Some Mathematical Problems in Biology (pp. 77–107). American Mathematical Society, Providence, RI.

MacArthur, R. H. and Wilson, E. O. (1963). An equilibrium theory of insular zoogeography. *Evolution*, **17**, 373–387.

MacArthur, R. H. and Wilson, E. O. (1967). The Theory of Island Biogeography. Princeton University Press, Princeton, NJ.

Mandle, L. and Ticktin, T. (2012). Interactions among fire, grazing, harvest and abiotic conditions shape palm demographic responses to disturbance. *Journal of Ecology*, **100**, 997–1008.

Maron, J. L. and Crone, E. (2006). Herbivory: effects on plant abundance, distribution and population growth. *Proceedings of the Royal Society B: Biological Sciences*, **273**, 2575–2584.

Massie, T. M., Weithoff, G., Kucklander, N., Gaedke, U., and Blasius, B. (2015). Enhanced Moran effect by spatial variation in environmental autocorrelation. *Nature Communications*, **6**, 5993.

Menges, E. S. (2000). Population viability analyses in plants: challenges and opportunities. *Trends in Ecology and Evolution*, **15**, 51–56.

Menges, E. S. and Quintana-Ascencio, P. F. (2004). Population viability with fire in *Eryngium cuneifolium*: deciphering a decade of demographic data. *Ecological Monographs*, **74**, 79–99.

Merow, C., Latimer, A. M., Wilson, A. M., McMahon, S. M., Rebelo, A. G.,and Silander, J. A. Jr. (2014). On using integral projection models to generate demographically driven predictions of species' distributions: development and validation using sparse data. *Ecography*, **37**, 1167– 1183.

Molowny-Horas, R., Suarez, M. L., and Lloret, F. (2017). Changes in the natural dynamics of *Nothofagus dombeyi* forests: population modeling with increasing drought frequencies. Ecosphere, **8**, e01708.

Moran, P. A. P. (1953). The statistical analysis of the Canadian lynx cycle. II Synchronization and meteorology. *Australian Journal of Zoology*, **1**, 291–298.

Moretti, M., Duelli, P., and Obrist, M. K. (2006). Biodiversity and resilience of arthropod communities after fire disturbance in temperate forests. *Oecologia*, **149**, 312–327.

Nicholson, A. J. (1933). Supplement: the balance of animal populations. *Journal of Animal Ecology*, **2**, 131–178.

Nicolè, F., Dahlgren, J. P., Vivat, A., Till-Bottraud, I., and Ehrlén, J. (2011). Interdependent effects of habitat quality and climate on population growth of an endangered plant. *Journal of Ecology*, **99**, 1211–1218.

O'Brien, T. G., Kinnaird, M. F., Nurcahyo, A., Prasetyaningrun, M., and Iqbal, M. (2003). Fire, demography and the persistence of siamang (*Symphalangus syndactylus*: hylobatidae) in a Sumatran rainforest. *Animal Conservation*, **6**, 115–121.

Ooi, M. K. J., Whelan, R. J., and Auld, T. D. (2006). Persistence of obligate-seedling species at the population scale: effects of fire intensity, fire patchiness and long fire-free intervals. *International Journal of Wildland Fire*, **15**, 261–269.

Palmer, T. M., Doak, D. F., Stanton, M. L., et al. (2010). Synergy of multiple partners, including freeloaders, increases host fitness in a multispecies mutualism. *Proceedings of the National Academy of Sciences*, **107**, 17234–17239.

Paradis, E., Baillie, S. R., Sutherland, W. J., and Gregory, R. D. (1998). Patterns of natal and breeding dispersal in birds. *Journal of Animal Ecology*, **67**, 518–536.

Parker, I. M. (1997.) Pollinator limitation of *Cytisus scoparius* (Scotch broom), an invasive exotic shrub. *Ecology*, **78**, 1457–1470.

Pascarella, J. B. and Horvitz, C. C. (1998). Hurricane disturbance and the population dynamics of a tropical understory shrub: megamatrix elasticity analysis. *Ecology*, **79**, 547–563.

Pérez-Jorge, S., Gomes, I., Hayes, K., et al. (2016). Effects of nature-based tourism and environmental drivers on the demography of a small dolphin population. *Biological Conservation*, **197**, 200–208.

Prevéy, J. S., Germino, M. J., and Huntly, N. J. (2010). Loss of foundation species increases population growth of exotic forbs in sagebrush steppe. *Ecological Applications*, **20**, 1890–1902.

Pulliam, H. R. (1988). Source, sinks, and population regulation. *The American Naturalist*, **132**, 652–661.

Quintana-Ascencio, P. F., Koontz, S. M., Smith, S. A., Sclater, V. L., David, A. S., and Menges, E. S. (2018). Predicting landscape level distribution and abundance: integrating demography, fire, elevation and landscape habitat configuration. *Journal of Ecology*, **106**, 2395–2408.

Quintana-Ascencio, P. F., Koontz, S. M., Ochocki, B. M., et al. (2019). Assessing the roles of seed bank, seed dispersal and historical disturbances for metapopulation persistence of a pyrogenic herb. *Journal of Ecology*, **107**, 2760–2771.

Ranta, E., Kaitala, V., Lindstrom, J., and Linden, H. (1995). Synchrony in population dynamics. *Proceedings of the Royal Society B: Biological Sciences*, **1364**, 113–118.

Rayfield, B., Fortin, M. -J., and Fall, A. (2011). Connectivity for conservation: a framework to classify network measures. *Ecology*, **92**, 847–858.

Reid, C. (1899). The Origin of the British Flora. Dulau & Co., London.

Ricketts, T. H. (2001). The matrix matters: effective isolation in fragmented landscapes. *The American Naturalist*, **158**, 87–99.

Rodenhouse, N. L., Sherry, T. W. and Holmes, R. T. (1997). Site-dependent regulation of population size: a new synthesis. *Ecology*, **78**, 2025–2042.

Rodriguez, R. J., White Jr, J. F., Arnold, A. E., and Redman, R. S. (2009). Fungal endophytes: diversity and functional roles. *New Phytologist*, **182**, 314–330.

Roy, B. A., Coulson, T., Blaser, W., et al. (2011). Population regulation by enemies of the grass *Brachypodium sylvaticum*: demography in native and invaded ranges. *Ecology*, **92**, 665–675.

Salguero-Gómez, R., Jones, O. R., Archer, C. R., et al. (2015). The COMPADRE plant matrix database: an open online repository for plant demography. *Journal of Ecology*, **103**, 202–218.

Scanga, S. E. (2014). Population dynamics in canopy gaps: nonlinear response to variable light regimes by an understory plant. *Plant Ecology*, **215**, 927–935.

Shefferson, R. P, Hutchings, M. J., Selosse, M. -A., et al. (2018). Drivers of vegetative dormancy across herbaceous perennial plant species. *Ecology Letters*, **21**, 724–733.

Skellam, J. G. (1951). Random dispersal in theoretical populations. *Biometrika*, **38**, 196–218.

Skellam, J. G. (1952). Studies in statistical ecology: i. Spatial pattern. *Biometrika*, **39**, 346–362.

Sletvold, N., Dahlgren, J. P., Øien, D. I., Moen, A., and Ehrlén, J. (2013). Climate warming alters effects of management on population viability of threatened species: results from a 30-year experimental study on a rare orchid. *Global Change Biology*, **19**, 2729–2738.

Smith, M. A. and Green, D. M. (2005). Dispersal and metapopulation paradigm in amphibian ecology and conservation: are all amphibian populations metapopulations? *Ecography*, **28**, 110–128.

Smucker, K. M., Hutto, R. L., and Steele, B. M. (2005). Changes in bird abundance after wildfire: importance of fire severity and time since fire. *Ecological Applications*, **15**, 1535–1549.

Stevens, J. T. and Latimer, A. M. (2015). Snowpack, fire, and forest disturbance: interactions affect montane invasions by non-native shrubs. *Global Change Biology*, **21**, 2379–2393.

Stevens, V. M., Turlure, C., and Baguette, M. (2010). A meta-analysis of dispersal in butterflies. *Biological Reviews*, **85**, 625–642.

Szczys, P., Oswald, S. A., and Arnold, J. M. (2017). Conservation implications of long-distance migration routes: regional metapopulation structure, asymmetrical dispersal, and population declines. *Biological Conservation*, **209**, 263–272.

Tansley, A. G. (1923). Practical Plant Ecology: A Guide for Beginners in the Field Study of Plant Communities. Allen and Unwin, London.

Teller, B. J., Adler, P. B., Edwards, C. B, Hooker, G., and Ellner, S. P. (2016). Linking demography with drivers: climate and competition. *Methods in Ecology and Evolution*, **7**, 171–183.

Trenham, P. C., Koenig, W. D., and Shaffer, H. B. (2001). Spatially autocorrelated demography and interpond dispersal in the salamander, *Ambystoma californiense*. *Ecology*, **82**, 3519–3530.

Tye, M. R., Menges, E. S., Weekley, C., Quintana-Ascencio, P. F., and Salguero-Gómez, R. (2016). A demographic ménage à trois: interactions between disturbances both amplify and dampen population dynamics of an endemic plant. *Journal of Ecology*, **104**, 1778–1788.

Urban, D. L., Minor, E. S., Treml, E. A., and Schick, R. S. (2009). Graph models of habitat mosaics. *Ecology Letters* **12**, 260–273.

Vardi, T., Williams, D., and Sandin, S. A. (2012). Population dynamics of threatned elkhorn oral in the northern Florida Keys, USA. *Endangered Species Research*, **19**, 157–1969.

Williams, A. L., Wills, K. E., Janes, J. K., Schoor, J. K. V., Newton, P. C. D. and Hovenden, M. J. (2007). Warming and free-air CO_2 enrichment alter demographics in four co-occurring grassland species. *New Phytologist*, **176**, 365–374.

Winter, S., Jung, L. S., Eckstein, R. L., Otte, A., Donath, T. W., and Kriechbaum, M. (2014). Control of the toxic plant *Colchicum autumnale* in semi-natural grasslands: effects of cutting treatments on demography and diversity. *Journal of Applied Ecology*, **51**, 524–533.

Yule, K. M., Miller, T. E. X., and Rudgers, J. A. (2013). Costs, benefits, and loss of vertically transmitted symbionts affect host population dynamics. *Oikos*, **122**, 1512–1520.

PART 2

Data- and Research Question-Driven Methods

CHAPTER 7

Abundance-based approaches

Jonas Knape and Andreas Lindén

7.1 Aims of abundance analyses

Data describing the abundance of populations across time or space, but lacking information on demographic structure, are widely available. Abundance data of this type are generally available at greater resolution, and at larger scales in time and space, than are data on demographic rates. In many cases abundance data are the only data available about populations at national, regional or landscape scales, while we may, at best, have estimates of vital rates from a handful of local and more detailed studies, often of short duration. Abundance data therefore form a backbone for the study of a large number of species and populations in the wild.

Abundance, N, is a natural aggregate property of populations that plays a fundamental role in applied and theoretically oriented ecological studies. Often abundance itself is of major interest, and aims may be directed to how abundance has changed historically, how it varies in space, and what the current levels of abundance are or are predicted to be in the future. For instance, estimating historical abundance is the focus of analyses of population trends from monitoring programmes, which may be used for red-listing of species or as indicators of the state of the environment, and estimates of current abundance are used to set fisheries quotas or hunting caps, sometimes in combination with short-term predictions on how harvesting might affect population abundance.

Abundance is also used to study patterns in dynamics. Classical questions addressed in animal ecology concern the level of synchrony of populations separated in space (Ranta et al., 1995; Liebhold et al., 2004), the existence of cycles in population dynamics (Barraquand et al., 2017), and how abundance varies with environmental characteristics such as across an altitudinal gradient or with weather (Ives, 1995) or climate (Ådahl et al., 2006).

Beyond focus on population levels and patterns, abundance may be used to learn about basic population-level parameters to better understand why populations change. The most fundamental population parameter available from abundance data, apart from abundance itself, is the population growth rate. In ecology, the growth rate is studied to understand patterns of density dependence and regulation, or how populations respond to environmental factors, including abundances of interacting species, while in evolutionary studies the growth rate may be used as a measure of fitness.

For populations growing continuously in time the per capita rate of increase may be defined as the derivative of population size with respect to time divided by current population size. That is, the instantaneous rate of additions of individuals per individual present in the population:

$$r(t) = \frac{dN}{dt}/N$$

The population will then grow when $r(t)$ is positive, decrease when it's negative, and be stable when it is zero. Alternatively, in discrete time, we may instead define the multiplicative growth rate via

$$\lambda_t = \frac{N_{t+1}}{N_t}$$

where λ_t is the per capita increase in population size from time t to $t+1$. Commonly discrete growth is studied through the logarithm of λ_t, $r_t = \log(\lambda_t)$ instead of directly. On the logarithmic scale multiplicative growth processes become additive, making it better suited for many population analytical approaches. The similarity in notation between $r(t)$ and r_t is no coincidence since if time steps are small, r_t is approximately equal to $r(t)$, and the discrete time population will similarly increase from time t to $t+1$ if r_t is positive and decrease when it is negative. From this perspective, the

Jonas Knape and Andreas Lindén, *Abundance-based approaches*. In: *Demographic Methods Across the Tree of Life*.
Edited by Roberto Salguero-Gómez and Marlène Gamelon, Oxford University Press.
© Oxford University Press (2021). DOI: 10.1093/oso/9780198838609.003.0007

aim of many studies of abundance is to understand how and why the function $r(t)$ or r_t varies over time, in space, and as a function of population size or external factors.

7.2 Measures of abundance

What do we mean by abundance? In section 7.1 we were deliberately vague about how abundance is defined, and in practice the meaning of abundance varies depending on the context. The most intuitive descriptors of abundance are the total number of individuals in the population, or the number of individuals per unit area, known as population density. These measures are commonly applied for several types of organisms, including most terrestrial vertebrates and arthropods. Other measures of abundance are in use and may be preferred for biological or practical reasons. For example, since tree growth depends on tree size it can be easier to predict the growth of a population of trees in a forest using its biomass rather than the number of trees. Fish also grow largely indeterminately and with size dependence, which partly explains the common use of biomass for studying the dynamics of fish populations. There can also be practical reasons for using fish biomass over the number of individuals. When populations are assessed using trawl surveys or by fishery vessels, it may require less effort to weigh the catch than it does to count it. For plants with clonal growth it may be difficult to separate individuals in the field, and counting ramets instead of individuals may be preferred and better represent abundance (Bullock, 1999). Another example of alternative measures of abundance is areal cover, which is used to describe the size of some populations of plants. Ultimately, the aims of the study in combination with practical considerations should be used as a guide when deciding on the definition of abundance for a population. If the aims are not properly considered, the data to be collected may not be able to address the questions that are asked of them (Reynolds et al., 2016).

Apart from determining what to measure, an important consideration when defining abundance for a population is to decide when and how often abundance should be measured. For some species and purposes, a single abundance measure per year may be sufficient, and this temporal scale of sampling is common for vertebrates in particular but also for other organisms. For organisms with fast life histories, perhaps with several generations in the course of a single year, more frequent sampling will be necessary for detailed studies

of changes in abundance. Similarly, some organisms including many plants and insects have strongly seasonal abundance patterns which makes the timing of sampling critical. Butterfly monitoring programmes, for example, may sample abundance on 20 or more occasions each year to cover the full seasonal cycle. For studying long-term changes in abundance these seasonal samples may be aggregated in some way to annual indexes of abundance, for example by estimating the area under or the maximum of the seasonal curve (Dennis et al., 2013). Similar measures may be applied to organisms with very short generation time, such as bacteria or phytoplankton. Considering seasonal patterns has become increasingly important for many organisms due to accelerating climate change and rapidly changing phenology in a wide range of taxa. In some cases, failing to account for changing phenology in abundance at the time of sampling may lead to biased estimates of population abundance.

It is further important to define the area across which the population is to be studied. Considering the type of organism and the spatial scale of the questions asked are fundamental to this definition. Delimiting the spatial extent of a population is typically easier for sessile organisms than it is for mobile ones. For the latter a complicating factor is that it is typically impossible to estimate emigration and immigration rates from abundance data. Therefore it is sometimes convenient to define populations over a large enough area so that migration does not play a major role in its dynamics.

7.2.1 Sampling abundance

Sampling abundance in the wild is in general a difficult and arduous task. A wide range of sampling methods exists, and the methods differ greatly among taxa. Sampling may involve counting individuals, counting animal tracks along a predefined transect, mapping territories, marking and recapturing individuals, sampling areal cover using pin-quadrats or photographs, camera traps, acoustic monitoring, electro fishing in streams, or trawl surveys in the sea. For an overview of these and other basic methods for obtaining abundance data we refer the reader to section (5.2.1) and Sutherland (2006).

The sampling of abundance almost invariably involves some degree of observation error, implying that we obtain an estimate of abundance rather than abundance itself. Observation error has repercussions for several types of analyses, and it is generally important to think of the consequences that it may have on analyses of abundance data. One obvious consequence of observation error is that estimated abundance will tend to be more variable than abundance itself. Failing to consider this could, for example, lead to overestimation of extinction risks. Observation error also weakens signals in population dynamics, making inference about population processes harder and more uncertain.

Samples of abundance may further be biased. A commonly encountered problem is that individuals may be hard to find, so that measured abundance tends to be smaller than actual abundance, sometimes by a considerable margin (see section 5.2.1). Some sampling schemes are designed to provide information of such bias due to detection errors. These include mark–recapture methods (McCrea and Morgan, 2015) and removal and distance sampling (Buckland et al., 2015). Other sampling schemes make no such attempts and instead provide an index of abundance. For example, an index of current abundance may be defined as the proportional abundance relative to a reference year, or relative indices of abundance may be computed for comparisons among sites.

Under certain conditions, an estimate of the number of individuals corrected for detection can be obtained from repeated counts at many sites, without any other dedicated sampling design (Royle, 2004; Kéry et al., 2005). This method is sensitive to model assumptions and can give misleading estimates of abundance if the assumptions, some of which cannot be checked without other types of data, are not met. It has therefore sparked controversy (Barker et al., 2018). It tends to work best when individuals are distributed uniformly in space and are not too mobile, detection rates are high, and abundances are small (Knape et al., 2018).

Even with sampling designs intended to quantify detection and other observation errors, estimates of total abundance will often be uncertain and biased. Obtaining useful relative estimates of abundance can be substantially easier. Relative estimates contain no information about total abundance, for example on how many individuals there are in a population, but they may be sufficient for inference on population growth rates (r_t). Time series of relative abundance can be applied, for instance, to estimate how fast a population is decreasing. How appropriate relative measures of abundance are in this case hinges on whether and how observation errors vary in time.

The large majority of methods discussed in this chapter are general enough that they may, perhaps with some minor adaptations, be applied to most types of abundance measures.

7.3 Analytic approaches

7.3.1 Regression-type models

One approach to analysing abundance data is to rely on standard regression techniques from statistics, such as linear models, generalised linear models, mixed models, smoothing methods such as additive models, or sparse regression methods. These approaches are used in several different contexts of abundance analyses. An example is the use of regression for estimating temporal changes in abundance from monitoring data. Several countries have implemented national monitoring programmes to track the status of species over time, primarily for birds, other terrestrial vertebrates, and butterflies. Their results are used in conservation settings for red-listing of species that are in decline, as general environmental indicators, or for applied research purposes to study relationships between environmental changes and population trends.

The basic premise for trend estimation from abundance data is in the form of a regression of an abundance measure x_t against time:

$$x_t = a + bt + \epsilon_t \tag{7.1}$$

where a is an intercept, b is the trend or change in population size, and ϵ_t is a term denoting sampling error. Typically x_t is an estimate of the logarithm of population size rather than the population size at the arithmetic scale, so that $x_t = \log(N_t)$. If this is the case, the exponent of the slope coefficient, e^b, is interpreted as the proportional change in the population per unit time, and could be reported in the form of a percentage. This parameter is one of the most typically reported statistics from monitoring studies and often applied in evaluation of red-listing criteria.

The above regression for estimating population trends applies to data on a single population but can be extended in many ways. For example, if surveys are conducted over multiple sites the model may be extended to study how population trends vary in space or with environmental variables in space and time (Oedekoven et al., 2017). Vanhatalo et al. (2017) used one such extension including spatiotemporal effects to study outbreak patterns in starfish at the Great Barrier Reef. Regression models may also be used for impact assessments to estimate the effects of infrastructure development (Pépino et al., 2012) or extreme natural events (Rochester et al. 2010; Chevalier et al. 2019) on abundance.

In regression-based approaches a statistical model is used to analyse the abundance data without necessarily explicitly specifying a model of population dynamics. In general, the focus of inference in regression-based approaches is more towards identifying patterns in and correlatives of abundance and growth rates, rather than to learn about the dynamic mechanisms behind the population processes.

Example: trends in the common cuckoo

The common cuckoo (*Cuculus canorus*) is a migrant bird with a widespread summer breeding range in Europe and Asia. It is a brood parasite, laying its eggs in the nests of mainly small-sized passerines (Cramp 1985). The Swedish breeding bird survey consists of standardised annual transect counts of breeding birds, including the cuckoo, at around 700 sites laid out on a grid across Sweden (Tayleur et al., 2015). The aim is to follow the change in the national population sizes of common breeding birds.

Counts at each site are conducted once a year during the breeding season, yielding data in the form of the number of seen or heard cuckoos, y_{it}, for site i in year t. To estimate a smooth trend in the total population size we may use an extension of eqn. (7.1)

$$y_{it} \sim \text{NegBin}(\exp(a_i + s(t) + \epsilon_t), \theta)$$

where $s(t)$ is now a smooth and possibly nonlinear trend over time, a_i is a site-specific random effect, ϵ_t is a random year effect, and the counts are assumed to follow a negative binomial distribution with overdispersion parameter θ. The random effects are assumed to be normally distributed and independent among, respectively, sites and years. The site effects account for variation in abundance level and detection rates among sites, while the year effects account for common annual variation in the data and hence modelling spatial synchrony in the unexplained variation (Knape, 2016).

The estimated national trend is shown in Figure 7.1. It suggests a slight increase in the cuckoo population in early 2000 followed by a slight decrease. We may ask whether this national trend is consistent across Sweden. To address this, we use the average length of the growing season as a proxy for gross spatial variation in habitat and climate over Sweden. The growing season is short in the northern alpine regions of Sweden, increases a bit in the northern and mid-latitude forested areas, and is longest in the southern, largely agricultural, regions. We then fit a model similar to the previous one but with a two-dimensional smooth term:

$$y_{it} \sim \text{NegBin}(\exp(a_i + s(t, g) + \epsilon_t), \theta).$$

The smooth interaction term $s(t, g)$ allows trends in time to vary in space as a function of the length of the

growing season g. Figure 7.1 shows the results from fitting this model. They indicate that population change varied in space with fairly stable population size in the south, a strong increase followed by a decline in central Sweden, and a recent decline from a stable level in the north.

7.3.2 Dynamic abundance models

An alternative to regression-based approaches is to use population ecological theory as the starting point for models and apply flexible statistical techniques for their assessment. For example, fishery biologist often rely on model-based approaches to guide sustainable management and harvest of fish stocks (Quinn and Deriso, 1999).

Mathematical models of population dynamics have a long history in ecology. They have been used in purely theoretical capacities to better understand the foundations of population growth, regulation, or outbreak risks. A classic example is May's examination of chaotic dynamics in simple mathematical population models when growth rates are high (May 1976). Population models are also commonly used and fitted to data in empirical studies. An early example is Gause's study of population limitation in yeast via a logistic model (Gause, 1932).

Observation error models

Perhaps the easiest way of linking data to a population model is to assume that the dynamics are deterministic but that abundance is observed with error. This leads to a regression problem that can be relatively easy to solve numerically. While the assumption of deterministic dynamics is often a gross simplification, it can make it possible to fit mathematically complex models that could otherwise pose problems.

We have in the previous section already encountered an example of an observation error population dynamics model, although it was not cast as such. The regression model of log-population size against time may be reformulated as a model of this type. The model can be expressed as

$$\begin{aligned} N_t &= N_0 \exp(rt) \\ y_t &= N_t \exp(\epsilon_t) \end{aligned} \tag{7.2}$$

Here we have changed the parametrisation of the model in eqn. (7.1) by using the initial population size N_0 at time 0 in place of a, linked via $N_0 = e^a$, and the slope b has been renamed to r to clarify its relation to the growth rate. The first equation is that of a deterministic exponential growth model. This model describes the dynamics of a population with constant per-capita

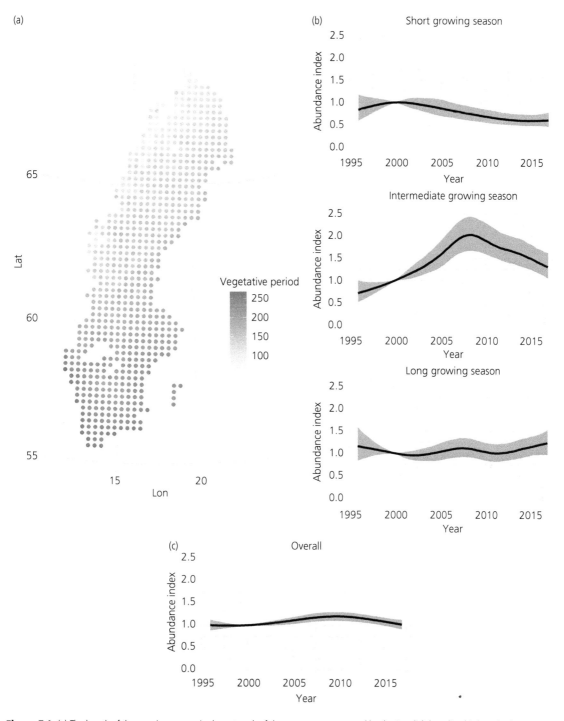

Figure 7.1 (a) The length of the growing season in days at each of the survey routes covered by the Swedish breeding bird monitoring programme. (b) The long-term population trends of common cuckoo are nonlinear in shape and vary across Sweden, here approximated as a function of the length of the growing season. The trends in abundance over years are illustrated as abundance indices, scaled relative to the reference year 2000. This is shown for areas with short, intermediate, and long growing season. (c) Abundance indices when the trend is assumed to be the same across all of Sweden. Shaded regions show approximate 90% confidence intervals for the population change relative to the reference year.

growth rate and is the solution to the differential equation

$$\frac{dN}{dt}\frac{1}{N} = r.$$

The second component of the model in eqn. (7.2) relates the observed or measured abundance, y_t, to the abundance N_t via a model for observation error. The random error term ε_t is often assumed to be normally distributed. The relation then describes the log-normal distribution, which can often be an acceptable choice for abundance data in the form of population numbers or biomass of large populations. The equivalence between this model and the linear regression model in eqn. (7.1) means that standard linear regression software can be used to fit it to data by putting log(yt) as the response variable and t as the explanatory variable.

The model in eqn. (7.2) highlights the distinguishing feature of observation models. They can be partitioned into a deterministic model of abundance, here the deterministic exponential growth model, and a stochastic observation error model of how observed abundance arises from the 'true' abundance, here a log-normal error distribution. In place of the log-normal distribution, other models for observation errors may sometimes be more appropriate. Binomial, Poisson, or negative binomial distributions could be used for discrete count data (Lindén and Mäntyniemi, 2011), a beta distribution for continuous cover data, or a Tweedie distribution for biomass data (Dunstan et al., 2013). For the exponential growth model these observation error distributions lead to a generalised regression problem, and generic GLM software can be used to fit such models to data. Most population dynamics models cannot, however, be framed as linear, and fitting them to data in an observation error framework then requires nonlinear regression techniques.

A population cannot grow indefinitely according to an exponential growth model because it would, sooner or later, either go extinct or consume all its available resources. To prevent this some form of negative density dependence, a decrease in per-capita growth rate as populations become large, must be at play (Hixon et al., 2002). The importance of regulation was realised early in the study of population ecology, and the classic logistic population model is built around this idea by simply extending the exponential growth model in eqn. (7.2) such that the per-capita growth rate decreases linearly with abundance:

$$\frac{dN}{dt}\frac{1}{N} = r(1 - N/K)$$

When the population is small the ratio N/K is negligible and we recover the exponential growth model. Thus, a population following a logistic model initially grows exponentially at a rate r, the intrinsic growth rate, if it starts from a small abundance. As the population approaches the carrying capacity K the growth rate decreases towards zero.

This model, which is nonlinear, can be fit to data assuming that deviations between model predictions and data are due to observation error only. To do so, it helps to solve the logistic differential equation to get a mathematical expression for how abundance changes with time under this model. This can be done using basic techniques of calculus and the solution is

$$N(t) = \frac{KN(0)e^{rt}}{K - N(0) + N(0)e^{rt}}$$

If we assume that the logistic model is observed with a log-normally distributed observation error at times $t_1, t_2, \ldots, t_m$, that is $\log y_i \sim \text{Normal}(N(t_i), \sigma^2)$, the log-likelihood for the data can be constructed from the density of the normal distribution as

$$l(r, K, N(0), \sigma; y) = -\frac{m}{2}\log(2\pi\sigma^2)$$
$$-\frac{1}{2\sigma^2}\sum_{i=1}^{m}(\log y_i - N(t_i))^2$$

The model can then be fit to data by maximising the likelihood over the parameters r, K, $N(0)$, and σ. Generic optimisation routines are available to solve this task (Bolker, 2008). Due to the normally distributed error (on the log scale), it is sufficient to numerically minimise the sum of squares over r, K, and $N(0)$, which is a nonlinear least squares problem.

In general, the most important requirement for making it technically possible to fit an observation error model to data is that we can predict the population trajectory, analytically or numerically, over time (and/or space) for all relevant values of the parameters. Since the population model is deterministic, population size is a function of time, t, and the parameter values ϕ (e.g. $\phi = (r, K, N(0))$ in the logistic model). Technically, the requirement therefore means that we can compute this function, $N(t, \phi)$, for various values of t and ϕ.

The population trajectory can then be coupled to data using a probability density for the observations that is appropriate for the data. For example if the abundance data are counts, a discrete observation error model such as a negative binomial distribution may fit the data better than the log-normal distribution used above. Formally, the link between data and population trajectory may be described via the probability density or probability mass function of the observation model, $p(y_t|N(t,\phi),\theta)$, where θ contains the parameters of the observation model. From this we may construct a general version of the log-likelihood,

$$l(\theta,\phi|y) = \sum \log p(y_i; N(t_i,\phi),\theta),$$

which can then be maximised numerically or alternatively could be used in a Bayesian framework.

This recipe can make it possible to fit very advanced deterministic models to data. For instance, $N(t,\phi)$ may be implicitly defined as the solution to a differential equation or a system of differential equations that describe population dynamics (Nisbet and Gurney 1982). Even if we cannot solve the differential equation analytically, as we could for the logistic and exponential models, we may be able to solve it numerically and compute a close approximation to $N(t,\phi)$ that can be plugged into the likelihood equation above. Nonlinear discrete time models may also be used. Even deterministic individual-based models can in principle be fitted this way. That is, one could simulate a deterministic model of individuals in a population over time, compute population size from the total number of individuals in the simulation, and then plug the simulated deterministic population size into the likelihood. In practice, care needs to be taken to avoid having more parameters than can be reasonably informed by the abundance data, and careful consideration of parameter identifiability would be necessary.

While the principle of estimating observation error models this way is straightforward, the likelihood surface can for some models be very complex, making estimation difficult in practice. Models with complex dynamics, for instance, can have erratic likelihood surfaces that are difficult to deal with using standard frequentist or Bayesian methods (Wood, 2010).

Intuitively, fitting observation error models can be viewed as fitting a population trajectory curve, specified by a population model, through the data. Before the widespread availability of computing power, the relative simplicity in adapting them to data sometimes made them the only viable option for estimating parameters of a population model. Even today, this simplicity is the main advantage of observation error models, and in cases where fitting a deterministic model is sufficient to address the question at hand, they are convenient to work with.

Example: damselflies in newly established ponds

Dragonflies were counted annually immediately following the establishment of a number of small ponds

at a site in England in 1962 (Moore, 1991). We consider data for the azure damselfly (*Coenagrion puella*) between 1963 and 1989. Counts analysed here are the annual maximum number of individuals recorded over multiple visits each year. The counts of this species initially increased slowly after the creation of the ponds before they possibly stabilised around some average level.

In management and conservation applications one may need indicators of habitat quality or suitability, which could be measured in terms of population dynamical characteristics. For example, the growth potential, or how many damselflies ponds could host on average in a saturated population, could indicate pond quality and suitability. We therefore estimate the intrinsic growth rate (r) and the carrying capacity (K) by fitting the logistic model for population growth. We assume that the annual counts are negatively binomially distributed due to observation errors, with an expected value equal to the population trajectory ($N(t)$). Model uncertainty is assessed using parametric bootstrap with 5,000 resamples, and also more specifically for parameter K, by plotting its profile likelihood function.

The model fit and residual diagnostics are shown in Figure 7.2. The parameter estimates (and 90% confidence intervals) are: initial population size $N(1) = 2.7$ $(1.1, 4.7)$, $r = 0.23$ $(0.16, 0.35)$, $K = 210$ $(105, 10^8)$, and overdispersion parameter $\theta = 0.55$ $(0.27, 0.76)$. The huge uncertainty of K, particularly its flat likelihood for values larger than the estimate (Figure 7.2b), reflects that it is unclear whether the population has yet approached an asymptote. Apart from the problems with uncertainty, analyses of the Dunn–Smyth residuals suggest a reasonable model fit (Figure 7.2c,d).

The results suggest that an exponential growth model fit the data approximately equally well and that these data cannot be used to obtain a reliable estimate of K. Even a 27-year time series, with several data points intuitively suggesting that population growth may have ceased, is not necessarily enough for this task. Proper uncertainty assessment is crucial in this kind of analysis, while a realistic point estimate ($K = 210$) and reasonable model fit are no guarantee for a reliable result. At the same time, we have good information on other parameters. Regarding population characteristics that could indicate habitat quality, our only option is to concentrate on the population growth potential (via the estimate of r). The situation could be improved even for K if there would be less variability in the counts and possibly if new data for

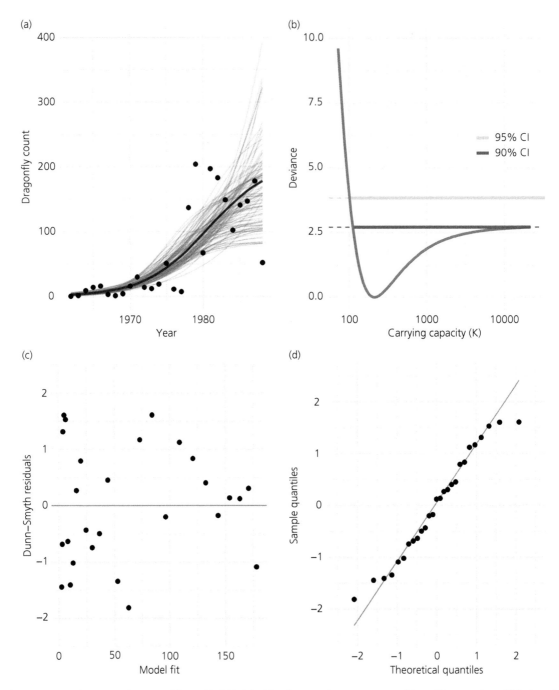

Figure 7.2 (a) A deterministic logistic model curve (thick black line) fitted to annual maximum counts of damselflies in newly established ponds (black points) in England. Green lines show 100 bootstrap samples, illustrating the uncertainty of the model fit. (b) The model deviance (i.e. difference in −2 log-likelihood) is a shallow function of carrying capacity (K) on the right side of the minimum (K = 210), barely reaching the critical value for an approximate 90% confidence interval. The upper limit of a corresponding 95% confidence interval goes to infinity, illustrating that we cannot reliably estimate K. (c) The Dunn–Smyth residuals show no obvious pattern when plotted against the model expectation, and (d) a QQ-plot of the Dunn–Smyth residuals indicate no apparent deviations from a normal distribution, suggesting a reasonable model fit.

a few years more would become available. Less variability in count data may be achieved by improving the monitoring practices or by somehow combining the information available from the multiple visits.

Stochastic population models

Variation plays a central role in the dynamics of populations. When using deterministic models, properties of population dynamics that are inherently stochastic cannot be studied adequately. For instance, if we want to estimate the risk of extinction for some species, a stochastic model is a natural starting point. The majority of analyses of abundance data therefore use population models that incorporate variability in the dynamics.

In the study of stochastic population dynamics one usually distinguishes between two main types of variation denoted environmental and demographic stochasticity (Engen et al., 1998). Demographic stochasticity operates at the level of individuals and is due to random variation in outcomes of life history events such as survival and reproduction. Environmental stochasticity, on the other hand, operates at the level of the population and is due to fluctuations in the environment affecting all individuals but varying through time. Demographic stochasticity therefore concerns variation in outcomes in life history events, while environmental stochasticity concerns variability in the average *rates* that govern those outcomes. For example, demographic stochasticity in survival is the variation caused by some individuals surviving and others not, while environmental stochasticity in survival is the among-year variation in annual survival probabilities.

Demographic stochasticity has the largest relative effects in small populations where the fate of single individuals can be crucial for population persistence. In large populations demographic stochasticity to a large extent averages out and becomes less important. In contrast, environmental stochasticity affects populations of all sizes. With abundance data, it is typically difficult to quantify the contribution of the demographic and environmental variation to the total variation in the dynamics. An approach often taken is to assume that demographic variation is negligible and to instead focus on environmental variation.

In applications to data, it is common to study stochastic population dynamics at a discrete timescale. A simple and widely used approach to build a stochastic population dynamics model with environmental variation is to start with a deterministic discrete time model and add a multiplicative error term to it (Brännström and Sumpter, 2006). For instance, we may take the deterministic discrete time exponential growth model as the starting point and multiply the multiplicative growth rate λ by an error term $\exp(\eta_t)$, where η_t is normally distributed with mean zero and variance σ^2. The stochastic exponential growth model then is

$$N_t = N_{t-1} \exp(r + \eta_t) \tag{7.3}$$

This model is often used in population viability analyses to assess extinction risks (Dennis et al., 1991; Holmes, 2001; Jacquemyn et al., 2007; Ramula et al., 2008). Because the model is stochastic, extinction risk is nonzero even if the average logarithmic growth rate r is positive.

The same approach to adding environmental variation is being used for other deterministic population models. The deterministic Ricker model, a discrete time variant of logistic growth, is

$$N_{t+1} = N_t \exp\left(r\left(1 - \frac{N_t}{K}\right)\right)$$

Environmental variation can be added to this model in the same way as for exponential growth by adding a multiplicative error term, resulting in the model

$$N_{t+1} = N_t \exp\left(r\left(1 - \frac{N_t}{K}\right) + \eta_t\right) \tag{7.4}$$

While adding a normally distributed error term to the exponent of the growth rate is common, it does not always result in a realistic description of growth. Many populations have an upper bound to how much they can increase over a single time period, and such a bound is not strictly honoured by the additive normally distributed error term. Upper bounds could be enforced by instead adding a multiplicative error term to the carrying capacity (Ruokolainen et al., 2009), resulting in little variation when the population is far from the equilibrium, or by replacing the exponential function for a logistic one (Sköld and Knape, 2018).

Autoregressive models

An alternative, but related, approach to the analysis of stochastic population dynamics is to rely on statistical methods for time series analysis. In population dynamics, this was popularised by Royama (1992), Turchin (1995), and others who used linear autoregressive (AR) models of the annual log growth rate. The simplest model of this type is the AR(1) model, an autoregressive model with one time lag:

$$X_{t+1} = a_0 + a_1 X_t + \eta_t \qquad (7.5)$$

where a_0 is an intercept, a_1 is the first order autoregressive coefficient and η_t is a normally distributed error term with zero mean. In studying population dynamics, X_t is usually taken to be the logarithm of abundance at time t (i.e. $N_t = \exp(X_t)$). On the arithmetic scale, the model then can be written

$$N_{t+1} = N_t \exp(a_0 + (a_1 - 1)\log(N_t) + \eta_t)$$

Similarly to the stochastic Ricker model above, this model has an annual growth rate that depends on abundance and a random error term representing environmental stochasticity. It is sometimes referred to as the Gompertz model and has been frequently used to study density dependence (Dennis and Taper 1994) via the autoregressive parameter a_1. When a_1 is 1 the dependence of the growth rate on abundance disappears, implying density independence. Values of a_1 less than 1 imply a negative effect of increased density on the growth rate. If a_1 is also larger than -1, N_t will reach an equilibrium value around which the population will fluctuate. The equilibrium value is $\exp(a_0/(1-a_1))$ and may be interpreted as a stochastic carrying capacity in a manner similar to K in the stochastic Ricker model in eqn. (7.4).

By adding further autoregressive terms to the model, delayed density dependence can be studied. Delayed density dependence occurs when population size at previous time steps affects current growth rates. This can happen through age structure in the population (Fromentin et al., 2001), through interactions with other species (Turchin, 2003), or due to effects of unknown autocorrelated environmental variables (Ripa and Ives, 2007). In the first two cases it may be argued that fundamental biological processes are missing from the model, but adding delayed density dependence can be seen as a phenomenological short-cut for modelling the desired dynamical properties without making the resulting model overly complex. The simplest model with delayed density dependence is the AR(2) model

$$X_{t+1} = a_0 + a_1 X_t + a_2 X_{t-1} + \eta_t$$

The second order autoregressive coefficient a_2 represents effects of the abundance two time steps previously, not captured by the effect of abundance in the previous time step. This model can, in contrast to the AR(1) model, produce quasi-cycles in dynamics (stochastic cycles which get out of phase over time). It has been used to study cycles and spatial synchrony (Buonaccorsi et al., 2001) in birds (Watson et al., 2000) and small

mammal dynamics (Cornulier et al., 2013), often in combination with effects of environmental variables on population growth rates (Stenseth et al., 2002). Another application is studying insect outbreak dynamics (Peltonen et al., 2002).

Autoregressive models of the logarithm of abundance may be interpreted directly as population models as above. Alternatively, they may be seen as linearisations of nonlinear models around an equilibrium point (Månsson et al., 2007). Linearisation can be useful for an improved understanding of nonlinear models or as an approximation to nonlinear dynamics for small deviations from an equilibrium (Ripa and Heino, 1999).

Fitting stochastic population models to data under no sampling error

In some simple cases, parameter estimates of stochastic populations models can be obtained fairly easily. For example, when the exponential growth model with environmental stochasticity is assumed to be observed without error, the logarithms of the annual growth rates are independent and normally distributed

$$\log(N_{t+1}/N_t) \sim \text{Normal}(r, \sigma^2).$$

Estimates of r and σ^2 can therefore be obtained from, respectively, the sample mean and the sample variance of the logarithm of the observed annual growth rates (Dennis et al., 1991; Morris and Doak, 2002). Most stochastic models, however, cannot be transformed to a set of independently observed quantities in this way. One then has to be aware of the temporal statistical dependence among abundances, induced by the stochastic population model. Handling this dependence when estimating stochastic population models, and dynamic statistical models in general, requires some care. Software specifically adapted for time series is available for certain models and greatly simplifies model estimation. In particular, there are several software options available for fitting AR models. This is also the case for ARMA models that are an extension of AR models that include moving average (MA) terms. The MA terms in ARMA models can be useful for implicitly accounting for measurement error (in contrast to the more explicit approach described in the section 7.3.3), or for accounting for interactions with other, unobserved, species (Ives et al., 2010). For nonstandard models for which dedicated software is not readily available, common approaches are to fit models using maximum likelihood (ML) or restricted maximum likelihood (REML) methods in a frequentistic framework, or Markov chain Monte Carlo (MCMC) in a Bayesian framework. Likelihoods for stochastic population models observed without error can often be computed without too much trouble. For example, a likelihood for the stochastic discrete Ricker model is similar to the likelihood of a linear regression of $\log(N_{t+1}/N_t)$ against N_t. Therefore, ML estimates for this and similar, models can be obtained using standard regression software. These estimates, however, do not necessarily behave in the same way as they do in an ordinary regression problem. In contrast to standard regression, the temporal dependence in time series can cause estimates to be biased in small samples, which can be problematic in practice since most ecological time series are short, and uncertainty estimates obtained from regression are typically not valid in the time series setting. One way to explore these issues is to use simulations (Bence 1995).

7.3.3 State-space models

The two approaches to estimation of dynamic models from abundance data we have seen so far assumed either that the population process is deterministic or that abundance was observed without error. Both of these assumptions are usually unsatisfactory, as in reality we expect considerable unpredictability in the population process, as well as difficulties in accurately measuring abundance. Moreover, observation error has been shown to have the capacity to severely bias estimates of population parameters and processes in dynamic models when data are assumed to be observed without error. Biased estimates under these circumstances have been reported for density dependence (Shenk et al., 1998; Lebreton and Gimenez, 2013), extinction risk, sustainable harvesting strategies (Walters and Ludwig, 1981), and effects of autocorrelated environmental covariates (Lindén and Knape, 2009). State-space models deal simultaneously with observation and population process errors and may be used to try to overcome these problems and to provide more realistic descriptions of data.

A simple state-space model is the exponential growth state-space model (Lindley 2003; Holmes *et al.* 2007)

$$
\begin{aligned}
N_{t+1} &= N_t \exp(r + \eta_t) \\
y_t &= N_t \exp(\epsilon_t)
\end{aligned}
\tag{7.6}
$$

The model combines the stochastic exponential growth model from eqn. (7.3) with the observation model from the exponential observation error model from eqn. (7.2). It illustrates the fundamental structure of population state-space models, with a dynamic model

for the unobserved population 'states' N_t and an observation error model for which observations y_t at different points in time are independent given the population process. Thus, the temporal statistical dependence in the data is assumed to arise from the population process only.

This basic model can be extended in several ways. One example is the Gompertz state-space model with log-normal observation error (Dennis et al., 2006). The model is obtained by replacing the exponential growth model by a Gompertz model in eqn. (7.6). Taking the logarithm of abundance and observations, this model may be written

$$
\begin{aligned}
X_{t+1} &= a_0 + a_1 X_t + \eta_t \\
\log(y_t) &= X_t + \epsilon_t
\end{aligned}
\tag{7.7}
$$

The Gompertz model, which includes the exponential model as a special case ($a_1 = 1$), is in the form of a linear Gaussian state-space model. Essentially, this means that both the observation and process equations are linear with Gaussian errors. State-space models that are linear and Gaussian are relatively easy to work with because efficient computational tools are available for them, in particular the Kalman filter. Two of the most influential statistical books on state-space models are almost exclusively devoted to models of this kind (Harvey, 1990; Durbin and Koopman, 2012). Linear state-space models can have multivariate observations as well as multivariate states, which allows joint modelling of spatially separated populations, inclusion of age structure, species interactions, incorporating lagged density dependence, seasonal effects, dealing with missing covariates and covariates observed with error (Almaraz et al., 2012), or dynamic factor analysis to look for common trends among multiple population time series (Zuur et al., 2003). In short, linear state-space models offer many possibilities and a lot of flexibility.

For some types of data a log-normal or Gaussian observation distribution may not be appropriate, and it is necessary to go beyond the linear state-space model framework. The observation equation may then be changed, for example to a negative binomial distribution for counts or to a beta-binomial distribution for pin-point cover data (Damgaard, 2012). Similarly, nonlinear dynamics such as the Ricker model can be accommodated by modifying the population process model. State-space models may also be used to combine a population dynamics model with an observation model derived from more complex sampling designs within a single model. For instance, a model of population dynamics may be combined with a model for distance sampling data and be jointly estimated. It is also possible to plug in existing estimates of variances in population abundances, along with the abundance estimates themselves, into state-space models in a two-stage approach (Newman et al., 2014).

Fitting state-space models to data can be done by specialised algorithms, by numerical ML estimation in cases where the likelihood can be readily computed (notably the linear state-space model), in Bayesian settings using MCMC, or by various kinds of approximation methods.

The joint treatment of observation and process errors makes state-space models compelling, and, for many reasons, they are often to be preferred over methods assuming only process error, or only observation error. However, they also bring substantial computational and complexity costs. Estimating a state-space model often takes orders of magnitude more time than a simpler alternative. For small or moderately sized data sets, model fitting can be reasonably quick despite longer computation times than alternative methods, but it can be more problematic for larger data sets. Furthermore, it is easy to overparameterise state-space models by including parameters for which there is not enough information in the data to estimate them well (Knape, 2008). Such identification problems can be particularly acute in the state-space model setting because there is usually considerable uncertainty about the observation and process error variances, unless one includes explicit information on either of them. For example, explicit information about observation error may be included from replicate measurements of the same population within short time periods and can aid the estimation of the observation error variance. In general, problems with identification are not always intuitive or easy to diagnose, but tools aimed at checking estimability can be useful (Lele et al., 2007).

Example: thistles in wheat fields

At Rothamsted in the United Kingdom, long-term experiments with wheat fields have been running since the late 1850s (Moss et al., 2004). As part of these experiments, data on weed species are collected at wheat fields that are not treated with herbicides. Here we consider annual data on the perennial creeping thistle (*Cirsium arvense*) (Tiley, 2010) from 1991 to 2018 at 18 plots that consist of small fields lined up next to each

other, each plot receiving a different fertilisation regime.

One may ask to what extent there is synchrony in the dynamics among the plots. Are the populations in the different plots behaving independently, in strong synchrony, for example due to shared external (e.g. weather) or internal (e.g. seed dispersal) drivers, or somewhere in between where both local features of the plots and shared drivers contribute to the dynamics. To address these questions we use a state-space model to deal with observation error and to analyse the structure of the temporal variation. Handling of observation error is important in this context since it will tend to dilute any synchrony in dynamics among the plots.

For each of the 18 plots the data consist of presence of the species in 25 random 0.1 m^2 quadrats. The quadrats are small relative to the size of the plots. It is therefore reasonable to assume that the observed species presences in the quadrats are independent samples of the occurrence in each plot. We therefore model observation error, which in this case is due to sampling variation, using a binomial model:

$$y_{it} \sim \text{Bin}(25, p_{it})$$

where y_{it} is the number of observed occurrences in plot i in year t and p_{it} is the probability of the species occurring in a random quadrat. This probability may be viewed as an index of abundance, observed with error, whose dynamics we are interested in. Since the index p_{it} is a probability it is restricted to the interval from 0 to 1, and we model it at the logit scale to satisfy this restriction. Hence, we define the dynamic process model in terms of $x_{it} = \text{logit}(p_{it})$. For this we use the AR(1) process in eqn. (7.5) as a starting point, but add some components to it.

First, potential synchrony could be caused by exposure to common fluctuating environmental variables. We include the average amount of rainfall in June, July, and August as one such potential variable. Additionally, in 1994, 2001, 2008, and 2015–2016 all plots were fallowed and no data were collected. To account for effects of the fallows we include a separate parameter for each of the four fallow periods that allow for instant changes in abundance. Second, the error terms η in the AR(1) model in eqn. (7.5) represent process error, in other words unexplained variation in the growth rates among years. To study synchrony in addition to that potentially caused by the measured covariates, rain and fallow, we allow the process errors to be correlated among plots. Specifically, we assume that the variance of the process errors consists of two components, σ_n^2 and σ_c^2, with a total variance $\sigma_n^2 + \sigma_c^2$. The first component is the variance due to variation in local dynamics and corresponds to errors that are independent among plots. The second component is variance that is shared among plots in such a way that the covariance is allowed to decline exponentially as a function of distance with $\mathrm{cov}(\eta_i, \eta_j) = \sigma_c^2 \exp(-\phi \cdot \mathrm{distance}(i, j))$ for plot i and j. In the special case that ϕ is equal or close to zero the dependence of the correlation on distance disappears and the correlation is the same regardless of distance. This is often referred to as compound symmetry and would suggest that factors that are shared among all plots are driving synchrony. On the other hand, the correlation dropping with distance would suggest that factors operating at a smaller spatial scale are driving synchrony.

The resulting process model at the logit scale is described by the equation

$$x_{it+1} = a_{0i} + a_1 x_{it} + c \cdot \mathrm{rain}_t + d_{f_t} 1(\text{fallow in year } t) + \eta_{it}$$

where c is the slope for the effect of rain and d_{f_t} are effects for the fallow years, with $1(\cdot)$ an indicator function being 1 if there was a fallow in year t and zero otherwise. We assume that the autoregressive parameter, the effect of rainfall, and the response to the fallowing are identical among plots at the logit scale. Intercepts a_{0i} are allowed to vary among plots to allow for different rates of growth due to the different fertilisation regimes. The initial logit scale indices at year 1, x_{i1}, are treated as unknown parameters to be estimated.

We fit a Bayesian version of the model to data using a custom implementation in the statistical programming language Stan (Carpenter et al., 2017). Figure 7.3 shows the estimated dynamics and parameters, and how the correlation in the dynamics among plots changes as a function of the distance between the plots. The results suggest that a large part of the synchrony in dynamics is mainly driven by local features that extend only a few plots away, with only some potential synchrony remaining between plots at opposite ends. We further see that there is some evidence for a slight positive effect of rainfall and that populations were reduced after the fallows in 2001, 2008, and 2015–2016 but not after the fallow in 1994.

7.4 Conclusions

When analysing population dynamics, it is common that the only available data are estimates of population sizes or indices of relative abundance. This chapter focused on such situations where demographically explicit data are not available or not used.

Multiple demographic mechanisms can give rise to very similar patterns of abundance. The lack of direct information about demographic processes in abundance data implies that we cannot make strong inference about the demographic drivers behind abundance patterns. Instead, the main aim of most abundance analyses is to reveal the patterns themselves. Typical analyses done for abundance data are trend estimation, indicating the state and direction of population status, assessment of population characteristics such as the growth potential, linking population change to environmental variables, estimating density dependence or cycles, or assessing spatial synchrony and other spatiotemporal patterns in population fluctuations.

Consistent with the information content in the data, analytic models for abundance data usually contain little detail about demographic mechanisms. Rather, joint effects of demographic processes are combined into a fairly simple process model. Stochastic components of the analytic models can be divided into observation errors and process errors. Observation errors are due to randomness in the counting or estimation uncertainty, that is, variability in the data which does not exist in the actual population. Process error, in turn, is true variation in the population dynamics, which cannot be predicted by the model and is therefore treated as random unexplained variation. Process errors may be due to unmeasured or missed environmental variables, variation in population structure, or so-called demographic stochasticity. Simple statistical models of abundance data presume that all randomness is either observation errors or process error, but in the real world both are inevitably present to some extent. State-space models are a group of models that simultaneously include both sources of stochasticity and may often provide better inference, but they are technically more challenging to fit and they require more from the data.

Analyses of population abundance are, arguably, most informative and useful when the data consist of long time series and have spatial replication. This often allows drawing conclusions about abundance patterns at scales in time and space that are otherwise not possible, or practical, to study.

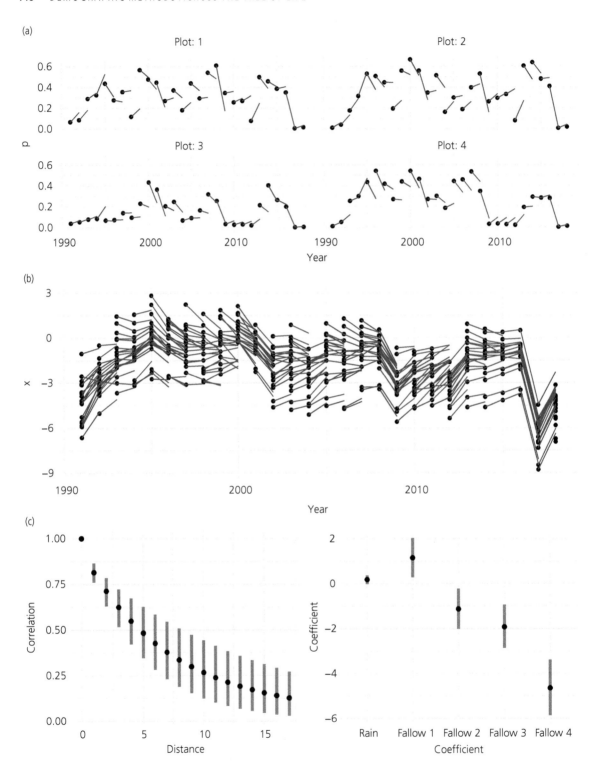

Figure 7.3 (a) One step ahead predictions of the abundance index at the first four plots. Black points show the estimated abundance index p in each year. Red lines connect these estimated indices to the abundance index predicted by the model in the following year. (b) Same as (a) but for all 18 plots together and on the logit scale. (c) Estimated correlation in process errors among the plots as a function of distance between them. (d) Estimates of coefficients (at the logit scale) for rainfall and each of the four fallows. In (c) and (d) green lines show 90% credible intervals.

References

Almaraz, P., Green, A.J., Aguilera, E., Rendón, M.A. & Bustamante, J. (2012). Estimating partial observability and nonlinear climate effects on stochastic community dynamics of migratory waterfowl. *Journal of Animal Ecology*, **81**, 1113–1125.

Ådahl, E., Lundberg, P. & Jonzén, N. (2006). From climate change to population change: The need to consider annual life cycles. *Global Change Biology*, **12**, 1627–1633.

Barker, R.J., Schofield, M.R., Link, W.A. & Sauer, J.R. (2018). On the reliability of N-mixture models for count data. *Biometrics*, **74**, 369–377.

Barraquand, F., Louca, S., Abbott, K.C., et al. (2017). Moving forward in circles: Challenges and opportunities in modelling population cycles. *Ecology Letters*, **20**, 1074–1092.

Bence, J.R. (1995). Analysis of short time series: correcting for autocorrelation. *Ecology*, **76**, 628–639.

Bolker, B.M. (2008). *Ecological Models and Data in R*. Princeton University Press.

Brännström, Å. & Sumpter, D.J.T. (2006). Stochastic analogues of deterministic single-species population models. *Theoretical Population Biology*, **69**, 442–451.

Buckland, S.T., Rexstad, E.A., Marques, T.A. & Oedekoven, C.S. (2015). *Distance Sampling: Methods and Applications*. Springer.

Bullock, J. (1999). *Plants. Ecological Census Techniques: A Handbook*. Cambridge University Press Cambridge.

Buonaccorsi, J.P., Elkinton, J.S., Evans, S.R. & Liebhold, A.M., et al. (2001). Measuring and testing for spatial synchrony. *Ecology*, **82**, 1668–1679.

Carpenter, B., Gelman, A., Hoffman, M.D., et al. (2017). Stan: A probabilistic programming language. *Journal of Statistical Software*, **76**, 1–32.

Chevalier, M., Russell, J.C. & Knape, J., et al. (2019). New measures for evaluation of environmental perturbations using Before-After-Control-Impact analyses. *Ecological Applications*, **29**, e01838.

Cornulier, T., Yoccoz, N.G., Bretagnolle, V., et al. (2013). Europe-wide dampening of population cycles in keystone herbivores. *Science*, **340**, 63–66.

Cramp, S. (1985). *Handbook of the Birds of Europe, the Middle East, and North Africa: Volume IV Terns to Woodpeckers*. Oxford University Press, Oxford.

Damgaard, C. (2012). Trend analyses of hierarchical pinpoint cover data. *Ecology*, **93**, 1269–1274.

Dennis, E.B., Freeman, S.N., Brereton, T. & Roy, D.B. (2013). Indexing butterfly abundance whilst accounting for missing counts and variability in seasonal pattern. *Methods in Ecology and Evolution*, **4**, 637–645.

Dennis, B., Munholland, P.L. & Scott, J.M. (1991). Estimation of growth and extinction parameters for endangered species. *Ecological Monographs*, **61**, 115–143.

Dennis, B., Ponciano, J.M., Lele, S.R., Taper, M.L. & Staples, D.F. (2006). Estimating density dependence, process noise, and observation error. *Ecological Monographs*, **76**, 323–341.

Dennis, B. & Taper, M.L. (1994). Density dependence in time series observations of natural populations: estimation and testing. *Ecological Monographs*, **64**, 205–224.

Dunstan, P.K., Foster, S.D., Hui, F.K.C. & Warton, D.I. (2013). Finite mixture of regression modeling for high-dimensional count and biomass data in ecology. *Journal of Agricultural, Biological, and Environmental Statistics*, **18**, 357–375.

Durbin, J. & Koopman, S.J. (2012). *Time Series Aanalysis by State Space Methods*. Oxford University Press.

Engen, S., Bakke, Ø. & Islam, A. (1998). Demographic and environmental stochasticity-concepts and definitions. *Biometrics*, **54**, 840–846.

Fromentin, J.M., Myers, R.A., Bjornstad, O.N., et al. (2001). Effects of density-dependent and stochastic processes on the regulation of cod populations. *Ecology*, **82**, 567–579.

Gause, G.F. (1932). Experimental studies on the struggle for existence: I. Mixed population of two species of yeast. *Journal of Experimental Biology*, 9, 389–402.

Harvey, A.C. (1990). *Forecasting, Structural Time Series Models and the Kalman Filter*. Cambridge University Press.

Hixon, M.A., Pacala, S.W. & Sandin, S.A. (2002). Population regulation: Historical context and contemporary challenges of open vs. closed systems. *Ecology*, **83**, 1490–1508.

Holmes, E.E. (2001). Estimating risks in declining populations with poor data. *Proceedings of the National Academy of Sciences of the United States of America*, **98**, 5072–5077.

Holmes, E.E., Sabo, J.L., Viscido, S.V. & Fagan, W.F. (2007). A statistical approach to quasi-extinction forecasting. *Ecology Letters*, **10**, 1182–1198.

Ives, A.R. (1995). Predicting the response of populations to environmental change. *Ecology*, **76**, 926–941.

Ives, A.R., Abbott, K.C. & Ziebarth, N.L. (2010). Analysis of ecological time series with ARMA(p,q) models. *Ecology*, **91**, 858–871.

Jacquemyn, H., Brys, R., Hermy, M. & Willems, J.H. (2007). Long-term dynamics and population viability in one of the last populations of the endangered *Spiranthes spiralis* (Orchidaceae) in the Netherlands. *Biological Conservation*, **134**, 14–21.

Kéry, M., Royle, J. & Schmid, H. (2005). Modeling avian abundance from replicated counts using binomial mixture models. *Ecological Applications*, **15**, 1450–1461.

Knape, J. (2016). Decomposing trends in Swedish bird populations using generalized additive mixed models. *Journal of Applied Ecology*, **53**, 1852–1861.

Knape, J. (2008). Estimability of density dependence in models of time series data. *Ecology*, **89**, 2994–3000.

Knape, J., Arlt, D., Barraquand, F., Berg, Å., Chevalier, M., Pärt, T., Ruete, A. & Żmihorski, M. (2018). Sensitivity of binomial N-mixture models to overdispersion: The importance of assessing model fit. *Methods in Ecology and Evolution*, **9**, 2102–2114.

Lebreton, J.-D. & Gimenez, O. (2013). Detecting and estimating density dependence in wildlife populations. *Journal of Wildlife Management*, **77**, 12–23.

Lele, S.R., Dennis, B. & Lutscher, F. (2007). Data cloning: Easy maximum likelihood estimation for complex ecological models using Bayesian Markov chain Monte Carlo methods. *Ecology Letters*, **10**, 551–563.

Liebhold, A., Koenig, W.D. & Bjørnstad, O.N. (2004). Spatial synchrony in population dynamics. *Annual Review of Ecology, Evolution, and Systematics*, 35, 467–490.

Lindén, A. & Knape, J. (2009). Estimating environmental effects on population dynamics: Consequences of observation error. *Oikos*, **118**, 675–680.

Lindén, A. & Mäntyniemi, S. (2011). Using the negative binomial distribution to model overdispersion in ecological count data. *Ecology*, **92**, 1414–1421.

Lindley, S.T. (2003). Estimation of population growth and extinction parameters from noisy data. *Ecological Applications*, **13**, 806–813.

May, R.M. (1976). Simple mathematical models with very complicated dynamics. *Nature*, **261**, 459–467.

Månsson, L., Ripa, J. & Lundberg, P. (2007). Time series modelling and trophic interactions: Rainfall, vegetation and ungulate dynamics. *Population Ecology*, **49**, 287–296.

McCrea, R. & Morgan, B. (2015). *Analysis of Capture-Recapture Data*. Taylor & Francis, Boca Raton.

Moore, N.W. (1991). The development of dragonfly communities and the consequences of territorial behaviour: A 27 year study on small ponds at Woodwalton Fen, Cambridgeshire, United Kingdom. *Odonatologica*, **20**, 203–231.

Morris, W.F. & Doak, D.F., et al. (2002). *Quantitative Conservation Biology*. Sinauer.

Moss, S.R., Storkey, J., Cussans, J.W., Perryman, S.A.M. & Hewitt, M.V. (2004). The Broadbalk long-term experiment at Rothamsted: What has it told us about weeds? *Weed Science*, **52**, 864–873.

Newman, K.B., Buckland, S.T., Morgan, B.J.T., et al. (2014). Modelling population dynamics using closed-population abundance estimates. *Modelling Population Dynamics: Model Formulation, Fitting and Assessment Using State-Space Methods* (eds K.B. Newman, S.T. Buckland, B.J.T. Morgan, R. King, D.L. Borchers, D.J. Cole, P. Besbeas, O. Gimenez & L. Thomas), pp. 123–145. Methods in Statistical Ecology. Springer.

Nisbet, R.M. & Gurney, W. (1982). *Modelling Fluctuating Populations*. John Wiley & Sons.

Oedekoven, C.S., Elston, D.A., Harrison, P.J., et al. (2017). Attributing changes in the distribution of species abundance to weather variables using the example of British breeding birds. *Methods in Ecology and Evolution*, **8**, 1690–1702.

Peltonen, M., Liebhold, A.M., Bjørnstad, O.N. & Williams, D.W. (2002). Spatial synchrony in forest insect outbreaks: roles of regional stochasticity and dispersal. *Ecology*, **83**, 3120–3129.

Pépino, M., Rodríguez, M.A. & Magnan, P. (2012). Impacts of highway crossings on density of brook charr in streams. *Journal of Applied Ecology*, **49**, 395–403.

Quinn, T.J. & Deriso, R.B. (1999). *Quantitative Fish Dynamics*. Oxford University Press.

Ramula, S., Puhakainen, L., Suhonen, J. & Vallius, E. (2008). Management actions are required to improve the viability of the rare grassland herb Carlina biebersteinii. *Nordic Journal of Botany*, **26**, 83–90.

Ranta, E., Kaitala, V., Lindstrom, J. & Linden, H., et al. (1995). Synchrony in population dynamics. *Proceedings: Biological Sciences*, **262**, 113–118.

Reynolds, J.H., Knutson, M.G., Newman, K.B., Silverman, E.D. & Thompson, W.L. (2016). A road map for designing and implementing a biological monitoring program. *Environmental Monitoring and Assessment*, **188**, 399.

Ripa, J. & Heino, M. (1999). Linear analysis solves two puzzles in population dynamics: The route to extinction and extinction in coloured environments. *Ecology Letters*, **2**, 219–222.

Ripa, J. & Ives, A.R. (2007). Interaction assessments in correlated and autocorrelated environments. *The Impact of Environmental Variability on Ecological Systems* (eds D.A. Vasseur & K.S. McCann), pp. 111–131. Springer Netherlands.

Rochester, C.J., Brehme, C.S., Clark, D.R., et al. (2010). Reptile and amphibian responses to large-scale wildfires in southern California. *Journal of Herpetology*, **44**, 333–351.

Royama, T. (1992). *Analytical Population Dynamics*. Springer Netherlands.

Royle, J.A. (2004). N-mixture models for estimating population size from spatially replicated counts. *Biometrics*, **60**, 108–115.

Ruokolainen, L., Linden, A., Kaitala, V. & Fowler, M.S. (2009). Ecological and evolutionary dynamics under coloured environmental variation. *Trends in Ecology & Evolution*, **24**, 555–563.

Shenk, T.M., White, G.C. & Burnham, K.P. (1998). Sampling-variance effects on detecting density dependence from temporal trends in natural populations. *Ecological Monographs*, **68**, 445–463.

Sköld, M. & Knape, J. (2018). Bounding reproductive rates in state-space models for animal population dynamics. *Ecosphere*, **9**, e02215.

Stenseth, N.C., Mysterud, A., Ottersen, G., et al. (2002). Ecological effects of climate fluctuations. *Science*, **297**, 1292–1296.

Sutherland, W.J. (Ed.). (2006). *Ecological Census Techniques: A Handbook*, 2nd edn. Cambridge University Press.

Tayleur, C., Caplat, P., Massimino, D., et al. (2015). Swedish birds are tracking temperature but not rainfall: Evidence from a decade of abundance changes. *Global Ecology and Biogeography*, **24**, 859–872.

Tiley, G.E.D. (2010). Biological flora of the British Isles: *Cirsium arvense* (L.) Scop. *Journal of Ecology*, **98**, 938–983.

Turchin, P. (2003). *Complex Population Dynamics: A Theoretical/Empirical Synthesis*. Princeton University Press.

Turchin, P. (1995). Population regulation: old arguments and a new synthesis. *Population Dynamics: New Approaches and Synthesis* (eds N. Capuccino & P. Price), pp. 19–40. Elsevier.

Vanhatalo, J., Hosack, G.R. & Sweatman, H. (2017). Spatiotemporal modelling of crown-of-thorns starfish outbreaks on the Great Barrier Reef to inform control strategies (M. Bode, Ed.). *Journal of Applied Ecology*, **54**, 188–197.

Walters, C.J. & Ludwig, D. (1981). Effects of measurement errors on the assessment of stock-recruitment relationships. *Canadian Journal of Fisheries and Aquatic Sciences*, **38**, 704–710.

Watson, A., Moss, R. & Rothery, P. (2000). Weather and synchrony in 10-Year population cycles of rock ptarmigan and red grouse in Scotland. *Ecology*, **81**, 2126–2136.

Wood, S.N. (2010). Statistical inference for noisy nonlinear ecological dynamic systems. *Nature*, **466**, 1102–1104.

Zuur, A.F., Tuck, I.D. & Bailey, N. (2003). Dynamic factor analysis to estimate common trends in fisheries time series. *Canadian Journal of Fisheries and Aquatic Sciences*, **60**, 542–552.

CHAPTER 8

Life tables: construction and interpretation

Owen R. Jones

8.1 Introduction

There is remarkable diversity in the life history strategies employed by different species and among populations. What unites these diverse strategies is the fundamental balancing equation for population growth: $N_{t+1} = N_t + B_t - D_t + I_t - E_t$, where the population's size at time $t+1$ (N_{t+1}) is calculated from the population size at time t (N_t), and where B represents the number of offspring produced, D represents deaths, and I and E represent immigration and emigration respectively between time t and $t+1$. Where births and immigration exceed deaths and emigration, the population will grow. It is convenient to consider only closed populations so that $N_{t+1} = N_t + B_t - D_t$.

These equations break down population growth into processes of birth and death (and migration) that occur to individuals. However, the equations assume that all individuals have identical probabilities of death and of producing offspring, which is of course usually not true. A major source of demographic heterogeneity is age (or developmental stage). For example, over their life course both birds and mammals tend to experience bathtub-shaped mortality trajectories (Siler 1979) with relatively high infant and juvenile mortality that declines with age towards the age of maturity, and eventually increases with the march of time and effects of senescence (Jones et al. 2008). Fertility also varies through the mammalian life course: In humans, fertility is unimodal, increasing to a peak and then declining with menopause, while in most other mammals there is no such menopausal fertility decline, but rather an increase to a plateau after maturity (Johnstone and Cant 2019). These patterns are not universal and demographic trajectories show remarkable diversity among species (Jones et al. 2014;

Jones and Vaupel 2017), within species between the sexes (Clutton-Brock and Isvaran 2007; Bonduriansky et al. 2009; Maklakov et al. 2009; Griffin et al. 2017), or depending on environmental conditions, especially during early development (Descamps et al. 2008; Reed et al. 2008; Bouwhuis et al. 2010; Boonekamp et al. 2014).

Life tables have been a fundamental tool for exploring this diversity by showing how various demographic metrics, chief among them the probability of death and fertility, change with advancing age. In this chapter, I aim to provide broad coverage of the calculation and use of life tables, with a particular focus on nonhuman animals.

The use of life tables has a long history. They originate in John Graunt's bills of mortality published in 1662 (Sutherland 1963) and have been extensively used in all aspects of human demography for centuries (Preston et al. 2000). Their use has not been restricted to humans, but serious use in nonhumans began only in the twentieth century. Perhaps the earliest nonhuman life tables were those published in a series of articles led by Raymond Pearl exploring mortality patterns using fruit fly (*Drosophila melanogaster*) beginning with Pearl and Parker (1921). A later article in the series broadened the taxonomic scope by presenting life tables for other 'lower organisms', including various other types of *Drosophila*, the freshwater *Hydra fusca* (now *Hydra oligactis*), the slug *Agriolimax agrestis* (now *Deroceras agreste*), the Oriental cockroach *Blatta orientalis*, and mice (*Mus musculus*), with the aim of comparing survivorship curves to understand how mortality varies with age and among species (Pearl and Miner 1935). Pearl and Miner used these comparisons to develop a categorisation of survivorship

Owen R. Jones, *Life tables: construction and interpretation*. In: *Demographic Methods Across the Tree of Life*.
Edited by Roberto Salguero-Gómez and Marlène Gamelon, Oxford University Press.
© Oxford University Press (2021). DOI: 10.1093/oso/9780198838609.003.0008

curves based on the distribution of mortality risk through the life course. Deevey (1947) further used life tables to explore mortality in species as diverse as barnacles (*Semibalanus balanoides*), Dall's sheep (*Ovis dalli*), blackbird (*Turdus merula*), and rotifer (*Floscularia conifera*) and redefined Pearl and Miner's (1935) survivorship categorisation as the now-familiar type I, II, and III survivorship, which result from increasing, constant, and decreasing mortality risk respectively (Pearl and Miner 1935; Deevey 1947; Jones 2016). Later work by Caughley (1966) used life tables to compare mortality trajectories among mammals, with the central finding that, despite great variation in typical life span, mammals generally have U- or 'fish-hook'-shaped mortality trajectories.

Since then, life tables have been put to a diversity of uses including studying sex differences in mortality (e.g. Clutton-Brock and Isvaran 2007), testing theories of life history evolution (Purvis and Harvey 1995) and human evolution (Burger et al. 2012; Colchero et al. 2016; e.g. Levitis et al. 2013), and the effects of caloric restriction on mortality (e.g. Carey et al. 2002), among many others. Although most nonhuman life tables have been used for animal populations, they have also been used to study plant demography, from herbs (Leverich and Levin 1979; Hegazy 1990) to trees (Van Valen 1975; Harcombe 1987). Life tables form the basis for age-structured population modelling, including the application of the Euler–Lotka equation (Lotka 1907), which is perhaps the most important equation in mathematical demography. Life tables have obvious similarities with age-based (Leslie) matrix population models (MPMs) (Leslie 1945; Caswell 2001), and producing an MPM from a life table that includes both survival and fertility information is straightforward (and vice versa). It is also possible to approximate an age-based life table from a stage-based (Lefkovitch) MPM (Cochran and Ellner 1992; Caswell 2001). MPMs are covered in detail in Chapter 9.

In this chapter, I focus on single decrement life tables. That is, life tables where individuals have only one possible exit from a state (i.e. death). Multiple decrement life tables allow more than one exit (e.g. multiple mutually exclusive causes of death) and are uncommon in nonhuman studies (see Chapter 4 in Preston et al. (2000) for excellent coverage of this topic).

8.2 Cohort and period life tables

The central aim of most life tables is to describe how the risk of death, and sometimes fertility, changes with age. The life table, therefore, makes an accounting of the number of individuals that enter a particular age

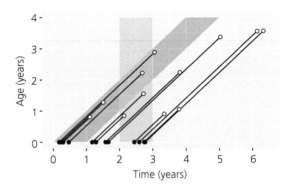

Figure 8.1 Lexis diagram illustrating the difference between cohort and period approaches. The figure shows the fate of three cohorts of four individuals. Time is represented by the horizontal axis and age by the vertical axis. Each line represents an individual born at the point marked with a filled circle and dying at the time marked with an open circle. The height of the open circle on the vertical axis thus represents age at death. The pink area highlights the cohort born in the interval 0–1. The cohort can be followed until all four individuals have died. The green area highlights the period in the interval 2–3. There are six individuals that enter the period at time=2 at various ages, and three of these individuals die. Four new individuals are born, which all survive until the end of the period. To calculate mortality risk, the period life table methods must account for the number of individuals 'at risk' during the period.

interval (i.e. the population at risk) and the number that die and the number of offspring produced within this interval in order to correctly calculate mortality and fertility. The calculation of life tables can be divided into approaches—cohort-based and period-based (also known as static or vertical life tables)—where, in both cases, the rows correspond to age intervals and the first column indicates the age intervals over which the life table is calculated. The choice of interval largely depends on the preferences of the researcher and on the life history of the species. It is often a year, but not necessarily so, and can vary within the life table. For example, in human life tables the age intervals at the start of life are commonly 1 year, then 4 years, and then broaden to 5 years in later life.

As their names imply, cohort-based methods aim to track the fate of a set of individuals born within a particular time-frame (e.g. in a particular year), while period life tables focus on the fate of a population of mixed ages during a particular time period such as a year (Figure 8.1). These differences can be important, and we should expect measures derived from cohort and period life tables to align only if the environment has been constant and unchanging. There is no agreed universal standard for the life table layout or nomenclature, but in this chapter (see Tables 8.1, 8.2, and 8.3) I follow the nomenclature of Preston et al. (2000). Here subscripts n and x define the length and start of the age interval respectively (i.e. $_np_x$ is the probability of survival from age x to $x+n$). Where the age interval is consistently the same (e.g. 1 year), the n subscript is normally omitted.

In the published literature, life table-like data are reported in diverse ways: Sometimes data are presented as numbers of individuals entering an age interval and/or dying in the interval, while at other times data may be presented as the proportion surviving to a particular age (i.e. survivorship standardised to begin at 1); or data may be reported as age-specific probability of survival or (less-commonly) hazard or death rate of individuals in the interval (the slope of the survivorship curve multiplied by –1). The close mathematical relationships among columns in the life table mean that usually a complete actuarial life table can be derived from these initial data, but knowledge of sample size (e.g. initial cohort size or the number of individuals in an age interval) can be useful to infer confidence for the estimated demographic quantities. In sections 8.2.1 and 8.2.2, I describe cohort and period approaches and illustrate them with examples.

8.2.1 Cohort life tables

Cohort life tables focus on a group of individuals born within a particular time interval, for example a year,

and follow them until (ideally) all individuals in the group are dead (Figure 8.1). The raw data used for the construction of these life tables are thus the age at death of individuals, which may also be recorded as the number of individuals entering each age interval, and the methods assume that all individuals in the population are equally likely to be sampled (Caughley 1977; Schwartz and Armitage 1998; Preston et al. 2000). I illustrate a cohort life table by calculating a life table for tree swallows (*Tachycineta bicolor*) using cohort-based data obtained from the DATLife database (Max-Planck Institute for Demographic Research [Germany] 2018), which sources the original data from Butler (1988) (Table 8.1). The first column (x) represents the exact age at the start of the interval, and the length of the interval (n) is the difference between the values of x in consecutive rows. Then, l_x is the number of individuals entering the interval at age x, with the first entry, therefore, being the number of individuals in the entire cohort and subsequent entries being the number surviving to each age (x). From this information, I can calculate the number dying between ages x and $x+n$ ($_nd_x$) as the difference between l_x and l_{x+1} and the probability of dying ($_nq_x$) or surviving ($_np_x$) the interval as $_nd_x/l_x$ and $1-(_nd_x/l_x)$, respectively.

Before proceeding towards the calculation of age-specific death rate, it is first useful to consider what is meant by a 'rate' and why this differs from a

Table 8.1 Cohort life table for tree swallows (*Tachycineta bicolor*) calculated for a cohort of 320 individuals from x, l_x, and $_na_x$ data obtained from DATLife. Deaths are assumed to be distributed uniformly during the interval, so $_na_x = 0.5$. Values in columns marked by asterisks have been rounded to 3 decimal places (dp).

Exact age	Number left alive at age x	Number dying between ages x and $x+n$	Probability of dying between ages x and $x+n$ *	Probability of surviving between ages x and $x+n$ *	Person-years lived between ages x and $x+n$	Person-years lived above age x	Life expectancy from age x *	Death rate in the cohort between ages x and $x+n$ *	Average number of years lived in the time interval by those dying in the time interval *
x	l_x	$_nd_x$	$_nq_x$	$_np_x$	$_nL_x$	T_x	e_x	$_nm_x$	$_na_x$
0	320	253	0.791	0.209	193.5	272.0	0.850	1.308	0.500
1	67	40	0.597	0.403	47	78.5	1.172	0.851	0.500
2	27	16	0.593	0.407	19	31.5	1.167	0.842	0.500
3	11	7	0.636	0.364	7.5	12.5	1.136	0.933	0.500
4	4	2	0.500	0.500	3	5.0	1.250	0.667	0.500
5	2	1	0.500	0.500	1.5	2.0	1.000	0.667	0.500
6	1	1	1.000	0.000	0.5	0.5	0.500	2.000	0.500

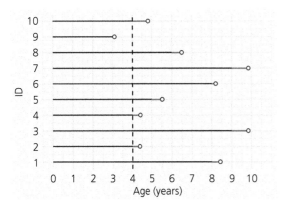

Figure 8.2 Understanding life expectancy (e_0) and the average length of time that those that die in the interval survive within the interval (a_x). The a_x is used to estimate L_x, the person-years lived during the interval. The horizontal lines represent life lines for ten individuals, with age at death marked with an open circle. Life expectancy from any particular age is the average remaining length of the lines. For example, life expectancy from birth (age = 0) is the average length of the lines (6.503 years) and life expectancy from age 4 is the average length of the lines conditional on reaching the vertical dashed line (2.882 years). In this case, individual number 9 is excluded because the individual dies before reaching 4 years. In most data sets, the precise age at death is not recorded and instead it is only recorded that the individual did not survive to the end of the age interval. However, dying individuals usually survive through at least some of their 'final' interval (here marked with a red segment), and this is accounted for by the a_x value which is the average length of time survived within this final interval, for each age interval. Here, the overall a_x is the average length of the red segments (0.503 years), but this could vary with age: typically, a_x is smaller for younger ages.

probability. In demography, a rate refers to the occurrence of an event in relation to the amount of exposure to the risk of occurrence of that event. The timing of the event during the interval is therefore important: Individuals that survive the interval are exposed to risk for the entire duration of the interval, while those that die are only exposed to risk until they die. I therefore calculate the death rate as the number of deaths occurring during a given time interval (e.g. a particular year of life), divided by the number of 'person'-years lived during that interval ($_nL_x$). The difference in denominator between the calculations for probability of death and death rate mean that these two quantities are distinct.

The number of 'person'-years lived during that interval ($_nL_x$) is calculated as the number of individuals entering the interval at age x (l_x), multiplied by their probability of survival ($_np_x$), plus the average length of time that those that die in the interval survive within the interval ($_na_x$) (Figure 8.2): $_nL_x = (_np_x \cdot l_x) +$

($_na_x \cdot {_nd_x}$). Estimation of $_na_x$ requires accurate time of death data that is normally not available for non-humans. A simplifying assumption is that deaths are distributed uniformly through the age interval such that $_na_x = 0.5n$. However, if deaths are concentrated early during the interval, for example, due to high infant mortality in the first interval, a smaller $_na_x$ value could be appropriate for that interval. An $_na_x$ value may also be informed by strong seasonality in deaths. In the animal demography literature, it is often implicitly assumed that individuals entering the interval survive the entire interval (i.e. $_na_x = 1$). Whether assumptions about a_x, and consequently L_x, make an appreciable difference to analytical outcomes depend on the life history of the organism concerned and on the questions being addressed.

From $_nL_x$ one can calculate the number of person-years lived beyond age $x(T_x)$ as the sum of $_nL_x$ from x onwards (i.e. $T_x = \sum_{a=x}^{\infty} {_nL_a}$). The age-specific death rate (hazard) ($_nm_x$) is calculated as the number of deaths in the age interval divided by the number of person-years lived in the age interval (i.e. $m_x = {_nd_x}/{_nL_x}$). Finally, life expectancy from age x (e_x^0 is calculated as T_x/l_x (Figure 8.2).

8.2.2 Period life tables

As their name suggests, period life tables focus on a particular time period, such as a year (Figure 8.1), and are also known as static or vertical life tables. They aim to model what would happen to an imaginary ('synthetic') cohort that experienced the mortality conditions of that period. They are sometimes favoured over cohort life tables because following an entire cohort until all members are dead can be impractical, especially for longer-lived organisms. The main assumptions of this approach are that the age distribution is stationary (i.e. the proportion of individuals in each age class is constant, and population growth rate is constant) and that every individual in the population has an equal probability of being sampled (Caughley 1977; Schwartz and Armitage 1998; Preston et al. 2000). These are strong assumptions that are likely rarely met in wild animals and, although 'moderate' deviations are probably acceptable (Stearns 1992, p. 25), the use of period life tables in fluctuating populations has been robustly cautioned against (Anderson et al. 1981; Menkens and Boyce 1993).

Nevertheless, understanding period life tables is useful and I will illustrate their construction using data for the European starling (*Sturnus vulgaris*) from the DATLife database (Max-Planck Institute

Table 8.2 Period life table for European starling (*Sturnus vulgaris*) calculated using x, $_nN_x$, $_nD_x$, and $_na_x$ data from DATLife (original data sourced from Deevey 1947) and with l_0 set to 1.000. See text for explanation of nomenclature. Noninteger values are rounded to 3 dp.

x	$_nN_x$	$_nD_x$	$_na_x$	$_nm_x$	$_nq_x$	$_np_x$	l_x	$_nd_x$	$_nL_x$	T_x	e_x
0	203	99	0.500	0.488	0.392	0.608	1.000	0.392	0.804	2.051	2.051
1	166	69	0.500	0.416	0.344	0.656	0.608	0.209	0.503	1.247	2.052
2	97	51	0.500	0.526	0.416	0.584	0.399	0.166	0.316	0.744	1.866
3	46	27	0.500	0.587	0.454	0.546	0.233	0.106	0.180	0.428	1.840
4	19	8	0.500	0.421	0.348	0.652	0.127	0.044	0.105	0.248	1.953
5	11	5	0.500	0.455	0.370	0.630	0.083	0.031	0.068	0.143	1.728
6	6	4	0.500	0.667	0.500	0.500	0.052	0.026	0.039	0.076	1.450
7	2	1	0.500	0.500	0.400	0.600	0.026	0.010	0.021	0.037	1.400
8	1	1	0.000	1.000	1.000	0.000	0.016	0.016	0.016	0.016	1.000

for Demographic Research (Germany) 2018) and originally sourced from Deevey (1947). The starting point for constructing a period life table is the calculation of age-specific death rate (hazard) ($_nm_x$) from the number of individuals alive in the *middle* of an age interval ($_nN_x$) and the number of those individuals that die in the observation period ($_nD_x$): $_nm_x = {}_nD_x/{}_nN_x$. In this case (Table 8.2) there are 203 individuals within the age interval 0–1, and 99 of them die; thus, the death rate is 99/203 = 0.488. (Note that I am assuming that these $_nN_x$ values are for the middle of the age interval.)

It is worth emphasising again here that death rates are distinct from probabilities of death and can exceed 1. This is clear if you recall that the numerator in the relevant equation is the number alive in the *middle* rather than the *start* of the interval and that the mortality rate is the slope of the survivorship curve multiplied by –1. The effect of this difference is best illustrated with an example: In an animal population with 1,000 individuals alive at the start of the interval and very high mortality risk, one may see 400 dying and 600 remaining alive by the time they reach the middle of the interval and then a further 400 dying during the second half of the interval (a total of 800 dying), producing a mortality rate ($_nm_x$) of 800/600 = 1.333. But how can we calculate the age-specific probability of death ($_nq_x$) from $_nm_x$? This relies on knowledge of how deaths are distributed within the age interval: If deaths are concentrated early in the interval, $_nq_x$ is larger. Thus, $_na_x$ and $_nm_x$ can be used to calculate $_nq_x$ as follows: $_nq_x = (n \cdot {}_nm_x)/(1 + (n - {}_na_x) \cdot {}_nm_x)$. For example, using the example from above, where $_nm_x$ is 1.333, $_na_x$ is 0.5, and n = 1, we can check that we obtain a sensible result: 1.333/(1 + 0.5 × 1.333) = 0.800. This corresponds

perfectly with the intuitive result obtained by dividing the number dying in the interval (800) by the number entering the interval (1,000). The derivation of this $_nq_x$ formula can be seen in Preston (2000, p. 43).

The age-specific probability of survival ($_np_x$) is $1 - {}_nq_x$. The final age interval is a special case because it extends from x to infinity. It is, therefore, appropriate to assume that all individuals die in this interval, that is, $_\infty q_x = 1.000$ and $_\infty p_x = 0.000$.

At this point, we can choose a population size to start our synthetic cohort (l_0), the 'radix', and calculate l_{x+n} as $l_x \cdot {}_np_x$. In human life tables, the radix is often set as 1 million or 100,000, though 1.000 is also convenient because then the l_x values are survivorship, or the probability of survival to age x. The values of the subsequent columns will vary in proportion to the value chosen for the radix. The number of individuals dying in a time interval ($_nd_x$) is $l_x - l_{x+1}$, and the number of person-years lived in the interval ($_nL_x$) is $n \cdot l_{x+n} + {}_na_x \cdot {}_nd_x$ for all but the final row. For the final row, which has an open interval, $_\infty L_x = l_x/{}_\infty m_x$. Person-years lived above age x (T_x) is the sum of $_nL_x$ from x onwards (i.e. $T_x = \sum_{a=x}^{\infty} {}_nL_a$) and life expectancy from age x (e_x^0) is T_x/l_x.

In addition to the cohort and period life table methods described above, life tables based on 'snapshots' of the standing age structure of the population or on the ages of death of individuals across multiple cohorts have been used extensively in nonhumans (Udevitz and Ballachey 1998; Udevitz and Gogan 2012). As with the period methods described above, these approaches normally assume that populations are stable and stationary (Caughley 1977), though this assumption can be relaxed if a time series of age structures are available (Hoenig and Gedamke 2007; Udevitz and Gogan 2012). Other methods exist to estimate life tables from nonstable populations if the population growth rate (Udevitz and Ballachey 1998) or mortality rates (Fryxell 1986) are known. When the stationarity assumptions are met, the age structure of the population will be the same as the standardized survivorship (l_x, where l_0 is standardised to 1), and it is straightforward to calculate the other columns in the life table from this starting point using the mathematical relationships among them.

8.2.3 Fertility

Until now I have focused on mortality trajectories, but information on how fertility changes with age is necessary to address many fundamental topics. For non-humans, life tables that include fertility are generally based on the female fraction of the population because measuring male reproductive output is hard. Methods for estimating fertility vary depending on the nature of the organism but usually involve carefully monitoring a sample of females to observe them with confirmed offspring (or not). More recently, genetic sampling methods are being used to confirm parenthood (see Chapter 1). Like death, fertility can be expressed as a rate by dividing the number of births in an interval by the number of person-years lived by individuals entering the interval. Nevertheless, in many studies on wild animals, fertility is expressed as the average number of female offspring born within an age interval to females that enter the interval (i.e. assuming that $a_x = 1$).

Assessment of fertility is complicated by the fact that the number of births can be hard to estimate in wild animals and may be ambiguous: Do we refer to live births or all births (including, for example, stillbirths and miscarriages)? Do we include offspring of both sexes or focus only on the female fraction of the population? The fact that very early mortality of offspring makes detection difficult means that part of a fertility estimate may include some neonatal survival. Moreover, the denominator ($_nL_x$) can also be ambiguous: Do we include all individuals in our calculation of person-years? Just the females? Only sexually active females? And so on. These decisions must be made with reference to the specific research question but are important when making comparisons across populations or species and have consequences for modelling population dynamics.

Notwithstanding the above caveats, Table 8.3 illustrates the construction of a life table that includes age-specific fertility rate ($_nb_x$) for Bali myna (*Leucopsar rothschildi*). This life table is based on data from the DATLife database (Max-Planck Institute for Demographic Research [Germany] 2018) which was originally sourced from Ricklefs et al. (2003). I assume that the fertility reported in DATLife, in this case, is based on the number of females entering the interval, rather than person-years lived during the interval (i.e. DATLife fertility = total number of births/N_x). I therefore calculate $_nb_x$ as a rate by multiplying the DATLife-reported fertility with N_x, to calculate the number of offspring produced by the sampled population, then dividing by L_x to calculate the fertility rate. This fertility information includes offspring of both sexes, so I further divide by 2 to obtain an estimate of females only (i.e. I assume a 50:50 sex ratio at birth). In practice, with wild animal data, it is unlikely that decisions about using N_x or L_x (and the values used for a_x in the L_x calculation) will make an appreciable difference for most use cases, but users should be aware of the potential impacts.

The data for this life table end at age 18 when l_x is still 0.116. I have arbitrarily closed the life table by setting $_nD_x$ to equal $_nN_x$ for this interval. This has consequences for the calculation of some of the subsequent derived columns (e.g. life expectancy will be underestimated) and emphasises the difficulty of collecting data for cohort life tables. Nevertheless, for some important metrics that require fertility information, this is not a problem because fertility ($_nb_x$) has dropped to zero at these advanced ages. For example, the calculation of net reproductive rate (R_0), generation time (T), and intrinsic population growth rate (r) are unaffected by this truncation. Net reproductive rate (R_0), or the average number of females produced by a female during her life, is calculated as the sum of the l_xb_x series ($R_0 = \sum l_xb_x$), which in this case is 4.674 female offspring per female. Generation time (G), the average age of the parents of all offspring produced by a cohort, is a particularly useful quantity (Gaillard et al. 2005) that can be calculated by weighting the sum of the production of offspring by the age at which they are produced and dividing by R_0 ($G = \sum l_xb_xx / \sum l_xb_x$). In this example, the generation time is 4.662 years. Population growth rate can then be estimated by dividing the natural logarithm of R_0 by generation time ($r \approx \ln(R_0)/G$, which is 0.331 in this case). Another important quantity, reproductive value, which is an individual's fitness expressed in terms of expected contribution to future population growth, is calculated as the number of offspring produced by individuals from age x onwards divided by the number of individuals

Table 8.3 Period life table for female Bali myna (*Leucopsar rothschildi*) calculated for a cohort of 644 individuals from x, N_x, $_na_x$, and $_nb_x$ data obtained from DATLife. Deaths are assumed to be distributed uniformly during the interval, so $_na_x = 0.5$, except for the final interval where it has been set to 0 and where q_x is set to 1 to arbitrarily complete the life table. Noninteger values have been rounded to 3 dp Nomenclature is the same as in Table 8.2, with the addition of fertility rate ($_nb_x$). Fertility values reported in DATLife are for offspring of both sexes and I assume the sex ratio is 50:50.

x	N_x	$_nD_x$	l_x	$_nq_x$	$_np_x$	$_nL_x$	T_x	e_x	$_nm_x$	$_na_x$	$_nb_x$
0	644	93	1.000	0.144	0.856	597.500	5757.500	8.940	0.156	0.500	0.050
1	551	45	0.856	0.082	0.918	528.500	5160.000	9.365	0.085	0.500	0.530
2	506	20	0.786	0.040	0.960	496.000	4631.500	9.153	0.040	0.500	0.895
3	486	9	0.755	0.019	0.981	481.500	4135.500	8.509	0.019	0.500	0.855
4	477	13	0.741	0.027	0.973	470.500	3654.000	7.660	0.028	0.500	1.000
5	464	54	0.720	0.116	0.884	437.000	3183.500	6.861	0.124	0.500	0.620
6	410	38	0.637	0.093	0.907	391.000	2746.500	6.699	0.097	0.500	0.540
7	372	37	0.578	0.099	0.901	353.500	2355.500	6.332	0.105	0.500	0.465
8	335	35	0.520	0.104	0.896	317.500	2002.000	5.976	0.110	0.500	0.375
9	300	40	0.466	0.133	0.867	280.000	1684.500	5.615	0.143	0.500	0.395
10	260	15	0.404	0.058	0.942	252.500	1404.500	5.402	0.059	0.500	0.300
11	245	11	0.380	0.045	0.955	239.500	1152.000	4.702	0.046	0.500	0.320
12	234	24	0.363	0.103	0.897	222.000	912.500	3.900	0.108	0.500	0.190
13	210	33	0.326	0.157	0.843	193.500	690.500	3.288	0.171	0.500	0.235
14	177	29	0.275	0.164	0.836	162.500	497.000	2.808	0.178	0.500	0.230
15	148	24	0.230	0.162	0.838	136.000	334.500	2.260	0.176	0.500	0.130
16	124	25	0.193	0.202	0.798	111.500	198.500	1.601	0.224	0.500	0.000
17	99	24	0.154	0.242	0.758	87.000	87.000	0.879	0.276	0.500	0.000
18	75	75	0.116	1.000	0.000	0.000	0.000	0.000		0.000	0.000

of age x (assuming a stable age distribution). Further development of these relationships and measures can be found in any good textbook on population biology (e.g. Gotelli 2001; Neal 2018).

8.3 The richness of life table information

From the life tables described it is possible to derive a dizzying array of metrics, some of which I have already mentioned. Many of these metrics can be considered as part of the life table since they vary through time within the life table and apply to particular age intervals (e.g. life expectancy, reproductive value, life span distribution [the empirical density function of the age at death], indicators of the force of selection [Baudisch 2005]). Others are single-value summaries such as population growth rate, generation time, or measures of the 'shape' of mortality such as entropy or life span equality (Demetrius 1978; Baudisch 2011; Wrycza et al.

2015; Colchero et al. 2016). These metrics can be considered life history traits of an organism living in a particular place at a particular time and are thus useful in a comparative evolutionary context. For example, using life tables derived from MPMs, researchers have explored the evolution of the pace and shape of senescence in plants (Baudisch 2011; Baudisch et al. 2013) and have explored the structure and evolution of life history across animals and plants using a large number of life-table-derived metrics (Salguero-Gómez et al. 2016; Salguero-Gómez and Jones 2017; Healy et al. 2019).

8.4 Complications, correlations, and limitations

The life table construction methods described are designed to use ideal data, where individuals in the population have perfect detectability and, for period-type life tables, where the population structure is stationary. They also rely on cohorts being followed until (almost) all individuals are dead or where the period-data include sufficient sample sizes across all ages. Such data are relatively easy to obtain from laboratory-based studies of model species like medflies (e.g. Carey et al. 2002) or fruit flies (e.g. Pearl and Parker 1921) or even naked mole rats (Sherman and Jarvis 2002). However, following the fate of most wild animals is difficult, and alternative approaches such as using standing age structure (to produce 'static life tables') are common (Udevitz and Gogan 2012). Other examples include the use of data from marine mammal strandings (e.g. Stolen and Barlow 2003), bird ringing (banding) recoveries (e.g. Lack 1943), skull recovery (Gonzalez and Crampe 2001; e.g. Simmons et al. 1984), fish otolith growth rings (e.g. Carlson et al. 2003), and fossil dinosaurs (Erickson et al. 2006; Erickson et al. 2009), where age can be estimated. This small selection of citations highlights the variety of data that can be coerced into the life table framework, but it is clear that there may sometimes be nuances in these data that complicate estimation of demographic trajectories.

Developments in the statistical field of survival analysis help researchers estimate demographic trajectories and quantify the effect of covariates, while accounting for the various nuances in the available data. These nuances include censoring, where not all individuals have experienced the event (death, in the present context) by the end of the study, and truncation, where the time of birth (and therefore age) is unknown. In addition, observational bias leading to the violation of the assumption that all individuals have an equal chance of observation can be problematic. The statistical approaches designed to cope with these issues include long-standing methods such as bivariate nonparametric analysis (Kaplan–Meier estimates: Kaplan and Meier 1958), proportional hazard approaches (Cox 1972), and accelerated failure time models (Wei 1992). In addition, and especially for wild animals where individual-based data are available, capture–recapture modelling (Lebreton et al. 1992; Lebreton and Gaillard 2016) has become increasingly popular. Some of these methods explicitly model mortality over continuous age with a functional form (e.g. Gompertz, Weibull, Siler) (Colchero and Clark 2012; Colchero et al. 2012). These approaches are covered in detail in Chapter 13.

Difficulties in estimating age in wild animals can also be a problem, and so researchers have often opted to use stage- rather than age-structured approaches like Lefkovitch MPMs (Lefkovitch 1965; Caswell 2001), even though age might be more appropriate under ideal conditions (see Chapter 9). Until recently, researchers were satisfied with a simplified (st)age structure (e.g. the two age classes of 'Juvenile' and 'Adult') that assumes constant mortality and fertility within those classes. However, vital rates can vary throughout adulthood (Jones et al. 2014), and senescence is commonplace, at least in birds and mammals (Jones et al. 2008; Nussey et al. 2013). Failure to account for these age-varying vital rates can have consequences for understanding population dynamics and in some cases can influence estimates of extinction risk (Robert et al. 2015; Colchero et al. 2019).

Furthermore, many organisms are naturally structured by something other than age, for example, distinct developmental stages (e.g. eggs, larvae, and adults in insects), and in some groups, notably plants, life histories can be complex, including both progression and retrogression (e.g. shrinkage) (e.g. Salguero-Gómez and Casper 2010). For these cases, MPMs are a flexible and powerful modelling tool (Caswell 2001; see Chapter 9), and in cases where life histories are structured by quantitative traits (e.g. body size, stem diameter, plant height) IPMs are similarly useful alternatives (Ellner and Rees 2006; Coulson 2012; see Chapter 10).

8.5 Conclusions

I have given a broad overview of the construction and use of life tables, with a focus on nonhuman studies. The use of life tables has a long history and they are still ubiquitous in human demographic studies and in controlled laboratory studies. Numerous metrics of importance to population biology can be calculated from them, including generation time, population growth rate, reproductive value, and entropy measures. Their use has been somewhat superseded in field-based demographic studies on nonhuman animals, largely because of the difficulty in meeting the required assumptions. Alternatives such as MPMs (Chapter 9), IPMs (Chapter 10), and Bayesian capture–recapture methods (Chapter 13) have become popular. Nevertheless, an understanding of life tables and the estimates derived from them is useful to appreciate and understand much of life history theory.

Data and code accessibility

Data and code to calculate the life tables and associated quantities and to produce the plots in this chapter are available at Zenodo: https://doi.org/10.5281/zenodo.3860588.

References

Anderson, D. R., Wywialowski, A. P., & Burnham, K. P. (1981). Tests of the assumptions underlying life table methods for estimating parameters from cohort data. *Ecology*, 62(4), 1121–1124.

Baudisch, A. (2005). Hamilton's indicators of the force of selection. *Proceedings of the National Academy of Sciences*, 102(23), 8263–8268.

Baudisch, A. (2011). The pace and shape of ageing. *Methods in Ecology and Evolution/British Ecological Society*, 2(4), 375–382.

Baudisch, A., Salguero-Gómez, R., Jones, O. R., et al. (2013). The pace and shape of senescence in angiosperms. *Journal of Ecology*, 101(3), 596–606.

Bonduriansky, R., Maklakov, A., Zajitschek, F., & Brooks, R. (2009). Sexual selection, sexual conflict and the evolution of ageing and life span. *Functional Ecology*, 22(3), 443–453.

Boonekamp, J. J., Salomons, M., Bouwhuis, S., Dijkstra, C., & Verhulst, S. (2014). Reproductive effort accelerates actuarial senescence in wild birds: an experimental study. *Ecology Letters*, 17(5), 599–605.

Bouwhuis, S., Charmantier, A., Verhulst, S., & Sheldon, B. C. (2010). Individual variation in rates of senescence: natal origin effects and disposable soma in a wild bird population. *Journal of Animal Ecology*, 79(6), 1251–1261.

Burger, O., Baudisch, A., & Vaupel, J. W. (2012). Human mortality improvement in evolutionary context. *Proceedings of the National Academy of Sciences*, 109(44), 18210–18214.

Butler, R. W. (1988). Population dynamics and migration routes of tree swallows, *Tachycineta bicolor*, in North America. *Journal of Field Ornithology*, 59(4), 395–402.

Carey, J. R., Liedo, P., Harshman, L., et al. (2002). Life history response of Mediterranean fruit flies to dietary restriction. *Aging Cell*, 1(2), 140–148.

Carlson, J. K., Cortés, E., & Bethea, D. M. (2003). Life history and population dynamics of the finetooth shark (*Carcharhinus isodon*) in the northeastern Gulf of Mexico. *Fishery Bulletin*, 101(2), 281–292.

Caswell, H. (2001). Matrix Population Models. Sinauer Associates, Sunderland, MA.

Caughley, G. (1966). Mortality patterns in mammals. *Ecology*, 47(6), 906–918.

Caughley, G. (1977). Analysis of Vertebrate Populations. John Wiley & Sons, London.

Clutton-Brock, T. H., & Isvaran, K. (2007). Sex differences in ageing in natural populations of vertebrates. *Proceedings. Biological Sciences/The Royal Society*, 274(1629), 3097–3104.

Cochran, M. E. & Ellner, S. (1992). Simple methods for calculating age-based life history parameters for stage-structured populations. *Ecological Monographs*, 62(3), 345–364.

Colchero, F. & Clark, J. S. (2012). Bayesian inference on age-specific survival for censored and truncated data. *Journal of Animal Ecology*, 81(1), 139–149.

Colchero, F., Jones, O. R., & Rebke, M. (2012). BaSTA: an R package for Bayesian estimation of age-specific survival from incomplete mark—recapture/recovery data with covariates. *Methods in Ecology and Evolution/British Ecological Society*, 3(3), 466–470.

Colchero, F., Rau, R., Jones, O. R., et al. (2016). The emergence of longevous populations. *Proceedings of the National Academy of Sciences*, 113(48), E7681–E7690.

Colchero, F., Jones, O. R., Conde, D. A., et al. (2019). The diversity of population responses to environmental change. *Ecology Letters*, 22(2), 342–353.

Coulson, T. (2012). Integral projections models, their construction and use in posing hypotheses in ecology. *Oikos*, 121(9), 1337–1350.

Cox, D. R. (1972). Regression models and life-tables. *Journal of the Royal Statistical Society. Series B, Statistical Methodology*, 34(2), 187–202.

Deevey, E. S. (1947). Life tables for natural populations of animals. *Quarterly Review of Biology*, 22(4), 283–314.

Demetrius, L. (1978). Adaptive value, entropy and survivorship curves. *Nature*, 275(5677), 213–214.

Descamps, S., Boutin, S., Berteaux, D., & Gaillard, J. M. (2008). Age-specific variation in survival, reproductive success and offspring quality in red squirrels: evidence of senescence. *Oikos*, 117(9), 1406–1416.

Ellner, S. P. & Rees, M. (2006). Integral projection models for species with complex demography. *The American Naturalist*, 167(3), 410–428.

Erickson, G. M., Currie, P. J., Inouye, B. D., & Winn, A. A. (2006). Tyrannosaur life tables: an example of nonavian dinosaur population biology. *Science*, 313(5784), 213–217.

Erickson, G. M., Makovicky, P. J., Inouye, B. D., Zhou, C.-F., & Gao, K.-Q. (2009). A life table for *Psittacosaurus lujiatunensis*: initial insights into ornithischian dinosaur population biology. *Anatomical Record*, 292(9), 1514–1521.

Fryxell, J. M. (1986). Age-specific mortality: an alternative approach. *Ecology*, 67, 1687–1692.

Gaillard, J.-M., Yoccoz, N. G., Lebreton, J.-D., et al. (2005). Review of *Generation time: a reliable metric to measure life-history variation among mammalian populations*. *The American Naturalist*, 166(1), 119–123; discussion 124–128.

Gonzalez, G. & Crampe, J.-P. (2001). Mortality patterns in a protected population of isards (*Rupicapra pyrenaica*). *Canadian Journal of Zoology*, 79(11), 2072–2079.

Gotelli, N. J. (2001). A Primer of Ecology. Sinauer Associates, Sunderland, MA.

Griffin, R. M., Hayward, A. D., Bolund, E., Maklakov, A. A., & Lummaa, V. (2017). Sex differences in adult mortality rate mediated by early-life environmental conditions. *Ecology Letters*, 21(2), 235–242.

Harcombe, P. A. (1987). Tree life tables. *Bioscience*, 37(8), 557–568.

Healy, K., Ezard, T. H. G., Jones, O. R., Salguero-Gómez, R., & Buckley, Y. M. (2019). Animal life history is shaped by the pace of life and the distribution of age-specific mortality and reproduction. *Nature Ecology & Evolution*, 3(8), 1217–1224.

Hegazy, A. K. (1990). Population ecology and implications for conservation of *Cleome droserifolia*: a threatened xerophyte. *Journal of Arid Environments*, 19, 269–282.

Hoenig, J. M. & Gedamke, T. (2007). A simple method for estimating survival rate from catch rates from multiple years. *Transactions of the American Fisheries Society*, 136(5), 1245–1251.

Johnstone, R. A. & Cant, M. A. (2019). Evolution of menopause. *Current Biology*, 29(4), R112–R115.

Jones, O. R. (2016). Life history patterns. In R. M. Kliman (ed.), *Encyclopaedia of Evolutionary Biology* (pp. 366–371). Academic Press, Oxford.

Jones, O. R. & Vaupel, J. W. (2017). Senescence is not inevitable. *Biogerontology*, 18(6), 965–971.

Jones, O. R., Gaillard, J.-M., Tuljapurkar, S., et al. (2008). Senescence rates are determined by ranking on the fast-slow life-history continuum. *Ecology Letters*, 11(7), 664–673.

Jones, O. R., Scheuerlein, A., Salguero-Gómez, R., et al. (2014). Diversity of ageing across the tree of life. *Nature*, 505(7482), 169–173.

Kaplan, E. L. & Meier, P. (1958). Nonparametric estimation from incomplete observations. *Journal of the American Statistical Association*, 53, 457–481.

Lack, D. (1943). The age of the blackbird. *British Birds; an Illustrated Magazine Devoted to the Birds on the British List*, 36(9), 166–175.

Lebreton, J.-D. & Gaillard, J.-M. (2016). Wildlife demography: population processes, analytical tools and management applications. In R. Mateo, B. Arroyo, & J. T. Garcia (eds), *Current Trends in Wildlife Research* (pp. 29–54). Springer, Cham.

Lebreton, J.-D., Burnham, K. P., Clobert, J., & Anderson, D. R. (1992). Modeling survival and testing biological hypotheses using marked animals: a unified approach with case studies. *Ecological Monographs*, 62(1), 67–118.

Lefkovitch, L. P. (1965). The study of population growth in organisms grouped by stages. *Biometrics*, 21(1), 1.

Leslie, P. H. (1945). On the use of matrices in certain population mathematics. *Biometrika*, 33, 183–212.

Leverich, W. J. & Levin, D. A. (1979). Age-specific survivorship and reproduction in *Phlox drummondii*. *The American Naturalist*, 113(6), 881–903.

Levitis, D. A., Burger, O., & Lackey, L. B. (2013). The human post-fertile lifespan in comparative evolutionary context. *Evolutionary Anthropology: Issues, News, and Reviews*, 22(2), 66–79.

Lotka, A. J. (1907). Relation between birth rates and death rates. *Science*, 26(653), 21–22.

Maklakov, A. A., Bonduriansky, R., & Brooks, R. C. (2009). Sex differences, sexual selection, and ageing: an experimental evolution approach. *Evolution; International Journal of Organic Evolution*, 63(10), 2491–2503.

Max-Planck Institute for Demographic Research (Germany). (2018). DATLife—the demography across the

Tree of Life—database. https://datlife.org/ (accessed 9 July 2019).

Menkens, G. E. & Boyce, M. S. (1993). Comments on the use of time-specific and cohort life tables. *Ecology*, 74(7), 2164–2168.

Neal, D. (2018). Introduction to Population Biology. Cambridge University Press, Cambridge.

Nussey, D. H., Froy, H., Lemaitre, J.-F., Gaillard, J.-M., & Austad, S. N. (2013). Senescence in natural populations of animals: widespread evidence and its implications for bio-gerontology. *Ageing Research Reviews*, 12(1), 214–225.

Pearl, R. & Miner, J. R. (1935). Experimental studies on the duration of life. XIV. The comparative mortality of certain lower organisms. *Quarterly Review of Biology*, 10(1), 60–79.

Pearl, R. & Parker, S. L. (1921). Experimental studies on the duration of life. I. Introductory discussion of the duration of life in *Drosophila*. *The American Naturalist*, 55(641), 481–509.

Preston, S., Heuveline, P., & Guillot, M. (2000). Demography: Measuring and Modeling Population Processes. Blackwell, Oxford.

Purvis, A. & Harvey, P. H. (1995). Mammal life-history evolution: a comparative test of Charnov's model. *Journal of Zoology*, 237(2), 259–283.

Reed, T. E., Kruuk, L. E. B., Wanless, S., Frederiksen, M., Cunningham, E. J. A., & Harris, M. P. (2008). Reproductive senescence in a long-lived seabird: rates of decline in late-life performance are associated with varying costs of early reproduction. *The American Naturalist*, 171(2), E89–E101.

Ricklefs, R. E., Scheuerlein, A., & Cohen, A. (2003). Age-related patterns of fertility in captive populations of birds and mammals. *Experimental Gerontology*, 38(7), 741–745.

Robert, A., Chantepie, S., Pavard, S., Sarrazin, F., & Teplitsky, C. (2015). Actuarial senescence can increase the risk of extinction of mammal populations. *Ecological Applications: A Publication of the Ecological Society of America*, 25(1), 116–124.

Salguero-Gómez, R. & Casper, B. B. (2010). Keeping plant shrinkage in the demographic loop. *Journal of Ecology*, 98(2), 312–323.

Salguero-Gómez, R. & Jones, O. R. (2017). Life history trade-offs modulate the speed of senescence. In R. P. Shefferson, O. R. Jones, & R. Salguero-Gómez (eds), *The Evolution of Senescence in the Tree of Life* (pp. 403–421). Cambridge University Press, Cambridge.

Salguero-Gómez, R., Jones, O. R., Jongejans, E., et al. (2016). Fast–slow continuum and reproductive strategies structure plant life-history variation worldwide. *Proceedings of the National Academy of Sciences*, 113(1), 230–235.

Schwartz, O. A. & Armitage, K. B. (1998). Empirical considerations on the stable age distribution. *Oecologia Montana*, 7(1–2), 1–6.

Sherman, P. W. & Jarvis, J. U. M. (2002). Extraordinary life spans of naked mole-rats (*Heterocephalus glaber*). *Journal of Zoology*, *258*(3), 307–311.

Siler, W. (1979). A competing-risk model for animal mortality. *Ecology*, *60*(4), 750–757.

Simmons, N. M., Bayer, M. B., & Sinkey, L. O. (1984). Demography of Dall's sheep in the MacKenzie Mountains, Northwest Territories. *Journal of Wildlife Management*, *48*(1), 156–162.

Stearns, S. C. (1992). *The Evolution of Life Histories*. Oxford University Press, Oxford.

Stolen, M. K. & Barlow, J. (2003). A model life table for bottlenose dolphins (*Tursiops truncatus*) from the Indian River Lagoon system, Florida, U.S.A. *Marine Mammal Science*, *19*(4), 630–649.

Sutherland, I. (1963). John Graunt: a tercentenary tribute. *Journal of the Royal Statistical Society. Series A*, *126*(4), 537.

Udevitz, M. S. & Ballachey, B. E. (1998). Estimating survival rates with age-structure data. *Journal of Wildlife Management*, *62*(2), 779–792.

Udevitz, M. S. & Gogan, P. J. P. (2012). Estimating survival rates with time series of standing age-structure data. *Ecology*, *93*(4), 726–732.

Van Valen, L. (1975). Life, death, and energy of a tree. *Biotropica*, *7*(4), 259–269.

Wei, L. J. (1992). The accelerated failure time model: a useful alternative to the cox regression model in survival analysis. *Statistics in Medicine*, 11, 1871–1879.

Wrycza, T. F., Missov, T. I., & Baudisch, A. (2015). Quantifying the shape of aging. *PLOS One*, *10*(3), e0119163.

CHAPTER 9

Introduction to matrix population models

Yngvild Vindenes, Christie Le Coeur, and Hal Caswell

9.1 Introduction

Population growth, extinction risk, and changes in trait distributions over time are examples of population-level processes that are studied in ecology and evolutionary biology. They are also intricately connected, since population growth ultimately depends on the demographic processes of individuals (such as survival, growth, and reproduction), which in turn depend on the state of individuals (the individual state, or i-state). The i-state may reflect properties such as age, life cycle stage, or morphological traits. Structured population models play a key role in linking such individual variation in state variables to population-level processes (Caswell and John 1992). Matrix population models (MPMs) represent a class of structured population model which is widely applied in ecology and evolutionary biology. This chapter will introduce MPMs and present a few of their applications.

The state of a population (or p-state) is a distribution of individuals in the i-state categories. This can be written as a vector $\mathbf{n}(t)$, and the transformation that maps the population from one time to the next can be written as a matrix: hence the name 'matrix population models'. Doing so opens a host of powerful mathematical methods to extract the population consequences of the individual life cycle. There is misapprehension that matrix mathematics is difficult. It is not. Because so much of ecology involves relations between multi-dimensional variables, and because matrices exist precisely to describe such relations, the basics of matrix algebra belong in the mathematical knowledge of every population ecologist. In this chapter, we will also write out some simple matrix expressions as summations and multiplications, to aid in understanding. This is definitely not the way to actually carry out computations, as matrix algebra is much more efficient.

MPMs are discrete time population models where individuals are classified into discrete stages (Caswell, 2001). Individuals within each stage are treated as identical, in the sense that they are assumed to experience the same vital parameters describing demographic processes and stage transitions. Indeed, the goal of defining stages is to accomplish this as closely as possible. MPMs were introduced by Leslie (1945, 1948) for age-structured populations and later extended to developmental stages (Lefkovitch, 1965). Today, MPMs are one of the most widely applied structured population models, because of many computational as well as analytical advances. A comprehensive presentation of MPMs, their analysis and applications, is provided by Caswell (2001). The purpose of this chapter is to give an introduction to age structured and stage structured MPMs and their key components. We also show how to calculate common model results using different empirical examples.

A key defining property of MPMs and other structured population models is the incorporation of individual differences, which provides the link between individual- and population-level processes (Caswell and John, 1992). Individual differences can arise from a range of properties, like age, developmental stage, morphological, and behavioural traits, and external factors like location (Bolnick et al., 2003; Dall et al., 2004; Caswell, 2009; Vindenes and Langangen, 2015). Understanding individual variation has always been a central problem in evolutionary biology, because trait variation is one of the prerequisites for evolution. MPMs, and their life table ancestors, have long been part of population ecology (Deevey, 1947).

Yngvild Vindenes, Christie Le Coeur, and Hal Caswell, *Introduction to matrix population models*. In: *Demographic Methods Across the Tree of Life*. Edited by Roberto Salguero-Gómez and Marlène Gamelon, Oxford University Press. © Oxford University Press (2021). DOI: 10.1093/oso/9780198838609.003.0009

There also exists a tradition, dating back to Lotka and Volterra, based on models in which all individuals are treated as equal. While such models yield important theoretical insights of population dynamics, they ignore the underlying mechanisms acting on individuals. By providing the crucial link from such individual processes (survival, development, growth, maturation, reproduction, migration, etc.) to the population dynamics, structured population models like MPMs have been central to our understanding of ecology and life history evolution (Caswell, 2001).

The number of potential state variables to include in an MPM is potentially infinite, and an important part of constructing an MPM for a given population is to choose the state variables. As is true for other modelling choices, this choice will depend on the study system, study questions, and available data and may involve trade-offs with other modelling choices such as analytical tractability. The chosen state variables will affect the model results, and the consequences of choosing a certain structure may be assessed along with other assumptions of the model. Statistical methods exist to evaluate choices of state variables (Caswell 2001; Chapter 12). In general, a state variable can represent any property or combination of properties relevant to describe the demographic processes of individuals. Age is commonly used, going back to life table analyses (Pearl, 1923; Deevey, 1947) and the early development of MPMs (Leslie, 1945, 1948). Classical life history theory was also developed for age-structured populations (Fisher, 1930; Roff, 1992; Charlesworth, 1994). Since several physiological and developmental processes change with age, age is often a good predictor of demographic processes. However, this is not always the case, and other variables, such as size, may be more important in many species, in particular with indeterminate growth. Some state variables are naturally discrete (e.g. distinct life cycle stages), while others are continuous, such as size, and must be discretised in a matrix model. Integral projection models (IPM; Chapter 10) are developed to incorporate continuous state variables, but in the end the underlying theory and the eventual analyses are very similar to MPMs (Ellner and Rees, 2006).

In the following sections, we first describe an age structured MPM and then extend this to a general stage structured MPM, dissecting the main model elements. This is followed by a brief introduction to some common results from analysis of an MPM, including parameters describing the long-term population dynamics and mean life history of the individuals in the population. We then present three case studies representing different kinds of demographic structure and demonstrate how to construct and

analyse an MPM for each case. The R code, together with more detailed information for each example, are provided in the online supplementary material (www.oup.com/companion/SalgueroGamelonDM). Our examples include a mammal, a bird, and an invertebrate. However, we emphasise that MPMs (and other demographic models) and their analyses depend on the structure of the life cycle, and the rates applying to individuals within that life cycle, and not at all on the taxon to which they are applied. This is part of the power of demographic analyses.

This chapter describes basic aspects of demographic models formulated in terms of matrices and vectors. We assume that the reader is familiar with basic matrix operations (addition and multiplication of matrices and vectors, inverses of matrices, eigenvalues, and eigenvectors). If you are not familiar with these, there are many resources available, online and in books (e.g. Caswell, 2001).

9.2 An age-structured MPM

We start by describing an age structured MPM, before extending this to a general case of stage structure. We will assume a constant environment and no density dependence, and for simplicity consider a female-based model with annual time-steps (other time intervals are also possible). Age-structured MPMs are widely used in demography and life history theory and are closely linked to life tables (Chapter 8).

9.2.1 Life table for age structured population

As an example, consider a perennial plant that reproduces at age 2 and 3 and where all individuals die before reaching age 4. The life table for this fictive population is given in Table 9.1.

Here, p_j denotes the annual survival probability from age j to age $j+1$, l_j denotes the cumulative survival from age 0 to age j (by definition $l_0 = 1$), and m_j is the fecundity, that is, the number of seeds (age 0) produced at exact age j (we assume a pulse reproduction). The last column shows the product of $l_j m_j$.

The life table can be used to calculate various life history characteristics of the population. For the plant

Table 9.1 Life table of plant example.

Age j	Annual survival p_j	Cumulative survival l_j	Fecundity m_j	$l_j m_j$
0	0.2	1	0	0
1	0.9	0.2	0	0
2	0.6	0.18	3	0.54
3	0	0.108	6	0.648

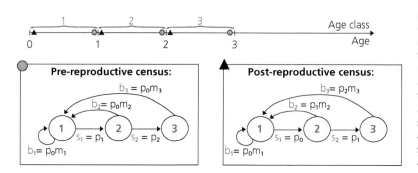

Figure 9.1 Relationship between discrete age class j and continuous age x (upper figure, modified from fig. 2.1 of Caswell, 2001), and the life cycle graphs for a pre- and post-reproductive census. Census time is indicated on the age-timeline using grey circles and black triangles, respectively. Parameters applying to age class (s_j and b_j) are shown in red, and parameters applying to actual age / birthday j (m_j and p_j) are shown in black. For simplicity only three age classes are shown.

example, the expected lifetime reproduction (net reproductive rate) is, for instance, given by $R_0 = \sum_{j=0}^{k} l_j m_j = 0+0+0.54+0.648 = 1.188$, and the cohort generation time is given by $G = \frac{1}{R_0} \sum_{j=0}^{k} j l_j m_j = (2 \cdot 0.54 + 3 \cdot 0.648)/1.188 \approx 2.5$. Finally, the long-term population growth rate λ is the solution of the Euler–Lotka equation,

$$\sum_{j=0}^{k} \lambda^{-j} l_j m_j = 1$$

$$0.54\lambda^{-2} + 0.648\lambda^{-3} = 1.$$

This equation can only be solved numerically, and for the plant example we find $\lambda \approx 1.07$. In a matrix model, all of these parameters (R_0, G, λ) and others can also be calculated directly from the projection matrix and its main components, as shown in section 9.4. First, we describe the age structured MPM using the plant example.

9.2.2 Projecting growth of age structured population

To describe how an age structured population is growing from one time step to the next, we need to keep track of the number of individuals in each of the age classes. Following Caswell (2001), we always denote the first age class as age class 1, irrespective of census time (see Figure 9.1; note that some authors use 0 to denote the first age class in the case of a post-reproductive census). Using the plant example described above, the population vector $\mathbf{n}_t = [n_1, n_2, n_3]'_t$ (the symbol $'$ indicates the transpose) describes the number of individuals in each age class at time t. Let s_j describe the survival probability of an individual in age class j to next year and b_j the number of offspring produced by an individual in age class j that enter next year's population (fertility coefficient). With these parameters we can describe the growth of each age class

$$n_1(t + 1) = b_1 n_1(t) + b_2 n_2(t) + b_3 n_3(t), \quad (9.1)$$

$$n_2(t + 1) = s_1 n_1(t), \quad (9.2)$$

$$n_3(t + 1) = s_2 n_2(t). \quad (9.3)$$

This is a linear system of equations, which can be written in matrix form:

$$\mathbf{n}_{t+1} = \mathbf{A} \mathbf{n}_t \quad (9.4)$$

$$\begin{bmatrix} n_1 \\ n_2 \\ n_3 \end{bmatrix}_{t+1} = \begin{bmatrix} b_1 & b_2 & b_3 \\ s_1 & 0 & 0 \\ 0 & s_2 & 0 \end{bmatrix} \begin{bmatrix} n_1 \\ n_2 \\ n_3 \end{bmatrix}_t.$$

where $\mathbf{A}$ is the population projection matrix; this age-classified case is called the Leslie matrix (named after early works of Leslie, 1945, 1948). The first row contains the fertilities, while the subdiagonal contains the survival probabilities. Next we consider how these are defined from the underlying age-dependent parameters of the life table.

9.2.3 Defining Leslie matrix elements for pre- and post-reproductive censuses

The fertility coefficients b_j and the survival probabilities s_j in the Leslie matrix describe transitions among the age classes from one time step to the next. However, because matrix models are defined in discrete time, while age is an inherently continuous variable, linking these parameters to the underlying age-specific parameters requires some careful definitions regarding (1) census time (when the population is counted during each time interval), and (2) the mode of reproduction (birth flow or birth pulse) of the study population. We have, for simplicity, assumed a birth-pulse population where individuals reproduce on their birthday (for birth flow populations see e.g. Caswell, 2001, Chapter 2.4.1). Regarding census time we will consider the two most common examples, where census is either right before (pre-reproductive) or right after (post-reproductive) the pulse reproduction (Caswell,

(a)
Pre-reproductive census

(b)
Post-reproductive census

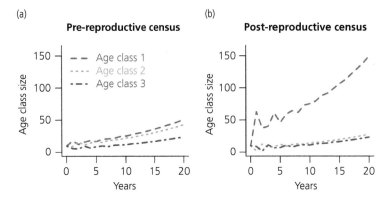

Figure 9.2 Projected growth in the age structured plant example, assuming a pre-reproductive census (a) or post-reproductive census (b). In both cases the asymptotic population growth rate is $\lambda = 1.07$.

2001, Chapter 2.4.2). The Leslie matrix elements will then depend on the age-specific parameters as follows:

1. With a *pre-reproductive census*, offspring (seedlings) will be nearly 1 year old when counted, and age classes correspond to actual age at counting (e.g. individuals in age class 1 are 1 year old when counted), so that $s_j = p_j$. The fertility coefficients are then given by $b_j = m_j p_0$, the product of the fecundity m_j and the first-year survival probability p_0 (Figure 9.1), so that the resulting Leslie matrix for the plant example is given by

$$\mathbf{A} = \begin{bmatrix} b_1 & b_2 & b_3 \\ s_1 & 0 & 0 \\ 0 & s_2 & 0 \end{bmatrix} = \begin{bmatrix} m_1 p_0 & m_2 p_0 & m_3 p_0 \\ p_1 & 0 & 0 \\ 0 & p_2 & 0 \end{bmatrix}$$

$$= \begin{bmatrix} 0 & 0.6 & 1.2 \\ 0.9 & 0 & 0 \\ 0 & 0.6 & 0 \end{bmatrix}.$$

2. With a *post-reproductive census*, the first age class will consist of newborn individuals (seeds), and age class does not correspond to actual age (individuals in age class j are counted at age $j-1$), so in this case $s_j = p_{j-1}$. The fertility in age class j is then given by $b_j = m_j p_{j-1} = m_j s_j$, the product of the parental survival probability until the next age, and the fecundity at the next age (fig. 9.1). For the plant example, the Leslie matrix for a post-reproductive census is then given by

$$\mathbf{A} = \begin{bmatrix} b_1 & b_2 & b_3 \\ s_1 & 0 & 0 \\ 0 & s_2 & 0 \end{bmatrix} = \begin{bmatrix} m_1 p_0 & m_2 p_1 & m_3 p_2 \\ p_0 & 0 & 0 \\ 0 & p_1 & 0 \end{bmatrix}$$

$$= \begin{bmatrix} 0 & 2.7 & 3.6 \\ 0.2 & 0 & 0 \\ 0 & 0.9 & 0 \end{bmatrix}.$$

Figure 9.2 shows the projected growth of the three age classes in the plant example, using a pre- and post-reproductive census. In the latter, the offspring age class consists of seeds and is therefore larger compared to the offspring class with pre-reproductive census. Note that errors in the definition of projection matrix elements seem to be a frequent issue in the literature where matrix models are used, in particular for models with a post-reproductive census (Kendall et al., 2019). With the exception of the asymptotic growth rate λ, all model results from an MPM will generally depend on the census time. This is not surprising, since populations are changing also within each time interval, so it matters at which time point during the time interval we make the census. This is also important to keep in mind when doing comparative analyses of MPM results from different studies—such comparisons should be made for models with similar census time.

9.3 Stage-structured MPMs

Age is a very special kind of population structure: individuals age exactly 1 year for every year that passes, it is impossible to move backwards in age, and reproduction always produces individuals in the first age class. We will now turn to a more general kind of structure, considering a population classified into k discrete stages, with no *a priori* restriction on what kind of transitions are permitted. The change in population vector from one time step to the next is still given by eqn. (9.4), but the projection matrix $\mathbf{A}$ is now a general stage-structured projection matrix.

It is helpful to write this matrix in terms of components that capture the different demographic processes of survival, reproduction, and transitions. To reduce the size of expressions, suppose that the life cycle includes $k = 3$ stages. Then write

$$\mathbf{A} = \begin{bmatrix} g_{11} & g_{12} & g_{13} \\ g_{21} & g_{22} & g_{23} \\ g_{31} & g_{32} & g_{33} \end{bmatrix} \begin{bmatrix} s_1 & 0 & 0 \\ 0 & s_2 & 0 \\ 0 & 0 & s_3 \end{bmatrix}$$

$$+ \begin{bmatrix} q_{11} & q_{12} & q_{13} \\ q_{21} & q_{22} & q_{23} \\ q_{31} & q_{32} & q_{33} \end{bmatrix} \begin{bmatrix} b_1 & 0 & 0 \\ 0 & b_2 & 0 \\ 0 & 0 & b_3 \end{bmatrix} \quad (9.5)$$

$$= \mathbf{G}\mathrm{Diag}(\mathbf{s}) + \mathbf{Q}\mathrm{Diag}(\mathbf{b}) \quad (9.6)$$

$$= \mathbf{U} + \mathbf{F} \quad (9.7)$$

This general projection matrix depends on four main demographic processes: (1) Survival, governed by the stage-specific survival probabilities s_j; (2) transitions among stages (conditional on survival), governed by the probabilities g_{ij} that an individual from stage j enters stage i next time step ($\sum_{i=1}^{k} g_{ij} = 1$); (3) reproduction, governed by the fertility coefficients b_j; and (4) distribution of new offspring among stages, governed by the probabilities q_{ij} that an offspring produced in stage j enters stage i ($\sum_{i=1}^{k} q_{ij} = 1$). We refer to the q_{ij} as offspring transition rates and the g_{ij} as individual transition rates.

We have now written $\mathbf{A}$ in terms of four component matrices, which capture each of these four demographic processes, and then as the sum of a matrix $\mathbf{U}$ that describes the survival and transitions among stages of extant individuals, and a matrix $\mathbf{F}$ that describes production of offspring from and to the various stages. These component matrices appear in any MPM, and by adjusting $\mathbf{G}$ and $\mathbf{Q}$, different kinds of demographic structure can be obtained. The age structured Leslie matrix from equation 9.4 is a special case of the general projection matrix: With age structure, surviving individuals will get one year older each time step, so that $g_{ij} = 1$ for $i = j + 1$ and zero otherwise. Similarly, all offspring enter the first age class, so that $q_{ij} = 1$ for $i = 1$ and zero otherwise. We show examples with different kinds of transition matrix structures in section 9.5.

Although we have written the projection matrix $\mathbf{A}$ as a constant matrix, depending on the situation it can be time dependent (stochastic or deterministic), dependent on environmental factors, or nonlinear (density- or frequency-dependent). Depending on the nature of the model, different kinds of analyses can be done based on the projection matrices, analytical or numerical. In any MPM, the most straightforward way to use the projection matrix (or matrices) is for numerical projection of the population vector over time by iterating eqn. (9.4), to see how the number and distribution of individuals in each stage is changing. Such projection is

now routine using a computer program like R (R Development Core Team, 2020) or MATLAB, for which several packages also exist. We show some examples using R in the supplementary material.

9.4 Basic analytical results from MPMs

MPMs can be used to calculate a number of properties of the dynamics and life history of the population and to investigate consequences of changing parameters (Caswell, 2001). These analyses go far beyond the simple numerical projection of the population over time, using eqn. (9.4). It is convenient to divide analytical results into those describing transient, short-term dynamics (Chapter 11) and those describing asymptotic, long-term dynamics. The latter include measures of population growth and structure, reproductive value, invasion fitness, selection gradients, equilibria, resilience, and sensitivity and elasticity analysis of these properties. Here we describe some basic analytical results for the general stage structured model defined above. Analysis of stochastic MPMs is described in Chapter 11. An overview of the variables and parameters defined here is given in Table 9.2.

9.4.1 Asymptotic properties from the projection matrix

Properties of the asymptotic population growth and structure are calculated in terms of the eigenvalues and eigenvectors of $\mathbf{A}$ (Caswell, 2001, Chapter 4.5). These include i) the long-term population growth rate λ, which is the dominant eigenvalue of $\mathbf{A}$, ii) the stable stage structure $\mathbf{w} = [w_1, ..., w_k]'$, which is the right eigenvector associated with λ and scaled so that $\sum_{j=1}^{k} w_j = 1$, and iii) the stage-specific reproductive values $\mathbf{v} = [v_1, ..., v_k]$, which is the left eigenvector associated with λ. This vector can be scaled in various ways; here we scale it so that $\sum_{j=1}^{k} w_j v_j = \mathbf{w}'\mathbf{v} = 1$. These parameters can also be found by projecting the population vector over a long time interval using $\mathbf{A}$. Regardless of the initial population vector, after a period of transient fluctuations the population will gradually reach the stable stage structure $\mathbf{w}$ where each stage grows with rate λ. In the context of life history theory, λ is also a measure of fitness of the phenotype described by $\mathbf{A}$ (see Metz et al. (1992) for discussion).

The reproductive values describe the relative contribution of individuals in each stage to future population size, compared to other stages. Fisher (1930) introduced the concept of reproductive value for a continuous-time age structured model and demonstrated that the total reproductive value, which

Table 9.2 Overview of variables and parameters for a general stage-structured matrix model. Model results assume no density dependence and a constant environment.

Parameter	Explanation
$N_t = \sum_{j=1}^{k} n_{j,t}$	Population size at time t.
k	Number of discrete stages.
$'$	Transpose of matrix or vector.
$\mathrm{Diag}(\mathbf{x})$	Diagonal matrix with elements of vector $\mathbf{x}$ on the main diagonal, and 0 elsewhere.
$\mathbf{s} = [s_1, ..., s_k]'$	Stage-specific survival probabilities.
$\mathbf{b} = [b_1, ..., b_k]'$	Stage-specific fertility coefficients (next year's offspring number).
g_{ij}	Transition probability from stage j to stage i.
q_{ij}	Probability that offspring produced in stage j is assigned to stage i.
$\mathbf{A} = \mathbf{G}\mathrm{Diag}(\mathbf{s}) + \mathbf{Q}\mathrm{Diag}(\mathbf{b})$	Projection matrix, dimension $k \times k$.
$\mathbf{G}$	Transition matrix conditional on survival, elements g_{ij}.
$\mathbf{U} = \mathbf{G}\mathrm{Diag}(\mathbf{s})$	Survival and transition matrix.
$\mathbf{Q}$	Offspring transition matrix conditional on survival, elements q_{ij}.
$\mathbf{F} = \mathbf{Q}\mathrm{Diag}(\mathbf{b})$	Offspring production matrix.
λ	Long-term (asymptotic) growth rate, dominant eigenvalue of $\mathbf{A}$.
$\mathbf{w} = [w_1, ..., w_k]'$	Stable stage structure, right eigenvector associated with λ.
$\mathbf{v} = [v_1, ..., v_k]$	Reproductive values, left eigenvector associated with λ.
$\mathbf{N} = (\mathbf{I} - \mathbf{U})^{-1}$	Fundamental matrix, elements n_{ij}.
$\eta_j = \sum_i n_{ij}$	Life expectancy in stage j.
$\mathbf{R} = \mathbf{F}(\mathbf{I} - \mathbf{U})^{-1}$	Generation to generation projection matrix.
R_0	Net reproductive rate, dominant eigenvalue of $\mathbf{R}$.
$T_M = \frac{\lambda}{\mathbf{vFw}}$	Mean age of parent at stable growth.

in our notation would be $V(t) = \sum_{i=1}^{k} n_i(t)v_i = \mathbf{v}'\mathbf{n}(t)$, grows exponentially with rate λ, even when the population is not currently at the stable structure $\mathbf{w}$. This result applies also to the discrete time stage-structured models described here (Engen et al., 2007; Vindenes et al., 2012) as well as to integral projection models (Vindenes et al., 2011). The stable structure and reproductive value vectors are key properties of the population dynamics of MPMs and also occur in other calculations such as sensitivity analysis of population growth rate.

It is important to note that asymptotic properties such as λ are consequences of the life cycle and the environment that determines the demographic rates. They show what would happen *if* the environment remains constant, but they do not require any assumption that the environment *will* remain constant.

9.4.2 Sensitivity analysis of population growth rate

Sensitivity analysis asks, and answers, the question, how would some result change if the parameters from which it is calculated were to change (Caswell, 2001,

2019b). One of the simplest and most commonly used analyses finds the sensitivities of λ to the elements of $\mathbf{A}$. Recent developments using matrix calculus have greatly extended the range of sensitivity analysis (Caswell, 2019b), but we do not address these here.

The sensitivity of λ can be calculated directly from the stable structure and reproductive values (using our scaling of v above; Caswell, 2001, Chapter 9),

$$\frac{d\lambda}{da_{ij}} = v_i w_j. \tag{9.8}$$

The sensitivity matrix, with sensitivity of λ to each matrix element, is thus given by $\mathbf{vw}'$.

In population genetics, these sensitivities are interpreted as stage-specific selection gradients linking the life history (as defined by $\mathbf{A}$) to fitness measured by λ (Lande, 1982). It is also possible to calculate sensitivities of λ (or any other output from an MPM) with respect to other parameters related to the life history (Caswell, 2001, Chapter 9). Sensitivities of λ to underlying parameters, such as the vital rates (survival probability, transition probability, etc.), can be found by applying the chain rule:

$$\frac{d\lambda}{dx} = \sum_{ij} \frac{d\lambda}{da_{ij}} \frac{da_{ij}}{dx}. \qquad (9.9)$$

The elasticity of λ is the proportional rate of change in response to a proportional change in x and is given by $\frac{d \ln \lambda}{d \ln x} = \frac{x}{\lambda}\frac{d\lambda}{dx}$.

9.4.3 Life history characteristics

The matrices $\mathbf{F}$ and $\mathbf{U}$ (elements of $\mathbf{A}$) contain information that can be used to calculate different characteristics of the underlying life history, such as the (stage-specific) lifetime reproductive success, the generation time, and stage-specific life expectancies.

The lifetime reproductive success, or net reproductive rate, R_0 describes the mean generation-to-generation population growth and the expected lifetime reproduction of an individual (Caswell, 2001). The generation-to- generation projection matrix is given by $\mathbf{R} = \mathbf{F}(\mathbf{I} - \mathbf{U})^{-1}$, and its dominant eigenvalue corresponds to R_0 (Section 5.3.4 in Caswell, 2001). See Caswell (2011) and van Daalen and Caswell (2017) for a deeper analysis that provides not only the mean lifetime reproductive success, but also the variance, skewness, and other measures of variability among individuals.

Generation time is a key life history characteristic, for which several measures exist. One common measure is the mean age of parents of offspring in a population at the stable stage distribution (Caswell, 2001). Using a Markov chain definition of the population dynamics in a general stage structured population, Bienvenu and Legendre (2015) demonstrated that this value can be calculated as $T_M = \frac{\lambda}{\mathbf{vFw}}$ (here adapted to use our scaling of $\mathbf{v}$).

The life expectancy in each stage can be calculated from the fundamental matrix $\mathbf{N} = (\mathbf{I} - \mathbf{U})^{-1}$ (Section 5.3.41 in Caswell, 2001). In this matrix, element n_{ij} is the expected number of time-steps an individual currently in stage j will spend in stage i before it dies. The remaining life expectancy, conditional on current stage, is thus given by the column sums of the fundamental matrix, $\eta_j = \sum_i n_{ij}$. In models with only one offspring stage, the life expectancy at birth is given by the column sum for this stage.

In the models discussed so far, all differences among individuals are due to a single demographic variable, that we have called 'stage'. However, it is possible to add additional variables that capture heterogeneity beyond that contributed by stage. The techniques for this are beyond this Chapter (Caswell et al., 2018),

but they make it possible to decompose the variance among individuals into each component (Hamel et al., 2016; Hartemink et al., 2017). Understanding the consequences of heterogeneity as well as its relative contribution to the variance in life history parameters is an important area of research (Vindenes et al., 2008; Tuljapurkar et al., 2009; Cam et al., 2016; Caswell et al., 2018; van Daalen and Caswell, 2020). In Box 9.1 we discuss how the use of matrix models has contributed to this field.

9.5 Case studies with different demographic structures

By adjusting the transition matrices $\mathbf{G}$ and $\mathbf{Q}$ of the general model, different kinds of demographic structure can be included (Vindenes et al., 2012). In the following we present some of these, with numerical examples. R code for calculations is provided as supplementary material.

9.5.1 Case study 1: Age structure in Asian elephants

In an age-structured model, transitions among the stages (in this case the age classes) are simple: surviving individuals always get one time step older, and will enter the next age class with probability 1 (thus $g_{ij} = 1$ for $i = j + 1$ and zero otherwise). All offspring enter the population in the first age class (thus $q_{ij} = 1$ for $i = 1$ and zero otherwise).

As an example of an age structured MPM we consider the female-based model for Asian elephants (*Elephas maximus*) developed by Goswami et al. (2014) to study potential consequences of conflict-induced mortality. Here we consider their baseline model with no such extra mortality. The main model parameters are the survival probabilities $\mathbf{s}$ and fertility coefficients $\mathbf{b}$ from age class 1 to 60 (Figure 9.3a; supplementary material). Goswami et al. (2014) assumed a post-reproductive census, so that offspring in age class 1 are newborn (actual age 0), and the survival from age class i to $i + 1$ then corresponds to the survival probability of actual age $i - 1$, $s_i = p_{i-1}$ (for instance, $s_1 = p_0$ is the survival probability from actual age 0 to age 1). The fertility of age class i is given by $b_i = m_i s_i$ and defines next year's number of female offspring produced by a female currently in age class i.

The long-term growth rate is $\lambda \approx 1.05$ for this model, indicating an annual growth of about 5% in the absence of extra mortality and density-dependent reproduction (Goswami et al., 2014). The stable age structure $\mathbf{w}$

Box 9.1 Effects of individual heterogeneity

Individual heterogeneity refers to variation among individuals that results in their experiencing different demographic rates. Heterogeneity is a topic of interest in different biological fields, including statistical ecology, behavioural biology, and conservation biology (e.g. Vaupel et al., 1979; Dall et al., 2004; Vindenes et al., 2008; Caswell, 2009; Tuljapurkar et al., 2009; Bolnick et al., 2011; Cam et al., 2016; Hamel et al., 2016; Jenouvrier et al., 2018). These fields have focused on different questions, from identifying unobserved heterogeneity in demographic data to evaluating its consequences for model results.

It is seldom realised however, that MPMs, and other demographic models, are essentially an approach to incorporate and account for individual heterogeneity. The choice to structure a model by age, size, stage, habitat, or something else is a judgement of what kind of heterogeneity is most important. The entries in the population projection matrix capture the differences in rates experienced by individuals of different ages, sizes, and so on. In other words, individual heterogeneity is nothing new. A main concern, however, is estimating and evaluating the consequences of heterogeneity that is not included in the description of the life cycle, perhaps because it is observed but neglected, or perhaps because it is invisible or unobserved. Try as you may to pick the most important state variables, there will always be residual variation among individuals.

The identification and estimation of unobserved heterogeneity in demographic data, so that it can be included in a model, is difficult. It is no surprise that estimating something without observing it is a difficult statistical problem. Recent advances, such as finite mixture models, sometimes make it possible to identify groups of individuals that share the same demographic rates, but where the difference between the groups is not explained by any apparent trait or property (Hamel et al., 2016; Hartemink and Caswell, 2018; Jenouvrier et al., 2018). Once heterogeneity is included in a model, whether as observed or unobserved state variables, the two main areas of research concern are (1) evaluating the potential consequences of ignoring heterogeneity for various model results, and (2) evaluating the relative contribution of heterogeneity to the variance in life history parameters, compared to other sources of variation among individuals.

One approach to studying consequences of ignoring heterogeneity is to compare results from models including the relevant state variable capturing heterogeneity or not (Vindenes et al., 2008). For instance, in the black-headed gull example, habitat structure is a source of observed heterogeneity. If we ignore habitat type, the resulting model becomes purely age structured (supplementary material). This model has the same value of $\lambda = 1.00$ as the model including habitat structure, but other model results differ, such as the mean lifetime reproduction and the age-specific reproductive values. For instance, the generation time would be 7.67 years rather than 7.69 (supplementary material). In this case, the habitat heterogeneity had rather small effects only on fertility, so ignoring it had small effects on model results. However, if we in addition to habitat also ignored age, which has much larger effects on fertility, the estimated generation time would be 5.51 years, underestimating by 28%. A similar comparison approach has been used to evaluate effects of ignoring heterogeneity on other parameters, such as parameters characterising stochastic population dynamics (demographic and environmental variance, e.g. Vindenes et al., 2008; Vindenes and Langangen, 2015).

Another important question is the relative contribution of heterogeneity to the variance among individuals in life history outcomes, compared to other sources of individual differences, including random variation. These methods are described at length in Caswell et al. (2018) and van Daalen and Caswell (2020), and make it possible to quantify the fraction of the variance in some outcome that is due to heterogeneity. The approach has so far been applied to longevity (Hartemink et al., 2017; Hartemink and Caswell, 2018; Seaman et al., 2019) and lifetime reproductive output (Jenouvrier et al., 2018; van Daalen and Caswell, 2020). Hartemink et al. (2017) considered effects of hidden heterogeneity ('frailty') on longevity in humans and found that most of the variation could be explained by individual stochasticity (random variation within each heterogeneity group). Jenouvrier et al. (2018) detected three heterogeneity groups in a population of southern fulmars (*Fulmarus glacialoides*) and found that this heterogeneity explained 3.7% of the variance in age at first reproduction, 5.9% in longevity, and 22% in lifetime reproductive success.

Theory for using demographic models like MPMs to estimate and evaluate consequences of heterogeneity is still in development, and there are several other questions where this theory and methods will be relevant (Cam et al., 2016), including senescence (McDonald and Fitzpatrick, 1996; Knape et al., 2011), extinction risk (Conner and White, 1999), and patterns of selection gradients (Caswell and Salguero-Gómez, 2013; Hernández et al., 2020).

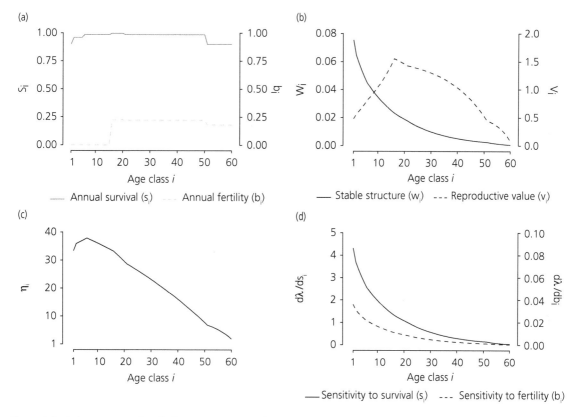

Figure 9.3 Vital parameters and results from an age-structured female-based model for Asian elephants. (a) Survival probability s_i and fertility b_i across age class i. (b) Stable age structure w_i and reproductive value v_i across age class i. (c) Life expectancy η_i in age class i. (d) Sensitivity of λ to survival probability and fertility.

declines monotonically with age class (Figure 9.3b), which is always true for growing age-structured populations. The reproductive value increases with age class at first and peaks around or shortly after the age at first reproduction and thereafter declines with age class (Figure 9.3b)—another common pattern for such life histories.

The remaining life expectancy in each age class is shown in figure 9.3c. Because the survival probability increases slightly during the first years of life (Figure 9.3a), the life expectancy also increases towards a peak in early life and then declines with age (Figure 9.3c). In contrast to life expectancy, generation time depends on both survival and fertility, and for this elephant model, the mean age of mothers is $T_M = \frac{\lambda}{\mathbf{vFw}} \approx$ 28 years.

With a post-reproductive census, the sensitivity of λ to survival s_j in age class j, taking into account effects on both age class transitions and reproduction, is given by

$$\frac{\partial \lambda}{\partial s_j} = v_{j+1}w_j + v_1 w_j m_j. \qquad (9.10)$$

These sensitivities are shown in Figure 9.3d, together with the sensitivities to fertility in each age class, which are given by the first row of the sensitivity matrix. These patterns are also universal for growing age-structured populations, with the sensitivities being highest at young age classes and then declining with age (Hamilton, 1966).

9.5.2 Case study 2: Life cycle stages in an isopod

Many species are characterised by discrete life history stages, reflecting ontogenetic development (Lefkovitch, 1965). A commonly used model for such stages (the 'standard size-classified model' of Caswell (2001)) proposes that, at each time step, surviving individuals in stage j will either make a transition to stage $j+1$, with probability γ_j, or remain in their current stage with probability $(1-\gamma_j)$. The transition matrix thus takes the values $g_{j+1,j} = \gamma_j$ and $g_{jj} = (1-\gamma_j)$ for $j = 1, ..., k-1$, $g_{kk} = 1$ for the final stage and zero elsewhere. As in the age structured model, all offspring enter the first stage 1 (thus $q_{ij} = 1$ for $i = 1$ and zero otherwise).

As an example of a species with distinct life cycle stages we consider the isopod *Porcellio scaber* (common rough woodlouse). In a toxicology study, Kammenga et al. (2001) defined a matrix model for this species, with monthly time-steps and seven distinct life history stages: 1) oocytes, 2) eggs/offspring (in marsupium), 3) juveniles, 4) first reproductive stage, 5) non-reproductive adult, 6) second reproductive stage, 7) senescent non-reproductive adult. Transition probabilities γ_j were calculated as the reciprocal of the average stage duration (Kammenga et al., 2001).

The average number of oocytes produced by reproductive individuals over a month is $m_4 = m_6 = 12$ for both reproductive stages (Kammenga et al., 2001). With a post-reproductive census, the fertility coefficients in stage 4 and 6 are

$$b_4 = 0.5m_4[\gamma_3 s_3 + (1 - \gamma_4)s_4] \tag{9.11}$$

$$b_6 = 0.5m_6[\gamma_5 s_5 + (1 - \gamma_6)s_6]. \tag{9.12}$$

The factor 0.5 represents an assumption of equal sex ratio of fertilised eggs. These fertility coefficients contain one component corresponding to reproduction by females that survive and make a transition from the preceding stage and one component from females that survive and remain in the reproductive stage. The resulting projection matrix for the isopods is given by

$$\mathbf{A} = \begin{bmatrix} (1 - \gamma_1)s_1 & 0 & 0 & 0.5m_4[\gamma_3 s_3 + (1 - \gamma_4)s_4] & 0 & 0.5m_6[\gamma_5 s_5 + (1 - \gamma_6)s_6] & 0 \\ \gamma_1 s_1 & (1 - \gamma_2)s_2 & 0 & 0 & 0 & 0 & 0 \\ 0 & \gamma_2 s_2 & (1 - \gamma_3)s_3 & 0 & 0 & 0 & 0 \\ 0 & 0 & \gamma_3 s_3 & (1 - \gamma_4)s_4 & 0 & 0 & 0 \\ 0 & 0 & 0 & \gamma_4 s_4 & (1 - \gamma_5)s_5 & 0 & 0 \\ 0 & 0 & 0 & 0 & \gamma_5 s_5 & (1 - \gamma_6)s_6 & 0 \\ 0 & 0 & 0 & 0 & 0 & \gamma_6 s_6 & s_7 \end{bmatrix}. \tag{9.13}$$

Vital rate values for this model are shown in Figure 9.6 as well as in supplementary information, and are taken from the control model of Kammenga et al. (2001) without any toxicity effects, and the corresponding life cycle graph for this model is given in figure 9.4. Life cycle graphs are useful tools for constructing MPMs, in particular for more complex stage structured models with various forms of transitions among the stages. Figure 9.5 shows the number of individuals in each stage over time, starting from a population consisting of 10 oocytes. After an initial period of transient fluctuations, the population reaches a stable stage structure, and all stages grow steadily at rate λ.

For this model the long-term growth rate is $\lambda \approx 1.07$, indicating a monthly population growth of around 7%. The stable stage structure is shown in figure 9.6c; it contains a high proportion of oocytes, eggs, and juveniles compared to adults. The reproductive value is higher for adults, especially in the two reproductive stages (Figure 9.6d). The last stage consists of non-reproducing senescent individuals with no contribution to future reproduction; thus the reproductive value is zero for this stage. If we were only concerned with λ, this stage could be omitted from the model. But if we want to know the entire population's structure, or if the model included

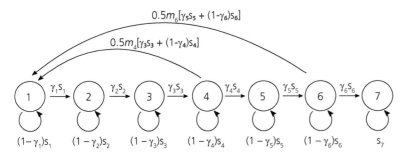

Figure 9.4 Life cycle graph for the isopod example of case study 2, with monthly transitions corresponding to the projection matrix of eqn. (9.13). Stages are 1) Oocytes, 2) Eggs, 3) Juveniles, 4) First reproductive stage, 5) Non-reproductive adults, 6) Second reproductive stage, 7) Senescent. The final stage does not contribute to population growth rate, λ.

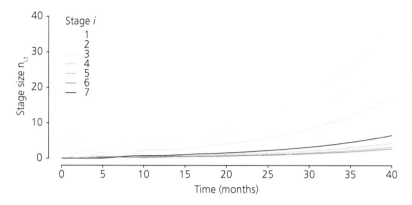

Figure 9.5 Projection of stage-specific abundance in the isopod case study, starting from a population of 10 oocytes and using the projection matrix in eqn. (9.13). Stages are 1) Oocytes, 2) Eggs, 3) Juveniles, 4) First reproductive stage, 5) Non-reproductive adults, 6) Second reproductive stage, 7) Senescent.

Table 9.3 Sensitivities in the isopod model, for stages 1 to 6. For the senescent stage 7 the sensitivity to survival is 0.

Stage j	To survival, $\frac{\partial \lambda}{\partial s_j}$	To transition, $\frac{\partial \lambda}{\partial \gamma_j}$	To fecundity, $\frac{\partial \lambda}{\partial m_j}$
1	$v_1 w_1(1 - \gamma_1) + v_2 w_1 \gamma_1$	$w_1(v_2 - v_1)s_1$	
2	$v_2 w_2(1 - \gamma_2) + v_3 w_2 \gamma_2$	$w_2(v_3 - v_2)s_2$	
3	$v_3 w_3(1 - \gamma_3) + v_4 w_3 \gamma_3 + 0.5 v_1 w_4 \gamma_3 m_4$	$w_3(v_4 - v_3)s_3 + 0.5 v_1 w_4 m_4 s_3$	
4	$v_4 w_4(1 - \gamma_4) + v_5 w_4 \gamma_4 + 0.5 v_1 w_4(1 - \gamma_4)m_4$	$w_4(v_5 - v_4)s_4 - 0.5 v_1 w_4 m_4 s_4$	$0.5 v_1 w_4 [\gamma_3 s_3 + (1 - \gamma_4)s_4]$
5	$v_5 w_5(1 - \gamma_5) + v_6 w_5 \gamma_5 + 0.5 v_1 w_6 \gamma_5 m_6$	$w_5(v_6 - v_5 s_5) + 0.5 v_1 w_6 m_6 s_5$	
6	$v_6 w_6(1 - \gamma_6) + v_7 w_6 \gamma_6 + 0.5 v_1 w_6(1 - \gamma_6)m_6$	$w_6(v_7 - v_6)s_6 - 0.5 v_1 w_6 m_6 s_6$	$0.5 v_1 w_6 [\gamma_5 s_5 + (1 - \gamma_6)s_6]$

density dependence, this stage would have to be included.

The expressions for sensitivities of λ to stage-specific survival probability s_j, transition rate γ_j and fecundity m_j are given in table 9.3, and the resulting values are shown in figure 9.7, together with the corresponding elasticities. These are found by applying the chain rule, as in eqn. (9.9). Population growth rate λ is most sensitive and most elastic to the survival and the growth probabilities of stages 2 and 3, which also contain a high proportion of the population (Figure 9.6). Overall, the sensitivities of λ to survival and transition probabilities are much higher than those to fecundity, while the elasticities are more similar in scale. Note that in the reproductive stages 4 and 6, the sensitivity to transition probability is negative. In stage 4, individuals moving to stage 5 lose the opportunity to reproduce and have to wait at least one more time step to enter the next reproductive stage, resulting in a lower reproductive value in stage 5 than in 4 and a corresponding negative sensitivity to the transition. Thus, this sensitivity analysis shows that increasing the probability of remaining in the first reproductive stage will increase λ, even though there is another reproductive stage later.

Regarding other life history parameters for this case model, the net reproductive rate is $R_0 \approx 1.77$, life expectancy in stage 1 is $\eta_1 = 5.44$ months, while the generation time is $T_m \approx 7.67$ months.

9.5.3 Case study 3: Age and habitat structure in black-headed gulls

Individuals rarely differ from each other by just one state variable. Multistate MPMs describe the dynamics of populations structured by more than one state variable, and age by stage models are an important example. Here, the stage can represent any of the familiar i-state variables (size, developmental stage, physiological condition, spatial location, habitat quality, or genotype), and the inclusion of age permits the stage-specific transitions, fertilities, and survival probabilities to vary with age. To analyse multistate models, the population vector must contain all combinations of the states, and the population projection matrix has a block structure corresponding to the arrangement of stages. We show a simple example here, but the possibilities are very general (Caswell et al., 2018).

Lebreton (1996) developed an age by stage MPM for black-headed gulls (*Chroicocephalus ridibundus*), where

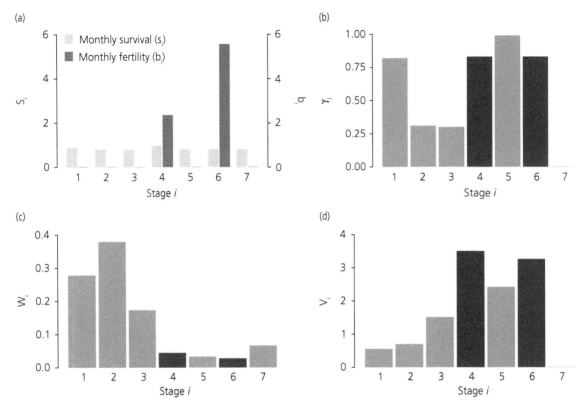

Figure 9.6 Vital rates and results from stage-structured MPM for isopods in case study 2. Stages are 1) Oocytes, 2) Eggs, 3) Juveniles, 4) First reproductive stage, 5) Non-reproductive adults, 6) Second reproductive stage, 7) Senescent. Results representing the two reproductive stages are shown in dark colour. (a) Stage-specific survival probability s_j and fertility b_j (fertility depends on other vital rates, see eq. (9.11) and eq. (9.12)). (b) Stage-specific probability γ_j of making a transition to the next stage. (c) Stable stage structure. (d) Stage-specific reproductive values.

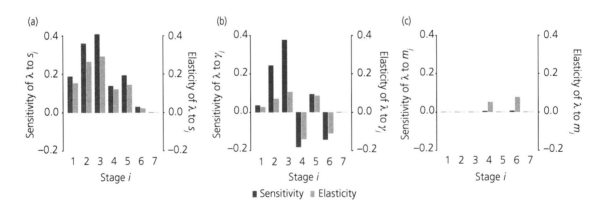

Figure 9.7 (a) Sensitivities and elasticities of λ to underlying vital rates of stage specific survival probabilities s_j (b) transition probabilities γ_j (c) and oocyte production m_j (c), for the isopod example in case study 2. Stages are 1) Oocytes, 2) Eggs, 3) Juveniles, 4) First reproductive stage, 5) Non-reproductive adults, 6) Second reproductive stage, 7) Senescent.

the stage represents breeding habitat quality. This model was also used by Hunter and Caswell (2005) in a demonstration of the vec-permutation matrix approach to constructing such models. This goes beyond the scope of this chapter but is described in detail in Caswell et al. (2018). The model has two habitat types ('good' and 'poor' quality) and five age classes, where the last class contains individuals of age 5 and above. Only the 1-year-olds disperse among the two habitats.

Including age within habitat, we get 10 stages, where the first five correspond to ages 1–5 in the good habitat and the last five correspond to ages 1–5 in the poor habitat. The resulting 10×10 projection matrix for black-headed gulls is given by

$$
\mathbf{A} =
\left[
\begin{array}{ccccc|ccccc}
b_1 & b_2 & b_3 & b_4 & b_5 & 0 & 0 & 0 & 0 & 0 \\
(1-p)s_1 & 0 & 0 & 0 & 0 & qs_6 & 0 & 0 & 0 & 0 \\
0 & s_2 & 0 & 0 & 0 & 0 & 0 & 0 & 0 & 0 \\
0 & 0 & s_3 & 0 & 0 & 0 & 0 & 0 & 0 & 0 \\
0 & 0 & 0 & s_4 & s_5 & 0 & 0 & 0 & 0 & 0 \\
\hline
0 & 0 & 0 & 0 & 0 & b_6 & b_7 & b_8 & b_9 & b_{10} \\
ps_1 & 0 & 0 & 0 & 0 & (1-q)s_6 & 0 & 0 & 0 & 0 \\
0 & 0 & 0 & 0 & 0 & 0 & s_7 & 0 & 0 & 0 \\
0 & 0 & 0 & 0 & 0 & 0 & 0 & s_8 & 0 & 0 \\
0 & 0 & 0 & 0 & 0 & 0 & 0 & 0 & s_9 & s_{10}
\end{array}
\right].
$$

$$(9.14)$$

Here, the upper left block of the matrix describes age-specific survival and reproduction in the good habitat, while the lower right matrix describes the same processes for the poor habitat. The upper right matrix describes transition from the poor to the good habitat (only possible in age 1), which occurs with probability q, while the lower left matrix describes transition from the good to the poor habitat, which occurs with probability p. These probabilities can be varied to study the consequences of different dispersal patterns (Lebreton, 1996; Hunter and Caswell, 2005).

According to Lebreton (1996), the age-specific survival probabilities are the same in the two habitats, while age-specific fecundities differ, leading to higher fertility in the good quality habitat (Figure 9.8b; parameter values in the supplementary material). For the special case without any dispersal among the habitats ($p = q = 0$), we get a separate age-structured MPM for each habitat. The long-term growth rate is then $\lambda \approx 1.01$ in the good habitat and $\lambda \approx 0.97$ in the poor habitat. When dispersal is included ($p = 0.25$, $q = 0.375$ in the baseline model), the long-term growth rate for the entire population, coupling the good and poor habitats, is $\lambda \approx 1.00$. The stable structure is skewed towards more individuals in the good habitat, because of higher reproduction combined with a larger probability of remaining in than leaving this habitat (Figure 9.8c). Reproductive values are higher in the good quality

habitat (Figure 9.8d), in particular for ages 2–5 after dispersal has occurred.

Hunter and Caswell (2005) calculated the sensitivities of λ to the dispersal probabilities p and q. Using our notation and scaling, these are given by

$$
\begin{aligned}
\frac{d\lambda}{dp} &= \sum_{i,j} \frac{\partial \lambda}{\partial a_{ij}} \frac{\partial a_{ij}}{\partial p} = -v_2 w_1 s_1 + v_7 w_1 s_1 \\
&= w_1 (v_7 - v_2) s_1
\end{aligned}
$$

$$(9.15)$$

and

$$
\frac{d\lambda}{dq} = v_2 w_6 s_6 - v_7 w_6 s_6 = w_6 s_6 (v_2 - v_7).
$$

$$(9.16)$$

With the baseline values of $p = 0.25$ and $q = 0.375$ these are $\frac{d\lambda}{dp} \approx -0.04$ and $\frac{d\lambda}{dq} \approx 0.02$. Thus, as expected, increased transition out of the good quality habitat has a negative effect on λ, while increased transition out of the poor quality habitat is positive. The sensitivities of λ to age-specific survival and fecundity in the two habitats are shown in Figure 9.8e,f, confirming that the vital parameters of the good quality habitat have a higher influence on the population growth.

9.6 Conclusions

Computational and analytical advances have opened new application of MPMs in a range of theoretical and applied fields. In conservation biology, for instance, sensitivity analysis can be used to identify vulnerable stages and predict responses to environmental change (e.g., Jenouvrier et al., 2009; Hunter et al., 2010; van de Pol et al., 2010). For harvested populations or in pest control, MPMs can be used to optimise the management effort and strategies with respect to different goals and risks (e.g., Bieber and Ruf, 2005). MPMs are widely used in ecotoxicology to evaluate the population consequences of pollutant effects on individual life cycles (e.g., Li et al., 2014; Miller et al., 2020). Because the essence of MPMs is the incorporation of individual differences through stages, they also represent a natural framework to study potential consequences of ignoring such differences (Box 9.1; Vindenes and Langangen 2015). In combination with modern statistical methods for identifying unobserved heterogeneity, MPMs represent a powerful tool to decompose sources of variation in life history characteristics (Caswell, 2009; Hamel et al., 2016; Jenouvrier et al., 2018).

We have introduced matrix models through a time-invariant stage-structured MPM and shown how different kinds of demographic structure can be obtained by adjusting the transition probabilities. The ability of matrices to mathematically describe transformations of

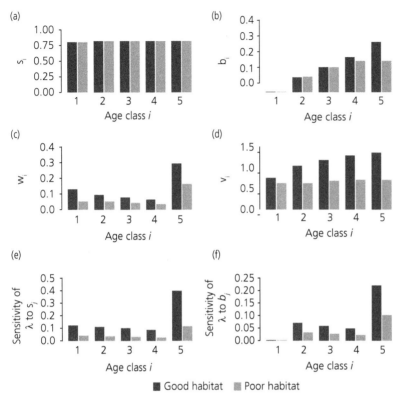

Figure 9.8 Vital parameters and model results from the age by stage structured model for black-headed gulls, case study 3. The last age class includes ages ≥ 5. (a) Survival probability s_i (same in the two habitats). (b) Fertility b_i. (c) Stable stage structure. (d) Reproductive values. (e) Sensitivity of λ to survival probability. (f) Sensitivity of λ to fertility.

multiple variables, like the population structure vector, makes MPMs the most flexible of any of the frameworks for structured population models. Our three case studies only scratch the surface of all the possible kinds of demographic structure that can be considered. Developments and applications of MPMs go far beyond the linear, time-invariant models here. The list is long, but we should draw attention to important recent developments of MPMs in **(1)** stochastic models (e.g., Haridas and Tuljapurkar, 2005; Tuljapurkar and Haridas, 2006; Steinsaltz et al., 2011), **(2)** nonlinear models for density dependence (Cushing et al., 2002; Caswell, 2008), **(3)** nonlinear two-sex models for frequency dependence (Shyu and Caswell, 2018), **(4)** stage-structured epidemic models (Klepac and Caswell, 2011; Metcalf et al., 2012), **(5)** models for species invasion dynamics (Neubert and Caswell, 2000; Neubert and Parker, 2004), **(6)** models including explicit population genetics and selection (de Vries and Caswell, 2019a, 2019b), **(7)** models for second-order systems, where individual fate depends not only on present stage but also on previous stage (de Vries and Caswell, 2018), **(8)** models that can incorporate any number of i-state dimensions (Roth and Caswell, 2016;

Caswell et al., 2018), and **(9)** models describing individuals in terms of their close relatives (Pavard and Branger, 2012) and entire kinship networks (Caswell, 2019a, 2020). Coupled with these developments has been the introduction of a new, powerful, and much more general approach to the sensitivity analysis of every aspect of population dynamics (transient and asymptotic, deterministic and stochastic, linear and nonlinear) using the methodology of matrix calculus (Caswell, 2019b).

A particularly important development has been theory and applications that couple MPMs explicitly to the stochastic movement of individuals through their life cycle. This leads to a rich connection with Markov chain methods, which have been used to analyse longevity and occupancy times (e.g., Tuljapurkar and Horvitz, 2006; Horvitz and Tuljapurkar, 2008; Caswell, 2009; Roth and Caswell, 2018) and have led to a complete stochastic theory for variation among individuals in lifetime reproductive output (Caswell, 2011; van Daalen and Caswell, 2017), and measures of population health (Caswell and Zarulli, 2018). The stochastic fates of individuals within an MPM provide the variability that leads to demographic stochasticity in population dynamics and to connections between MPMs, multitype branching processes, and diffusion models (Vindenes et al., 2008; Caswell and Vindenes, 2018).

MPMs are increasingly applied also in comparative analyses (Chapter 18). The growing open access databases COMPADRE (2016) and COMADRE (2017) contain published projection matrices from the literature; they cover an increasing number of species of plants and animals. The metadata for these studies provide information about the study system, location, phylogeny, source publication, and other factors useful for comparative analyses. MPMs can now be used to study patterns of variation in demographic parameters across the Tree of Life. For instance, Che Castaldo et al. (2018) used plant MPMs from COMPADRE (2016) to evaluate the extent to which demographic results from one species can be extrapolated to related or biologically similar species, as commonly advocated in conservation when demographic data are lacking. Other examples of applications include studies aiming to identify life history strategies in variable environments (McDonald et al., 2017), to understand evolution of senescence in plants (Caswell and Salguero-Gómez, 2013), or to identify diverging patterns of senescence across the Tree of Life (Jones et al., 2014).

This chapter has provided only a glimpse of possible applications of MPMs, but even a model assuming time-independent vital rates yields much information about the underlying life history and population dynamics. Through the methods provided in this chapter and in the cited references, MPMs allow a deeper understanding of the underlying mechanisms shaping ecological and evolutionary processes.

References

Bieber, C., and T. Ruf. 2005. Population dynamics in wild boar *Sus scrofa*: ecology, elasticity of growth rate and implications for the management of pulsed resource consumers. J. Appl. Ecol. **42**: 1203–1213.

Bienvenu, F., and S. Legendre. 2015. A new approach to the generation time in matrix population models. Am. Nat. **185**: 834–843.

Bolnick, D. I., R. Svanback, J. A. Fordyce, et al. 2003. The ecology of individuals: Incidence and implications of individual specialization. Am. Nat. **161**: 1–28.

Bolnick, D. I., P. Amarasekare, M. Araújo, et al. 2011. Why intraspecific trait variation matters in community ecology. Trends Ecol. Evol. **26**: 183–192.

Cam, E., L. M. Aubry, and M. Authier. 2016. The conundrum of heterogeneities in life history studies. Trends Ecol. Evol. **31**: 872–886.

Caswell, H. 2001. Matrix population models. 2nd edition. Sinauer Associates, Sunderland, Massachusetts.

Caswell, H. 2008. Perturbation analysis of nonlinear matrix population models. Demographic Research **18**: 59–116.

Caswell, H. 2009. Stage, age and individual stochasticity in demography. Oikos **118**: 1763–1782.

Caswell, H. 2011. Beyond R0: demographic models for variability of lifetime reproductive output. PLOS One **6**:e20809.

Caswell, H. 2019a. The formal demography of kinship: A matrix formulation. Demographic Res. **41**: 679–712.

Caswell, H. 2019b. Sensitivity Analysis: Matrix Methods in Demography and Ecology. Demographic Research Monographs, Springer Nature Switzerland, Cham.

Caswell, H. 2020. The formal demography of kinship. II. Multistate models, parity, and sibship. Demographic Res. **42**: 1097–1144.

Caswell, H., and A. M. John, 1992. From the individual to the population in demographic models. Pages 36–61 in D. L. DeAngelis and L. J. Gross, editors. Individual-based models and approaches in ecology. Chapman & Hall, New York.

Caswell, H., and R. Salguero-Gómez. 2013. Age, stage and senescence in plants. J. Ecol. **101**: 585–595.

Caswell, H., and Y. Vindenes. 2018. Demographic variance in heterogeneous populations: Matrix models and sensitivity analysis. Oikos **127**: 648–663.

Caswell, H., and V. Zarulli. 2018. Matrix methods in health demography: a new approach to the stochastic analysis of healthy longevity and DALYs. Population Health Metrics **16**: 8.

Caswell, H., C. de Vries, N. Hartemink, et al. 2018. Age x stage-classified demographic analysis: a comprehensive approach. Ecol. Monogr. **88**: 560–584.

Charlesworth, B. 1994. Evolution in age-structured populations. 2nd edition. Cambridge University Press, Cambridge.

Che Castaldo, J. P., C. C. Che Castaldo, and M. C. Neel. 2018. Predictability of demographic rates based on phylogeny and biological similarity. Conservation **32**: 1290–1300.

COMPADRE, 2016. Plant Matrix Database. Available at www.compadre-db.org. Max Planck Institute for Demographic Research (Germany), version 4.0.1, accessed 5 October 2016.

COMADRE, 2017. Animal Matrix Database. Available at www.comadre-db.org. Max Planck Institute for Demographic Research (Germany), version 2.0.1, accessed 3 September 2017.

Conner, M., and G. White. 1999. Effects of individual heterogeneity in estimating the persistence of small populations. Nat. Resource Model. **12**: 109–127.

Cushing, J. M., R. F. Costantino, B. Dennis, et al. 2002. Chaos in ecology: experimental nonlinear dynamics. Elsevier, New York.

Dall, S. R., A. I. Houston, and J. M. McNamara. 2004. The behavioural ecology of personality: consistent individual differences from an adaptive perspective. Ecol. Lett. **7**: 734–739.

de Vries, C., and H. Caswell. 2018. Demography when history matters: construction and analysis of second-order matrix population models. Theor. Ecol. **11**: 129–140.

de Vries, C., and H. Caswell. 2019a. Selection in two-sex stage-structured populations: Genetics, demography, and polymorphism. Theor. Popul. Biol. **130**: 160–169.

de Vries, C., and H. Caswell. 2019b. Stage-structured evolutionary demography: linking life histories, population genetics, and ecological dynamics. Am. Nat. **193**: 545–559.

Deevey, E. S. J. 1947. Life tables for natural populations of animals. Quart. Rev. Biol. **22**: 283–314.

Ellner, S. P., and M. Rees. 2006. Integral projection models for species with complex demography. Am. Nat. **167**: 410–428.

Engen, S., R. Lande, B.-E. Sæther, et al. 2007. Using reproductive value to estimate key parameters in density-independent age-structured populations. J. Theor. Biol. **244**: 308–317.

Fisher, R. A. 1930. The genetical theory of natural selection. Clarendon Press, Oxford.

Goswami, V. R., D. Vasudev, and M. K. Oli. 2014. The importance of conflict-induced mortality for conservation planning in areas of human–elephant co-occurrence. Biol. Cons. **176**: 191–198.

Hamel, S., N. G. Yoccoz, and J.-M. Gaillard. 2016. Assessing variation in life-history tactics within a population using mixture regression models: a practical guide for evolutionary ecologists. Biol. Rev. **92**: 754–775.

Hamilton, W. D. 1966. The moulding of senescence by natural selection. J. Theor. Biol. **12**: 12–45.

Haridas, C. V., and S. Tuljapurkar. 2005. Elasticities in variable environments: properties and implications. Am. Nat. **166**: 481–495.

Hartemink, N., and H. Caswell. 2018. Variance in animal longevity: contributions of heterogeneity and stochasticity. Popul. Ecol. **60**: 89–99.

Hartemink, N., T. I. Missov, and H. Caswell. 2017. Stochasticity, heterogeneity, and variance in longevity in human populations. Theor. Popul. Biol. **114**: 107–116.

Hernández, C. M., S. F. van Daalen, H. Caswell, et al. 2020. A demographic and evolutionary analysis of maternal effect senescence. Proc. Natl. Acad. Sci. U.S.A. **117**: 16431–16437.

Horvitz, C. C., and S. Tuljapurkar. 2008. Stage dynamics, period survival, and mortality plateaus. Am. Nat. **172**: 203–215.

Hunter, C. M., and H. Caswell. 2005. The use of the vec-permutation matrix in spatial matrix population models. Ecol. Model. **188**: 15–21.

Hunter, C. M., H. Caswell, M. C. Runge, et al. 2010. Climate change threatens polar bear populations: a stochastic demographic analysis. Ecology **91**: 2883–3897.

Jenouvrier, S., H. Caswell, C. Barbraud, et al. 2009. Demographic models and IPCC climate projections predict the decline of an emperor penguin population. Proc. Natl. Acad. Sci. U.S.A. **106**: 1844–1847.

Jenouvrier, S., L. M. Aubry, C. Barbraud, et al. 2018. Interacting effects of unobserved heterogeneity and individual stochasticity in the life history of the southern fulmar. J. Anim. Ecol. **87**: 212–222.

Jones, O. R., A. Scheuerlein, R. Salguero-Gómez, et al. 2014. Diversity of ageing across the tree of life. Nature **505**: 169–173.

Kammenga, J. E., C. A. M. v. Gestel, and E. Hornung. 2001. Switching life-history sensitivities to stress in soil invertebrates. Ecol. Appl. **11**: 226–238.

Kendall, B. E., M. Fujiwara, J. Diaz-Lopez, et al. 2019. Persistent problems in the construction of matrix population models. Ecol. Model. **406**: 33–43.

Klepac, P., and H. Caswell. 2011. The stage-structured epidemic: linking disease and demography with a multi-state matrix approach model. Theor. Ecol. **4**: 301–319.

Knape, J., N. Jonzén, M. Sköld, et al. 2011. Individual heterogeneity and senescence in Silvereyes on Heron Island. Ecology **92**: 813–820.

Lande, R. 1982. A quantitative genetic theory of life history evolution. Ecology **63**: 607–615.

Lebreton, J. D. 1996. Demographic models for subdivided populations: the renewal equation approach. Theor. Popul. Biol. **49**: 291–313.

Lefkovitch, L. P. 1965. The study of population growth in organisms grouped by stages. Biometrics **21**: 1–18.

Leslie, P. H. 1945. On the use of matrices in certain population mathematics. Biometrika **33**: 183–212.

Leslie, P. H. 1948. Some further notes on the use of matrices in population mathematics. Biometrika **35**: 213–245.

Li, W.-H., Y.-R. Ju, C.-M. Liao, et al. 2014. Assessment of selenium toxicity on the life cycle of *Caenorhabditis elegans*. Ecotoxicology **23**: 1245–1253.

McDonald, D. B., and J. W. Fitzpatrick. 1996. Actuarial senescence and demographic heterogeneity in the Florida scrub jay. Ecology **77**: 2373–2381.

McDonald, J. L., M. Franco, S. Townley, et al. 2017. Divergent demographic strategies of plants in variable environments. Nat. Ecol. Evol. **1**: 0029.

Metcalf, C. J. E., J. Lessler, P. Klepac, et al. 2012. Structured models of infectious disease: inference with discrete data. Theor. Popul. Biol. **82**: 275–282.

Metz, J. A. J., R. M. Nisbet, and S. A. H. Geritz. 1992. How should we define 'fitness' for general ecological scenarios? Trends Ecol. Evol. **7**: 198–202.

Miller, D. H., B. W. Clark, and D. E. Nacci. 2020. A multidimensional density dependent matrix population model for assessing risk of stressors to fish populations. Ecotoxicol. Environ. Safety **201**: 110786.

Neubert, M. G., and H. Caswell. 2000. Demography and dispersal: calculation and sensitivity analysis of invasion speed for structured populations. Ecology **81**: 1613–1628.

Neubert, M. G., and I. M. Parker. 2004. Projecting rates of spread for invasive species. Risk Analysis **24**: 817–831.

Pavard, S., and F. Branger. 2012. Effect of maternal and grandmaternal care on population dynamics and human life-history evolution: A matrix projection model. Theor. Popul. Biol. **82**: 364–376.

Pearl, R. 1923. Introduction to medical biometry and statistics. W. B. Saunders, Philadelphia.

R Development Core Team, 2020. R: A Language and Environment for Statistical Computing. R Foundation for Statistical Computing, Vienna.

Roff, D. A. 1992. The evolution of life histories: theory and analysis. Chapman & Hall, New York, U.S.A.

Roth, G., and H. Caswell. 2016. Hyperstate matrix models: extending demographic state spaces to higher dimensions. Methods Ecol. Evol. **7**: 1438–1450.

Roth, G., and H. Caswell. 2018. Occupancy time in sets of states for demographic models. Theor. Popul. Biol. **120**: 62–77.

Seaman, R., T. Riffe, and H. Caswell. 2019. Changing contribution of area-level deprivation to total variance in age at death: a population-based decomposition analysis. BMJ Open **9**:e024952.

Shyu, E., and H. Caswell. 2018. Mating, births, and transitions: a flexible two-sex matrix model for evolutionary demography. Popul. Ecol. **60**: 21–36.

Steinsaltz, D., S. Tuljapurkar, and C. Horvitz. 2011. Derivatives of the stochastic growth rate. Theor. Popul. Biol. **80**: 1–15.

Tuljapurkar, S., and C. V. Haridas. 2006. Temporal autocorrelation and stochastic population growth. Ecol. Lett. **9**: 324–334.

Tuljapurkar, S., and C. C. Horvitz. 2006. From stage to age in variable environments: life expectancy and survivorship. Ecology **87**: 1497–1509.

Tuljapurkar, S., U. K. Steiner, and S. H. Orzack. 2009. Dynamic heterogeneity in life histories. Ecol. Lett. **12**: 93–106.

van Daalen, S. F., and H. Caswell. 2017. Lifetime reproductive output: individual stochasticity, variance, and sensitivity analysis. Theor. Ecol. **10**: 355–374.

van Daalen, S. F., and H. Caswell. 2020. Variance as a life history outcome: sensitivity analysis of the contributions of stochasticity and heterogeneity. Ecol. Model. **417**: 108856.

Vaupel, J. W., K. G. Manton, and E. Stallard. 1979. The impact of heterogeneity in individual frailty on the dynamics of mortality. Demography **16**: 439–454.

Vindenes, Y., S. Engen, and B.-E. Sæther. 2008. Individual heterogeneity in vital parameters and demographic stochasticity. Am. Nat. **171**: 455–467.

van de Pol, M., Y. Vindenes, B.-E. Sæther, et al. 2010. Effects of climate change and variability on population dynamics in a long-lived shorebird. Ecology **91**: 1192–1204.

Vindenes, Y., S. Engen, and B.-E. Sæther. 2011. Integral projection models for finite populations in a stochastic environment. Ecology **92**: 1146–1156.

Vindenes, Y., B.-E. Sæther, and S. Engen. 2012. Effects of demographic structure on key properties of stochastic density-independent population dynamics. Theor. Popul. Biol. **82**: 253–263.

Vindenes, Y., and Ø. Langangen. 2015. Individual heterogeneity in life histories and eco-evolutionary dynamics. Ecol. Lett. **18**: 417–432.

CHAPTER 10

Integral projection models

Edgar J. González, Dylan Z. Childs, Pedro F. Quintana-Ascencio, and Roberto Salguero-Gómez

10.1 Introduction

When the dynamics of a population is modelled in discrete time, but the variable that structures the population is continuous (e.g. size), rather than discrete (e.g. developmental stage) as in the case of matrix population models (MPMs, Chapter 9), an integral projection model (IPM) represents an ideal approach to describe the behaviour of the population; it also takes advantage of such nature to make the model more robust to smaller data sets (Ramula et al. 2009) and better incorporate other explanatory variables (e.g. environment [Merow et al. 2014a] and density dependence [Metcalf et al. 2008]). In contrast, when a continuous state variable is discretised, as required by an MPM, a subjective number of classes and class boundaries have to be introduced, a decision that can influence outputs derived from such models (Salguero-Gómez and Plotkin 2010).

IPMs closely relate to MPMs, and thus all the properties and analyses that have been developed for them (see Chapter 9) can also be applied to IPMs. However, IPMs also inherit the assumptions of MPMs. In its simplest form, an IPM assumes that the population is surveyed in discrete time periods (ideally coinciding with or close to the pulse of reproduction), the population is density independent (i.e. the population changes over time in an exponential manner), the population has a deterministic behaviour (i.e. the dynamics of the population is strictly determined by the vital rates, which in turn are strictly determined by the state variables), the population dynamics is constant over time (i.e. the dynamics does not change over time), and the state of a population at time $t+1$ strictly depends on its state at time t. IPMs that consider more complex scenarios where one or more of these assumptions are eliminated have been developed (e.g.

those that include density dependence [Metcalf et al. 2008]; stochastic dynamics [Childs et al. 2004; Ellner and Rees 2007]; both continuous and discrete state variables [Childs et al. 2003; Ellner and Rees 2006]; or time-lagged dynamics Kuss et al. 2008]). In the case studies presented here, we will consider a simple IPM with a continuous state variable and a more complex one with an individual and two environmental variables.

An important difference between IPMs and MPMs is that the former necessarily require the construction of statistical models to describe the vital rates as a function of the state variable, whereas the latter, although largely benefitted by their use, can be constructed without them (but see Box 9.1 in Chapter 9). Thus, this chapter aims at providing an introduction to IPMs, emphasising the statistical aspects relevant for the construction of the vital-rate models and exemplifying them in an animal and a plant system. For a thorough explanation of the logic, analyses, and applications of IPMs, we refer the reader to Ellner et al. (2016).

10.2 Constructing a basic IPM

As with MPMs, state variables, which determine differential demographic behaviours among individuals, include characteristics both of the individual itself and of the environment it inhabits. The former includes genetic, ontogenic, physiological, or behavioural attributes (Ellner et al. 2016; Rees and Ellner 2016), whilst the latter involve biotic or abiotic conditions, measured at the individual, subpopulation, or population levels (Hegland et al. 2010; Merow et al. 2014b). In the construction of an IPM, one assumes that an individual in a particular state performs demographically similarly to other individuals in this

Edgar J. González et al., *Integral projection models*. In: *Demographic Methods Across the Tree of Life*.
Edited by Roberto Salguero-Gómez and Marlène Gamelon, Oxford University Press.
© Oxford University Press (2021). DOI: 10.1093/oso/9780198838609.003.0010

same state. Also, one has to distinguish between dynamic and static state variables, depending on whether an individual can or cannot change from one state of the variable to another during its lifetime. For example, individuals may change in size over their life, whilst they may not change their genetic configuration or the edaphic environment they experience. This distinction is important, as in the former case, one has to consider in the IPM a function describing the probabilities among states when describing their life cycles. Here, we will construct an IPM using a dynamic state variable: individual size.

To start, let us consider a closed population (i.e. with no immigration or emigration) structured by individual size and identify the relevant vital processes. In a census performed at a discrete time $t+1$, the population consists of the individuals that existed when census at time t was performed and survived to time $t+1$ (n_P in Figure 10.1) and those that, through reproduction, entered the population between these two time points and survived to time $t+1$ (n_B in Figure 10.1). The size of the individuals that survived to time $t+1$ might have increased, remained, or decreased; the individuals born between the two time points must have a size at time $t+1$. Thus, in our case, the set of vital processes that we must consider to adequately describe the life cycle are: survival (s), change in size of existing individuals (G), and number (b) and size (Q) of newborns.

Once the relevant vital processes are identified, these must be integrated in the kernel, the function that describes the dynamics of the population of interest between times t and $t+1$. The dynamics of the pre-existing individuals and that of the newborns describe the dynamics of the whole population. Let us focus on the former. At time t there is a probability density function $d_t(z)$ that describes the probability an individual in the population has of having a size z. If at this moment the population consists of N_t individuals, with their sizes distributed according to d_t, we can define the function as

$$n_t(z) = N_t d_t(z) \tag{10.1}$$

Note that $\int_Z n_t(z)\mathrm{d}z = N_t \int_Z d_t(z)\mathrm{d}z = N_t$, as $d_t(z)$ is a probability density and integrates to one over the set of possible size values Z. Thus, although $n_t(z)$ is not a probability density, it behaves as one, in the sense that it is continuous, and, if normalised (i.e. divided by its integral), it integrates to one. Therefore, for the remainder of the chapter we will refer to this type of function as *density functions*, restricting the term *probability density function* to those that strictly integrate to one.

If $s(z)$ denotes the survival function, which gives us the probability an individual of size z has of surviving to time $t+1$, and if at time t we had a density function $n_t(z)$, at time $t+1$ we will have a density $s(z) \cdot n_t(z)$ of the surviving individuals of size z. Note that by making s to depend only on z and not on N_t, we are assuming that population size does not impact survival—that is, that survival is density independent.

Next, a surviving individual must have a size at time $t+1$; the function that describes the probability an individual that at time t had a size z has of having a size z' at time $t+1$ is denoted by $G(z', z)$ and should have the property that $\int_Z G(z', z)\mathrm{d}z' = 1$ for all z values. Therefore, the density of the surviving individuals of size z at time t that reached a size z' at time $t+1$ is

$$n_P(z', z) = G(z', z) \cdot s(z) \cdot n_t(z) \tag{10.2}$$

To estimate the density of the individuals of size z' at time $t+1$, irrespective of their original size at time t, we need to integrate this function over all possible original z sizes. Thus,

$$n_P(z') = \int_Z G(z', z) \cdot s(z) \cdot n_t(z)\, dz \tag{10.3}$$

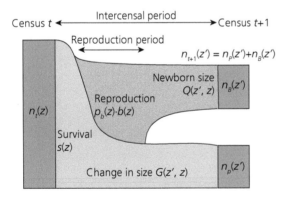

Figure 10.1 Components of an IPM. The individuals in the population at census time t (left rectangle) are structured according to $n_t(z)$. During the intercensal period, a fraction of them survive according to $s(z)$ and change to a size z' at time t+1 with probability $G(z', z)$. The surviving individuals at time t+1 will distribute according to $n_P(z')$. The reproduction period occurs over part of the intercensal period, and a fraction $p_b(z)$ of the individuals bear $b(z)$ newborns, which at time t+1 will have a size z' with probability $Q(z', z)$. At time t+1 there will be newborns that survived to this moment and are distributed as $n_B(z')$. The density of individuals at time t+1 results from the addition of $n_P(z')$ and $n_B(z')$.

As with eqn. 10.1, if we now integrate over all possible z' sizes, $\int_Z n_P(z')dz' = N_P$ is the total number of individuals that existed at time t and survived to time $t+1$.

Let us next focus on the newborns. In a closed population, the number of newborns will strictly depend on the number of reproductive individuals. Given a density $n_t(z)$ of existing individuals in the populations, a size-dependent fraction of them, $p_b(z)$, will reproduce. Thus, we will have a density $p_b(z) \cdot n_t(z)$ of reproductive organisms. Each of these individuals will produce a size-dependent number of newborns, $b(z)$, that will be alive at time $t+1$; that is, $b(z) \cdot p_b(z) \cdot n_t(z)$ newborns will be alive at time $t+1$. Again, note that b does not depend on N_t, and thus we are assuming that fecundity is density independent.

Finally, given that the size of the newborn at time $t+1$ is relevant, we need to know the probability $Q(z', z)$ that a newborn has of having a size z' at time $t+1$, given that its parent had a size z at time t; again, this function must satisfy that $\int_Z Q(z', z)dz' = 1$ for all z values. Thus, the density of the newborns of size z' that originated from parents of size z is given by

$$n_B(z', z) = Q(z', z) \cdot b(z) \cdot p_b(z) \cdot n_t(z) \qquad (10.4)$$

Again, we can integrate over all original parents of size z and get

$$n_B(z') = \int_Z Q(z', z) \cdot b(z) \cdot p_b(z) \cdot n_t(z)\, dz, \qquad (10.5)$$

the density of the newborns of size z' at time $t+1$, and $\int_Z n_B(z')dz' = N_B$ is the total number of newborns at time $t+1$. In constructing this function, we are relating the size of the parent with that of its descendant. This relation can be based on evidence from the census data (see case study 1, section 10.7.1); however, as we will see in case study 2 (section 10.7.2), we sometimes do not have evidence of such relation or we cannot associate parents with newborns, and thus we have to model Q only as a function of z'.

Given that the population at time $t+1$ consists of the individuals that existed at time t and survived to time $t+1$, plus the newborns present at time $t+1$, following eqns. 10.3 and 10.5, the density of the existing individuals of size z' at time $t+1$ is

$$n_{t+1}(z') = n_P(z') + n_B(z') = \int_Z [G(z', z) \cdot s(z)$$
$$+ Q(z', z) \cdot b(z) \cdot p_b(z)] \cdot n_t(z)\, dz \qquad (10.6)$$

If we define the *kernel* as

$$K(z', z) = G(z', z) \cdot s(z) + Q(z', z) \cdot b(z) \cdot p_b(z) \quad (10.7)$$

we have that

$$n_{t+1}(z') = \int_Z K(z', z) \cdot n_t(z)\, dz, \qquad (10.8)$$

and, as stated above,

$$\int_Z n_{t+1}(z')\, dz' = N_{t+1}. \qquad (10.9)$$

Eqns. 10.7 and 10.8 contain the structure of a basic IPM and describe how the density of individuals at time t, n_t, turns into that at time $t+1$, n_{t+1}, while considering the size-structured dynamics of the population. Evidently, alternative kernel structures can be conceived as each species, state variable, and census timing will present their particular modelling considerations; several structures are explored by Ellner et al. (2016). Also, not all vital processes may be easily quantifiable, in which case we can resort to inverse estimation (see Box 10.1).

As stated above, the kernel, $K(z', z)$, is the continuous equivalent of the transition matrix in the MPMs (eqn. 9.1). As with MPMs, we can extract the asymptotic properties from the kernel, such as the asymptotic population growth rate (λ), perform sensitivity and elasticity analyses (see section 10.5), and, by iterating eqn. 10.8, we can project the population structure (and size, using eqn. 10.9) over time to determine the trajectory a population will be expected to follow under the assumptions of the model.

Finally, IPMs are mechanistic models, in the sense that eqn. 10.6 describes the way in which the size structure changes over time as a function of the underlying processes responsible for such change. However, functions s, G, b, p_b, and Q must still be defined. These could in turn be described mechanistically and several efforts have been made to achieve this, for example, by including energy budget (Smallegange et al. 2017) or genetic (Coulson et al. 2011) dynamics as drivers of individual performance. In the case when no mechanism was considered in studying the population, one can resort to phenomenological (i.e. nonmechanistic) relations between the state variable and the vital processes; such would be a semimechanistic IPM (Ellner et al. 1998). Here, to provide a general framework that does not depend on the identity of the state variable used, we will present modelling options to establish phenomenological relationships between state variables and vital rates.

Box 10.1 Dealing with missing life cycle parts: an inverse modelling approach

The lack of information on a particular stage of the life cycle of the species of interest or the inability to follow individual dynamics through time are common situations in demographic research. For instance, seed dispersal and germination in many plants, death in long-lived trees, and migration and hibernation in animals are processes that can be hard to record; our inability to do so will render imperfect censuses. Under these scenarios, many researchers can rely on off-field experiments to estimate transition probabilities over these unobserved processes (e.g. Williams et al. 2010; Zambrano and Salguero-Gómez 2014), but these experiments can hardly reproduce the vagaries of natural conditions. An alternative to this challenge is to recognise that the output of an IPM, that is, the population size and structure (n and N in eqns. 10.8 and 10.9, respectively), contains information on these unobserved processes and that we can use it to our advantage. In an inverse modelling approach, we use the output of a model as its input to infer the value of the parameters associated to these missing processes (González et al. 2016). It is worth noting that IPMs are better suited for this approach than MPMs, because the former depend usually on fewer parameters than the latter, and the larger the number of parameters, the harder it is to estimate them inversely.

Let us assume that in the Soay sheep example (see section 10.7.1) we do not have information on the weight-dependent number of newborns each ewe has per year. This would be the case if surveying could not be performed during the breeding season and newborns could not be associated to their parents. Still, we know that the size structure of the population fluctuates over time and thus the number of reproductive females; if reproduction were weight dependent, we would also expect a correlated fluctuation in newborns. However, this correlation is blurred by the survival and change in weight of ewes, as the population size structure reflects all these processes. Thus, we have to use the survival and weight data, a traditional input of an IPM, along with the population size and structure, the IPM output, to estimate the p_b and b functions in eqn. 10.7. However, these functions involve a large number of parameters (associated to both fixed and random variables), which cannot be estimated from the limited information that population size and structures provide.

Thus, we must fit a simpler fecundity function: if we know that ewes can have up to two newborns per year (Clutton-Brock and Pemberton 2004), a linear logistic model with an upper asymptote of two would involve the estimation of only an intercept and a size slope, and would

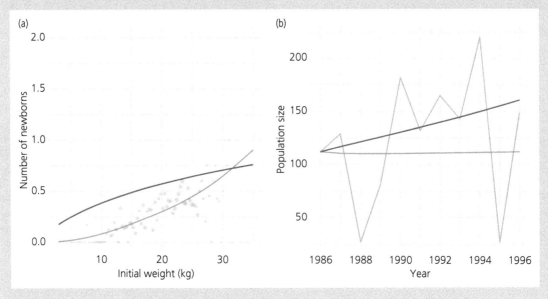

Boxed Figure 10.1 Inverse estimation of fecundity in Soay sheep. (a) Number of newborns produced as a function of ewe weight. Green indicates directly estimated number, corresponding to $b(z) \cdot p_b(z)$ in eqn. 10.7, using individual fecundity data (dots); red indicates inversely estimated fecundity, corresponding to $f(z)$ in eqn. 10.B1, using population data (i.e. population structures and sizes). (b) Population sizes observed during the study period. Green indicates projected sizes using individual data; red indicates projected sizes using individual and population data.

approximate the number of newborns produced by a ewe. As with $b(z)$ in eqn. 10.32, we multiply this model by 0.5 to consider only female lambs. This model can be formulated as

$$f(z) = 0.5 \cdot 2 \cdot \text{logistic}(\beta_0 + \beta_1 \cdot z), \qquad (10.B1)$$

and approximates the product $b(z) \cdot p_b(z)$ in eqn. 10.7. We want to find the values of these two parameters that maximise the likelihood of both the individual-level (survival, weight of existing individuals and newborns) and population-level (population size and structure) data. As shown in the code provided in the Appendix (please go

to www.oup.com/companion/SalgueroGamelonDM), we use differential evolution adaptive Metropolis (implemented in the dream package; Vrugt et al. 2009) to estimate the parameters. A comparison between the fecundities estimated through direct and inverse modelling is shown in Boxed Figure 10.1a.

Finally, we have to emphasise that, given that inverse modelling aims at estimating vital-rate parameters that translate into population sizes and structures that match those observed, this approach translates into an IPM that better predicts population-level patterns (Boxed Figure 10.1b).

10.3 Modelling vital rates in terms of the state variable

We have several options on how to relate the census data on the state variable and those on the vital rates. First, we have to recognise that usually the response variables (i.e. the vital rates) will be measurements (G and Q in eqn. 10.7), probabilities (s and p_b), or counts (b). To model measurement variables, linear models (LMs) are commonly used, whilst generalised linear models (GLMs) are needed to model probabilities and counts. In the following subsections we will introduce these two types of models, starting with LMs.

10.3.1 Linear models

The basic LM associates a response variable with an explanatory one, assuming independence among data points, linearity, normality, and homoscedasticity. Thus, if y is a continuous (measurement) variable that can take any real value and x is an explanatory variable, we can resume the above as

$$y \sim \text{Normal}(\beta_0 + \beta_1 \cdot x, \sigma_\varepsilon). \qquad (10.10)$$

This equation states that y follows a normal distribution with mean $\beta_0 + \beta_1 \cdot x$ (the mean of y changes linearly with respect to x) and standard deviation σ_ε; β_0, β_1, and σ_ε are parameters that have to be estimated from data. If we remember the equation of the normal probability density function, we can express the above equation as

$$f(y, x) = \frac{1}{\sqrt{2\pi}\sigma_\varepsilon} e^{-\frac{[y-(\beta_0+\beta_1 x)]^2}{2\sigma_\varepsilon^2}} \qquad (10.11)$$

Note that, stated in this way, we show that this is a bivariate function. Thus, if y and x are sizes at time $t+1$ and t, respectively (i.e. z' and z in eqn. 10.6), and

we assume that, on average, they relate linearly, we can use eqn. 10.11 to construct our change-in-size and size-of-newborn functions, $G(z', z)$ and $Q(z', z)$, respectively. Given that size can only take positive values, we usually log-transform this variable to make it span all real numbers, as required by the normal distribution, or use strictly positive distributions such as the log-normal or gamma.

10.3.2 Generalised linear models

As stated, vital processes may describe probabilities or counts. Since such variables do not (and cannot be made to) span all real numbers, and normality cannot be achieved without such range, we need models that allow for different ranges of the response variable; these are known as GLMs (Zuur et al. 2013).

The probabilities we have to model come from binary events (0: dead, 1: alive for the s function in eqn. 10.7; and 0: nonreproductive, 1: reproductive for the p_b function); such events can be described with a binomial distribution. Thus, if we substitute the normal distribution in eqn. 10.10 for this distribution, we have the correct value range to model these events. However, changing the distribution is not the only modification we have to make; given that a line spans all real numbers, we have to transform it to a more appropriately scale; in the case of probabilities, the logistic function is a transformation that restricts a line to the [0, 1] interval. Similar to eqn. 10.10, we can describe mathematically this GLM as

$$y \sim \text{Binomial}(N_t, \text{logistic}(\beta_0 + \beta_1 \cdot z)), \qquad (10.12)$$

where $N_t(z)$ is the number of events involved and logistic($\beta_0 + \beta_1 \cdot z$) is the size-dependent probability of success (i.e. of having a value of 1); β_0 and β_1 are parameters that have to be estimated from the data. This size-dependent probability can be used to describe the mean behaviour of our binary events s and p_b. Thus,

$$s(z) = \text{logistic}(\beta_0 + \beta_1 \cdot z) \text{ and}$$
$$p_b(z) = \text{logistic}(\beta_0 + \beta_1 \cdot z). \tag{10.13}$$

For count variables (in our case b in eqn. 10.7), the negative binomial distribution is the most commonly used one, and the exponential function is a transformation that restricts a line to positive values. This GLM can be expressed as

$$b \sim \text{Negative binomial}(\exp(\beta_0 + \beta_1 \cdot z), \theta), \tag{10.14}$$

where θ is a scale parameter. The mean of this distribution is

$$b(z) = \exp(\beta_0 + \beta_1 \cdot z). \tag{10.15}$$

Therefore, eqns. 10.11, 10.13, and 10.15 fully define the functions involved in the construction of the kernel in eqn. 10.7.

10.4 Combining the vital functions in the IPM

Once the vital processes have been modelled as a function of the state variables, we would like to use the associated functions, s, G, p_b, b, and Q, through eqn. 10.6, to combine them in the kernel. Although eqn. 10.6 clearly states how the functions combine, its computational implementation requires a few steps.

First, we must establish the (closed) interval of values of the state variable $Z = [L, U]$ we will assume an individual can take. We set Z as a closed interval because it makes biological sense that individuals can only display state values within a certain interval. Evidently, L and U must be chosen so that no potential individual in the population is excluded. However, as we will see next, if these two values make relatively good biological sense, the dynamics will be correctly modelled.

Second, by restricting our state variable to take values only within the $[L, U]$ interval, we have to take into account that, if we originally considered a different scale to model a vital process, we have to correct the scale. For example, if we log-transformed size we are assuming that this variable can take any real value and that a normal distribution (eqn. 10.11) correctly describes the variance around the mean in G and Q. To correct the scale of the variable, we can assume that any individual with size $z > U$ or $z < L$ at time $t+1$ will actually behave demographically as one having a size equal to U or L, respectively. Accordingly, we

have to reassign the probability in the tails of the normal probability outside of the $[L, U]$ interval to these extreme values. By doing this we are avoiding the accidental eviction of individuals from the population due to their too small/large size (see more details in Williams et al. 2012). To achieve this we can redefine the bivariate function in eqn. 10.10 as a restricted function (Easterling et al. 2000),

$$F_r(z', z) = F(z', z), \quad \text{if } L \leq z' \leq U,$$
$$= F(L, z), \quad \text{if } z' < L, \text{ and} \tag{10.16}$$
$$= F(U, z), \quad \text{if } z' > U.$$

Third, to implement eqn. 10.8, we would require to analytically integrate $K(z', z) \cdot n_t(z)$ over Z. However, this is computationally impractical, and one usually resorts to numerical methods to approximate this integral (Easterling et al. 2000). A common approach is the midpoint rule (Figure 10.2; see Zuidema et al. 2010 and Ellner et al. 2016 for alternatives), where the kernel is evaluated on a partition of the $[L, U]$ interval into m classes of equal width $h = (U - L)/m$. Thus, if $z_i = L + (i - 0.5)h$ are the midpoints of these classes, the integral at midpoint z'_j is approximated as

$$n_{t+1}(z'_j) = \int_L^U K(z'_j, z) \, n_t(z) \, dz \approx \sum_{i=1}^m h K(z'_j, z_i) \, n_t(z_i). \tag{10.17}$$

As the error in this approximation decreases (quadratically) with m, we will get better approximations with a large m-value ($m > 100$; Ellner et al. 2016).

As stated above, we can use the theory developed for MPMs to analyse an IPM if we note that $h \cdot K(z'_j, z_i)$ is an $m \times m$ matrix. Thus, if we denote $\mathbf{K} = h \cdot K(z'_j, z_i)$ and define the matrices $\mathbf{G} = h \cdot G(z'_j, z_i)$ and $\mathbf{Q} = h \cdot Q(z'_j, z_i)$, and the vectors $\mathbf{s} = s(z_i)$, $\mathbf{b} = b(z_i)$ and $\mathbf{p}_b = p_b(z_i)$, we can express eqn. 10.7 with the notation of Chapter 9 as

$$\mathbf{K} = \mathbf{G}\text{Diag}(\mathbf{s}) + \mathbf{Q}\text{Diag}(\mathbf{b})\,\text{Diag}(\mathbf{p}_b). \tag{10.18}$$

The only difference with that chapter is that the $\mathbf{F}$ matrix is $\mathbf{Q}\text{Diag}(\mathbf{b})\text{Diag}(\mathbf{p}_b)$. Therefore, $\mathbf{K}$ can be used as the transition matrix in the previous chapter.

Finally, we want to emphasise that resorting to discretisation is not an advocacy for MPMs. The main difference is that the vital-rate models were constructed without any categorisation, and thus, for example, the $\mathbf{s}$ vector does not contain the average survival of the observed individuals within each category, but the estimated one following a model of the relationship between the continuous state variable and survival.

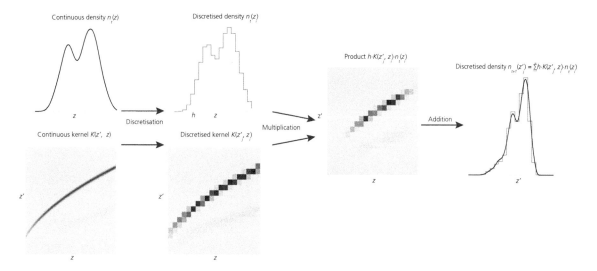

Figure 10.2 Midpoint rule integration procedure to iterate an IPM. (a) Starting with the continuous population size density at time t, $n_t(z)$ (black), we discretise it (grey); here the partition has 20 classes. (b) Under the same partition, we discretize the kernel, $K(z', z)$, (c) multiply it by the discrete size structure, and (d) obtain a discrete size structure at time $t+1$ (black), closely resembling the corresponding continuous size density at time $t+1$, $n_{t+1}(z')$ (grey).

10.5 Perturbation analysis: sensitivity and elasticity

Perturbation analyses evaluate the effect that a perturbation of the kernel or some other underlying part of it has on the measures derived from the IPM; in the remainder of the section, we will focus on the population growth rate (λ), as it is a commonly examined metric—but we emphasise that perturbations can be done with regards to any outcome of the model. For some metrics perturbations, analytical formulae exist (see Caswell 2001), and for others, numerical methods are necessary ('brute force' *sensu* Morris and Doak 2002).

In evaluating the effect of a perturbation, we might be interested in its absolute effect on λ, or on its effect relative to its original state; sensitivity and elasticity analyses evaluate these absolute and relative effects, respectively. An important assumption we make in performing these analyses is that vital processes are independent from one another, and thus perturbations do not propagate to other processes; for a discussion on this matter in MPMs see van Tienderen (1995) and De Kroon et al. (2000), and in IPMs see Griffith (2017).

10.5.1 Perturbation on the kernel

A perturbation on the kernel addresses the question: How much does λ change when we perturb $K(z', z)$

at some point (z'_0, z_0)? Ellner et al. (2016) show that this change can be calculated in the same way as with MPMs (eqn. 9.2). Therefore, we can calculate the sensitivity (S) of λ to a change of $K(z', z)$ at point (z'_0, z_0) as

$$S_K(z'_0, z_0) = \frac{\partial \lambda}{\partial K}(z'_0, z_0) = v(z'_0) \cdot w(z_0), \quad (10.19)$$

where $v(z')$ are the reproductive values and $w(z)$ the stable (size) structure associated to $K(z', z)$. Note that, as in Chapter 9, v and w are scaled so that $\int_Z v(z')\mathrm{d}z' = \int_Z w(z)\mathrm{d}z = 1$.

In turn, the elasticity (E) of λ to a change of $K(z', z)$ at point (z'_0, z_0) is

$$E_K(z'_0, z_0) = \frac{\frac{\partial \lambda}{\partial K}(z'_0, z_0)}{\frac{\lambda}{K(z'_0, z_0)}} = \frac{K(z'_0, z_0)}{\lambda} \cdot \frac{\partial \lambda}{\partial K}(z'_0, z_0)$$

$$= \frac{K(z'_0, z_0)}{\lambda} \cdot S_K(z'_0, z_0). \quad (10.20)$$

10.5.2 Perturbation on the pre-existing and newborn sections of the kernel

As we described in section 10.2 and as is shown in Figure 10.1 (blue vs. orange areas), the dynamics described by $K(z', z)$ can be decomposed into the pre-existing and newborn sections (**U** and **F** matrices in Chapter 9)—that is, $K(z', z) = U(z', z) + F(z', z)$, where $U(z', z) = G(z', z) \cdot s(z)$ and $F(z', z) = Q(z', z) \cdot b(z) \cdot p_b(z)$.

We may be interested in evaluating the sensitivity (or elasticity) of λ to a change in one of these components of the system. Using the chain rule and some derivative properties, we can show that the sensitivity of $U(z', z)$ at (z'_0, z_0) is equal to the sensitivity of $K(z', z)$:

$$S_U = \frac{\partial \lambda}{\partial U} = \frac{\partial \lambda}{\partial K} \cdot \frac{\partial K}{\partial U} = \frac{\partial \lambda}{\partial K} \cdot \frac{\partial [U+F]}{\partial U}$$

$$= \frac{\partial \lambda}{\partial K} \cdot \left[\frac{\partial U}{\partial U} + \frac{\partial F}{\partial U}\right] = \frac{\partial \lambda}{\partial K} \cdot [1+0] = S_K. \quad (10.21)$$

Note that, for simplification, we are omitting the point of evaluation (z'_0, z_0) in the equation. The same procedure shows that S_F, the sensitivity of $F(z', z)$, is equal to S_K.

The elasticities E_U and E_F, however, are not the same, as they differ in their denominators:

$$E_U = \frac{\frac{\partial \lambda}{\partial U}}{\frac{\lambda}{U}} = \frac{U}{\lambda} \cdot S_K \quad \text{and} \quad E_F = \frac{\frac{\partial \lambda}{\partial F}}{\frac{\lambda}{F}} = \frac{F}{\lambda} \cdot S_K.$$

$$(10.22)$$

Considering that the elasticity of λ to a change in a demographic component f is always the sensitivity of λ to a change in f divided by the quotient λ/f (i.e. $E_f = S_f/(\lambda/f)$), for the remainder of the section we will focus only on the sensitivities, leaving to the reader the calculation of this division.

10.5.3 Sensitivities on the vital rates

If we are interested in perturbing a particular vital rate, we have to remember that we have bivariate and univariate vital rate functions (i.e. $G(z', z)$ and $Q(z', z)$, and $s(z)$, $b(z)$, and $p_b(z)$, respectively). For the former, and using $G(z', z)$ to exemplify the procedure, the sensitivity of λ to a perturbation of $G(z', z)$ at a point (z'_0, z_0) can be derived using again the chain rule, the product rule, and the results of eqn. 10.21 as

$$S_G = \frac{\partial \lambda}{\partial G} = \frac{\partial \lambda}{\partial U} \cdot \frac{\partial U}{\partial G} = S_K \cdot \frac{\partial G \cdot s}{\partial G}$$

$$= S_K \cdot \left[\frac{\partial G}{\partial G} \cdot s + G \cdot \frac{\partial s}{\partial G}\right] = S_K \cdot s. \quad (10.23)$$

Similarly, the sensitivity of λ to a perturbation of $Q(z', z)$ at point (z'_0, z_0) is

$$S_Q = S_K.b.p_b. \quad (10.24)$$

For the univariate vital rates, and using $s(z)$ to exemplify the procedure, a perturbation of $s(z)$ at a point z_0

is a perturbation of $s(z)$ at a point z_0 and all points z'; that is,

$$S_s(z_0) = \frac{\partial \lambda}{\partial s}(z_0) = \int_L \frac{\partial \lambda}{\partial s}(z_0)\,dz'.$$

Again, using derivative rules and the results of eqn. 10.21 we obtain

$$\frac{\partial \lambda}{\partial s} = \frac{\partial \lambda}{\partial U} \cdot \frac{\partial U}{\partial s} = S_K \cdot \frac{\partial G \cdot s}{\partial s} = S_K \cdot \left[\frac{\partial G}{\partial s} \cdot s + G \cdot \frac{\partial s}{\partial s}\right]$$

$$= S_K \cdot G. \quad (10.25)$$

Thus,

$$S_s(z_0) = \int_L \frac{\partial \lambda}{\partial s}(z_0)\,dz' = \int_L S_K(z', z_0) \cdot G(z', z_0)\,dz'.$$

$$(10.26)$$

Under the same procedure, we can show that

$$S_b(z_0) = \int_L S_K(z', z_0) \cdot Q(z', z_0) \cdot p_b(z_0)\,dz' \quad \text{and}$$

$$S_{p_b}(z_0) = \int_L S_K(z', z_0) \cdot Q(z', z_0) \cdot b(z_0)\,dz'. \quad (10.27)$$

Note that these are the sensitivities of λ to a perturbation of a vital rate at a particular point z_0. If we were interested in the sensitivity to a perturbation of the entire vital rate, we only have to integrate over all points z. Thus, for example, perturbing the entire survival would correspond to

$$S_s = \int_L S_s(z)\,dz. \quad (10.28)$$

10.5.4 Sensitivities to vital rate parameters

Finally, we may be interested in how a perturbation in a parameter θ of a vital rate function f impacts λ. Now, vital rate functions can take many forms different from those described in eqns. 10.11, 10.13, and 10.15 (see e.g. section 10.6.1). Thus, the sensitivity of λ to a perturbation on a particular parameter has no general expression. However, due to the chain rule, we only have to focus on the derivative of the vital rate function with respect to the parameter to correctly calculate the sensitivity function:

$$S_\theta = \frac{\partial \lambda}{\partial \theta} = \frac{\partial \lambda}{\partial f} \cdot \frac{\partial f}{\partial \theta} = S_f \cdot \frac{\partial f}{\partial \theta}. \qquad (10.29)$$

To exemplify this, we can calculate the sensitivity of λ to a perturbation in β_1 of s in eqn. 10.13 at point z_0. Thus,

$$
\begin{aligned}
S_{\beta_1} = S_s \cdot \frac{\partial s}{\partial \beta_1} &= S_s \cdot \frac{\partial \left[\frac{1}{1+e^{-(\beta_0 + \beta_1 z)}} \right]}{\partial \beta_1} \\
&= S_s \left(z_0 \right) \cdot \left[s \left(z_0 \right) \right]^2 \cdot e^{-(\beta_0 + \beta_1 z_0)} \cdot z_0.
\end{aligned}
$$
$$(10.30)$$

The interpretation of parameter-level sensitivities is not always straightforward. For instance, in the above example, the sensitivity to a perturbation of β_1 will be non-negative for all z values (since all factors are non-negative); that is, a change in the survival of individuals will affect λ in the same direction irrespective of their size. However, if survival is described by a nonlinear function, a perturbation in β_1 can produce both negative and positive sensitivities on λ. Consider the survival equation s in the case study 1 (eqn. 10.32), which includes a nonlinear function: a spline (i.e. a function of interconnected polynomials); in this example the sensitivity to a perturbation in β_1 at size z_0 is

$$S_{\beta_1} = S_s \left(z_0 \right) \cdot \left[s \left(z_0 \right) \right]^2 \cdot e^{-[\beta_0 + \beta_1 \cdot \mathrm{spline}(z_0)]} \cdot \mathrm{spline}(z_0).$$
$$(10.31)$$

Since $S_s(z_0)$, $[s(z_0)]^2$, and $\exp\{-[\beta_0 + \beta_1 \cdot \mathrm{spline}(z_0)]\}$ are non-negative, the sign of the sensitivity is determined by the sign of the spline at z_0. In our case, the spline takes positive values for small and large individuals and negative for medium-size ones (Figure 10.3a). Therefore, the sensitivity to a perturbation in β_1 will be negative for these latter individuals, and thus an increase in β_1 in medium-size individuals will translate into a lower λ.

10.6 Further considerations in the construction of vital-rate models

Having presented the theory around IPMs, the construction of the vital-rate models, and their integration in the kernel, we can now consider two important subjects for the construction of the vital-rate models: (1) the extensions of the LM that allow the use of data that violate its assumptions, and (2) the selection procedure for a set of competing vital-rate models.

10.6.1 Extensions of the linear model

The statistical theory on LMs includes alternative models when the data violate some or all of the assumptions of a traditional LM: independence among data points, linearity in the relation between the explanatory and the response variables, and homoscedasticity in the response variable around its mean. These alternatives provide a high degree of flexibility in the phenomenological models we can construct.

Nonindependence among data points occurs when these are considered as spatially or temporarily related. Data points may be spatially associated if, for example, several plots were set within the experimental design, and thus we need to consider all data within a plot as nonindependent; data are temporally related if they come from an individual that was measured more than once. In such cases, the LM framework allows for the inclusion of random factors (plot, individual, year, etc.), which can make any parameter of the model to vary between values of these factors. The models that allow including random factors are the so-called linear mixed models (Bolker et al. 2009; Zuur et al. 2013).

Nonlinearity usually arises when mechanistic relationships between response and explanatory variables are considered. However, we can also construct nonlinear phenomenological models within the previous framework. These are the generalised *additive* models (Ozgul et al. 2012; Hastie 2017; Wood 2017), which use splines to fit any nonlinear relationship (e.g. Figure 10.3a).

Finally, heteroscedasticity can be addressed by modelling σ_ε in eqn. 10.11, not as a constant but as a function of z, be it linear or not (Easterling et al. 2000; Rigby et al. 2017).

10.6.2 Model selection

If we measured either one or more individual characteristic or environmental factor or a combination of both, it is likely that we may want to construct a kernel that considers the relevance of these variables in describing the vital rates. Under such a scenario, we will first consider multiple models regarding how well (or poorly) the variables describe each vital rate; in addition to linear vs. nonlinear relations between each state variable and the vital rate, we have to consider the potential interactions that may exist between state variables in the effect they have on each vital rate. Every option will translate into a model, and we will end up with a set of models from which we have to select

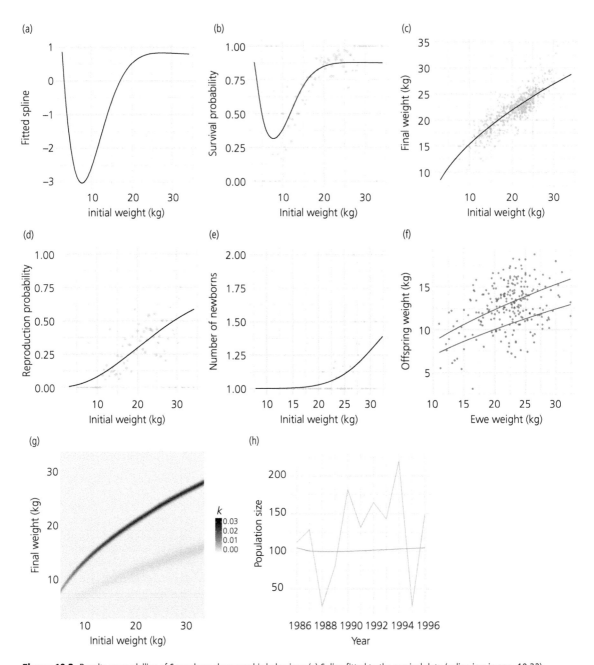

Figure 10.3 Results on modelling of Soay sheep demographic behaviour. (a) Spline fitted to the survival data (spline in *s* in eqn. 10.32); (b) weight-dependent survival; (c) individual change in weight between consecutive time points; (d) weight-dependent probability of reproduction; (e) weight-dependent fecundity; (f) weight of newborns as a function of parent weight (blue and red lines indicate single and twin pregnancies, respectively); (g) heat map of the kernel (eqn. 10.7) as a function of individual size at two consecutive time points; (h) population sizes observed during the study period (green indicates projected sizes using the IPM).

the best one. Information criteria, among which the most commonly used is the sample-corrected Akaike information criterion (AICc), allow the comparison of models taking into account their goodness-of-fit, their complexity, and the amount of information available (i.e. sample size). We consider the best-supported model as the one having the smallest value of the criterion. For a full description of the theory around model selection see Burnham and Anderson (2002).

10.7 Case studies

Here we exemplify IPMs with two study systems: an animal and a plant population. The R code associated to the construction of the IPMs is included in the Appendix (please go to www.oup. com/companion/SalgueroGamelonDM).

10.7.1 Case study 1: Soay sheep population structured by weight

We will start with a relatively simple case study on an animal system: The database on the Soay sheep (*Ovis aries*). The data come from a survey of uniquely marked female sheep inhabiting the Village Bay of Hirta Island, St Kilda archipelago, Scotland, United Kingdom, between 1986 and 1996 (Clutton-Brock and Pemberton 2004; Coulson 2012). The survey year ran from 1 August to 31 July, and the state variable was the weight of the sheep. Females reach a maximum longevity of 16 years and can conceive at around 7 months of age. A maximum of two newborns are born per female sheep in April of each year, so they are 4 months old by the time the population is surveyed. Because the sex of the newborn was not recorded, we assume that there is a 50% chance that it is female.

To model the vital rates, we first have to recognise that, since an individual can live for more than a year and all individuals in a given year experienced the same environmental conditions, we have to take into account the nonindependence between the data associated to the individual and the year of study; we can also consider nonlinear relations between weight (z, kg) and each vital rate. Additionally, to model reproduction we have to modify function b in eqn. 10.15, as the number of newborns is restricted to two; thus, a binary event (0 = single newborn, 1 = two newborns), as in eqn. 10.12, is a better modelling approach. Finally, we have to modify function Q to account for the fact that it is known that in this species the weight of the newborn (z') depends both on the weight of the ewe (z) and on whether it was the product of a single or twin pregnancy (b) (Clutton-Brock and Pemberton

2004); thus, for this function we consider the binary explanatory variable b. The Appendix (please go to www.oup.com/companion/SalgueroGamelonDM) presents the set of models considered and the model selection procedure followed. Modelling was performed in R (R Core Team 2020) using packages lme4 (for generalised linear mixed models; Bates et al. 2015), gamm4 (for generalised additive mixed models; Wood and Scheipl 2017), and AICcmodavg (for model selection; Mazerolle 2017).

The (marginal) structures of the selected models were:

$$s(z) = \text{logistic}(\beta_0 + \beta_1 \cdot \text{spline}(z)),$$
$$G(z', z) = \frac{1}{2\sqrt{\pi}\sigma_\varepsilon} e^{-\frac{[z'-(\beta_0+\beta_1 z)]^2}{2\sigma_\varepsilon^2}},$$
$$p_b(z) = \text{logistic}(\beta_0 + \beta_1 \cdot z), \quad (10.32)$$
$$b(z) = 0.5 \cdot (1 + \text{logistic}(\beta_0 + \beta_1 \cdot z), \text{and}$$
$$Q(z', z, b) = \frac{1}{2\sqrt{\pi}\sigma_\varepsilon} e^{-\frac{\{z'-[\beta_0+\beta_1 z+\beta_2 b]^2\}}{2\sigma_\varepsilon^2}}.$$

The graphs associated to these models are presented in Figure 10.3b–f. Note that survival is described nonlinearly (through a spline, Figure 10.3a) by the state variable, while $p_b(z)$ is linear, both in the logistic scale (Figure 10.3b,d). $G(z', z)$ and $Q(z')$ follow a structure similar to eqn. 10.11; since $Q(z', z)$ depends on b, single pregnancies produce heavier newborns than twin ones (Figure 10.3f; blue and red lines, respectively). Note that in $b(z)$ we add 1 to the logistic model, which ranges in the [0, 1] interval, to make it range in the [1, 2] newborns interval and then multiply it by 0.5 to obtain the number of newborn females (Figure 10.3e). These functions were integrated into the kernel following eqn. 10.7 (Figure 10.3g; note that the darker line reflects the product of the s and G functions and the lighter one reflects the product of p_b, b, and Q). The model is then iterated following eqn. 10.8 (Figure 10.2), producing estimated population sizes that reflect the average (marginal) dynamics of the population over the 10-year study period (green line in Figure 10.3h).

Finally, Box 10.1 demonstrates how, in a scenario where information on the weight-dependent number of newborns produced by each ewe were not available, an inverse estimation procedure could be used to approximate the function (b) describing it.

10.7.2 Case study 2: Perennial herb population structured by height, elevation, and time since fire

We will now model the dynamics of a more complex study system. *Hypericum cumulicola* (Hypericaceae) is a

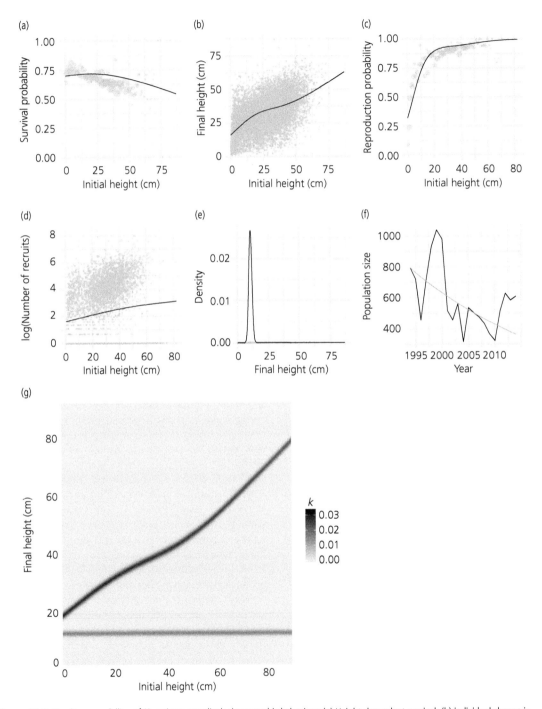

Figure 10.4 Results on modelling of *Hypericum cumulicola* demographic behaviour. (a) Height-dependent survival; (b) individual change in height between consecutive years; (c) height-dependent probability of reproduction; (d) height-dependent fecundity; (e) density function of the height of newborns; (f) population sizes observed during the study period (green indicates projected sizes using the IPM); (g) heat map of the kernel (eqn. 10.7) as a function of individual size at two consecutive years. Graphs a–d and g present the models evaluated over the individual state variable and at mean values of two environmental variables (relative altitude and time since fire).

small perennial herbaceous species endemic to central Florida, United States (Estill and Cruzan 2001). Flowering occurs mostly between mid-summer and mid-autumn. Every year, starting in spring, established individuals may develop multiflowered reproductive branches that senesce in the winter. Aboveground individuals are killed by fire (Quintana-Ascencio and Morales-Hernández 1997), but seeds in the soil can survive it (Quintana-Ascencio et al. 1998). *H. cumulicola* is mostly limited to gaps between vegetation patches in Florida rosemary scrub (Menges et al. 2017). The patches occur at variable elevation, thus affecting *H. cumulicola* demography probably due to differential water availability and community composition (Quintana-Ascencio et al. 2018). We estimate the vital rates using a 21-year-long annual census database across 12 sites, using as state variables an individual characteristic, height (z, cm), and two environmental factors—relative elevation (e, m) and time since fire (h, years), measured at the site level. We also consider the nonindependence of data points due to the study year and the spatial arrangement of data points into sites, gaps, and the 0.5×1 m study quadrats placed within each gap.

Given that we are dealing with more than one state variable, we have to construct a large set of models and perform model selection. Due to space limit, we refer the reader to the Appendix of this chapter (please go to www.oup.com/companion/SalgueroGamelonDM), where, for each vital rate, we construct all possible models and perform model selection. The (marginal) structures of the selected models for each vital rate were

$$s\,(z, e, h) = \text{logistic}(\beta_0 + \beta_1 \cdot \text{spline}\,(z, e, h)),$$

$$G\,(z', z, e, h) \;=\; \frac{1}{2\sqrt{\pi}\sigma_\varepsilon} e^{-\frac{\left\{z' - [\beta_0 + \beta_1 \cdot \text{spline}(z, e, h)]\right\}^2}{2\sigma_\varepsilon^2}},$$

$$p_b\,(z, e, h) = \text{logistic}(\beta_0 + \beta_1 \cdot \text{spline}\,(z, e, h)),$$

$$b\,(z, e, h) = \exp(\beta_0 + \beta_1 \cdot \text{spline}\,(z)$$
$$+\, \beta_2 \cdot e + \beta_3 \cdot h + \beta_4 \cdot e \cdot h)\text{and}$$

$$Q\,(z') \;=\; \frac{1}{2\sqrt{\pi}\sigma_\varepsilon} e^{-\frac{(z' - \beta_0)^2}{2\sigma_\varepsilon^2}}. \qquad (10.33)$$

$s(z, e, h)$, $G(z', z, e, h)$, $p_b(z, e, h)$, and $b(z, e, h)$ have more complex structures compared with eqn. 10.11 as they included a spline that relates nonlinearly the explanatory variables with the vital rates. $Q(z')$ in turn has a simpler structure, as we could not relate the size of the parent (z) with that of its descendant (z') and thus

had to assume that newborns have the same size distribution irrespective of the size of their parents. The graphs associated to these models, evaluated at the mean values of e and h, are presented in Figure 10.4a–e. We also include the kernel evaluated at these mean values (Figure 10.4g) and the population sizes derived from the model (green line in Figure 10.4f). Although seed dormancy has a significant role for *H. cumulicola*, in this implementation we ignore this state for heuristic purposes.

References

Bates, D., Maechler, M., Bolker, B. and Walker, S., 2015. Fitting linear mixed-effects models using lme4. Journal of Statistical Software, **67**(19), pp. 1–48.

Bolker, B. M., Brooks, M. E., Clark, C. J. et al., 2009. Generalized linear mixed models: a practical guide for ecology and evolution. Trends in Ecology and Evolution, **24**(3), pp. 127–135.

Burnham, K. P. and Anderson, D. R. 2002. Model Selection and Multimodel Inference: A Practical Information-Theoretic Approach, 2nd edition. Springer, New York.

Caswell, H. 2001. Matrix Population Models: Construction, Analysis, and Interpretation, 2nd edition. Sinauer Associates, Sunderland, MA.

Childs, D. Z., Rees, M., Rose, K. E., Grubb, P. J. and Ellner, S. P., 2003. Evolution of complex flowering strategies: an age–and size–structured integral projection model. Proceedings of the Royal Society of London. Series B: Biological Sciences, **270**(1526), pp. 1829–1838.

Childs, D. Z., Rees, M., Rose, K. E., Grubb, P. J. and Ellner, S. P., 2004. Evolution of size–dependent flowering in a variable environment: construction and analysis of a stochastic integral projection model. Proceedings of the Royal Society of London. Series B: Biological Sciences, **271**(1537), pp. 425–434.

Clutton-Brock, T. H. and Pemberton, J. M. (eds), 2004. *Soay Sheep: Dynamics and Selection in an Island Population*. Cambridge University Press, Cambridge.

Coulson, T., 2012. Integral projections models, their construction and use in posing hypotheses in ecology. Oikos, **121**(9), pp. 1337–1350.

Coulson, T., MacNulty, D. R., Stahler, D. R., Wayne, R. K. and Smith, D. W., 2011. Modeling effects of environmental change on wolf population dynamics, trait evolution, and life history. Science, **334**(6060), pp. 1275–1278.

De Kroon, H., Van Groenendael, J. and Ehrlén, J., 2000. Elasticities: a review of methods and model limitations. Ecology, **81**(3), pp. 607–618.

Easterling, M. R., Ellner, S. P. and Dixon, P. M., 2000. Size-specific sensitivity: applying a new structured population model. Ecology, **81**(3), pp. 694–708.

Ellner, S. P. and Rees, M., 2006. Integral projection models for species with complex demography. The American Naturalist, 167(3), pp. 410–428.

Ellner, S. P. and Rees, M., 2007. Stochastic stable population growth in integral projection models: theory and application. Journal of Mathematical Biology, 54(2), pp. 227–256.

Ellner, S. P., Bailey, B. A., Bobashev, G. V., Gallant, A. R., Grenfell, B. T. and Nychka, D. W., 1998. Noise and nonlinearity in measles epidemics: combining mechanistic and statistical approaches to population modeling. The American Naturalist, 151(5), pp. 425–440.

Ellner, S. P., Childs, D. Z. and Rees, M., 2016. Data-Driven Modelling of Structured Populations. Springer, Cham.

Estill, J. C. and Cruzan, M. B. 2001. Phytogeography of rare plant species endemic to the southeastern United States. Castanea, 1, pp. 3–23.

González, E. J., Martorell, C. and Bolker, B. M., 2016. Inverse estimation of integral projection model parameters using time series of population-level data. Methods in Ecology and Evolution, 7(2), pp. 147–156.

Griffith, A. B., 2017. Perturbation approaches for integral projection models. Oikos, 126(12), pp. 1675–1686.

Griffith, A. B., Salguero-Gómez, R., Merow, C. and McMahon, S., 2016. Demography beyond the population. Journal of Ecology, 104(2), pp. 271–280.

Hastie, T. J., 2017. Generalized additive models. In: Wood, S. N. (ed.), Statistical Models in S (pp. 249–307). Routledge, Abingdon-on-Thames.

Hegland, S. J., Jongejans, E. and Rydgren, K., 2010. Investigating the interaction between ungulate grazing and resource effects on Vaccinium myrtillus populations with integral projection models. Oecologia, 163(3), pp. 695–706.

Kuss, P., Rees, M., Ægisdóttir, H. H., Ellner, S. P. and Stöcklin, J., 2008. Evolutionary demography of long-lived monocarpic perennials: a time-lagged integral projection model. Journal of Ecology, 96(4), pp. 821–832.

Mazerolle, M. J., 2017. AICcmodavg: model selection and multimodel inference based on (Q)AIC(c). R package version 2, pp. 1–1.

Menges, E. S., Crate, S. J. H. and Quintana-Ascencio, P. F. 2017. Dynamics of gaps, vegetation, and plant species with and without fire. American Journal of Botany 104, pp. 1825–1836.

Merow, C., Dahlgren, J. P., Metcalf, C. J. E. et al., 2014a. Advancing population ecology with integral projection models: a practical guide. Methods in Ecology and Evolution, 5(2), pp. 99–110.

Merow, C., Latimer, A. M., Wilson, A. M., McMahon, S. M., Rebelo, A. G. and Silander Jr, J. A., 2014b. On using integral projection models to generate demographically driven predictions of species' distributions: development and validation using sparse data. Ecography, 37(12), pp. 1167–1183.

Metcalf, C. J. E., Rose, K. E., Childs, D. Z., Sheppard, A. W., Grubb, P. J. and Rees, M., 2008. Evolution of flowering decisions in a stochastic, density-dependent environment. Proceedings of the National Academy of Sciences, 105(30), pp. 10466–10470.

Morris, W. F. and Doak, D. F. 2002. Quantitative Conservation Biology. Sinauer Associates, Sunderland, MA.

Ozgul, A., Coulson, T., Reynolds, A., Cameron, T. C. and Benton, T. G., 2012. Population responses to perturbations: the importance of trait-based analysis illustrated through a microcosm experiment. The American Naturalist, 179(5), pp. 582–594.

Quintana-Ascencio, P. F. and Morales-Hernández, M. 1997. Fire-mediated effects of shrubs, lichens and herbs on the demography of Hypericum cumulicola in patchy Florida scrub. Oecologia 112(2), pp. 267–271.

Quintana-Ascencio, P. F., Dolan, R. W. and Menges, E. S. 1998. Hypericum cumulicola demography in unoccupied and occupied Florida scrub patches with different time-since-fire. Journal of Ecology 86(4), pp. 640–651.

Quintana-Ascencio, P. F., Koontz, S. M., Smith, S. A., Sclater, V. L., David, A. S. and Menges, E. S., 2018. Predicting landscape-level distribution and abundance: integrating demography, fire, elevation and landscape habitat configuration. Journal of Ecology, 106(6), pp. 2395–2408.

R Core Team, 2020. R: A Language and Environment for Statistical Computing. R Foundation for Statistical Computing, Vienna.

Ramula, S., Rees, M. and Buckley, Y. M., 2009. Integral projection models perform better for small demographic data sets than matrix population models: a case study of two perennial herbs. Journal of Applied Ecology, 46(5), pp. 1048–1053.

Rees, M. and Ellner, S. P., 2016. Evolving integral projection models: evolutionary demography meets eco-evolutionary dynamics. Methods in Ecology and Evolution, 7(2), pp. 157–170.

Rigby, R. A., Stasinopoulos, D. M., Heller, G. Z. and De Bastiani, F., 2017. Distributions for modelling location, scale, and shape: using GAMLSS in R. www.gamlss.org (accessed 5 March 2018).

Salguero-Gomez, R. and Plotkin, J. B. 2010. Matrix dimensions bias demographic inferences: implications for comparative plant demography. The American Naturalist, 176(6), pp. 710–722.

Smallegange, I. M., Caswell, H. Toorians, M. E. M. and De Roos A. M. 2017. Mechanistic description of population dynamics using dynamic energy budget theory incorporated into integral projection models. Methods in Ecology and Evolution, 8(2), pp. 146–154.

Van Tienderen, P. H. 1995. Life cycle trade-offs in matrix population models. Ecology, 76(8), pp. 2482–2489.

Vrugt, J. A., Ter Braak, C. J. F., Diks, C. G. H., Robinson, B. A., Hyman, J. M. and Higdon, D., 2009. Accelerating Markov chain Monte Carlo simulation by differential evolution with self-adaptive randomized subspace sampling. International Journal of Nonlinear Sciences and Numerical Simulation, 10(3), pp. 273–290.

Williams, J. L., Auge, H. and Maron, J. L., 2010. Testing hypotheses for exotic plant success: parallel experiments in the native and introduced ranges. Ecology, **91**(5), pp. 1355–1366.

Williams, J. L., Miller, T. E. and Ellner, S. P., 2012. Avoiding unintentional eviction from integral projection models. Ecology, **93**(9), pp. 2008–2014.

Wood, S. N., 2017. *Generalized Additive Models: An Introduction with R*. Chapman and Hall/CRC Press, Boca Raton, FL.

Wood, S. and Scheipl, F., 2017. gamm4: generalized additive mixed models using 'mgcv' and 'lme4'. R package version 0. pp. 2–5. https://cran.r-project.org/web/packages/gamm4/index.html.

Zambrano, J. and SalgueroGómez, R., 2014. Forest fragmentation alters the population dynamics of a late-successional tropical tree. Biotropica, **46**(5), pp. 556–564.

Zuidema, P. A., Jongejans, E., Chien, P. D., During, H. J. and Schieving, F., 2010. Integral projection models for trees: a new parameterization method and a validation of model output. Journal of Ecology, **98**(2), pp.345–355.

Zuur, A. F., Hilbe, J. M. and Ieno, E. N., 2013. *A Beginner's Guide to GLM and GLMM with R: A Frequentist and Bayesian Perspective for Ecologists*. Highland Statistics, Newburgh.

CHAPTER 11

Transient analyses of population dynamics using matrix projection models

David N. Koons, David T. Iles, and Iain Stott

11.1 Introduction

In this chapter we address the burgeoning topic of transient population dynamics using matrix projection models (MPMs), which are introduced in Chapter 9 and widely used for examining the dynamics of populations structured by discrete life cycle stages (e.g. developmental phases, sex) or discretized continuous life history traits (e.g. age, size; Caswell 2001). The popularity of the MPM hails in part from its reliance on algebra and calculus, which are familiar to most quantitatively inclined ecologists given the early foundations built by pioneers such as Alfred Lotka (1924) and Robert May (1981). Vindenes et al. provide a lucid overview of how to conduct asymptotic analyses of MPMs in Chapter 9, with valuable explanation of terminology and extensions to a wide array of environmental conditions.

Integral projection models (IPMs) (Easterling et al. 2000), eloquently reviewed in Chapter 10 by González et al., are continuously structured population models in discrete time and akin to infinite-dimensional MPMs with a stage width of zero. Though transient analyses are more mature for MPMs, the limit–property connection between discrete classes of an MPM and continuous states of an IPM should allow many of the analytical tools developed for MPMs to be applied to IPMs, and when not, numerical simulation is often a sufficient replacement (Ellner et al. 2016).

It is useful to distinguish models for constant vs. time-varying environments; they have fundamentally different dynamics and require a different set of tools to study. MPMs for constant environments assume that each vital rate is described by a single numerical value that does not change over time. Although they are undoubtedly abstractions of nature, MPMs for constant environments are often used to characterise the general (or average) properties of particular populations or life histories, and a well-developed suite of analytical methods are available to understand their dynamics (Caswell 2001; see Chapter 9). Because of their analytical tractability, models for constant environments have been the focus of most studies that utilise MPMs (Boyce et al. 2006). In contrast, time-varying models allow vital rates to change over time as a result of density dependence, environmental covariates, unexplained stochastic variation, or a combination of variables (Caswell 2001). Time-varying models introduce additional layers of realism but require more sophisticated analytical tools, many of which are under continued development.

The vast majority of studies that use MPMs and IPMs for constant environments focus on the long-term, asymptotic dynamics once stage structure has stabilised. In reviewing these concepts, Vindenes et al. in Chapter 9 introduce us to the 'transient dynamics' or 'transients' that occur before asymptotic dynamics are reached as a result of an initially nonstable stage structure. This prompts the question of how influential a nonstable stage structure can be on transient population dynamics. The near-term dynamics resulting from an initial stage structure skewed toward individuals with low reproductive value will be very different compared to one skewed toward individuals with high reproductive value: intuition alone tells us that the former will initially perform more poorly than the latter (see section 2.2 for verification).

David N. Koons, David T. Iles, and Iain Stott, *Transient analyses of population dynamics using matrix projection models*.
In: *Demographic Methods Across the Tree of Life*. Edited by Roberto Salguero-Gómez and Marlène Gamelon, Oxford University Press.
© Oxford University Press (2021). DOI: 10.1093/oso/9780198838609.003.0011

Given the potential impact of nonstable stage structure on population dynamics, analyses of transients are becoming popular in an era of global change because of numerous situations that cause nonstable stage structures. For example, stage structure can be altered by sudden biotic forces (e.g. disease, novel predation pressure), abiotic forces (e.g. extreme weather events, fire), and anthropogenic forces (e.g. harvesting, manipulation of habitat) that quickly remove individuals of certain stages from the population (defined here as a demographic 'disturbance'). Alternatively, the vital rates of certain stages can change with less immediate impact on the number of individuals in each stage (defined here as a demographic 'perturbation'). Demographic disturbances affect the current stage structure, whilst demographic perturbations change the stable stage structure: both mechanisms create a mismatch between vital rates and stage structure (remember the fundamental projection equation from Chapter 9, $\mathbf{n}_{t+1} = \mathbf{A}\mathbf{n}_t$, and the IPM analogue in Chapter 10).

Disturbances may be common: for example, the leading edge of a plant or marine invertebrate invasion or colonisation consists entirely of dispersing propagules that later progress through the life cycle, whereas terrestrial animal and marine vertebrate colonisations originate from the (usually adult or subadult) stages most likely to disperse (Shigesada and Kawasaki 1997), and humans directly manipulate stage structure whenever they relocate, reintroduce, or stock individuals into a wild habitat. Human interventions often force prolonged perturbations, where populations are managed to boost or suppress particular vital rates. Prolonged environmental perturbations may also result from surpassing an environmental 'tipping point' or shifts to alternative stable states of an ecosystem (Jenouvrier et al. 2012). These are just a few examples of when transients might actually be of greater relevance to ecologists than asymptotic dynamics (Hastings 2004; Hastings et al. 2018).

Our goal in this chapter is to provide a gentle introduction to the concepts and analysis of transient dynamics using MPMs such that readers are prepared to dive into reviews on these topics (Ezard et al. 2010; Stott et al. 2011; Stott 2016) and then apply analyses of transients to relevant questions and problems of their own. We begin with the analysis of transient dynamics in constant environments and then highlight recent extensions to time-varying environments, providing comparisons to asymptotic analogues throughout. A large number of demographic tools for analysing asymptotic and transient dynamics

are referred to in the chapter, and we provide a concise overview of these tools in Table 11.1 to assist the reader.

11.2 Examples

11.2.1 Example life cycle models

We herein illustrate analyses of transient population dynamics using two published MPMs, but everything we demonstrate is readily applicable to population models for other types of organisms. We downloaded vital-rate data for both examples from COM(P)ADRE (Salguero-Gómez et al. 2015, 2016) with the assistance of the ShinyPop application (https://iainmstott.shinyapps.io/shinypop/). Our first example focuses on a study of the lady orchid (*Orchis purpurea*) in eastern Belgium. This study was conducted by Jacquemyn et al. (2010), and we focus specifically on their data for the third plot in a light environment, which was one of six study plots and monitored over 6 years. When averaged over time and rounded, the stage-classified MPM for the lady orchid in light plot 3 is

$$\mathbf{A}_{\text{orchid}} = \begin{bmatrix} 0 & 0 & 0 & 0 & 0 & 25.31 \\ 0.02 & 0 & 0 & 0 & 0 & 0 \\ 0 & 0.09 & 0 & 0 & 0 & 0 \\ 0 & 0 & 0.9 & 0.24 & 0 & 0 \\ 0 & 0 & 0 & 0.76 & 0.41 & 0.29 \\ 0 & 0 & 0 & 0 & 0.57 & 0.70 \end{bmatrix}$$

$$(11.1)$$

The six discrete stages were based on developmental phases: protocorms, tubers, seedlings, juveniles, nonflowering plants, and flowering plants (Jacquemyn et al. 2010). After each time step, surviving individuals either advanced to the next stage (subdiagonal of eqn. 11.1), remained in a stage (diagonal), or, for some flowering plants, reverted to a nonflowing plant (supradiagonal). Because the asymptotic growth rate of $\mathbf{A}_{\text{orchid}}$ is $\lambda_1 = 1.02$, we will use this population to illustrate how transients might be used to further enhance population growth of this beautiful and locally abundant orchid that is otherwise rare and threatened near the northern extent of the species' range (Jacquemyn et al. 2010).

Our second example focuses on a study of polar bears (*Ursus maritimus*) in the southern Beaufort Sea off the northern coast of Alaska, United States. This study was conducted by Hunter et al. (2010) and occurred over 5 years (though the study is ongoing). When averaged over time and rounded, the stage-classified MPM for polar bears is

Table 11.1 A concise overview of various ways to analyse MPMs that are discussed further within the text. When applicable, the key references refer to seminal books or review papers, which contain citations of the original work that developed the theoretical basis for a given demographic tool. To find applied examples of a tool, we highly recommend that readers use a search engine to find papers that have cited the original works or our key references here.

Demographic tool	Type of environment	Interpretation	Key references
General properties of model			
Asymptotic analysis of deterministic model	Constant(time-invariant)	Measures long-term population growth rate, (st)age structure, reproductive value, generation time, and more when population structure has stabilised	Caswell 2001 and references therein
Transient analysis of deterministic model	Constant	Measures properties of transient dynamics (including amplification, attenuation, duration, inertia, transient envelope, and other transient behaviour of specific disturbances) before population structure has stabilised	Ezard et al. 2010; Stott et al. 2011; Stott 2016; and references therein
Stochastic analysis	Time-varying(changing)	Measures long-term stochastic growth rate (λ_s) and other demographic outcomes in a time-varying environment once population structure has reached a multivariate stationary distribution	Tuljapurkar 1990 and references therein
Transient analysis of time-varying model	Time-varying	Measures realised population growth rates ($\lambda_{real,t}$), abundances, density dependence, population viability, and other demographic outcomes in any type of time-varying environment without assuming stationary conditions	Gotelli and Ellison 2006; Hunter et al. 2010; Schaub et al. 2013
Prospective perturbation analysis			
Asymptotic elasticities for deterministic models	Constant	Measure effects of proportional change in vital rates on asymptotic demographic outcomes, most commonly λ_1	Caswell 2001 and references therein
Transient elasticities for deterministic models	Constant	Measure effects of proportional change in vital rates or components of initial stage structure on transient demographic outcomes	Stott 2016 and references therein
Stochastic elasticities	Time-varying	Measure effects of proportional change in mean and variance of vital rates on λ_s	Tulajpurkar et al. 2003; Haridas and Tuljapurkar 2005; Haridas et al. 2009
Transient elasticities for time-varying models	Time-varying	Measure effects of proportional change in vital rates on transient demographic outcomes in a time-varying environment	Caswell 2007; Haridas et al. 2009
Retrospective perturbation analysis			
Asymptotic LTRE with random design	Constant	Measures the contribution of vital rates to variation in λ_1 among deterministic models	Caswell 1989; Horvitz et al. 1997
Stochastic LTRE	Time-varying	Measures the contribution of vital rates and stage structure to differences in λ_s between or among different populations	Caswell 2010; Davison et al. 2010
Transient LTRE	Time-varying	Measures the contribution of vital rates and stage structure to temporal variation (or other summaries) of $\lambda_{real,t}$ within a single population	Koons et al. 2016; Maldonado-Chaparro et al. 2017

$$
\mathbf{A}_{\text{bear}} = \begin{bmatrix}
0 & 0 & 0 & 0 & 0 & 0.46 \\
0.86 & 0 & 0 & 0 & 0 & 0 \\
0 & 0.86 & 0 & 0 & 0 & 0 \\
0 & 0 & 0.86 & 0.50 & 0.38 & 0.90 \\
0 & 0 & 0 & 0.40 & 0.07 & 0 \\
0 & 0 & 0 & 0 & 0.44 & 0
\end{bmatrix}
$$

$$(11.2)$$

The six discrete stages were respectively classified as age 2, 3, and 4 subadults, adult females available to breed without dependent young, adult females accompanied by one or more cubs of the year, and adult females accompanied by one or more yearling cubs. Equation 11.2 pertains to the female component of the polar bear population and assumes an equal sex ratio among surviving offspring (see Hunter et al. 2010 for details about the underlying lower-level vital rates that comprise each matrix entry). Because the asymptotic growth rate of $\mathbf{A}_{\text{bear}}$ is $\lambda_1 = 0.94$, we will use this population to illustrate how transients might be used to inform conservation strategies for declining or threatened and endangered species (Caughley 1994).

11.2.2 Transient analysis in constant environments

To demonstrate analyses of transient population dynamics in constant environments, we scaled matrices $\mathbf{A}_{\text{orchid}}$ and $\mathbf{A}_{\text{bear}}$ by their respective asymptotic population growth rates (i.e. divided each matrix by its dominant eigenvalue). This results in $\lambda_1 = 1$ for any standardised matrix, thereby removing its long-term trend but preserving other asymptotic properties (e.g. stable stage structure, elasticities). This facilitates comparison across populations and species and provides for clear interpretation of any transient index of population density as a multiple of asymptotic density. With a scaled matrix, indices >1 indicate a transient population density that is greater than asymptotic density produced by a stable stage structure, an index >0 but <1 indicates a density less than asymptotic density, and an index = 1 pertains to a population projected from its stable stage structure (Stott et al. 2011).

Using a standard eigen-analysis described in Chapter 9, we found that for the standardised matrix $\mathbf{A}_{\text{orchid}}$, the quantities of the stable stage structure (0.92, 0.02, 0.00, 0.00, 0.02, 0.04) and reproductive values (1, 46.24, 522.57, 590.62, 604.07, 638.01) were greatest for stages 1 and 6 respectively. To illustrate the transient behaviour that can occur when a population

is away from its stable stage structure, we projected the population dynamics of lady orchids over 10 years with all individuals initially in the first stage (i.e. $\mathbf{n}_0 = [1, 0, \ldots, 0]$), representing a hypothetical colonisation wavefront beginning solely with protocorms. In this case, the population experienced a phenomenon called 'attenuation' in which unfavourable short-term population growth rates led to lower densities (bottom solid line in Figure 11.1a) than a population with equivalent vital rates but initiated from the stable stage structure (a.k.a. the 'stable equivalent'; dashed line in Figure 11.1a). In particular, this initial stage structure yielded the largest possible transient attenuation (reduction) in population density, which likely occurred because stage 1 individuals had the lowest asymptotic reproductive value, but the formal link between the reproductive values associated with an initial stage structure and resulting transient dynamics remains to be proven. In contrast, when we initiated a lady orchid population with 100% of the individuals in the last stage (i.e. $\mathbf{n}_0 = [0, \ldots, 0, 1]$) with highest reproductive value, which could represent human introduction of ornamental flowering orchids, the population experienced an opposite phenomenon: favourable short-term population growth led to the largest possible transient 'amplification' (increase) in population density relative to a stable equivalent (top solid line in Figure 11.1a).

Upon closer inspection, there are interesting dynamics in these transient amplifications and attenuations that several researchers have suggested as key indices for comparison across populations and species. Across any of the possible initial stage structures, the maximum amplification and attenuation after the first time step are respectively called the 'reactivity' ($\bar{\rho}_1$; Neubert and Caswell 1997) and 'first time-step attenuation' ($\underline{\rho}_1$; Townley et al. 2007; Townley and Hodgson 2008). These instantaneous transient responses will always be transient in nature but are useful for gauging immediate population responses to demographic perturbations or disturbances. Limiting one's evaluation of transients to the first time interval, however, risks missing the full extent of transient dynamics, and there are other indices to consider. For example, the greatest magnitudes of amplification and attenuation that are possible at any time step, even if they are dissipated at a later time, are called the 'maximal amplification' ($\bar{\rho}_{\max}$) and 'maximal attenuation' ($\underline{\rho}_{\min}$, maximal in attenuated magnitude but minimal in numerical value; see Figure 11.1). Measurement of these maximal transient responses is effective for capturing the full extent

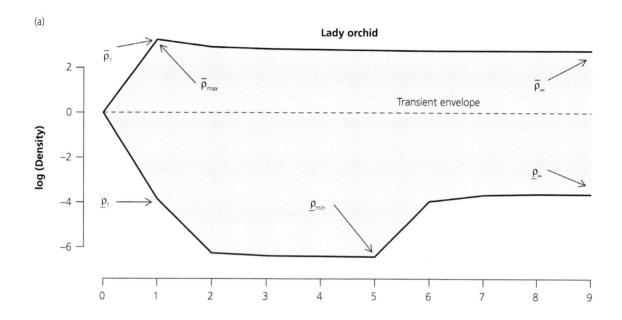

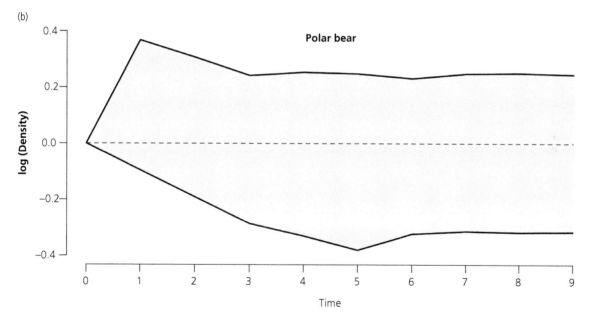

Figure 11.1 Standardised transient population dynamics, excluding the influence of asymptotic growth, for (a) the mean lady orchid MPM in a constant environment. Bold lines indicate extreme amplifications and attenuations of transient densities on the log scale for respective initial stage structures with 100% of the individuals in last stage (top bold line) or first stage (bottom bold line). Labelled points along the paths of transient amplification and attenuation are as described in the main text, and these extremities encapsulate the transient envelope (grey areas), which covers the range of possible transient dynamics for all other case-specific initial stage structures. The dashed line indicates a reference projection of population dynamics from an initially stable stage structure. (b) Analogous projections of standardised transient population dynamics for the polar bear example. Note the different behaviours of transient amplifications compared to attenuations over time.

of transient dynamics: Regardless of the initial stage structure, population density should always be smaller than maximal amplification and larger than maximal attenuation. These four near-term indices were all originally defined as transient bounds (Neubert and Caswell 1997; Townley et al. 2007; Townley and Hodgson 2008), but analogues also exist for projections from case-specific initial stage structures (i.e. any credible structure where individuals are not stacked into a single stage that yields maximal transients). Collectively, these capture the majority of variation in near-term transient responses that occur across different initial conditions and are coined the 'transient envelope' (grey areas in Figure 11.1; Stott et al. 2011).

An important long-term outcome of an initially nonstable stage structure and ensuing transients is that they affect asymptotic population density $\mathbf{n}_\infty$ but not asymptotic population growth, which means that unlike λ_1, density and abundance are not ergodic, meaning they are not independent of initial conditions (see Caswell 2001, chapter 4; see also Stott et al. 2010b for explanation of reducible vs. irreducible matrices and implications for transients). A generalisation of 'population momentum' (Keyfitz 1971), population inertia I measures the long-term ratio of the density of a population originating from any initial stage structure relative to a reference initial stage structure, with the reference most often being the stable equivalent (Tuljapurkar and Lee 1997; Koons et al. 2007). Though population inertia is not a transient, it nicely captures the net effect of transient dynamics on long-term population density. When the reference initial stage structure equals the stable structure, population inertia is defined as

$$I = \frac{\mathbf{e}^{\mathrm{T}} \left(\mathbf{v}_1^* \mathbf{n}_0 \right) \mathbf{w}_1}{\mathbf{e}^{\mathrm{T}} \mathbf{n}_0} \qquad (11.3)$$

where $\mathbf{e}$ is a vector with each entry equal to one, ^{T} denotes the transpose of a vector, * denotes the complex conjugate transpose, $\mathbf{n}_0$ is as defined above, and $\mathbf{w}_1$ and $\mathbf{v}_1$ are the dominant right and left eigenvectors of a projection model (Caswell 2001; Koons et al. 2007).

In introductory physics courses we are taught about the momentum/inertia of moving objects with mass, such as a maritime ship attempting to turn east at a fixed point from a northerly course. By the time the ship achieves a perfect easterly bearing, it will have drifted north of the intended turning point as a result of the ship's physical inertia. Although not often

thought of in this way, structured populations are moving objects with mass (structured density). Nonstable stage structure exerts an inertia on abundance, causing it to move in directions other than that exerted by the vital rates acting on their own (i.e. acting on a stable stage structure).

In an applied context, population inertia could help or hinder a manager's ability to achieve near- and long-term population goals (Koons et al. 2006a, 2006b). Consider the maximum bounds for amplified ($\bar{\rho}_\infty$) and attenuated inertia ($\underline{\rho}_\infty$) for lady orchids in Figure 11.1a. Since this population is predicted to grow slightly in the long run ($\lambda_1 = 1.02$), a nonstable stage structure that produces a large amplified inertia like $\bar{\rho}_\infty$ (resulting from an initial preponderance of flowering plants in the population) would be a hindrance for managers attempting to stabilise or reduce population density but a benefit to those wishing to more rapidly enhance population density. A large attenuated inertia (resulting from an initial population consisting predominantly of protocorms) would result in opposite outcomes relevant to management goals.

The outcomes of transient dynamics and population inertia can nevertheless be very different across populations and species (e.g. Koons et al. 2005, 2006b; Stott et al. 2010a; Gamelon et al. 2014; Iles et al. 2016). Take for example polar bears, for which entries of the stable stage structure (0.05, 0.05, 0.04, 0.51, 0.23, 0.11) and reproductive values (1, 1.10, 1.21, 1.33, 1.47, 1.76) associated with standardised matrix $\mathbf{A}_{\mathrm{bear}}$ were greatest for stages 4 and 6. Though a somewhat different life history pattern compared to lady orchids, an initial stage structure with 100% of the individuals in the first stage ($\mathbf{n}_0 = [1, 0, \ldots, 0]$) also produced the greatest attenuation in transient dynamics and long-term inertia in density relative to a stable equivalent (bottom solid line in Figure 11.1b). When we initiated a polar bear population with 100% of the individuals in the last stage ($\mathbf{n}_0 = [0, \ldots, 0, 1]$), the population experienced maximal amplification in both transient dynamics and inertia (top solid line in Figure 11.1b), again likely because the final stage has the greatest reproductive value. Relocations or colonisations by adults would therefore induce amplified transient dynamics and population inertia. But it is nevertheless important to note the vastly different magnitudes in standardised transients between the lady orchid and polar bear examples that result from their different life histories (Figure 11.1). R code is provided in Appendix 11.1 (please go to www.oup.com/companion/SalgueroGamelonDM) for building graphics like that in Figure 11.1.

11.2.3 Transient perturbation analysis in constant environments

Section 11.2.2 examined the effects of disturbances to stage structure, assuming vital rates remain unchanged. However, analysis of an MPM or IPM would be incomplete without also investigating the consequences of perturbations to the vital rates. Prospective perturbation analysis (see Chapters 9 and 10) answers the question: 'All else being equal, what would happen to the population dynamics if one-by-one, each life cycle vital rate (or entry of stage structure) were changed by the same amount?' These are highly valuable tools for managers of natural populations whose everyday business is 'changing demography' to achieve population objectives. A perturbation analysis can provide information about the life cycle stages and vital rates that can be targeted in management interventions to give maximum impact on the focal demographic outcome for minimum effort (while remembering that one should also consider the logistical and financial realities of actually inducing changes in demographic parameters; Nichols and Hines 2002; Baxter et al. 2006). Elasticities are prospective perturbation tools for comparing the demographic outcomes of proportional changes in demographic parameters (Caswell 2000, 2001), which can help identify the changes in demographic parameters that would be most effective at achieving desired demographic outcomes (e.g. a targeted change in population growth rate; see Heppell 1998).

A basic elasticity analysis of standardised $\mathbf{A}_{orchid}$ indicates that the asymptotic population growth rate for lady orchids in a constant environment is, not surprisingly, most sensitive to proportional changes in survival and stasis of individuals in the flowering plant stage:

$$\mathbf{E}^{\lambda_1}_{\mathbf{A}_{orchid}} = \begin{bmatrix} 0 & 0 & 0 & 0 & 0 & 0.02 \\ 0.02 & 0 & 0 & 0 & 0 & 0 \\ 0 & 0.02 & 0 & 0 & 0 & 0 \\ 0 & 0 & 0.02 & 0.01 & 0 & 0 \\ 0 & 0 & 0 & 0.02 & 0.13 & 0.16 \\ 0 & 0 & 0 & 0 & 0.19 & \mathbf{0.41} \end{bmatrix}$$

(11.4)

All else being equal, maintaining a large pool of flowering plants in the population therefore induces the greatest proportional increase in the asymptotic population growth rate for lady orchids. The asymptotic population growth rate for standardised $\mathbf{A}_{bear}$, however, was most sensitive to proportional changes in

survival of stage 4 adult polar bears without offspring (row 4, column 4 of eqn. 11.5) and their chance of surviving to produce and care for cubs (row 5, column 4):

$$\mathbf{E}^{\lambda_1}_{\mathbf{A}_{bear}} = \begin{bmatrix} 0 & 0 & 0 & 0 & 0 & 0.04 \\ 0.04 & 0 & 0 & 0 & 0 & 0 \\ 0 & 0.04 & 0 & 0 & 0 & 0 \\ 0 & 0 & 0.04 & \mathbf{0.26} & 0.09 & 0.10 \\ 0 & 0 & 0 & \mathbf{0.23} & 0.02 & 0 \\ 0 & 0 & 0 & 0 & 0.14 & 0 \end{bmatrix}$$

(11.5)

which if increased, should induce the greatest proportional increase in λ_1.

A strict focus on perturbation analysis of the asymptotic population growth rate could limit one's evaluation of population dynamics for informing actions to achieve near-term population objectives. Perturbations of several transient indices could be considered for informing near- to moderate-term management of population density or abundance, but population inertia is a valuable omnibus measure because it essentially combines the effects of transient dynamics over time, it correlates well with other transient indices, and it is also relevant to long-term abundance objectives (Stott et al. 2011). When initiating each of the example populations with 100% of the individuals in stage 1, we had previously shown that this induces a maximal attenuation of population inertia for lady orchids (Figure 11.1a), which could be problematic since the species is generally rare and threatened near the northern extent of the species' range (Jacquemyn et al. 2010). The same initial stage structure for polar bears also induced a maximal attenuation in population inertia, which would be problematic since the population is predicted to decline by 6% per time step in the long run.

Using perturbation methods developed by Koons et al. (2007), we found that in terms of a focus on life-cycle vital rates, the attenuated inertia for lady orchids was most sensitive to proportional changes in survival and transition of nonflowering plants to becoming flowering plants and also the survival and stasis of nonflowering plants:

$$\mathbf{E}^{l}_{\mathbf{A}_{orchid}} = \begin{bmatrix} 0 & 0 & 0 & 0 & 0 & 1.18 \\ 1.28 & 0 & 0 & 0 & 0 & 0 \\ 0 & 0.63 & 0 & 0 & 0 & 0 \\ 0 & 0 & 0.25 & -0.18 & 0 & 0 \\ 0 & 0 & 0 & -0.55 & -1.88 & 0.38 \\ 0 & 0 & 0 & 0 & -\mathbf{2.55} & 1.44 \end{bmatrix}$$

(11.6)

Note that these elasticities sum to zero, not one, and that negative values imply that, quite counterintuitively (but we promise it to be true!), increasing the matrix-level vital rate will decrease the numerical value of population inertia. Contrary to elasticity patterns for the asymptotic population growth rate (eqn. 11.4), managers wishing to dampen attenuated inertial effects of the initial stage structure on the eventual density of lady orchids should, all else being equal, try to 'decrease' the survival of nonflowering plants. This management action would reduce the fraction of nonflowering plants in the population. But we note that perpetually decreasing this vital rate will also decrease λ_1 and long-term population density, which highlights the care that must be taken when interpreting the transient and asymptotic impacts of manipulating a species' vital rates (Figure 11.2, light-grey line). Another management option might be to instead focus on increasing the survival and stasis of flowering plants. The elasticity of λ_1 to change in this life cycle transition was highly positive (eqn. 11.4), indicating that enhancements would significantly increase λ_1. The elasticity of population inertia to change in this vital rate was also positive (eqn. 11.6), and thus an enhancement would also increase the numerical value of inertia and dampen the degree of attenuation caused by a colonising stage structure on population density (Figure 11.2, orange line).

The attenuated inertia for polar bears was most sensitive to proportional changes in the survival of 2-year-old subadults (row 2, column 1 in eqn. 11.7):

$$
\mathbf{E}^I_{\mathbf{A}_{bear}} =
\begin{bmatrix}
0 & 0 & 0 & 0 & 0 & 0.13 \\
\mathbf{1.10} & 0 & 0 & 0 & 0 & 0 \\
0 & 0.43 & 0 & 0 & 0 & 0 \\
0 & 0 & -0.37 & \mathbf{-0.58} & -0.15 & 0.29 \\
0 & 0 & 0 & \mathbf{-0.55} & -0.03 & 0 \\
0 & 0 & 0 & 0 & -0.27 & 0
\end{bmatrix}
\tag{11.7}
$$

This elasticity is positive, indicating that increasing the survival of young subadults would be effective at increasing the numerical value of inertia and thus reducing its magnitude of attenuation. The inertia elasticity associated with survival of adults without offspring (row 4, column 4 in eqn. 11.7) and their chance of surviving to eventually produce and care for cubs (row 5, column 4) are fairly negative, however (eqn. 11.7), meaning that although an increase in these vital rates would have a strong positive effect on λ_1 (eqn. 11.5), it would also reduce the numerical value of population inertia and enhance the magnitude of

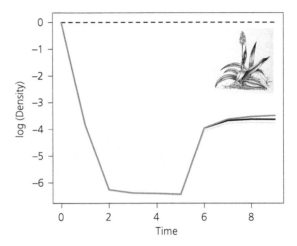

Figure 11.2 Projections of standardised population dynamics, excluding the influence of asymptotic growth, for the lady orchid MPM in a constant environment (note that the nonstandardised MPM yields population growth). The dashed line indicates a reference projection of population dynamics from an initially stable stage structure, and the solid black line indicates projection from a colonising stage structure without any perturbations to the vital rates. The light-grey line refers to projection from the colonising stage structure, with a permanent 10% decrement to the life cycle transition from stage 5 to 6 in an attempt to reduce attenuated inertial effects of the initial stage structure, but with an inadvertent effect of decreasing long-term population growth rate that also affects long-term abundance. The orange line refers to projection from the colonising stage structure, with a permanent 10% enhancement of survival and stasis of flowering plants within stage 6, which reduces the severity of attenuated population inertia and enhances the long-term population growth rate. Orchid image sampled from *Hardy Perennials and Old-fashioned Garden Flowers* (Wood, 1884, public domain).

attenuation in short-term density. For some populations, such perturbations could drive them to the brink of extinction, again highlighting the care that must be taken when interpreting the transient and asymptotic impacts of manipulating a species' vital rates.

Another option not presented here would be to consider direct disturbances to stage structure by, for example, (re)introducing individuals in key stages of development (Koons et al. 2007). R code is provided in Appendix 11.2 (please go to www.oup.com/companion/SalgueroGamelonDM) for conducting the asymptotic and transient analyses presented above.

11.2.4 Population dynamics in time-varying environments

Though highly useful, metrics of population dynamics derived from a time-invariant MPM cannot describe the effects of environmental variability (Cohen

1977a; Tuljapurkar and Orzack 1980). MPMs can be /nobreakextended to accommodate the reality that temporal variation in vital rates persistently changes stage structure, $\mathbf{n}_{t+1} = \mathbf{A}_t\mathbf{n}_t$, and that both the vital rates and stage structure contribute to changes in realised population growth rates over time: $\lambda_{real,t} = \frac{\|\mathbf{n}_{t+1}\|}{\|\mathbf{n}_t\|} = \frac{\|\mathbf{A}_t\mathbf{n}_t\|}{\|\mathbf{n}_t\|}$. Most of the theoretical and applied work on stochastic population dynamics concerns cases in which vital rates fluctuate with stationary distributions, leading to stationary (but fluctuating) distributions of stage structure (Figure 11.3a) and realised population growth rate (Figure 11.3b). The stochastic population growth rate (λ_s) measures the long-term geometric average of the realised growth rates once stage structure has reached a multivariate stationary distribution: $\lambda_s = \lim_{T\to\infty} \left(\prod_{t=0}^{T-1} \lambda_{real,t}\right)^{1/T}$, $\log \lambda_s = \lim_{T\to\infty} \frac{1}{T}\sum_{t=0}^{T-1} \log \lambda_{real,t}$ (Lewontin and Cohen 1969; Cohen 1976, 1977a, 1977b; Tuljapurkar and Orzack 1980). Analogous to the asymptotically stable stage structure of a deterministic MPM for constant environments, the stationary distribution of fluctuating population structure in a time-varying environment means that the proportion of individuals in each stage has a stable mean, variance, and other moments over time (Cohen 1976, 1977a, 1977b) (see

Figure 11.3a). A great deal of theoretical and applied work focuses on aspects of the environment and relationships with demographic parameters that affect λ_s (e.g. Tuljapurkar 1982; Doak et al. 2005; Boyce et al. 2006; Koons et al. 2009; Lawson et al. 2015; Iles et al. 2019), factors that contribute to differences in λ_s between life histories and populations (e.g. Koons et al. 2008; Davison et al. 2019), and the sensitivity of λ_s to proportional changes in both the mean and temporal variance of vital rates and underlying environmental drivers (e.g. Tuljapurkar et al. 2003; Lee 2017).

Lacking more appropriate methods for retrospective perturbation analysis in time-varying environments, past studies have applied traditional, asymptotic life table response experiments (LTREs) with a 'random design' to decompose vital-rate contributions to variance in the asymptotic λ_1 associated with a set of deterministic MPMs, that is, the dominant eigenvalues. Such analyses can be useful for comparisons of λ_1 across populations, such as spatial locations, but are not appropriate for retrospective analysis in a time-varying environment (Koons et al. 2016). Though traditional random-design LTREs have often been applied to time series of MPMs (reviewed by Bassar et al. 2010; see also Table 11.1), an analysis of the dominant

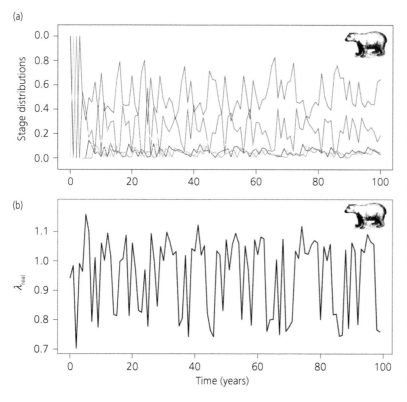

(a)

Stage distributions

(b)

λ_{real}

Time (years)

Figure 11.3 (a) Projections of polar bear population dynamics over time in a simulated stochastic environment, whereby the population was initiated with 100% of the individuals in stage 1 (black line in a), then projected over time with random selection of an MPM at each time step from one of the five MPMs published by Hunter et al. (2010). The distributions of abundance in stages 2 through 6 are shown in blue, orange, dark grey, firebrick red, and light grey. Note that eventually the means and variances of these distributions stabilise to 'stationary distributions'. (b) As this happens, the temporal mean and variance of the realised population growth rates ($\lambda_{real,t}$) also stabilise, providing the theoretical basis for computing λ_s, as described in the main text.

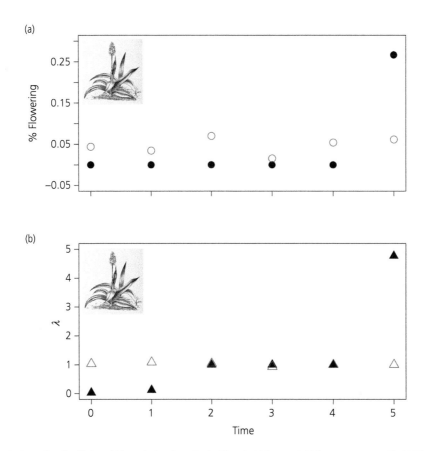

Figure 11.4 Projections of realised lady orchid population dynamics (solid symbols) from an initial stage structure with 100% of the individuals in stage 1 and the observed time series of MPMs (Jacquemyn et al. 2010; see www.oup.com/companion/SalgueroGamelonDM). (a) Realised fraction of the population in the flowering stage (solid circles) compared to the asymptotically stable proportion associated with each individual MPM (open circles). (b) Depicts the realised population growth rates (solid triangles) associated with the short-term projection of population dynamics compared to the asymptotic population growth rates associated with each individual MPM (open triangles), which are markedly different at most time steps. Orchid image sampled from *Hardy Perennials and Old-fashioned Garden Flowers* (Wood, 1884, public domain).

eigenvalues provides nonfactual inference about the demographic drivers of population dynamics in a time-varying environment because this essentially assumes that an asymptotically stable stage structure associated with the MPM at time t can transform into the stable structure associated with a new MPM at $t+1$ (Figure 11.4a, open circles). This simply does not occur when stage structure is appropriately projected by time-varying MPMs in a recursive fashion: $\mathbf{n}_{t+1} = \mathbf{A}_t\mathbf{n}_t$, (Figure 11.4a, closed circles), and the difference between λ_{real} and λ_1 at each time step can be substantial (Figure 11.4b).

Relatively recently, LTREs have been developed for the stochastic growth rate that appropriately accommodate the manner in which population dynamics operate in time-varying environments (Caswell 2010; Davison et al. 2010). These 'stochastic LTREs'

quantify the demographic contributions to 'differences' in λ_s between populations (Davison et al. 2019). But stochastic LTREs do not quantify the demographic determinants of change in $\lambda_{real,t}$ over time in a single population, which are of high interest to managers and conservationists (Williams et al. 2002).

The dependence of λ_s on a stationary stage structure moreover assumes that temporal distributions of vital rates (and the environmental drivers thereof) are stationary (e.g. Tuljapurkar 1990). Many contemporary environments are not stationary. For example, changes in climate, land use, and hydrology are directional, creating 'nonstationary' environments for much of the world's biota that are not likely to stabilise any time soon (Wolkovich et al. 2014). Nonstationary environments are defined by trends in average conditions, environmental variance, or both, which keep populations in a perpetual transient state because they never have a chance to converge to a stationary stage structure (Hastings 2004; Wolkovich et al. 2014). Fortunately, the same MPM projection equation that is used to study λ_s ($\mathbf{n}_{t+1} = \mathbf{A}_t\mathbf{n}_t$) can be used to study population dynamics in nonstationary environments (Gotelli and Ellison 2006). Alongside analyses of λ_s, Hunter et al. (2010) provide an elegant example of doing just this to project the impacts of forecasted climate change on polar bear population viability in the southern Beaufort Sea.

11.2.5 Transient perturbation analysis in time-varying environments

In many cases, the observed population dynamics will be transient because either the environment is not stationary or too little time will have elapsed for environmental conditions and demographic stage structure to have converged upon stationary distributions. Rather than assuming these properties and proceeding with a long-term stochastic demographic analysis, one can instead focus on the abundances and realised population growth rates produced by an observed sequence of time-varying environments and corresponding MPMs (see Figure 11.4). Only recently have perturbation analyses been developed for examining transients in time-varying environments. Caswell (2007) developed prospective sensitivity and elasticity tools for predicting the change in transient demographic outcomes that would result from any specified change in the parameters that determine those outcomes.

One outcome that is appealing for the analysis of transient dynamics in time-varying environments is the realised population growth rate ($\lambda_{real,t}$) because it accounts for the manner in which vital rates and stage structure jointly determine population growth in any type of environment, remembering that $\lambda_{real,t} = \frac{\|\mathbf{n}_{t+1}\|}{\|\mathbf{n}_t\|} = \frac{\|\mathbf{A}_t\mathbf{n}_t\|}{\|\mathbf{n}_t\|}$ (Haridas et al. 2009). Koons et al. (2016) developed a set of 'retrospective' transient LTREs that decompose an observed sequence of realised population growth rates into contributions from observed changes in vital rates and stage structure. Accordingly, transient LTREs provide deep insight into the demographic pathways through which populations change in potentially unstable time-varying environments. Maldonado-Chaparro et al. (2017) extended the methods of Koons et al. (2016) (1) to continuously structured IPMs, and (2) to measure explicit effects of past change in specific environmental variables (e.g. temperature)

on realised population growth rates. These collective developments of prospective and retrospective tools for examining transients in time-varying environments have relinquished practitioners from the need to use deterministic or stochastic perturbation analyses.

To illustrate the utility of transient LTREs for gaining a better understanding of population dynamics amidst changing environmental conditions, we again use the lady orchid and polar bear case studies. For consistency, we created scenarios with initial stage structures resembling colonisation of the youngest stage: $\mathbf{n}_0 = [1, 0, \ldots, 0]$. For each population, we then projected the initial stage structure forward in time using the empirically estimated sequence of MPMs over 6 or 5 years respectively. At each time step of projection we kept track of the normalised stage structure (i.e. the vector of relative stage-specific abundances that sum to one) and the realised annual population growth rate, $\lambda_{real,t}$. We used this information to implement a transient LTRE to measure the contributions of past change in both the vital rates and stage structure to temporal variance of the realised population growth rates var($\lambda_{real,t}$) (see Box 11.1). We also kept track of the asymptotic λ_1 associated with the (nonstandardised) MPM at each time step of study in order to compare results from the transient LTRE to the asymptotic LTRE with a random design. We should note that a minimum of 3 years (or other time step) of data are needed to implement either of these LTREs because one needs to be able to estimate the temporal process 'variance' of each demographic parameter.

The asymptotic LTRE with a random design for lady orchid indicated that variation in the rate of transition into and stasis in the flowering stage contributed (C) most to the variance in λ_1:

$$\mathbf{C}_{\mathbf{A}_{orchid}}^{var(\lambda_1)} = \begin{bmatrix} 0 & 0 & 0 & 0 & 0 & -0.0001 \\ 0.0009 & 0 & 0 & 0 & 0 & 0 \\ 0 & 0.0073 & 0 & 0 & 0 & 0 \\ 0 & 0 & 0.0002 & -0.0004 & 0 & 0 \\ 0 & 0 & 0 & 0.0004 & -0.0020 & -0.0018 \\ 0 & 0 & 0 & 0 & 0.0023 & 0.0021 \end{bmatrix}$$
(11.8)

However, the sum of these contributions severely underestimated the actual temporal variance of $\lambda_{real,t}$ (relative difference on log scale = −6.41; see Appendix 11.2 for calculation; please go to www.oup.com/companion/SalgueroGamelonDM). This discrepancy arises because asymptotic LTREs with a random design do not account for the highly unstable initial stage structure, nor the subsequent fluctuations in stage structure driven by temporal variation in the vital rates. In contrast, while the first-order approximation (see Box 11.1) of the transient LTRE was not perfect,

---

Box 11.1 The transient life-table response experiment for decomposing temporal variability in realised population growth rates into contributions from vital rates and components of stage structure.

To decompose var($\lambda_{real,t}$) over time, Koons et al. (2016) used methods derived from the definition of variance for a linear combination of variables (see Chapter 10 of Caswell 2001). Without affecting the measurement of $\lambda_{real,t}$ all $\mathbf{n}_t$ should be normalised to sum to one in order to avoid issues with numerical computation. All components of $\mathbf{n}_t$ and each parameter comprising the elements of the MPM should be placed into a vector Θ_t, and then the sensitivity of $\lambda_{real,t}$ to change in each underlying demographic parameter ($\partial\lambda_{real,t}/\partial\theta_{i,t}$) should be calculated using either symbolic calculus (which we demonstrate in Appendix 11.2; please go to www.oup.com/companion/SalgueroGamelonDM), numerical differentiation, or matrix calculus (see Caswell 2007). These sensitivities are then used along with covariance among all elements of Θ_t to obtain the first-order approximation of temporal variation in $\lambda_{real,t}$:

$$\text{var}(\lambda_{real,t}) \approx \sum_i \sum_j \text{cov}\left(\theta_{i,t}, \theta_{j,t}\right) \left. \frac{\partial\lambda_{real,t}}{\partial\theta_{i,t}} \frac{\partial\lambda_{real,t}}{\partial\theta_{j,t}} \right|_{\bar{\theta}_{ij}},$$

where the sensitivities are evaluated at the mean of Θ_t (as implied by the notation right of the vertical bar) across a realised time series (note that italic theta refers to a single element of Θ_t).

Each term in the equation above is the contribution to temporal variance in $\lambda_{real,t}$ from the covariance among each pair of parameters (which includes stage structure and vital rates). A measure of the contribution of variation in each θ_i to var($\lambda_{real,t}$) is obtained by focusing on θ_i specifically (i.e. dropping the first summation) and summing over the covariances with all other θ_j:

$$\chi_{\theta_i} \approx \sum_j \text{cov}\left(\theta_{i,t}, \theta_{j,t}\right) \left. \frac{\partial\lambda_{real,t}}{\partial\theta_{i,t}} \frac{\partial\lambda_{real,t}}{\partial\theta_{j,t}} \right|_{\bar{\theta}_{ij}}$$

The magnitude of χ_{θ_i} may be small because $\lambda_{real,t}$ is insensitive to changes in $\theta_{i,t}$, because θ_i exhibits little variability over time, or because negative covariation with other θ_j nullifies its direct effect. A transient analysis therefore reveals how strongly fluctuations in population growth rate have been driven by the direct effects of each vital rate, indirect effects channelled through components of stage structure, and sums of χ_{θ_i} across joint parameters of interest. It is nevertheless important to note that a decomposition of 'variance' in $\lambda_{real,t}$ would be most relevant to questions about population dynamics in stationary environmental conditions when one suspects that stage structure has not had enough time to reach a stationary distribution (Maldonado-Chaparro et al. 2017). Other properties of change in $\lambda_{real,t}$ over time would be more relevant in a nonstationary environment, and Koons et al. (2016) provide several, somewhat more complicated, transient LTREs for such conditions.

---

it provided a much more accurate decomposition of temporal variation in realised population growth rates over the 6-year study (relative difference on the log scale was only 0.75):

$$\mathbf{C}_{\mathbf{A}_{orchid}}^{\text{var}(\lambda_{real,t})} = \begin{bmatrix} 0 & 0 & 0 & 0 & 0 & 0.19 \\ -0.01 & 0 & 0 & 0 & 0 & 0 \\ 0 & -0.02 & 0 & 0 & 0 & 0 \\ 0 & 0 & -0.08 & 0.15 & 0 & 0 \\ 0 & 0 & 0 & -0.15 & 0.04 & 0.01 \\ 0 & 0 & 0 & 0 & -0.05 & -0.01 \end{bmatrix}$$

$$+ \begin{bmatrix} 0.20 \\ 0.94 \\ 0.27 \\ 0.08 \\ -0.67 \\ 6.21 \end{bmatrix}$$

(11.9)

The transient LTRE moreover revealed that the largest contributions to variance in realised population growth rates were caused by fluctuations in stage structure, especially the proportion of the population in the flowering stage (last entry of far-right vector in eqn. 11.9). The greatest direct effect of any vital rate was that of fecundity (eqn. 11.9, upper-right corner of matrix), thus indicating quite different parameters as being the dominant contributors to variation in population growth rates during the study compared to the asymptotic LTRE with a random design (eqn. 11.8). Not only do these differing results hinder understanding of past population dynamics, but also application of inappropriate retrospective tools in time-varying environments (i.e. the asymptotic LTRE with a random design) can impair near-term predictions about future population dynamics that are informed by the past (see below).

The asymptotic LTRE with a random design for polar bears indicated that variance in λ_1 was primarily driven by variation in the chance of adults without offspring surviving and then producing and caring for cubs (bold entry in eqn. 11.10):

$$
\mathbf{C}_{\mathbf{A}_{bear}}^{\mathrm{var}(\lambda_1)} = \begin{bmatrix} 0 & 0 & 0 & 0 & 0 & 0.0018 \\ 0.0010 & 0 & 0 & 0 & 0 & 0 \\ 0 & 0.0010 & 0 & 0 & 0 & 0 \\ 0 & 0 & 0.0010 & -0.0080 & -0.0014 & 0.0018 \\ 0 & 0 & 0 & \mathbf{0.0183} & 0.0015 & 0 \\ 0 & 0 & 0 & 0 & 0.0060 & 0 \end{bmatrix}
$$

$$(11.10)$$

Though the LTRE happened to identify the same matrix entry as being most influential on temporal variation in past population growth rates as the transient LTRE (eqn. 11.11), the sum of the contributions overestimated the actual variance of $\lambda_{real,t}$ over time (relative difference on log scale = −0.14). In contrast, the transient LTRE provided a highly accurate decomposition of the demographic parameter contributions to temporal variation in $\lambda_{real,t}$ (relative difference on the log scale was only −0.005):

$$
\mathbf{C}_{\mathbf{A}_{bear}}^{\mathrm{var}(\lambda_{real,t})} = \begin{bmatrix} 0 & 0 & 0 & 0 & 0 & 0.0004 \\ 0 & 0 & 0 & 0 & 0 & 0 \\ 0 & 0.0034 & 0 & 0 & 0 & 0 \\ 0 & 0 & 0.0034 & -0.0053 & -0.0003 & 0.0003 \\ 0 & 0 & 0 & \mathbf{0.0115} & 0.0003 & 0 \\ 0 & 0 & 0 & 0 & 0.0010 & 0 \end{bmatrix}
$$
$$
+ \begin{bmatrix} 0 \\ -0.0007 \\ -0.0007 \\ -0.0002 \\ 0.0001 \\ -0.0009 \end{bmatrix}
$$

$$(11.11)$$

With its full demographic accountability for decomposing retrospective variation in realised population growth rates, the transient LTRE is more reliable for identifying the most influential demographic parameters and is currently the preferred tool for retrospective analyses of transients in time-varying environments (Koons et al. 2016; Maldonado-Chaparro et al. 2017). R code is provided in Appendix 11.2 (please go to www.oup.com/companion/SalgueroGamelonDM) for conducting the LTRE analyses presented above and in Box 11.1.

The collective results above remind us of two questions we are often asked by students: (1) Why do the results from retrospective perturbation analyses differ from those of a prospective analysis? and (2) How can retrospective analyses be applied to conservation and management? To address the first question, prospective perturbation analyses compare the effects of 'equivalent changes' in demographic parameters on demographic outcomes (e.g. population growth rate, population inertia). The derivatives used in a prospective perturbation analysis also appear in retrospective perturbation analyses (see Box 11.1), but

the key difference is that retrospective analyses compare the effects of 'unequal changes' in demographic parameters on demographic outcomes observed in the past. On their own, retrospective analyses do not tell us anything about what will happen in the future or what should be changed to most efficiently affect future demographic outcomes (Caswell 2000). But when implemented in an iterative, adaptive manner, our response to the second question is that transient LTREs (a retrospective analysis) can serve as extremely valuable tools for making predictions about future population dynamics that are informed by the past. These predictions can be gauged against future realisations and iteratively honed until practitioners become sound ecological forecasters in the nonstationary world that we and other organisms currently live in (see Dietze 2017a, 2017b; Dietze et al. 2018).

As is often the case (e.g. Gaillard et al. 2000, 2003), the inference gleaned from our transient LTRE case studies was different than that attained from the prospective elasticities (compare eqns. 11.9 and 11.11 to corresponding prospective results in eqns. 11.4–11.7). This should not be surprising because (1) retrospective perturbation analyses do not force the perturbations to be equivalent across demographic parameters, and (2) the parameters with the greatest elasticity sometimes change the least in response to environmental change (i.e. are the most buffered; Boyce et al. 2006; Koons et al. 2009). When used carefully, the full arsenal of prospective and retrospective transient perturbation tools can be useful in preparing and planning for the future over near-term time-scales.

11.3 Conclusions

We have provided a gentle introduction to the concepts and analysis of transient dynamics with MPMs in both constant and time-varying environments. The study of transients that arise from environmental disturbances and perturbations in otherwise constant environments has matured nicely since Caswell's seminal book (2001), and the reader should now be prepared to dive into thorough reviews on these topics (e.g. Ezard et al. 2010; Stott et al. 2011; Stott 2016). But the study of transients in time-varying environments is in its infancy, and there are ample opportunities to build theoretical foundations for 'realised transients' in today's nonstationary world.

The large amount of work on transients of deterministic models provides a platform for developing needed theory in time-varying environments. There are notions of population inertia in slowly changing environments (Tulajpurkar and Lee 1997), but analogues to transient bounds of a transient envelop have not been developed (see Figure 11.1 for a deterministic example). Fluctuating stage structure in a time-varying environment plays an important role in either buffering or amplifying the effects of vital rate variation on population dynamics (Ellis and Crone 2013; Gamelon et al. 2016; Koons et al. 2016; McDonald et al. 2016). Continued efforts to determine how transients alter population sensitivity to ongoing environmental change are needed. Theoretical studies of what is possible across life histories and various environmental conditions would provide a beacon of light, much like the study of stochastic demography in stationary environments progressed from a practice of theoretical to empirical studies.

Perturbation analyses now exist for transients in time-varying environments but have been applied to only a handful of studies (e.g. Freckleton et al. 2018; Taylor et al. 2018; Layton-Matthews et al. 2019; Paquet et al. 2019). Transient LTREs are first-order approximations to decomposing past change in realised population growth rates that seem to work well for a variety of life histories and conditions (Koons et al. 2016; Maldonado-Chaparro et al. 2017), but we should not be surprised if they perform poorly in some situations. In rapidly changing environments there is a need for more accurate solutions to transient LTREs that incorporate skewness and higher moments of demographic time series (Rice et al. 2011). These solutions should be broadly applicable to any type of structuring variable (not just age structure; Brown and Alexander 1991) and other types of demographic outcomes (e.g. lifetime reproductive success) and be relatively easy to implement like the current first-order approximations. Fervent investment in the study of transients in time-varying environments will rejuvenate an otherwise mature field of population biology, and it will hone our abilities to forecast population dynamics in an ever-changing world (Dietze 2017a, 2017b; Dietze et al. 2018).

Acknowledgements

We thank A. Compagnoni and P. Capdevila for their helpful comments on an earlier draft of the chapter. DNK was supported by the James C. Kennedy Wetland & Waterfowl Conservation endowment. We dedicate this chapter to friend and colleague Markus Dyck, whose life was tragically snatched from this world while doing what he did best, studying polar bears in the Arctic.

References

Bassar RD, A López-Sepulcre, MR Walsh, MM Turcotte, M Torres-Mejia & DN Reznick. 2010. Bridging the gap between ecology and evolution: integrating density regulation and life-history evolution. Annals of the New York Academy of Sciences 1206:17–34.

Baxter PWJ, MA McCarthy, HP Possingham, PW Menkhorst & N McLean. 2006. Accounting for management costs in sensitivity analyses of matrix population models. Conservation Biology 20:893–905.

Boyce MS, CV Haridas & CT Lee. 2006. Demography in an increasingly variable world. Trends in Ecology and Evolution 21:141–148.

Brown D & N Alexander. 1991. The analysis of the variance and covariance of products. Biometrics 47:429–444.

Caswell H. 1989. The analysis of life table response experiments. I. Decomposition of treatment effects on population growth rate. Ecological Modelling 46:221–237.

Caswell H. 2000. Prospective and retrospective perturbation analyses: their roles in conservation biology. Ecology 81:619–627.

Caswell H. 2001. Matrix Population Models: Construction, Analysis and Interpretation, 2nd edition. Sinauer Associates, Sunderland, MA.

Caswell H. 2007. Sensitivity analysis of transient population dynamics. Ecology Letters 10:1–15.

Caswell H. 2010. Life table response experiment analysis of the stochastic growth rate. Journal of Ecology 98:324–333.

Caughley G. 1994. Directions in conservation biology. Journal of Animal Ecology 63:215–244.

Cohen JE. 1976. Ergodicity of age structure in populations with Markovian vital rates, I: countable states. Journal of the American Statistical Association 71:335–339.

Cohen JE. 1977a. Ergodicity of age structure in populations with Markovian vital rates, II: general states. Advances in Applied Probability 9:18–37.

Cohen JE. 1977b. Ergodicity of age structure in populations with Markovian vital rates, III: finite-state moments and growth rate; an illustration. Advances in Applied Probability 9:462–475.

Davison R, H Jacquemyn, D Adriaens, O Honnay, H deKroon & S Tuljapurkar. 2010. Demographic effects of extreme weather events on a short-lived calcareous grassland species: stochastic life table response experiments. Journal of Ecology 98:255–267.

Davison R, M Stadman & E Jongejans. 2019. Stochastic effects contribute to population fitness differences. Ecological Modelling 408:108760.

Dietze MC. 2017a. Ecological Forecasting. Princeton University Press, Princeton, NJ.

Dietze MC. 2017b. Prediction in ecology: a first-principles framework. Ecological Applications 27:2048–2060.

Dietze MC, A Fox, LM Beck-Johnson, JL et al. 2018. Iterative near-term ecological forecasting: needs, opportunities, and challenges. Proceedings of the National Academy of Sciences 115:1424–1432.

Doak DF, WF Morris, CP Fister, BE Kendall & EM Bruna. 2005. Correctly estimating how environmental stochasticity influences fitness and population growth. The American Naturalist 166:E14–E21.

Easterling MR, SP Ellner & PM Dixon. 2000. Size-specific sensitivity: applying a new structured population model. Ecology 81:694–708.

Ellis MM & EE Crone. 2013. The role of transient dynamics in stochastic population growth for nine perennial plants. Ecology 94:1681–1686.

Ellner SP, DZ Childs & M Rees. 2016. Data-Driven Modelling of Structured Populations. Springer, Cham.

Ezard THG, JM Bullock, HJ Dalgleish, et al. 2010. Matrix models for a changeable world: the importance of transient dynamics in population management. Journal of Applied Ecology 47:515–523.

Freckleton RP, HL Hicks, D Comont, et al. 2018. Measuring the effectiveness of management interventions at regional scales by integrating ecological monitoring and modelling. Pest Management Science 74:2287–2295.

Gaillard J-M & NG Yoccoz. 2003. Temporal variation in survival of mammals: a case of environmental canalization? Ecology 84:3294–3306.

Gaillard J-M, M Festa-Bianchet, NG Yoccoz, A Loison & C Toigo 2000. Temporal variation in fitness components and population dynamics of large herbivores. Annual Review in Ecology and Systematics 31:367–393.

Gamelon M, O Gimenez, E Baubet, T Coulson, S Tuljapurkar & J-M Gaillard. 2014. Influence of life-history tactics on transient dynamics: a comparative analysis across mammalian populations. The American Naturalist 184:673–683.

Gamelon M, J-M Gaillard, O Gimenez, T Coulson, S Tuljapurkar & E Baubet. 2016. Linking demographic responses and life history tactics from longitudinal data in mammals. Oikos 125:395–404.

Gotelli NJ & AM Ellison. 2006. Forecasting extinction risk with nonstationary matrix models. Ecological Applications 16:51–61.

Haridas CV & S Tuljapurkar. 2005. Elasticities in varying environments: properties and implications. The American Naturalist 166:481–495.

Haridas CV, S Tuljapurkar & T Coulson. 2009. Estimating stochastic elasticities directly from longitudinal data. Ecology Letters 12:806–812.

Hastings A. 2004. Transients: the key to long-term ecological understanding? Trends in Ecology and Evolution 19:39–45.

Hastings A, KC Abbott, K Cuddington, et al. 2018. Transient phenomena in ecology. Science 361:eaat6412.

Heppell S. 1998. Application of life-history theory and population model analysis to turtle conservation. Copeia 2:367–375.

Horvitz C, DW Schemske & H Caswell. 1997. The relative 'importance' of life-history stages to population growth: prospective and retrospective analyses. In: S Tuljapurkar and H Caswell (eds). Structured-Population Models in Marine, Terrestrial and Freshwater Systems (pp. 247–271). Chapman and Hall, New York.

Hunter CM, H Caswell, MC Runge, EV Regehr, SC Amstrup & I Stirling. 2010. Climate change threatens polar bear populations: a stochastic demographic analysis. Ecology 91:2883–2897.

Iles DT, R Salguero-Gómez, PB Adler & DN Koons. 2016. Linking transient dynamics and life history to biological invasion success. Journal of Ecology 104:399–408.

Iles DT, RF Rockwell & DN Koons. 2019. Shifting vital rate correlations alter predicted population responses to increasingly variable environments. The American Naturalist 193:E57–E64.

Jacquemyn H, R Brys & E Jongejans. 2010. Seed limitation restricts population growth in shaded populations of a perennial woodland orchid. Ecology 91:119–129.

Jenouvrier S, M Holland, J Stroeve, et al. 2012. Effects of climate change on an emperor penguin population: analysis of coupled demographic and climate models. Global Change Biology 18:2756–2770.

Keyfitz N. 1971. On the momentum of population growth. Demography 8:71–80.

Koons DN, JB Grand, B Zinner & RF Rockwell. 2005. Transient population dynamics: relations to life history and initial population state. Ecological Modelling 185:283–297.

Koons DN, RF Rockwell & JB Grand. 2006a. Population momentum: implications for wildlife management. Journal of Wildlife Management 70:19–26.

Koons DN, JB Grand & JM Arnold. 2006b. Population momentum across vertebrate life histories. Ecological Modelling 197:418–430.

Koons DN, RR Holmes & JB Grand. 2007. Population inertia and its sensitivity to changes in vital rates and population structure. Ecology 88:2857–2867.

Koons DN, CJE Metcalf & S Tuljapurkar. 2008. Evolution of delayed reproduction in uncertain environments: a life history perspective. The American Naturalist 172:797–805.

Koons DN, S Pavard, A Baudisch & CJE Metcalf. 2009. Is life-history buffering or lability adaptive in stochastic environments? Oikos 118:972–980.

Koons DN, DT Iles, M Schaub & H Caswell. 2016. A life-history perspective on the demographic drivers of structured population dynamics in changing environments. Ecology Letters 19:1023–1031.

Lawson CR, Y Vindenes, L Bailey & M van de Pol. 2015. Environmental variation and population responses to global change. Ecology Letters 18:724–736.

Layton-Matthews K, MJJE Loonen, BB Hansen, CFD Coste, B-E Sæther & V Grøtan. 2019. Density-dependent population dynamics of a high Arctic capital breeder, the barnacle goose. Journal of Animal Ecology 88:2181191–1201.

Lee CT. 2017. Elasticity of population growth with respect to the intensity of biotic or abiotic driving factors. Ecology 98:1016–1025.

Lewontin RC & D Cohen. 1969. On population growth in a randomly varying environment. Proceedings of the National Academy of Sciences USA 62:1056–1060.

Lotka AJ. 1924. Elements of Physical Biology. Williams and Wilkins, Baltimore, MD. (Reprinted 1956 by Dover Publications, New York as Elements of Mathematical Biology.)

Maldonado-Chaparro AA, DT Blumstein, KB Armitage & DZ Childs. 2017. Transient LTRE analysis reveals the demographic and trait-mediated processes that buffer population growth. Ecology Letters 21: 1693–1703.

May RM. 1981. Theoretical Ecology, Principles and Applications. Blackwell, Oxford.

McDonald JL, I Stott, S Townley & DJ Hodgson. 2016. Transients drive the demographic dynamics of plant populations in variable environments. Journal of Ecology 104:306–314.

Neubert MG & H Caswell. 1997. Alternatives to resilience for measuring the responses of ecological systems to perturbations. Ecology 78:653–665.

Nichols JD & JE Hines. 2002. Approaches for the direct estimation of λ, and demographic contributions to λ, using capture–recapture data. Journal of Applied Statistics 29:539–568.

Paquet M, D Arlt, J Knape, M Low, P Forslund & T Pärt. 2019. Quantifying the links between land use and population growth rate in a declining farmland bird. Ecology and Evolution 9:868–879.

Rice SH, A Papadopoulos & J Harting. 2011. Stochastic processes driving directional evolution. In: P. Pontarotti (ed.), Evolutionary Biology—Concepts, Biodiversity, Macroevolution and Genome Evolution (pp. 21–33). Springer, Berlin.

Salguero-Gómez R, OR Jones, CR Archer, et al. 2015. The COMPADRE Plant Matrix Database: an open online repository for plant demography. Journal of Ecology 103:202–218.

Salguero-Gómez R, OR Jones, CR Archer, et al. 2016. COMADRE: a global data base of animal demography. Journal of Animal Ecology 85:371–384.

Schaub M, H Jakober & W Stauber. 2013. Strong contribution of immigration to local population regulation: evidence from a migratory passerine. Ecology 94:1828–1838.

Shigesada N & K Kawasaki. 1997. Biological Invasions: Theory and Practice. Oxford University Press, Oxford.

Stott I. 2016. Perturbation analysis of transient population dynamics using matrix projection models. Methods in Ecology and Evolution 7:666–678.

Stott I, M Franco, D Carslake, S Townley & DJ Hodgson. 2010a. Boom or bust? A comparative analysis of transient population dynamics in plants. Journal of Ecology 98: 302–311.

Stott I, S Townley, D Carslake & DJ Hodgson. 2010b. On reducibility and ergodicity of population projection matrix models. Methods in Ecology and Evolution 1: 242–252.

Stott I, S Townley & DJ Hodgson. 2011. A framework for studying transient dynamics of population projection matrix models. Ecology Letters 14:959–970.

Taylor LU, BK Woodworth, BK Sandercock & NT Wheelwright. 2018. Demographic drivers of collapse in an island population of tree swallows. The Condor: Ornithological Applications 120:828–841.

Townley S & D Hodgson. 2008. Erratum et addendum: transient amplification and attenuation in stage-structured population dynamics. Journal of Applied Ecology 45:1836–1839.

Townley S, D Carslake, O Kellie-Smith, D McCarthy & D Hodgson. 2007. Predicting transient amplification in perturbed ecological systems. Journal of Applied Ecology 44:1243–1251.

Tuljapurkar S & SH Orzack. 1980. Population dynamics in variable environments I: long-run growth rates and extinction. Theoretical Population Biology 18:314–342.

Tuljapurkar S. 1982. Population dynamics in variable environments. II. Correlated environments, sensitivity analysis and dynamics. Theoretical Population Biology 21: 114–140.

Tuljapurkar S. 1990. Population Dynamics in Variable Environments. Springer, New York.

Tuljapurkar S & R Lee. 1997. Demographic uncertainty and the stable equivalent population. Mathematical and Computer Modelling 26:39–56.

Tuljapurkar S, CC Horvitz & JB Pascarella. 2003. The many growth rates and elasticities of populations in random environments. The American Naturalist 162: 489–502.

Williams BK, JD Nichols & MJ Conroy 2002. Analysis and Management of Animal Populations. Academic Press, San Diego, CA.

Wolkovich EM, BI Cook, KK McLauchlan & TJ Davies. 2014. Temporal ecology in the Anthropocene. Ecology Letters 17:1365–1379.

CHAPTER 12

Individual-based models

Viktoriia Radchuk, Stephanie Kramer-Schadt, Uta Berger, Cédric Scherer, Pia Backmann, and Volker Grimm

12.1 Introduction

Individual-based models (IBMs, also often referred to as agent-based models, ABMs) represent a broad class of simulation models that depict relevant processes at the individual level, thereby allowing higher-level processes such as population and community dynamics to emerge. They are used whenever there is reason to believe that one or more of the following aspects are critical to consider: intraspecific trait variation (e.g. regarding age, size, or personality), local interactions (in contrast to global interactions assumed in nonspatial models), heterogeneous environments, nonlinear feedbacks, and adaptive behaviour (DeAngelis and Grimm 2014).

In the 1990s, when computing power for running IBMs became generally available, IBMs started to be used in different disciplines dealing with autonomous agents, for example ecology, geography, economics, and social sciences (Vincenot 2018). Nowadays, IBMs/ABMs are established tools in virtually every discipline investigating systems comprised of agents. Earlier differences between IBMs, which focused on trait variation and local interactions, for example in forest models, and ABMs, which focused on adaptive behaviour, are fading away (Railsback and Grimm 2019), so that both terms can be used interchangeably.

Demography is an explicit aspect of IBMs that deal with the full life cycle of their agents. Such IBMs distinguish among agents by differentiating their age, traits, and other variables. They thus complement more traditional modelling approaches in demography and are particularly suitable for addressing the questions about the response of agent-based systems to some kind of change (e.g. in physical or social environment). In the following we first briefly characterise all elements of an IBM. Then we discuss in more detail when IBMs

can be important, or even mandatory, to use even though they require more effort than more aggregated models. We will then characterise the kind of data required for parameterising IBMs and present three examples to demonstrate how IBMs look, work, and help answering research questions.

12.2 What are individual-based models and how do we document them?

IBMs are simulation models that iteratively apply the rules describing individual-level processes in discrete time. For a long time the only precise and complete documentation of an IBM was the computer program implementing it, but since IBMs are implemented in all kinds of programming languages and on different computer types and operation systems, no standard way of communicating IBMs existed. This severely limited the use of IBMs as a scientific tool because repeatability is a key element of the scientific method.

To overcome this situation, a standard format for describing IBMs, the ODD (overview, design concepts, and details) protocol (Grimm et al. 2006, 2010), was proposed and is now widely used (Vincenot 2018). Here we use it to introduce seven main elements of an IBM in general. Examples of full ODD model descriptions can be found in the online supplements that are linked to our example IBMs described in Section 12.6.

(1) *Purpose and patterns*: IBMs are, like any other model, developed for a certain purpose. This purpose to a large degree determines which aspects of reality are included in an IBM. Additionally to the model purpose, a recent update of ODD (Grimm et al. 2020) requires listing the criteria that are used to judge a model to be realistic enough for

Viktoriia Radchuk et al., *Individual-based models*. In: *Demographic Methods Across the Tree of Life*.
Edited by Roberto Salguero-Gómez and Marlène Gamelon, Oxford University Press.
© Oxford University Press (2021). DOI: 10.1093/oso/9780198838609.003.0012

its intended purpose. For applied problems, this can be a set of observed specific patterns ('pattern-oriented modelling'; Grimm and Railsback 2012); for more theoretical models, it can be more generic patterns, for example the very fact that a population does not go extinct.

(2) *Entities, state variables, and scales*: An IBM has entities, which are characterised by a set of state variables. Typical entities are agents (e.g. plants, animals, humans), spatial units (e.g. grid cells of a certain size), and the overall environment. Agents are often characterised by age, size, wealth, or behavioural types; spatial units by soil type, moisture, plant cover, habitat suitability, or price; the environment by temperature, precipitation, or demand. IBMs have a certain temporal extent and resolution, that is, length of the time step by which the model proceeds. Spatially explicit IBMs are additionally characterised by a spatial extent and resolution (cell size).

(3) *Process overview and scheduling*: Processes constituting the core of the model are executed either each time step or depending on certain conditions. To understand how IBMs work it is important to realise that most of these processes are not executed by the modeller (also called the 'observer' in individual-based modelling) but by the entities of the model: for example, individuals grow, reproduce, and die. Thus, in the process overview or central schedule of an IBM, the question is: Who is doing what, at what time, and in what sequence?

(4) *Design concepts*: IBMs are developed according to important design concepts (Railsback 2001). The most important ones are: Which behaviours of the agents are imposed, and which emerge from their own decision-making ('emergence'; for details see Section 12.3.1)? What do agents want to achieve ('objectives')? What do agents sense and know about themselves and their environment ('sensing')? How do agents interact ('interaction')? How does the modeller observe the simulated system ('observation')?

(5) *Initialisation*: With what parameter settings is the model initialised?

(6) *Input data*: Are there any input data imported from external files or other models which describe environmental variables that drive the dynamics of the system?

(7) *Submodels*: How, in detail, are the processes listed in schedule (3) implemented in terms of algorithms and equations? What are the parameters, their values, their units, and their uncertainty, and where do they come from?

The ODD protocol facilitates understanding what a model is and does. It should include the factual description of its elements and the rationales behind the model's assumptions and design. ODD is independent of scientific discipline, model purpose and complexity, and implementation details. Still, like any verbal description, it can contain ambiguities that may prevent full replication. Therefore, ideally, ODDs would contain links to chunks of their corresponding computer codes; the rationale behind such interlinks is that at the level of a few code lines, most programming languages can be understood without being familiar with them. In addition to ODD, the communication of IBMs is facilitated by diagrams based on Unified Modelling Language (UML). UML diagrams are an approved ISO standard for the visualisation of process-based software. Although they are not yet commonly used for IBMs, this promising tool may facilitate the communication of IBMs in the future.

12.3 Why individual-based models?

IBMs are indispensable to analyse and predict the development of small populations (as required, for example, in population viability analyses [PVA]), in which the fate of each individual is of particular importance and to a large degree affected by demographic and environmental noise (Caughley 1994). It may seem, however, that IBMs are less needed when studying large populations and communities. In such systems, one would assume mean field approximations, differential equations, or matrix model notations to be more suitable because the price of describing 'averaged individuals' seems to be paid off by the tractability of the models. However, this is not generally true. Matrix models grouping individuals with similar characteristics into separate stages might be feasible in simple cases, but even those models become analytically intractable when the number of individual characteristics increases.

12.3.1 Emergent behaviour vs. imposed demographic rates

Matrix and many other aggregated models are based on 'demographic thinking' (Grimm and Berger 2016) where population growth is the result of demographic, or vital, rates, that is, fecundity and mortality. If these rates are known, for example from mark–recapture studies (see Chapter 13), they indeed allow us to predict population growth for the environmental conditions and population structure under which they were measured. However, if the task is to predict how a population responds to changes in the environment and population structure, it can become critical to let those demographic rates emerge from what the individual organisms are actually doing, that is, their behaviour and life cycle (Radchuk et al. 2019). In the Anthropocene, characterised by multiple environmental factors changing at unprecedented rates, predictions of response to such changes have become a mandatory task for ecological modelling.

Basing models on demographic rates means to *impose* the performance of individuals instead of letting it *emerge* from their adaptive behaviour. 'Full-fledged' IBMs therefore usually let one or a few key behaviours of the individuals, for example foraging, emerge from their adaptive decision-making, while processes to which the population dynamics is less sensitive are still imposed, for example some background mortality.

The key element of such 'next-generation' IBMs (Grimm and Berger 2016), adaptive behaviour, requires representing the trade-offs that individuals have to consider (DeAngelis and Diaz 2019), for example the trade-off between predation and starvation risk for animals or the allocation of resources to the shoots or roots for plants. Adaptive behaviour ensures individual and hence species persistence in a changing world and therefore is crucial to be accounted for in the modern research on demography.

For example, including adaptive habitat selection in a fish population dynamics model allowed assessing the effects of multiple environmental factors on the population persistence (Harvey and Railsback 2007). Results of such an assessment would have been biased had the model failed to consider adaptive behaviour, because individual responses to each environmental factor would have been imposed according to fixed phenomenologically measured relations, failing to consider that such relations may change as an organism adapts to environmental changes. Such adaptive responses may occur via microevolution, phenotypic plasticity, or both (Gienapp et al. 2008). Arguably, such eco-evolutionary dynamics is much easier to implement and interpret using IBMs compared to other modelling approaches (e.g. Dunlop et al. 2009).

IBMs of eco-evolutionary dynamics differ widely with regard to how they depict evolutionary processes in the model. Thus, some models directly hard-wire a mean phenotypic trait value that undergoes evolution due to mutation and drift, without explicit consideration of the underlying genetic structure. For example, this is the way in which ecological traits are evolving in the asexual version of the model developed by Dieckmann and Doebeli (1999) to investigate the potential of sympatric speciation. At the same time, other models

explicitly code multilocus systems for each trait that is transmissible. Such an approach was used to investigate the effects of climate change on the microevolution of Atlantic salmon (Piou and Prévost 2012a, 2012b) and of stocking and barriers on the population persistence of brown trout (Frank and Baret 2013). The level of detail with which genes are mapped to the traits or behaviours they code for depends on the specific question for which the model is developed.

In IBMs, density dependence is not imposed, for example by making a vital rate depend on population density, but emerges from the individuals competing for resources. For example, Stephens et al. (2002) showed that four alternative model versions, varying in how they implemented density dependence of alpine marmots, differed greatly in the resulting population dynamics. All model versions, including a matrix model (Chapter 9), resulted in similar population dynamics around the equilibrium densities. However, only a behaviour-based IBM version, in which density dependence at the population level emerged from the lower-level individual interactions, could capture the processes at lower population densities, including positive density effects.

12.3.2 Local interactions and nonlinear feedbacks

Another key element of most IBMs is local interactions, which often have important implications for the demography of the focal species and the ones sharing resources with it. Such interactions are easy to capture in IBMs, but it is difficult to do so using mean field approaches, especially when space is heterogeneous. Donalson and Nisbet (1999) investigated the effects of relaxing two main assumptions of a classical Lotka–Volterra model: (1) a well-mixed population and (2) the absence of demographic stochasticity. The results of population persistence were strikingly different, even in homogeneous environments.

Finally, the responses of individuals to environment and to individuals of con- and hetero-specifics are often nonlinear (Harvey and Railsback 2007). Such nonlinearities at the individual level can further propagate to the population level, resulting in unexpected or counterintuitive outcomes. This is especially the case when individuals experience the impact of multiple factors, as successfully demonstrated by an IBM applied to the population dynamics of cutthroat trout subjected to elevated turbidity, elevated stream temperature, and reduced pool frequency (Harvey and Railsback 2007). Similarly, an IBM of a freshwater detritivore revealed that multiple environmental stressors interacted in their effects on population viability, but implications were much stronger for the ecosystem functioning delivered by this detritivore (Galic et al. 2017).

12.4 The pros and cons of IBMs

Generally, IBMs are not bound to a particular model structure like some other model types (e.g. matrix models; Chapter 9) and instead are very flexible, allowing a focus on the processes that are especially important for answering a specific research question. For example, if individual growth and resource allocation are essential for understanding competition among plants, these processes can be included in the model (Backmann et al. 2019; see Section 12.6). Additionally to the research question, the choice of an IBM's structure is determined by the availability of data to parameterise the required processes (Radchuk et al. 2016b).

Thanks to their flexibility, IBMs are used for understanding a wide range of phenomena and systems, including bird flocking (Hemelrijk and Hildenbrandt 2015), viability of small populations (Beissinger and Westphal 1998), disease dynamics and persistence (Lange and Thulke 2017), functioning of socio-ecological systems (Schlüter et al. 2017), and the structure and dynamics of plant (May et al. 2009) and microbial communities (Hellweger et al. 2016) and even entire ecosystems (Fu et al. 2018).

However, this flexibility is also associated with costs. IBMs usually have richer structure than, for example,

matrix models. They are harder to develop, parameterise, and understand than more aggregated models. Understanding model behaviour, in particular, requires considerable time and effort. Indeed, a comparison of resources needed to construct and analyse a matrix model vs. an IBM used for the PVA of a butterfly species (Radchuk 2012) showed that IBMs require twice as much time for model development and analysis and have much longer running time (Table 12.1). Likewise, higher complexity implies more parameters (Table 12.1) and hence more data to parameterise them. Such data are not always available, so that parameters remain uncertain.

Careful exploration of the sensitivity of the model output to this uncertainty ('parameter uncertainty') is one of the key aspects of individual-based modelling. Similarly, it is critical that modellers assess the sensitivity of choices made about the structure of the included submodels ('structural uncertainty'; Railsback and Grimm 2019). Given the complexity of IBMs, it is essential that all the choices made during iterative model development and testing are well reported and documented (Grimm et al. 2014).

Together with sensitivity analysis, model validation is another key tool in model analysis. Validation in its strict sense reflects the ability of a model to reproduce the patterns in data that were not used for model development. Validation is thus the ultimate test of model credibility (Grimm et al. 2014). Importantly, in more aggregated models, such as matrix

Table 12.1 Comparison of the resources needed to construct and run two model types: a matrix model and an IBM. Both models were used to assess the population viability of a vulnerable butterfly species, *Boloria eunomia* (the bog fritillary butterfly), in the south of Belgium. This table is modified after Radchuk (2012).

Resources	Matrix model	IBM
Data analyses	*Capture–mark–recapture data:* 12 days × 7 years	*Capture–mark–recapture data:* 12 days × 7 years
	Survival experiments: Eggs: 5 working days Larva 1: 7 working days Larva 2: 12 working days Pupa: 10 working days *Fecundity experiments:* 8 working days *Habitat mapping:* 17 working days	*Survival experiments:* Eggs: 5 working days Larva 1: 7 working days Larva 2: 12 working days Pupa: 10 working days *Fecundity experiments:* 12 working days *Habitat mapping:* 17 working days
Model development and analysis	160 working days	380 working days
Running time for baseline scenario (10 runs)	10 min 27 s	12 h 24 min 19 s
Number of parameters	21 + 800 of average and SD dispersal rates[1]	47 + 800 of average and SD dispersal rates[1]
Skills	Knowledge of a programming language	Knowledge of a programming language

[1] Dispersal rates are drawn from the phenomenologically obtained empirical distributions, reflecting the probability of emigration in this 20-patch meta-population system.

models, validation is possible at the highest organisation level only—the population level—for example by hindcasting the population sizes to the past. In contrast, the hierarchical structure of IBMs that are based on individual-level processes from which the higher-level phenomena are emerging allows us to conduct validation at several organisational levels, given the respective data are available. For example, if demographic rates or emigration rates are emerging from the underlying individual behaviour, they can be used as patterns for validation, in addition to the distribution of the population sizes.

To sum up, IBMs, like other model types, have their pros and cons. They complement more aggregated modelling approaches by capturing the mutual and inseparable micro–macro link between the behaviour of individuals and the structure and dynamics of the systems they comprise.

12.5 Types of data used to build an IBM

The data used to build an IBM for demography can be grouped into two main classes: species data and environmental data (Radchuk et al. 2016b) (Figure 12.1). Types of species data will vary depending on the purpose of the model, the structure of the model, and the detail at which processes are represented.

The following categories of species data can be distinguished: (1) individual-level data (Chapter 5) may include rules describing the energy budget of an organism (e.g. consumption rates, allocation rates), vital rates, decision rules describing sociality (Chapter 3) and territoriality if necessary, and rules for habitat selection and mate choice; (2) movement and dispersal data may be represented by emigration and immigration rates (in the case of aggregated dispersal description) or by detailed movement rules and parameters (in the case of fine-grained movement description); (3) genetic data (Chapter 1) may include mutation rate, heritability, and so on. If a model represents more than one species, species data are needed for all species considered, in addition to the rules describing the interactions of hetero-specific individuals.

Environmental data (Chapter 6) can be categorised into two main classes: climate data (describing the climatic variables that are essential to the viability of the population) and landscape data represented by maps of habitat availability and quality (if several different resources are distinguished they should all be depicted by a map). Of course, only those environmental data that either modulate the effect of the habitat, for example soil moisture, on individuals, or have direct effects on individual processes should be considered.

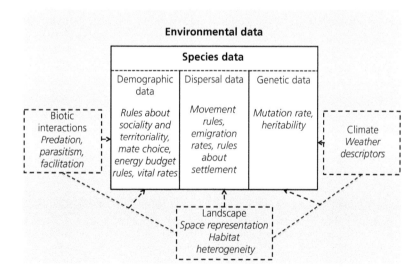

Figure 12.1 Schematic representation of the types of data that can be used in demographic IBMs (modified after Radchuk et al. 2016b). Species data are shown on the white background and can be categorised into three groups. Individuals of each species closely interact with their environment, and therefore species data are 'embedded' in a block representing environmental data (shown on the grey background). We distinguish three groups of environmental data. Examples of data for each group are given in italics. Environmental effects on species are shown with dashed arrows. Not only different environmental aspects can affect one or several aspects of species viability (i.e. effects on demography, dispersal, genetics), but also they may interact in their effects on species (depicted by dashed lines between aspects of environment and the arrows going from these lines).

If environmental data are included, there should also be rules and parameters defining the effect of the environment (climate and/or habitat) on the individual-level processes (e.g. behaviour, resource acquisition, metabolism, survival, fecundity, movement, and/or dispersal). For example, Meli et al. (2014) included linear regressions parameterising the effect of the toxicant's concentration on the survival and fecundity of individuals as well as their avoidance of the polluted grid cells.

Importantly, for the purpose of demographic IBMs, not all the above-mentioned data types have to be included in the model. As explained at the beginning, the data demand is driven by the model structure, which is dictated by the research question and the ecology of the species in consideration. Therefore, if the spatial aspect is considered unnecessary to answer the research question and to appropriately describe the demography of the studied species, the landscape data, as well as dispersal data, may be left out.

12.6 Case studies

Below we briefly present three IBMs as examples of different aspects and uses of IBMs. These address population cycles in rodents, the persistence of wildlife diseases in a given landscape, and the emergence of delayed chemical defence of plants against insect herbivores. The models are quite similar in terms of complexity, while much simpler but also much more complex IBMs exist. They are also all rooted in empirical patterns but still include a large proportion of stylised representations of processes; entirely stylised or highly detailed and realistic IBMs also exist. The ODD model descriptions and the corresponding codes for all three models are provided in Section 12.8 (Supplementary materials 1–5). Additionally, we uploaded the descriptions and codes of all models to the CoMSES model library. Demography is implicit in all these models, since age structure is important in all three cases. Moreover, sex differences and belonging to a certain family group are crucial in the models of vole cycles and disease persistence. Taken together, these individual-level differences generate differences in population structure that have importants implications for population dynamics.

12.6.1 Vole–mustelid model

Rodent population cycles have attracted much research attention. Predation, an extrinsic factor, was suggested as one of the main drivers behind population cycles,

although by itself it is insufficient to reproduce the cycles observed in natural populations. At the same time, empirical evidence shows that population cycles are affected by intrinsic factors, such as dispersal and sociality. The importance of interactions between intrinsic and extrinsic factors was highlighted a long time ago (Lidicker 1988), and recently a conceptual framework was proposed (Andreassen et al. 2013) that cycles are driven by a combination of two intrinsic (sociality and negative density-dependent dispersal) and one extrinsic (predation) factor.

The purpose of the vole–mustelid model presented here (Radchuk et al. 2016a) was to assess which factors (or their combinations) lead to the population cycles characteristic of natural vole populations. Radchuk et al. (2016a) contrasted alternative model versions by 'switching off/on' submodels to systematically investigate their ability to reproduce a set of observed patterns ('pattern-oriented modelling'; Grimm and Railsback 2012). An IBM was needed mainly because the vole life cycle, in particular aspects related to sociality, could not be accurately depicted with more aggregated models.

The modelled system imitates the experimental system at Evenstad Research Station (Norway) from which most of the parameters on root voles (*Microtus oeconomus*) were derived. The system consists of six habitat patches embedded into a hostile matrix (Andreassen et al. 2013). Individual voles are characterised by *age* (weeks), *stage* (weanling: <3 weeks old; subadult: 3–4 weeks old; and adult: >4 weeks), *sex* (female or male), *family* (ID of the family an individual belongs to), maturity state (*mature/not mature*), and reproductive state for mature individuals (*reproductive/not reproductive*). Model time step is 1 week. The duration of each simulation is 1,020 weeks, corresponding to 30 average generations across the prey and predator species. Since seasonality is important in the vole life cycle, we distinguish two seasons: summer (weeks 17–43) and winter (weeks 1–16 and 44–52).

The following processes are modelled: reproduction, survival, dispersal, maturation, and predation (Figure 12.2). Two processes occur only in summer, namely reproduction and dispersal. The number of weanlings is positively affected by the number of females in the patch (maximum 2), reflecting the kinship effect. For dispersal, the emigration probability is negatively density dependent and sex- and stage-specific. Density dependence, as here, is still often imposed in IBM but only when referring to spatial units that are small enough to assume that local

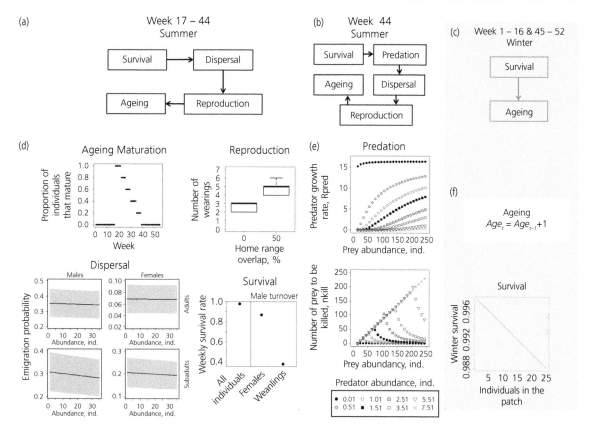

Figure 12.2 Schematic representation of the processes included in the vole-mustelid model and their schedule (reproduced from Radchuk et al. 2016a). (a)–(c) Schedules of processes taking place during different times of the year. Processes taking place in the summer season are shown on a white background, and those taking place in winter are shown on a grey background. (d)–(f) Details on parameterisation and functional relations used to implement the modelled processes: maturation, dispersal, survival, and reproduction in summer (d), predation occurring in week 44 (e), and ageing and survival in winter (f).

density indeed affects all involved individuals via local interactions. Each emigrant incurs dispersal mortality and ends up in one of the habitat patches that do not contain an adult of the same sex as its own. In winter, survival is negatively density dependent according to empirical findings (Aars and Ims 2002), and in summer it is fixed irrespective of the stage, age, and sex of individuals. If the sociality submodel is activated, then survival probability of a female and her weanlings declines in case of a male turnover.

The predation by a specialist predator (i.e. weasels) is modelled using the predator–prey model commonly used for voles (see Turchin and Hanski 1997). Predation is modelled once a year in week 44 and consists of two steps. In the first step, predator population density is updated using the predator population growth rate. In the second step, the number of prey to be killed by predator is determined according to the type 2 functional response only if the vole population

density is higher than the critical one. As for the most important design concept, emergence, virtually all behaviours of the individuals were imposed via empirical rules or rates. For the purpose of the model, explaining the emergence of population cycles in a given environment, this was sufficient.

Radchuk et al. (2016a) compared the population dynamics of empirically observed natural vole populations with those obtained using four alternative models representing studied combinations of extrinsic and intrinsic factors. Four metrics (so-called patterns in pattern-oriented modelling: Grimm and Railsback 2012) were used to describe the vole population cycles and to contrast the models with real populations: autumn population density, periodicity, amplitude, and population growth rate. Only the full model that incorporated predation *and* two intrinsic factors reproduced all four patterns the closest to the empirical ones (Figure 12.3). Periods were too short if one of

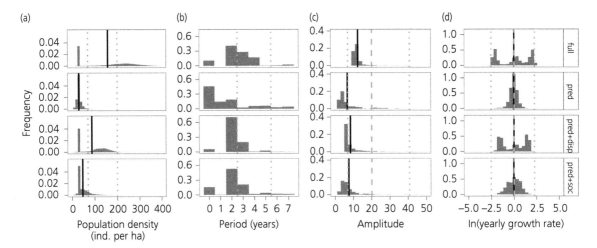

Figure 12.3 Performance of the four models against the four patterns summarising the empirical vole population cycles: (a) autumn population density, individuals (ind.) per ha; (b) cycle period; (c) amplitude; and (d) yearly population growth rates (natural-log transformed). Black lines denote the means of the patterns observed in simulations under each model; grey dashed lines show the means of the empirically observed patterns; and grey dotted lines show the 5 and 95 percentiles of the empirical distributions (except for (a), where dotted lines show the observed range of mean autumn population abundances across studies and locations). Models reflecting the prevailing hypotheses about the drivers of vole cycles are: full (predation and both intrinsic factors), pred (predation only), pred + disp (predation and dispersal), and pred + soc (predation and sociality). Only the full model version yields the amplitude and period of the cycles closest to those characterising natural populations. (This figure is adapted from Radchuk et al. (2016a))

the intrinsic factors was left out. Similarly, observed population densities were too low for the models with predation only or for a combination of predation and sociality. The model with predation only failed to capture the multimodal distribution of growth rates as observed in natural populations.

This example IBM demonstrates the interplay of empirical knowledge, data, and modelling. Direct observations to determine which mechanisms are responsible for the observed population cycles were virtually impossible. Still, each observed pattern contained information about the underlying mechanisms, and the IBM could be used to inversely 'decode' this information (Railsback and Grimm 2019). IBMs therefore are a powerful tool for inferring mechanisms from observed patterns.

12.6.2 Wild boar–classical swine fever (CSF) virus model

The ability of a pathogenic agent to successfully invade and avoid extinction—termed disease persistence—depends on both the transmission characteristics of the pathogen and the mixing and hence contact rates of hosts (Hagenaars et al. 2004). Heterogeneities in host mixing, namely spatiotemporal differences in host densities, arise from the demography of the host

species (e.g. seasonal changes in birth and survival rates) and the underlying landscape (Lloyd-Smith et al. 2005). The consideration of such spatiotemporal heterogeneities in host densities and their effects on disease persistence is facilitated by spatially explicit IBMs that allow for a detailed representation of (1) host demography, (2) the transmission characteristics of the pathogen, and (3) the underlying resource landscape. Further, such models enable making quantitative predictions, which are important for disease management (Eisinger and Thulke 2008).

The purpose of the host–pathogen model originally developed by Kramer-Schadt et al. (2009) is to understand how the interaction between virus transmission characteristics, host demography, and landscape features determine the persistence of a disease in a given landscape. The model is a combination of a spatially explicit, stochastic, individual-based demographic model of wild boars (*Sus scrofa* L.) and an epidemiological model for the classical swine fever (CSF) virus.

The model entities are individual wild boars and square spatial units or habitat cells. A wild boar is characterised by sex and age: piglet (<34 weeks), subadult (between 34 weeks and <1 year for females and <2 years for males), and adult. Further, it is described by an epidemiological status: susceptible, transiently infected, lethally infected, and immune. Females that are at least subadult may be assigned as breeders according to the habitat quality of their cell. Each grid cell is characterised by habitat quality that determines its maximum group size. Each cell represents an area of 2 km × 2 km encompassing an average home range area of a wild boar group. All processes take place on a spatial grid of 100 × 25 cells. The model proceeds in weekly time steps that correspond to the approximate CSF incubation time.

The IBM's demographic model mimics seasonal reproduction, natal dispersal, and mortality of wild boars, while the epidemiological model describes the effects of virus on the wild boar population density and structure via virus-induced mortality and litter size depression (Figure 12.4). The pathogen spread from infected to susceptible individuals is modelled via locally density-dependent within- and between-group infection rates. Infected individuals either die after a given infectious period (*mue*) or gain lifelong immunity depending on the case fatality ratio (*cfr*, defined as proportion of disease-induced deaths). The habitat quality of each cell is assigned randomly (Figure 12.5a). The virus is released in a random week of year 2 to a certain boar group, and simulations are run as long as any infected individuals are present.

Using a slightly adopted version of this model (Sciaini et al. 2018); Scherer (2019) demonstrated that disease persistence was more likely to occur in case of moderate case fatality ratios and long-lasting infectious periods (Figure 12.5b). Longer infectious period (e.g. mean of 9 weeks) led to higher number of infected individuals (Figure 12.5c, A+B) compared to shorter infectious periods (e.g. mean of 4 weeks; Figure 12.5c, C+D). Moderate case fatality ratio allowed for a large population of susceptible and immune individuals, thereby favouring a sustained transmission chain and annual cycles of infection due to sufficient contacts and the birth of susceptible piglets (Figure 12.5c, A+C). In contrast, high case fatality ratios caused a high probability of pathogen extinction after the initial spread due to overexploitation of the host by disease-induced deaths leading to stochastic fade-outs (Figure 12.5c, B+D). Insights obtained with such epidemiological IBMs are essential for the planning of diseasse prevention and control measures.

12.6.3 Wild tobacco–insect model (TIMELY)

Many plants can produce toxic chemicals to defend themselves against insect herbivores. Since producing these chemicals is costly and thus might reduce a plant's growth and, more generally, fitness, many species use induced defences, which are triggered by an actual herbivore attack. One might assume that plants should react as quickly as possible after the herbivore attack with the onset of chemical defence to prevent tissue loss to the hebivores (Karban et al. 1999). However, a wide variety of delays is observed in nature (Mathur et al. 2011). Backmann et al. (2019) hypothesised that, if the attacked plants grow in dense cohorts with strong competition, it may be beneficial for a plant to delay its chemical response until the herbivore larvae are mobile enough to move away and also to damage neighbouring plants, thus reducing the competition with the focal plant. Thus, delayed responses could reduce the competitive pressure on the focal plant (Figure 12.6).

The purpose of the model by Backmann et al. (2019) was to explore this hypothesis. The developed IBM depicted plant–herbivore interactions using as a model system interactions between the wild tobacco species *Nicotiana attenuata* and its specialised invertebrate herbivore, caterpillars of the moth *Manduca sexta*. This is one of the best-studied systems for induced defences, with a large amount of data collected both in the field and in the laboratory on the relevant aspects (from physiology and chemical responses to ecological dynamics), allowing for model parameterisation.

The IBM is based on an existing model of intraspecific plant competition (May et al. 2009) that was further developed to include plant defences, fitness costs, and detailed representation of the herbivore's demography. The model's entities are tobacco plants, larvae, and grid cells. Plants are characterised by their position, above- and belowground biomass, their defence level, and the delay time (time between a larva attacking the plant and the plant starting defence production). Larvae are characterised by their biomass, age, and position. The model represents an area of 250 × 250 square grid cells that are used to model zones of influence in a discretised way (May et al. 2009). Time step is 4 hours; simulations were run for 27 days, corresponding to a caterpillar's life cycle.

In each time step the following processes were scheduled for the plants: growth, competition for resources, production of defence compounds if attacked, and death as a result of being completely eaten by a caterpillar. The underlying model of plant competition is based on a two-layer zone of influence (ZOI) approach, in which one layer represents belowground competition of plants for nutrients and water (size-symmetric) and the second layer represents aboveground competition for light mainly

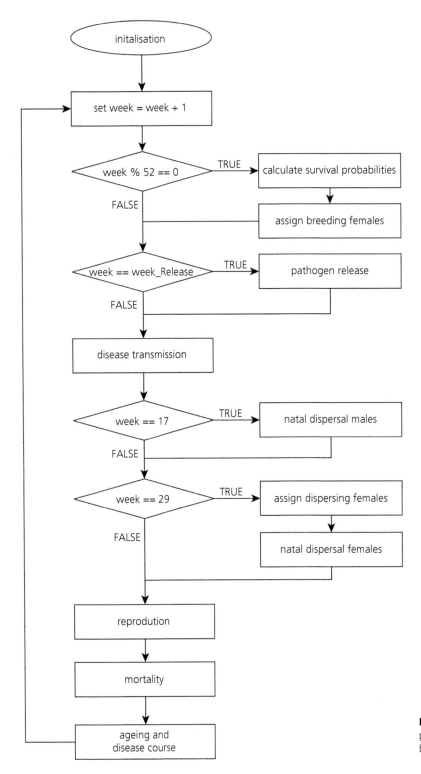

Figure 12.4 Flowchart depicting main processes and their timing in the wild boar—CSF virus model.

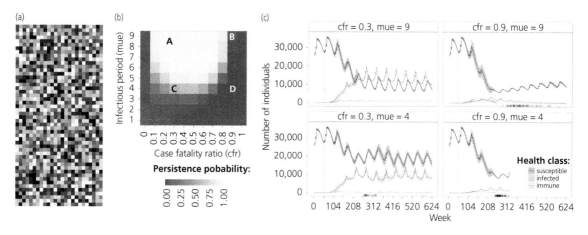

Figure 12.5 (a) Heterogeneous resource landscape depicting the mean number of reproducing wild boar females per raster cell (i.e. the darker the grey scale, the higher host density). (b) Effects of the infectious period (mue, in weeks) and case fatality ratio (cfr) on disease persistence and (c) exemplary temporal dynamics of the three health classes (i.e. susceptible, infected, immune) for specific parameter combinations. (b) Disease persistence was estimated using a 10-year threshold, that is, the disease was defined as 'persistent' only in those simulation runs in which any infected individuals were present 10 years after the pathogen was released to the naive population. Probability of persistence for each parameter combination was calculated as the proportion of runs that were classified as 'persistent' out of 200 simulation runs. Letters indicate the respective parameter combination used in (c). (c) Number of individuals per health class (mean ± sd), estimated for each week of the simulation run (i.e. as long as any infected individuals were present after the initial infection). One wild boar group was initially infected during the second year (indicated by the vertical dotted line). The rugs on the x-axis visualise pathogen extinction events.

(size-asymmetric). The radii of ZOI were allometrically scaled to the respective above- and belowground plant biomass. When, after an insect attack, a plant is induced it starts investing a proportion (30%) of its metabolic resources into defence production; the rest is invested into growth and reproduction.

Larvae initially stayed on their host plant, where they fed and grew. After having reached a certain age and weight, they became mobile enough to leave the host plant if it initiated chemical defence. Following processes are modelled for larvae in each time step: growth, mortality, and host switching. Caterpillars grew exponentially and consumed plant mass proportionally to their weight. When the larvae reached their maximum size, they pupated and were inactive in the model. Caterpillar growth and instar progression were negatively affected by the plant defence compounds. Caterpillars' per-time-step mortality scaled proportionally with their current host plant's defence level and inversely with the caterpillar's body mass. If the plant defence level reached a certain threshold, caterpillars could switch the host if they were heavy enough to move to another plant. The next host was chosen randomly among all plants within the movement radius of the larva. Larvae switching the hosts incurred higher mortality.

After testing the model by exploring the interaction of two plants, 400 plants were simulated. Either all 400 plants were assigned the same delay time or they were assigned random delay times. Plants were assigned to random positions. Larvae were placed on plants randomly, and several levels of larvae densities were tested. The mean number of larval host switches peaked at the intermediate time delay levels (3–5 days). Interestingly, by using a genetic algorithm, where the success of having a certain delay time in terms of the final biomass determined the proportion of this delay time in the following generation, the same delay times, 3–5 days, emerged as being dominant. These findings demonstrate that in contrast to former assumptions, under heavy competitive pressure, intermediate delay times are more favourable than an instantaneous defence reaction. Under high larval densities longer delay times were favoured, whereas under low densities the delay times decreased as the number of generations passed (Figure 12.7).

This model demonstrates several features of IBMs: the reuse of an existing model, which indicates a maturation of the IBM approach, as we increasingly can base IBMs on existing building blocks instead of always having to start from scratch (Thiele and Grimm 2015). Furthermore, IBMs can combine both realistic setting and highly stylised representations to explore ideas and test hypotheses. Moreover, IBMs can, in combination with genetic algorithms or neural networks, simulate the emergence of certain

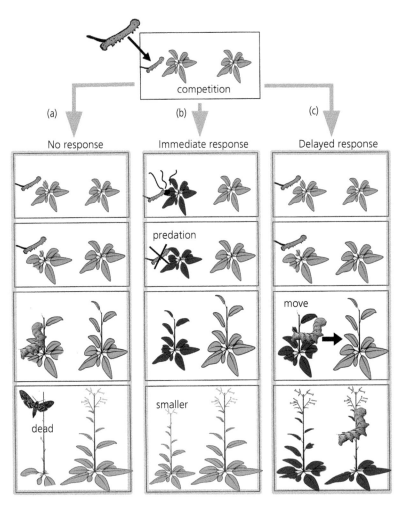

Figure 12.6 Schematic representation of the possible pathways in the TIMELY model. Plants with induced chemical defence are violet; plants without defence are green. Three possible ways a plant can react to the attack by a larva: (a) the plant does not produce defence compounds and as a result of herbivory it is dead; (b) the plant reacts immediately by producing defence compounds, which leads to the death of the larva but also lowers final biomass of the plant compared to its competitor; (c) the plant reacts after a certain delay time, causing the larva to move to the neighbouring plant and as a result having higher final biomass. Pathway (c) was the one that emerged in simulations based on genetic algorithm. (The figure is adapted from Backman et al. (2019))

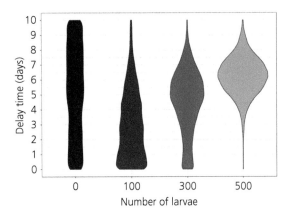

Figure 12.7 Violin plots of the final delay times obtained under different initial larvae number. The genetic algorithm started with randomly assigned delay times in the first generation and ran for 300 generations. (The figure is adapted from Backman et al. (2019))

behaviours and how they depend on environmental conditions and the trade-offs individual organisms should consider.

12.7 Conclusions

IBMs describing the demography of populations, that is, considering age-dependent vital rates, were developed for a wide variety of species differing in their ecology and life history (e.g. Stephens et al. 2002; Radchuk 2012). It was suggested, however, that IBMs are more beneficial for species with complex life histories, for example social or territorial species (e.g. wild dogs: Gusset et al. 2009; root voles: Radchuk et al. 2016a; marmots: Stephens et al. 2002) or complex movement behaviour (e.g. Eurasian lynx: Kramer-Schadt et al. 2004; lizard *Tiliqua rugosa*: Malishev et al. 2018; Fender's blue butterfly: McIntire et al. 2007).

On the other hand, species with a comparatively simple ecology can usually be modelled sufficiently well with matrix projection models (see e.g. Caswell 2002; Chapter 9) as long as local interactions and adaptive behaviour can be neglected. However, as we demonstrate in this chapter, this is rarely the case under natural conditions. Therefore, models that fail to consider the consequences of these processes for population dynamics may result in distorted projections of the population dynamics (Meli et al. 2014), and the IBM approach is often beneficial for making demographic projections.

In our examples of IBMs we focused mainly on animal systems, but it does not mean that these tools are used exclusively in zoology. In fact, IBMs investigating demography were applied to diverse taxa: plants (Berger and Hildenbrandt 2000; May et al. 2009), mammals (Stephens et al. 2002; Kramer-Schadt et al. 2004; Gusset et al. 2009; Radchuk et al. 2016a), birds (Green et al. 1996; Penteriani et al. 2005), reptiles (Malishev et al. 2018), microbes (Kreft et al. 2013; Banitz et al. 2015; Hellweger et al. 2016), fish (Harvey and Railsback 2007; Piou and Provost 2012a, 2012b; Frank and Baret 2013), and arthropods (McIntire et al. 2007; Radchuk 2012; Meli et al. 2014).

IBMs are indispensable to properly depict the demography of species characterised by complex interactions in time and space, to flexibly handle management scenarios, or if asking eco-evolutionary questions. IBMs are in general needed if a model is supposed to predict the response of a population to new conditions because they allow for emergence of demographic rates from the behaviour of individuals instead of imposing them (Radchuk et al. 2019). These advantages of IBMs, however, come with the costs associated with the data required to parameterise them and time needed to develop and properly test such complex models. These costs can be partially alleviated given the increase in computer power in recent decades, alternative ways to collect the data for model parameterisation (GPS tracking of many individuals, remote sensing, next generation sequencing), and if the modules of already developed models would be reused when developing a new one. Indeed, a large share of time that goes into model developing and testing can be saved if modellers would reuse the existing model blocks developed for the same processes in species with similar ecologies (Thiele and Grimm 2015).

Better understanding of the factors affecting the demography of species can be achieved by contrasting IBMs with other model types, for example integral-projection models (Chapter 12) or matrix projection models. By doing so systematically across species with different ecologies and for models differing in their purposes we would be able to derive a set of species characteristics and model purposes for which one or the other model type may be a better trade-off between the costs of model development and the insight gained. However, we are still behind in terms of such comparative analyses. Until then, given the necessary data are available, IBMs should be preferred if the system is characterised by nonlinear dynamics, when modelling the demography of small or declining populations, if inter- and intraspecific interactions in time and space are crucial for capturing the ecology of the species, and if a species is characterised by a non-negligible adaptive potential.

12.8 Note on Supplementary Comparison Supplementary list

The following supplementary materials for this chapter are available to download for free at the companion website:https://www.oup.com/companion/SalgueroGamelonDM

Supplementary Material 1. ODD of vole–mustelid model.

Supplementary Material 2. NetLogo code of vole–mustelid model.

Supplementary Material 3. ODD of wild boar–CSF virus model.

Supplementary Material 4. NetLogo code of wild boar–CSF virus model.

Supplementary Material 5. ODD of TIMELY model.

Supplementary Material 6. NetLogo code of TIMELY model.

NetLogo codes and ODDs of all models are uploaded to the CoMSES model library (https://www.comses.net/) and can be found at the following links: https://www.comses.net/codebases/4b484186-d8fb-4307-a710-fc05daa36afa/releases/1.0.0/ (vole–mustelid model),https://www.comses.net/codebases/82f2b53e-ae0e-4ac0-9e06-d541ecb4d1b6/releases/1.0.0/(wild boar–CSF virus model), and https://www.comses.net/codebases/a6ba9582-da40-4323-b4c9-3a45def89db1/releases/1.0.0/(TIMELY model).

References

Aars, J. & Ims, R.A. (2002) Intrinsic and climatic determinants of population demography: the winter dynamics of tundra voles. *Ecology*, **83**, 3449–56.

Andreassen, H.P., Glorvigen, P., Rémy, A., & Ims, R. A. (2013) New views on how population-intrinsic and community-extrinsic processes interact during the vole population cycles. *Oikos*, **122**, 507–15.

Backmann, P., Grimm, V., Jetschke, G., et al. (2019) Delayed chemical defense: timely expulsion of herbivores can reduce competition with neighboring plants. *The American Naturalist*, **193**, 125–39.

Banitz, T., Gras, A., & Ginovart, M. (2015) Individual-based modeling of soil organic matter in NetLogo: transparent, user-friendly, and open. *Environmental Modelling and Software*, **71**, 39–45.

Beissinger, S.R. & Westphal, M.I. (1998) On the use of demographic models of population viability in endangered species management. *Journal of Wildlife Management*, **62**, 821–41.

Berger, U. & Hildenbrandt, H. (2000) A new approach to spatially explicit modelling of forest dynamics: spacing, ageing and neighbourhood competition of mangrove trees. *Ecological Modelling*, **132**, 287–302.

Caswell, H. (2002) Matrix Population Models: Construction, Analysis and Interpretation. Sinauer Associates, Sunderland, MA.

Caughley, G. (1994) Directions in conservation biology. *Journal of Animal Ecology*, **63**, 215–44.

DeAngelis, D.L. & Diaz, S.G. (2019) Decision-making in agent-based modeling: a current review and future prospectus. *Frontiers in Ecology and Evolution*, **6**, 237.

DeAngelis, D.L. & Grimm, V. (2014) Individual-based models in ecology after four decades. *F1000Prime Reports*, **6**, 39.

Dieckmann, U. & Doebeli, M. (1999) On the origin of species by sympatric speciation. *Nature*, **400**, 354–7.

Donalson, D.D. & Nisbet, R.M. (1999) Population dynamics and spatial scale: effects of system size on population persistence. *Ecology*, **80**, 2492–507.

Dunlop, E.S., Heino, M., & Dieckmann, U.L.F. (2009) Eco-genetic modeling of contemporary life-history evolution. *Ecological Applications*, **19**, 1815–34.

Eisinger, D. & Thulke, H.-H. (2008) Spatial pattern formation facilitates eradication of infectious diseases. *Journal of Applied Ecology*, **45**, 1321–9.

Frank, B.M. & Baret, P.V. (2013) Simulating brown trout demogenetics in a river/nursery brook system: the individual-based model DemGenTrout. *Ecological Modelling*, **248**, 184–202.

Fu, C., Travers-Trolet, M., Velez, L., et al. (2018) Risky business: the combined effects of fishing and changes in primary productivity on fish communities. *Ecological Modelling*, **368**, 265–76.

Galic, N., Grimm, V., & Forbes, V.E. (2017) Impaired ecosystem process despite little effects on populations: modeling combined effects of warming and toxicants. *Global Change Biology*, **23**, 2973–89.

Gienapp, P., Teplitsky, C., Alho, J.S., Mills, J.A., & Merilä, J. (2008) Climate change and evolution: disentangling environmental and genetic responses. *Molecular Ecology*, **17**, 167–78.

Green, R.E., Pienkowski, M.W., & Love, J.A. (1996) Long-term viability of the re-introduced population of the white-tailed eagle *Haliaeetus albicilla* in Scotland. *Journal of Applied Ecology*, **33**, 357–68.

Grimm, V. & Berger, U. (2016) Structural realism, emergence, and predictions in next-generation ecological modelling: synthesis from a special issue. *Ecological Modelling*, **326**, 177–87.

Grimm, V. & Railsback, S.F. (2012) Pattern-oriented modelling: a 'multi-scope' for predictive systems ecology. *Philosophical Transactions of the Royal Society B: Biological Sciences*, **367**, 298–310.

Grimm, V., Berger, U., Bastiansen, F., et al. (2006) A standard protocol for describing individual-based and agent-based models. *Ecological Modelling*, **198**, 115–26.

Grimm, V., Berger, U., DeAngelis, D.L., Polhill, J.G., Giske, J., & Railsback, S.F. (2010) The ODD protocol: a review and first update. *Ecological Modelling*, **221**, 2760–8.

Grimm, V., Augusiak, J., Focks, A., et al. (2014) Towards better modelling and decision support: documenting model development, testing, and analysis using TRACE. *Ecological Modelling*, **280**, 129–39.

Grimm, V., Railsback, S.F., Vincenot, C.E., et al. (2020) The ODD protocol for describing agent-based and other simulation models: a second update to improve clarity, replication, and structural realism. *Journal of Artificial Societies and Social Simulation*, **23**, 7.

Gusset, M., Jakoby, O., Muller, M.S., Somers, M.J., Slotow, R., & Grimm, V. (2009) Dogs on the catwalk: modelling re-introduction and translocation of endangered wild dogs in South Africa. *Biological Conservation*, **142**, 2774–81.

Hagenaars, T.J., Donnelly, C.A., & Ferguson, N.M. (2004) Spatial heterogeneity and the persistence of infectious diseases. *Journal of Theoretical Biology*, **229**, 349–59.

Harvey, B.C. & Railsback, S.F. (2007) Estimating multi-factor cumulative watershed effects on fish populations with an individual-based model. *Fisheries*, **32**, 292–8.

Hellweger, F.L., Van Sebille, E., Calfee, B.C., et al. (2016) The role of ocean currents in the temperature selection of plankton: insights from an individual-based model. *PLOS One*, **11**, e0167010.

Hemelrijk, C.K. & Hildenbrandt, H. (2015) Diffusion and topological neighbours in flocks of starlings: relating a model to empirical data. *PLOS One*, **10**, e0126913.

Karban, R., Agrawal, A.A., Thaler, J.S., & Adler, L.S. (1999) Induced plant responses and information content about risk of herbivory. *Trends in Ecology and Evolution*, **14**, 443–7.

Kramer-Schadt, S., Revilla, E., Wiegand, T., & Breitenmoser, U. (2004) Fragmented landscapes, road mortality and patch connectivity: modelling influences on the dispersal of Eurasian lynx. *Journal of Applied Ecology*, **41**, 711–23.

Kramer-Schadt, S., Fernández, N., Eisinger, D., Grimm, V., & Thulke, H.H. (2009) Individual variations in infectiousness explain long-term disease persistence in wildlife populations. *Oikos*, **118**, 199–208.

Kreft, J.U., Plugge, C.M., Grimm, V., et al. (2013) Mighty small: observing and modeling individual microbes becomes big science. *Proceedings of the National Academy of Sciences*, **110**, 18027–8.

Lange, M. & Thulke, H.H. (2017) Elucidating transmission parameters of African swine fever through wild boar carcasses by combining spatio-temporal notification data and agent-based modelling. *Stochastic Environmental Research and Risk Assessment*, **31**, 379–91.

Lidicker, W.Z. (1988) Solving the enigma of Microtine cycles. *Journal of Mammalogy*, **69**, 225–35.

Lloyd-Smith, J.O., Schreiber, S.J., Kopp, P.E., & Getz, W.M. (2005) Superspreading and the effect of individual variation on disease emergence. *Nature*, **438**, 355–9.

Malishev, M., Bull, C.M., & Kearney, M.R. (2018) An individual-based model of ectotherm movement integrating metabolic and microclimatic constraints. *Methods in Ecology and Evolution*, **9**, 472–89.

Mathur, V., Ganta, S., Raaijmakers, C.E., Reddy, A.S., Vet, L.E.M., & Van Dam, N.M. (2011) Temporal dynamics of herbivore-induced responses in *Brassica juncea* and their effect on generalist and specialist herbivores. *Entomologia Experimentalis et Applicata*, **139**, 215–25.

May, F., Grimm, V., & Jeltsch, F. (2009) Reversed effects of grazing on plant diversity: the role of below-ground competition and size symmetry. *Oikos*, **118**, 1830–43.

McIntire, E.J.B., Schultz, C.B., & Crone, E.E. (2007) Designing a network for butterfly habitat restoration: where individuals, populations and landscapes interact. *Journal of Applied Ecology*, **44**, 725–36.

Meli, M., Palmqvist, A., Forbes, V.E., Groeneveld, J., & Grimm, V. (2014) Two pairs of eyes are better than one: combining individual-based and matrix models for ecological risk assessment of chemicals. *Ecological Modelling*, **280**, 40–52.

Penteriani, V., Otalora, F., Sergio, F., & Ferrer, M. (2005) Environmental stochasticity in dispersal areas can explain the 'mysterious' disappearance of breeding populations. *Proceedings of the Royal Society B: Biological Sciences*, **272**, 1265–9.

Piou, C. & Prévost, E. (2012a) A demo-genetic individual-based model for Atlantic salmon populations: model structure, parameterization and sensitivity. *Ecological Modelling*, **231**, 37–52.

Piou, C. & Prévost, E. (2012b) Contrasting effects of climate change in continental vs. oceanic environments on population persistence and microevolution of Atlantic salmon. *Global Change Biology*, **19**, 711–23.

Radchuk, V. (2012) *Dealing with Biological Complexity in Population Viability Analysis: Lessons from Two Endangered Butterfly Species*. Univertsité Catholique de Louvain, Louvain-la-Neuve.

Radchuk, V., Ims, R.A., & Andreassen, H.P. (2016a) From individuals to population cycles: the role of extrinsic and intrinsic factors in rodent populations. *Ecology*, **97**, 720–32.

Radchuk, V., Oppel, S., Groeneveld, J., Grimm, V., & Schtickzelle, N. (2016b) Simple or complex: relative impact of data availability and model purpose on the choice of model types for population viability analyses. *Ecological Modelling*, **323**, 87–95.

Radchuk, V., Kramer-Schadt, S., & Grimm, V. (2019) Transferability of mechanistic ecological models is about emergence. *Trends in Ecology & Evolution*, **34**, 487–8.

Railsback, S.F. (2001) Concepts from complex adaptive systems as a framework for individual-based modelling. *Ecological Modelling*, **139**, 47–62.

Railsback, S.F. & Grimm, V. (2019) *Agent-Based and Individual-Based Modelling: A Practical Introduction*. Princeton University Press, Princeton, NJ.

Scherer, C. (2019) *Infection on the Move: Individual Host Movement Drives Disease Persistence in Structured Landscapes*. University of Potsdam, Potsdam.

Schlüter, M., Baeza, A., Dressler, G., et al. (2017) A framework for mapping and comparing behavioural theories in models of social-ecological systems. *Ecological Economics*, **131**, 21–35.

Sciaini, M., Fritsch, M., Scherer, C., & Simpkins, C.E. (2018) NLMR and landscapetools: an integrated environment for simulating and modifying neutral landscape models in R. *Methods in Ecology and Evolution*, **9**, 2240–8.

Stephens, P.A., Frey-Roos, F., Arnold, W., & Sutherland, W.J. (2002) Model complexity and population predictions. The alpine marmot as a case study. *Journal of Animal Ecology*, **71**, 343–61.

Thiele, J. & Grimm, V. (2015) Replicating and breaking models: good for you and good for ecology. *Oikos*, **124**, 691–6.

Turchin, P. & Hanski, I. (1997) An empirically based model for latitudinal gradient in vole population dynamics. *The American Naturalist*, **149**, 842–74.

Vincenot, C.E. (2018) How new concepts become universal scientific approaches: insights from citation network analysis of agent-based complex systems science. *Proceedings of the Royal Society B: Biological Sciences*, **285**, 20172360.

CHAPTER 13

Survival analyses

Sarah Cubaynes, Simon Galas, Myriam Richaud, Ana Sanz Aguilar, Roger Pradel, Giacomo Tavecchia, Fernando Colchero, Sebastien Roques, Richard Shefferson, and Carlo Giovanni Camarda

13.1 Introduction

Fitness differences among individuals are the bedrock of ecological and evolutionary dynamics (Stearns 1992). Survival is without doubt one major component of fitness, which makes survival analyses a key tool for demographers, ecologists, and evolutionary biologists (Metcalf and Pavard 2007).

Assessing survival is not always an easy task. This is because individuals die only once in a lifetime, therefore precluding repeated measurements on the same individual, and death is rarely directly observed. However, a diversity of monitoring techniques exists to gather survival data for species across the Tree of Life, from short-living lab organisms and plants exhibiting dormancy, to long-living and elusive wild vertebrates (see Chapters 4 and 5). Each monitoring technique leads to a peculiar type of survival data, and a plethora of methods exist to estimate survival, from simple nonparametric estimators to more or less complex semiparametric and fully parametric models fitted in a continuous or in a discrete manner (reviewed in e.g. Wienke 2010; Miller 2011; Klein et al. 2016; Canudas-Romo et al. 2018; Cox 2018). The choice of method is guided by the type of survival data (e.g. collecting age at death or monitoring live individuals with perfect or imperfect detection), the species life history (e.g. single or numerous stages or ages), and the environment it experiences (e.g. controlled conditions vs. variable environments).

The aim of this chapter is not to provide another extensive review of the existing techniques for survival analyses, but rather to illustrate and contrast the most commonly used methods to estimate survival across the Tree of Life. The focus is placed on eight case studies of survival analyses in lab organisms, free-ranging animal and plant populations, and human populations. While it is relatively straightforward to gather survival data and assess survival using simple models under controlled lab conditions (Klein 2016; see section 13.2), monitoring survival in free-ranging populations often requires more sophisticated capture–mark–recapture (CMR) techniques to deal with imperfect detection of individuals (Williams et al. 2002; see section 13.3). Indeed, survival data are often 'incomplete' in free-ranging populations, timing and cause of death can be hard to assess, and multiple environmental factors are at play in influencing survival (see section 13.3; see also Chapters 4, 5, and 6). Humans are an exception, with the existence of several consequent databases with perfect knowledge of age and cause of death for several human populations, such as the Human Mortality Database (University of California, Berkeley (USA), and Max Planck Institute for Demographic Research (Germany) (2019). A great diversity of mortality models has been developed specifically to analyse age at death data in human demography (e.g. Canudas-Romo et al. 2018; see section 13.4).

13.2 What is survival analysis and why do we need it?

13.2.1 Time matters

Survival analysis is used to analyse the time up to a specific event that is going to occur. The event is generally death, but it can be of another kind (e.g. reproduction, migration, or exposure to a pathogen). Survival data usually involve following a set of individuals over

Sarah Cubaynes et al., *Survival analyses*. In: *Demographic Methods Across the Tree of Life*.
Edited by Roberto Salguero-Gómez and Marlène Gamelon, Oxford University Press.
© Oxford University Press (2021). DOI: 10.1093/oso/9780198838609.003.0013

a specific period of time and recording the time of occurrence of the event of interest.

13.2.2 Censoring and missing data

By nature, survival data have some particularities: survival cannot take a negative value, survival data are often censored, and missing data are frequent. Right-censoring occurs when an individual remains alive after the end of the study or drops out of the study, for example due to permanent emigration (Klein et al. 2006). Left-censoring occurs when we cannot observe the time when the event occurred: for example, an individual was already exposed to a pathogen before the study started (Klein et al. 2006). Other types of missing data are common, because individuals cannot always be observed at all occasions (see Chapter 5 and section 13.3). We need survival analyses to avoid introducing bias, because ordinary linear regression cannot effectively handle the censoring of observations and missing data (but see section 13.3).

13.2.3 Known fate data versus imperfect detection

In the presence of data referred to as 'known fate' data, survival can be modelled in a continuous manner using continuous distributions (e.g. Gompertz, Weibull, Makeham, Siler). The survival function $S(t)$ is the probability that an individual survives up to a certain time t. The hazard function $h(t)$ represents the instantaneous event rate for an individual who has already survived to time t. Both functions are related: the hazard relates to the death rate, while survival reflects its cumulative nonoccurrence. One might choose to model survival when the data involve counts of individuals alive at different points in time, while mortality models can be preferred to model age at death data.

Alternatively, CMR models are used to estimate survival when the data involve individuals missing at certain monitoring occasions and unknown time at death (Lebreton et al. 1992; see Chapter 5). CMR models involve estimating survival probability in between consecutive occasions as a function of a set of covariates, while accounting for the imperfect detection of individuals at each occasion (e.g. a marked individual might be present in the study area at a given occasion but missed). Fully parametric or semiparametric approaches are available to model survival, transition, or detection in CMR models (see examples in Section 13.3).

In both cases, survival or mortality can be modelled using a nonparametric (see example in section 13.2), semiparametric (see example in section 13.4), or fully parametric (see examples in sections 13.3 and 13.4) model.

13.3 Survival analyses in the lab: when it's (almost) all under control

13.3.1 Monitoring survival in the lab

In the lab, gathering survival data usually involves counting the number of individuals alive at different times or simply collecting the exact time at death. Collecting survival data is generally easier than in the wild because individuals are available for monitoring at all times during the study, and the influence of external variables is limited and often controlled. A common objective is to analyse differences in survival functions between groups of individuals that present different characteristics: for example various genotypes; groups of individuals exposed to various treatments; and age, sex, and life history differences.

13.3.2 Kaplan–Meier estimator and log-rank tests

Laboratory experiments generate data sometimes referred to as 'known fate' data, in which the probability of detection is 1 and thus does not need to be accounted for when estimating survival. The survival function $S(t)$ is usually estimated using the nonparametric Kaplan–Meier (KM) estimator (Kaplan and Meier 1958). For each time interval, it is calculated as the number of individuals surviving divided by the number of individuals at risk. Censored individuals who have dropped out of the study are not counted in the denominator. The cumulative survival probability is calculated by multiplying probabilities of surviving from one interval to the next. A plot of the KM survival probability against time provides a useful summary that can be used to estimate parameters such as median survival time (Figure 13.1). The smaller the time intervals, the smoother the survival curve.

Graphically, a vertical gap between the survival curves of different groups means that at a specific time point one group has a greater fraction of individuals surviving, while a horizontal gap means that it took longer for one group to experience a certain fraction of deaths (Figure 13.1). The log-rank test (Peto et al. 1977) is often used to compare the survival distributions of two or more groups. It is a nonparametric test based

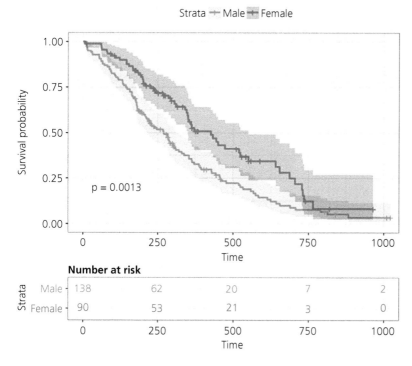

Figure 13.1 Example of KM survival curves for male and female patients with advanced lung cancer (data from Loprinzi et al. 1994). KM estimator (plain line) with 95% confidence intervals (shaded area) is provided for each group against time. The number of patients at risk at different times is given for each group. Vertical lines represent censored individuals. The p-value of the log-rank test indicates a significant difference between male and female patients' survival rates.

on a chi-square statistic, which makes no assumptions about the survival distributions. It assumes that the groups have the same survival as the null hypothesis. When the log-rank statistic is large, it is evidence for a significant difference in the survival times between the groups.

KM curves and log-rank tests are very useful in assessing whether a categorial covariate (e.g. treatment A vs. treatment B; males vs. females) affects survival. However, it does not allow investigating the effects of multiple or continuous covariates (e.g. weight, age) and to know how much more at risk one group is than another.

13.3.3 Cox's proportional hazards to adjust for covariates

An alternative method is Cox's proportional hazards regression analysis (Cox 1972). It is a semiparametric method which can be used to assess simultaneously the effect of several risk factors (both categorical and continuous covariates) on survival time. It is analogous to a multiple regression in which the response variable is the hazard measuring the instantaneous rate of the event. It assumes a constant proportional hazard across groups over time; that is, the ratio of the risk of dying at a particular point in time in one

group over another group is constant over time. Proportionality tests are used to evaluate this assumption (e.g. Miller 2011; see Supplementary Information S1 for an example; please go to www.oup.com/comp anion/SalgueroGamelonDM). Models allowing for different assumptions, such as accelerated failure time (AFT) models, are discussed for example in Kirkwood and Sterne (2003) and Klein et al. (2016).

13.3.4 Example of a stress assay in an unconventional resistant organism: the tardigrade

Tardigrades hold their own phylum that stands in between the phylum Arthropoda and Nematoda (Figure 13.2). They are renowned organisms for their ability to cope with the harshest environments, such as exposure to organic solvents, extreme temperatures (from −272 to 151°C), or high radiation doses, at any stage of their life (e.g. Jönsson et al. 2005). Most notably, some tardigrades have survived a 10-day flight in the vacuum of space and high-pressure equivalent to a depth of 180 km below the surface of the earth (Ono et al. 2016). However, it seems that the genetic 'toolbox' that ensures the uncommon resistance of these organisms to extreme stress is not shared by all tardigrade species. The resistance of tardigrade

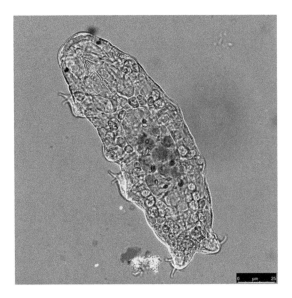

Figure 13.2 Laser confocal image of an adult tardigrade *Hypsibius exemplaris*. (Photo M. Richaud.)

species can be assessed and compared under controlled conditions in the lab.

In this example, we illustrate the use of the KM survival curve and log-rank test to evaluate the effect of the chemical stressor on tardigrades survival time. We use a Cox Proportional Hazard model to assess the effect of the chemical stressor while taking into account tardigrades age. The full procedure including data and R script to run the analysis using the R package *survival* is provided in Supplementary Information S1.

13.4 Survival analyses in the wild: dealing with uncertainty and variable environments

In free-ranging animal and plant populations, CMR field methods are generally used to gather survival data (procedure described in Chapter 5). A plethora of CMR models exist to estimate demographic parameters (including survival, access to reproduction, dispersal) while accounting for imperfect detection of individuals (e.g. Lebreton et al. 1992; Williams et al. 2002; Schaub et al. 2004; Pradel et al. 2005).

Hereafter, we introduce the principle of CMR analysis and the main types of CMR models. We then provide an overview of their use in animal and plant demography with step-by-step study cases.

13.4.1 Overview of CMR models

Principle

Within a typical CMR protocol designed to estimate survival probabilities, individuals are sampled on discrete occasions (often $\geq$ three occasions), at which they may be detected or not. Data collected in the field are then encoded into encounter histories organised by individual or cohort (see details in White and Burnham 1999). The simplest way to codify the encounter histories is by using binary codes '1' and '0' specifying if the individual has been detected or not. However, when additional information reflecting the state of the individual (e.g. breeder vs. nonbreeder, different resighting sites, epidemiological states) is collected, encounter histories may include additional codes (see multistate and multi-event capture–recapture frameworks below). Moreover, individual covariates can also be recorded and included at the end of the individual encounter histories either in order to categorise individuals (e.g. males and females, age class, different populations) or by indicating a particular quantitative individual trait (e.g. size) (White and Burnham 1999).

Models for the analysis of CMR data are classically based on multinomial distributions that describe on the one hand the biological processes and on the other hand the observational processes, conditional to the biological ones. In the simplest case, the probabilities involved are ϕ_i (probability of surviving the time interval i to $i+1$, and p_i (probability to detect a live individual at occasion i). More complex models (see sections 13.3.4.2 and 13.3.4.3) actuate multinomial rather than binomial biological processes and observational processes, but the basic structure remains the same.

Single-state models

The Cormack–Jolly–Seber (CJS) model was the first CMR model (with an earlier formulation: the Jolly–Seber (JS) model [Jolly 1965; Seber 1965]) to allow the estimation of demographic parameters under the assumption of an open population (i.e. open to birth, immigration, death, and emigration, a.k.a. BIDE. models). Under the CJS approach, apparent survival (ϕ_i) is the probability that an individual alive at time i will be alive at time $i+1$, while resighting (p_i) is the probability that an individual alive and present just before time i is seen (and marked) on that occasion (Lebreton et al. 1992). Apparent survival is generally not referred to simply as 'survival' because it may be confounded by permanent emigration of marked individuals out of the study site and by long-term vegetative dormancy in plants. Further, apparent survival is the probability to survive during the interval between two monitoring occasions and thus not at a specific monitoring occasion. In contrast, resighting is an estimator of detection of previously seen individuals during the monitoring occasion.

CMR models rely on several assumptions, the most important of which are that marked individuals are independent, tags are not lost, and the past history does not influence the fate of the individual (i.e. no trap response or negative effect of capture on survival). Pollock et al. (1990) developed a series of goodness-of-fit tests based on contingency tables for the CJS model to assess the validity of the assumptions. Later, directional tests for the detection of specific effects were derived (trap-dependence, Pradel 1993; transients, Pradel et al. 1997). The availability of informative goodness-of-fit tests makes the CJS model a common 'umbrella' model in model selection procedure (Lebreton et al. 1992). The CJS model can be easily expanded to age-dependent parameters when new animals are released at different ages or in multiple cohorts, or reduced assuming parameters are constant over time (examples in Lebreton et al. 1992).

The CJS model, and its extensions, have been, and still are, extensively used in the ecological literature, as they provide a suitable analytical framework to address multiple questions that tackle variability in survival over time (Lebreton et al. 1992), with external covariates (e.g. Grobois et al. 2008), changes in recruitment probabilities (e.g. Pradel and Lebreton 1999), recapture processes (e.g. Sanz-Aguilar et al. 2010), or evolutionary trade-offs (Tavecchia et al. 2001). However, the CJS model is based on capture–recapture data obtained from a single population and cannot explicitly frame observations of animals moving across multiple sites or between relevant biological states.

Multi-state models

Mutli-state models extend the CJS model by making the parameters state-specific (Arnason 1973; Schwartz 1993). Thus, apparent survival (ϕ_i^{jk}) is now the probability that an individual alive in state j on monitoring occasion i survives to occasion $i+1$ and, during the latter occasion, transits to state k. This parameter may be decomposed into two parameters unseen in the CJS model: state-specific survival (S_i^j) and state-transition (Ψ_i^{jk}). Here, state-specific survival (S_i^j) is the probability that an individual alive in state j at

monitoring occasion i survives to monitoring occasion $i+1$, irrespective of the state of the individual at the latter occasion. State-transition (Ψ_i^{jk}) is the probability, conditional on survival, of changing or moving from state j at time i to state k at time $i+1$, assuming that the individual survived the interval between the two occasions. Additionally, multistate models estimate resighting (p_i^j), defined as the probability that an individual alive and in state j at monitoring occasion i is also observed in that occasion.

This multisite–multistate formulation allows to address questions on survival and movement probabilities in metapopulation systems (e.g. Balkiz et al. 2010), but it also provides a suitable framework for the study of between-state transitions to study evolutionary trade-offs (Nichols and Kendall 1995) and recruitment probability (e.g. Jenouvrier et al. 2008) or to mix information of different types, that is, recoveries and recaptures (e.g. Lebreton et al. 1999). The multisite–multistate and robust design models have also been successfully applied to model dispersal to unobservable sites or transition to unobservable states—that is, places or states through which animals can move but in which they cannot be seen—or unobservable dormancy states in plants (Kendall and Nichols 2002). In the unobservable site/state, probability of detection is fixed to zero, and for this reason they are often referred to as 'ghost' sites/states (Jenouvrier et al. 2008; Balkiz et al. 2010). Grosbois and Tavecchia (2003) applied this idea to unobservable transitions. They considered the probability of dispersal as a two-step process, one accounting for the probability of leaving a given site and a second, conditional to this, incorporating the probability of settling into a new site. A similar approach was used by Schaub and Pradel (2004) to estimate the relative importance of different causes of death. However, multisite–multistate models have an important limitation: they assume that the state or the site in which an individual is observed is always certain.

Multi-event models

State uncertainty is a general problem in CMR models, but it might be particularly relevant in some studies. For example, the study of the evolutionary trade-off between survival and reproduction relies on the fact that the breeding state of the individuals observed is always determined correctly. This can be true in most cases, but sometimes it can be difficult to determine with certainty whether an animal or plant is breeding or not, is healthy or affected by a particular disease, or even if it is a male or a female. Pradel (2005) solved this problem by generalising the multisite–multistate model into a multi-event framework. In this new framework, individuals are still assumed to move across different states through survival and transition processes, but a new parameter, the initial state probability, appears. Usually, field observations are not fully informative of biological processes. By formally separating the 'real' state process from the observational process events, Pradel included a parameter to account for state uncertainty, that is, the probability to not assign or erroneously assign a given state to an individual. Multi-event models provide a solution to estimate sex- and/or age-dependent survival in species with cryptic or little age and/or sexual dimorphism (see section 13.3.4.1). Transients can be considered as a particular initial state and modelled directly using multi-event models (Genovart et al. 2012; Santidrian et al. 2017). More than uncertainty about the state of an individual, multi-event models allow to explicitly model unobserved heterogeneity among individuals. Indeed, latent state mixture models can be easily implemented within the multi-event framework, a model particularly suitable to frame individual heterogeneity or frailty (Gimenez et al. 2018a). Other approaches, for example using individual random effects, can be used to implement frailty (e.g. Cam et al. 2016; Hamel et al. 2018; see also section 13.3.2).

Multi-event and multistate models can be used to address many other ecological and evolutionary questions. If individual states can change over time, multi-event also allows modelling the transition dynamics between, for example, breeding states (Desprez et al. 2013) and epidemiological status (Benhaiem et al. 2018). Lagrange et al. (2014) developed multi-event models able to study dispersal among numerous sites for birds and amphibians. Tavecchia et al. (2012) modelled mortality due to different causes of mortality of radiotagged individuals, while accounting for the loss of the radio signal (see section 13.3.4.2). Multi-event models have been used to model survival when marks identifying individuals are lost (even totally; see Badia-Boher et al. 2019). The multi-event approach can be also used to exploit supplementary information and estimate survival, dispersal, and/or recruitment in partially monitored populations (Sanz-Aguilar et al. 2016; Tavecchia et al. 2016). Finally, trap-responses and memory effects can be modelled into the multi-event approach (Rouan et al. 2009; Pradel and Sanz-Aguilar 2012).

13.4.2 Environmental variability and individual heterogeneity in CMR models

Within a CMR model, survival, transition, and/or detection parameters can be assumed to be a function of external covariates (Lebreton et al. 1992). This formulation allows the inclusion of environmental effects in a regression-like framework, by including, for example, climatic variables, or individual time-invariant characters such as genotype. Multifactorial effects of environmental covariates can also be modelled using hierarchical models (King et al. 2009), for example to study evolutionary processes in the wild (Cubaynes et al. 2012). However, time-varying individual covariates, such as body weight, are tricky because when an individual is not detected, the value of the covariate is unknown. Inference can be based on a conditional likelihood approach using only the observed covariate values (trinomial approach; Catchpole et al. 2008), or missing values can be imputed from an underlying distribution (e.g. multiple imputation; Worthington et al. 2015). However, methods of imputation are sensitive to the underlying model and the number of missing values (Langrock and King 2013). One possibility is to discretise the covariate and use a multistate model (Fernández-Chacón et al. 2015; Gimenez et al. 2018a).

Unfortunately, we often do not measure all covariates influencing demographic parameters. Also individuals may react in different ways to environmental variation depending on unobservable individual states. This leads to unobserved (latent) individual heterogeneity (see Chapter 9). Ignoring latent individual heterogeneity may lead to flawed inference about the ecological or evolutionary processes at hand (Cam et al. 2016; Hamel et al. 2018), such as senescence patterns (Cam and Monnat 2000; Service 2000; Peron et al. 2010). Latent individual heterogeneity can be framed using finite mixtures or as individual random effects (see Gimenez et al. 2018a for a review about how to implement individual heterogeneity in CMR models).

13.4.3 Inference framework

Implementation of CMR models can be carried out using either a frequentist or Bayesian approach. A different philosophy stands behind each approach, and there is a long-standing debate about whether ecologists should use one or the other (e.g. Lele and Dennis 2009). While the frequentist approach may be faster, the Bayesian approach allows a great flexibility in the model writing, which can be useful to tackle analytical complexity, such as choosing underlying distributions for model parameters or fitting temporal random effects (e.g. Kery and Schaub 2011). Another appeal of the Bayesian approach is the possibility to include prior knowledge on biological parameters to facilitate the estimation, for example information on body weight or survival of a closely related species (MacCarthy et al. 2005). The Bayesian approach of CMR models uses the state-space modelling (SSM) formulation that clearly distinguishes the observation process (detection) from the underlying demographic process of interest (transition between states; e.g. Gimenez et al. 2007; Royle 2008; Kery and Schaub 2011), the observation process being conditional to the state process. The SSM formulation therefore allows to easily implement complex multifactorial observation processes and combine multiple sources of information (e.g. Buoro et al. 2012; see Supplementary Information S5 for an SSM formulation of the JS model).

Prior to model fitting, goodness-of-fit tests are generally performed to check the validity of the assumptions behind a CMR model, for example using the R2ucare package (Gimenez et al. 2018b). In the frequentist approach, model implementation can be carried out using program MARK and the widely used RMark package (White and Burnham 1999; Laake 2013), marked (Laake 2013), or E-SURGE (Choquet et al. 2009). Common tools for model comparisons include the Akaike information criterion (AIC) and its variants (AICc, QAIC, wAIC; Burnham and Anderson 2002), which serves to rank the models and calculate weights of evidence for each of them or for a particular effect (Burnham and Anderson 2002). Analysis of Deviance (Anodev) is also used to calculate the proportion of variance explained by a specific covariate (Grosbois et al. 2008). In the Bayesian approach, models can be implemented using program JAGS (Plummer 2003), R packages such as rjags (Plummer et al. 2018), or Bayesian Survival Trajectory Analysis (BaSTA) (Colchero et al. 2012). Posterior predictive checks can be used for performing model assessment (Chambert et al. 2014), and information criterion such as deviance information criterion (DIC) (Spiegelhalter et al. 2002) or wAIC are often used for model comparison (Hooten and Hobbs 2015). Further details about implementation in both a frequentist and Bayesian framework can be found in McCrea and Morgan (2014).

Hereafter, we develop study cases in animal and plant demography showing how to implement CMR models in a frequentist framework using program E-SURGE and R package marked (Laake et al. 2013) and in a Bayesian framework using package BaSTA and rjags.

13.4.5 Case studies in animal demography

Example 1: Estimating sex-dependent survival when sex assignment is uncertain: a multi-event model of the Balearic wall lizard (*Podarcis lilfordi*)

In this example, we consider the possibility of erroneous assignment of sex to newly captured individuals, a situation common to the monitoring of species, for example, with little sexual dimorphisms. In CMR analyses, erroneously assigning sex at the beginning of the capture history leads to bias in the estimated survival difference between the sexes. Here, our aim is to estimate sex-specific survival rates in the Balearic wall lizard, a small species endemic to the Balearic archipelago, Spain. Immature males are sometimes difficult to be sexed in the field and can be confounded with mature females of similar size. Sex-specific survival rates can be estimated by accounting for the uncertainty on sex assignment using a multi-event CMR model to separate the 'real' sex of the individuals (state) from the 'apparent' (observed) sex (event). In this model, we considered four events (type of observations) which code three states ('real' state of the individual). Data, together with step-by-step instructions to implement the models in E-SURGE and interpret the results, are provided in Supplementary Information S2.

Example 2: Survival and the issue of tag-loss: study case of the red kite (*Milvus milvus*)

Tag-loss is a common issue in wildlife monitoring of marked individuals and can lead to underestimated survival (Arnason and Mills 1981). To cope with this issue, ecologists have developed advanced methodological tools: from multiple-marking to advanced statistical methods to integrate tag-loss in the individual state (Cowen and Schwarz 2006; Tavecchia et al. 2012). Here is one case study to integrate the loss of a remote tracking device in the multi-event modelling framework.

The use of remote tracking devices (radio-satellite and GSM/GPS transmitters) to collect detailed individual history data is increasingly common in the ecological literature (see Chapters 3 and 5). A problem in estimating survival from tracking data is that the life span of the remote signal is commonly shorter than the life span of the individual that carries the device. In this case the survival probability refers to the life span of

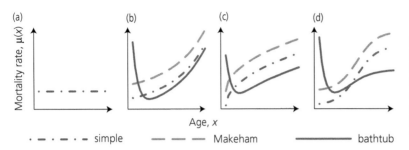

Figure 13.3 Mortality, $\mu_X(x|\theta)$, resulting from the four basic models included in BaSTA: (a) exponential; (b) Gompertz; (c) Weibull; and (d) logistic. The three different lines in each plot (except in a) show examples of the shapes that can be tested with BaSTA, namely 'simple', 'Makeham', and 'bathtub'.

the radio signal and not to the one of the animal. When animals are marked with tags or rings in addition to the radio device, their encounter history can follow the loss of the radio signal. In this example, we illustrate how multi-event models can accommodate the loss of the signal and provide unbiased estimate of survival in the presence of radio-loss or radio-failure using a real data set on red kite on the island of Mallorca (**data provided bt T. Muñoz, GOB-Mallorca**). Data, together with step-by-step instructions to implement the models in E-SURGE and interpret the results, are provided in Supplementary Information S3.

Bayesian implementation using the R package BaSTA

The R package BaSTA (Colchero et al. 2012) provides a set of tools that complement other CMR methods when users want to estimate age-specific mortality from CMR data sets where times of birth are known only for a few individuals (or none). Several parametric mortality models are available in BaSTA, including the exponential, Gompertz, Weibull, logistic, Makeham, and Gompert–Makeham models (Figure 13.3). In order to include all records in the analysis, BaSTA estimates the missing ages at birth and at death, which reduces the bias in the estimation of the mortality and cumulative survival functions. An example of implementation is provided in Supplementary Information S4.

Bayesian state-space formulation of the JS model to study stopover decisions of migratory birds using JAGS

Migratory birds cannot realise their journey between breeding and wintering areas in a single flight of thousands of kilometres and usually stop-over at places where they can replenish their energy reserves. At these stopover places individuals are not easy to detect. Studying the stopover decisions of migratory birds is a typical case where the detectability needs to be taken into account to be able to make strong ecological inference. In this example the survival between two

capture occasions (ϕ_i) is considered as the remaining probability at the stopover place, and thus $1-\phi_i$ is the departure probability between two occasions. As individuals may have arrived in the stopover area before first capture, the model needs to not be conditional on the first capture (as in the CJS model). Thus, we can use the JS model parametrised with entry probabilities noted as η_i for the probability to entry in the stopover area between time i and $i+1$ if not previously entered (Schwarz and Arnason 1996). This SSM formulation of the JS model allows an easy implementation in the Bayesian framework and a straightforward computation of the stopover duration (Lyons et al. 2016). We can also easily incorporate the effect of a weather covariate on the departure probability. The implementation of this example with R and JAGS is provided in Supplementary Information S5.

13.4.6 Case studies in plant demography

Plants do not move, but certain aspects of their ecologies, such as vegetative dormancy, variable sprouting times, and complex growth patterns, can make them just as challenging to work with as animals.

Example 1: Linear modelling of plant survival

The most common method to analyse survival using plant resighting data sets is using linear analysis under a logistic, generalised linear model (GLM) or generalised linear mixed model (GLMM) framework, with survival modelled assuming a binomial distribution (e.g. Salguero-Gómez et al. 2012). This method assumes that, at the very least, the resighting of previously observed individuals is nearly perfect, because any phenomenon decreasing redetection would be observed as mortality and yield-biased survival estimates. In cases where redetection is not perfect, some have argued that this approach is still useful provided that monitoring data sets are particularly long and large and that redetection is still above 90% (Shefferson et al. 2018). However, in studies of vegetative dormancy-prone perennials, dormancy will increasingly be confused with mortality as study length decreases. In the final year of a study, there will be no ability to differentiate the dead from the dormant. This suggests that the final 1–3 years of data in a study should be used simply to assign dormancy within the remaining data and that survival should not be estimated for those years. This loss of estimable years adds value to long data sets, particularly those over 10 years long.

We illustrate the use of linear modelling for survival analysis in plants using a case study on *Cypripedium parviflorum*, the North American small yellow lady's slipper, using data collected from 1994 to 2003 within a larger population from Illinois in the United States. These data were previously used in, for example, Shefferson et al. (2018). Using the R packages lme4 for model fitting and MuMIn for model comparison, we show that nonflowering plants have decreasing survival with increasing size, while flowering plants have increasing survival with increasing size. Full procedure including data and R script to run the analysis are provided in Supplementary Information S6.

Example 2: CMR survival analysis for plants

Plant population ecologists have long used field methods that may be considered in the same vein as mark–recapture methods in wildlife ecology. However, the application of CMR methods to plant population ecology is very recent. In one of the first studies to use mark–recapture analyses in plants, Alexander et al. (1997) faced all of these problems in a population of the Mead's milkweed, *Asclepias meadii*. This population consists of plants that grow in high densities and do not always produce aboveground tissue in a growing season. Closed population mark–recapture analysis allowed them to produce estimates of population size unbiased by these challenges. Expanding on this work, Shefferson et al. (2001) proposed the use of open population mark–recapture models to estimate annual survivorship in populations in which living individuals do not always sprout in a growing season. Since then, CMR studies have blossomed in plant population ecology, with extensions into the estimation of transition rates among life history stages (e.g. Shefferson et al. 2003), estimation of the demographic impacts of herbivory (e.g. Kéry and Gregg 2004), investigations into relationships among life history traits (e.g. Shefferson et al. 2003), tests of correlation with climatic factors (e.g. Shefferson and Tali 2007), estimation of minimal recruitment levels necessary to sustain populations (Slade et al. 2003), and theoretical papers inspired by

the problem of unobserved life stages (Kendall and Nichols 2002).

In this example, we also use the *Cypripedium parviflorum* data set, in combination with the *R* package marked (Laake et al. 2013), to investigate costs of reproduction using CMR models. Full procedure including data and R script to run the analysis are provided in Supplementary Information S7.

13.5 Mortality analysis in human populations

Modelling mortality in human populations is relatively easier than in nonhuman ones. For a given group of individuals, we often know their age-at-death, calendar year of the event, and their sex. Thanks to these reliable data sources, methodological advances have been produced since De Moivre (1725) and Gompertz (1825). These long-standing demographic and statistical developments have been often drawn by political, military, and economic reasons. In the following sections, we present a brief overview of the most common models used to describe mortality patterns over age and/or time on human mortality data.

13.5.1 Human data and assumptions

For a given sex, we usually have deaths and exposures to the risk of death arranged in two matrices, whose rows and columns are classified by age at death and year of death. The stochastic assumption behind mortality has a central role in modelling it. The most suitable distribution when we observe mortality data is the Poisson distribution. The aim of any mortality model is to seek for a parsimonious, yet satisfactory description of the so-called force of mortality $\mu_{i,j}$, given observed deaths $d_{i,j}$, and exposures $e_{i,j}$. One could estimate force of mortality in a fully nonparametric framework computing the death rates $\mu_{i,j} \approx \frac{d_{i,j}}{e_{i,j}}$. Simple plots of rates over age and/or time are good tools for a first presentation of mortality development. Commonly, rates are plotted on a logarithmic scale to better acknowledge differences (Figure 13.4). A relatively strong assumption behind this is that within the Poisson distribution mean and variance are equal. When the observed variance is larger than the theoretical one, we often attribute this feature either to overdispersion or to some hidden patterns in the data. Specific methods for coping with this issue have been proposed in the literature for all models below, though they will not be presented in this chapter. For a comprehensive

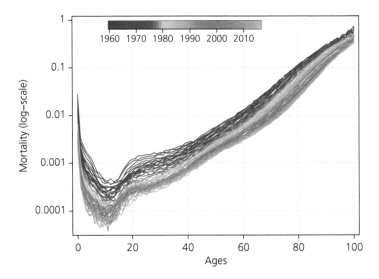

Figure 13.4 Death rates in log-scale over ages for all available years. Japanese females, 1960–2016, ages 0–100.

overview of them see, for example, Cameron and Trivedi (2013). Furthermore, binomial distribution could be used when we deal with probability of dying (deaths divided by persons-at-risk), and the multinomial distribution has been suggested for modelling mortality from a cohort perspective (Canudas Romo et al. 2018).

13.5.2 Parametric models over age

Parametric models for a suitable representation of the variation of mortality over age have been proposed since Gompertz (1825). He observed that after a certain age, a 'law of geometric progression pervades, in an approximate degree, large portions of different tables of mortality' (Gompertz 1825, p. 514). He thus suggested representing force of mortality as:

$$\mu_i = ae^{bi} \tag{13.1}$$

where a represents the mortality at time zero (usually age 30/40) and b is the rate of increase of mortality and is frequently used as a measure of the rate of ageing.

Makeham (1860) extended Gompertz's equation by adding a constant, an age-independent term, c >0, to account for risks of death that do not depend on age:

$$\mu_i = c + ae^{bi} \tag{13.2}$$

Human mortality often shows a levelling-off above certain ages (often 80) (Vaupel 1997). Logistic models have been proposed to portray this feature. Perks (1932) was the first to propose a logistic modification

of the Gompertz–Makeham models. A logistic function to model the late-life mortality deceleration can begiven by

$$\mu_i = c + \frac{ae^{bi}}{1 + \alpha e^{bi}} \tag{13.3}$$

where α captures the mortality deceleration at oldest ages. This law of mortality can be derived when heterogeneity is assumed in a proportional hazard setting. Commonly called the Gamma–Gompertz model, eqn. (13.3) is the hazard at a population level when standard mortality is described as a Gompertz and frailty values are assumed to be Gamma distributed (Wienke 2010).

A simplified version of the previous logistic law has been proposed by Thatcher et al. (1998):

$$\mu_i = c + \frac{ae^{bi}}{1 + ae^{bi}} \tag{13.4}$$

Heligman and Pollard (1980) derived a descriptive model, covering the whole age range. Here we propose a version for death rates:

$$\mu_i = A^{(i+B)^C} + De^{-E(lni - lnF)^2} + \frac{GH^i}{1 + GH^i} \tag{13.5}$$

where $A, B, \ldots, H$ are the eight parameters in the model. Each component aims to describe mortality at childhood, during young-adult ages, and at older ages. It is easy to see that such parameterisation can cause difficulties in the estimation procedure. Moreover, it would be hard to disentangle the physical meaning of each parameter (Booth and Tickle 2008).

Another three-component, competing-risk mortality model has been proposed by Siler (1983). Initially developed for animal survival data, this model has been recently used in human demography, especially for simulating possible scenarios in mortality developments (Canudas-Romo 2018). This model aims at portraying the whole of the age range with five parameters:

$$\mu_i = a_1 e^{-b_i i} + a_2 + a_3 e^{b_3 i} \qquad (13.6)$$

13.5.3 Overparametrised models: the example of the Lee–Carter model

Moving to a two-dimensional perspective and given the wealth of data, traditional demographic methods for analysing mortality surfaces, that is, data on deaths and exposures cross-classified by age and year of occurrence, tend to apply a high number of parameters, leading to all but parsimonious models. A typical example of this kind is the widely used model introduced by Lee and Carter (1992). In its original formulation, this approach reduces the complexity of the whole surface by introducing the following bilinear model for the log-death-rates:

$$\ln\left(m_{ij}\right) = \alpha_i + \beta_i \kappa_j + \varepsilon_{ij}, \quad i = l, \ldots, m \qquad (13.7)$$

$$j = l, \ldots, n$$

where αi, βi, and κj are vectors of parameters to be estimated, and εij represents the error term. Interpretation of the parameters is straightforward: αi and βi describe the general shape of mortality and the fixed rate of mortality improvement at age i respectively, and κj is a time-varying index which captures the general level of mortality. The variance εij in Lee and Carter (1992) is assumed to be constant for all i and j. As presented, the Lee–Carter (LC) model is underdetermined and requires additional constraints on the parameters to be successfully estimated. Usually, the model is centred by $\sum^{\beta_i} = 1, \sum^{\kappa_j} = 0$.

As pointed out in section 13.4.1, a Poisson assumption is more suitable for estimating mortality models, and a further development of the LC model was devoted to this issue (Brouhns et al. 2002). In Supplementary Information S7 we provide computational details for estimating the LC model within a Poisson framework.

In the last decades, further variants of this model have been proposed for enhancing several features of the model (e.g. Booth et al. 2002).

13.5.4 Semiparametric models: the example of P-splines

A compromise between simple parametric and overparametrised models could be found in the area of semiparametric statistics. Without the aim of producing estimated parameters with physical meanings and searching for a good fit to the data, smoothing approaches have been lately developed in the study of nonlinear phenomena. Among numerous options, we will mention here (and illustrate with more details in Supplementary Information S7) a methodology that is particularly suitable for the analysis of mortality developments: B-splines with penalties, known as P-splines. In a unidimensional setting, applications to mortality have been proposed by Currie et al. (2004) and Camarda (2008). Estimation of a model for the complete mortality surface will be achieved by a specific R package: MortalitySmooth (Camarda 2012). It is noteworthy that, being an extremely flexible tool, this methodology can be easily implemented for the analysis of demographic phenomena other than mortality. The main idea of the P-splines is to construct an intentionally overparametrised model and simultaneously to restrict, via a penalty, all redundant features for achieving a wisely parsimonious description of the data. Instead of smoothing a given structure, this approach used local supports such as equally spaced B-splines over ages and/or years and penalty term(s) on the associated coefficients.

13.5.5 Implementation example

We use data from the Human Mortality Database (2019) on Japanese females from the years 1960 to 2016 and from ages 0 to 100 to fit the models described above. Guidance with R script is provided in Supplementary Information S8.

13.6 Conclusions

This chapter has illustrated how the choice of a specific method for survival analysis is driven by the research question (e.g. comparing survival across groups vs. assessing the effects of environmental predictors), the species life history (more or less easy to monitor), and its living environment (controlled vs. stochastic environments), which determine the type of survival data (e.g. proportion of individuals alive, CMR data, or age at death).

With the increase in new monitoring techniques allowing the gathering of more and more detailed data at the individual level (see Chapter 5) has come the development of advanced statistical tools for survival analyses, including multi-event models and hierarchical models. More than simply assessing survival to feed population projection models (see Chapters 9, 10, 11, and 14), modern survival analyses have addressed questions about evolutionary trade-offs (e.g. Nichols and Kendall 1995), static versus dynamic heterogeneity in demographic parameters (e.g. Cam et al. 2016; Gimenez et al. 2018a), assessing senescence (e.g. Peron et al. 2010), and quantifying the heritability of demographic parameters in the wild (e.g. Papaïx et al. 2010). Addressing more and more complex questions has brought new challenges to the field of survival analyses. Among others, current methodological developments deal with methods to implement models with numerous states, predict age at death, and consider dependence among individuals when estimating survival.

References

Alexander, H.M., Slade, N.A. and Kettle, W.D. (1997). Application of mark–recapture models to estimation of the population size of plants. Ecology, 78, 1230–1237.

Arnason, A. N. (1973). The estimation of population size, migration rates and survival in a stratified population. Researches on Population Ecology, 15, 1–8.

Arnason, A. N., and Mills, K. H. (1981). Bias and loss of precision due to tag loss in Jolly–Seber estimates for mark–recapture experiments. Canadian Journal of Fisheries and Aquatic Sciences, 38(9), 1077–1095.

Badia-Boher, J. A., Sanz-Aguilar, A., De La Riva, M., et al. (2019). Evaluating European LIFE conservation projects: improvements in survival of an endangered vulture. Journal of Applied Ecology, 56(5), 1210–1219.

Balkız, O., Béchet, A., Rouan, L., et al. (2010). Experience-dependent natal philopatry of breeding greater flamingos. Journal of Animal Ecology 79, 1045–1056.

Benhaiem, S., Marescot, L., Hofer, H., et al. (2018). Robustness of eco-epidemiological capture–recapture parameter estimates to variation in infection state uncertainty. Frontiers in Veterinary Sciences 5, 197.

Booth, H., and Tickle, L. (2008). Mortality modelling and forecasting: a review of methods. Annals of Actuarial Science, 3(1–2), 3–43.

Booth, H., Maindonald, J., and Smith, L. (2002). Applying Lee–Carter under conditions of variable mortality decline. Population Studies 56, 325–336.

Brouhns, N., Denuit, M., and Vermunt, J. K. (2002). A Poisson log-bilinear regression approach to the construction of projected lifetables. Insurance: Mathematics and Economics 31, 373–393.

Buoro M., Prévost, E. and Gimenez, O. (2012). Digging through model complexity: using hierarchical models to uncover evolutionary processes in the wild. Journal of Evolutionary Biology 25(10), 2077–2090.

Burnham, K.P. and Anderson, D.R. (2002). Model Selection and Multimodel Inference: A Practical Information-Theoretic Approach, 2nd edition. Springer, New York.

Cam, E., and Monnat, J. Y. (2000). Apparent inferiority of first-time breeders in the kittiwake: the role of heterogeneity among age classes. Journal of Animal Ecology, 69(3), 380–394.

Cam, E., Aubry, L. M., and Authier, M. (2016). The conundrum of heterogeneities in life history studies. Trends in Ecology and Evolution, 31(11), 872–886.

Camarda, C. G. (2008). Smoothing Methods for the Analysis of Mortality Development. PhD thesis, Programa de Doctorado en Ingeniería Matemática. Universidad Carlos III, Departamento de Estadística, Madrid.

Camarda, C. G. (2012). MortalitySmooth: an R package for smoothing Poisson counts with P-splines. Journal of Statistical Software 50, 1–24.

Cameron, A. C., and Trivedi, P. K. (2013). Regression Analysis of Count Data (Vol. 53). Cambridge University Press, Cambridge.

Canudas-Romo, V., Mazzuco, S., and Zanotto, L. (2018). Measures and models of mortality. In: C.R. Rao (ed.), Handbook of Statistics (Vol. 39, pp. 405–442). Elsevier, Amsterdam.

Catchpole, E. A., Morgan, B. J., and Tavecchia, G. (2008). A new method for analysing discrete life history data with missing covariate values. Journal of the Royal Statistical Society: Series B (Statistical Methodology), 70(2), 445–460.

Chambert, T., Rotella, J. J., and Higgs, M. D. (2014). Use of posterior predictive checks as an inferential tool for investigating individual heterogeneity in animal population vital rates. Ecology and Evolution, 4(8), 1389–1397.

Choquet, R., Rouan, L., and Pradel, R. (2009). Program E-SURGE: a software application for fitting multi-event models. In: D.L. Thomson, E.G. Cooch, M.J. Conroy, (eds), Modeling Demographic Processes in Marked Populations (pp. 845–865). Springer, Boston, MA.

Colchero, F., Jones, O. R., and Rebke, M. (2012). BaSTA: an R package for Bayesian estimation of age-specific survival from incomplete mark–recapture/recovery data with covariates. Methods in Ecology and Evolution, 3(3), 466–470.

Cowen, L., and Schwarz, C. J. (2006). The Jolly–Seber model with tag loss. Biometrics, 62(3), 699–705.

Cox, D. R. (1972). Regression models and life-tables. Journal of the Royal Statistical Society: Series B (Methodological), 34(2), 187–202.

Cox, D. R. (2018). Analysis of Survival Data. Chapman and Hall/CRC Press, Boca Raton.

Cubaynes, S., Doutrelant, C., Grégoire, A., Perret, P., Faivre, B., and Gimenez, O. (2012). Testing hypotheses in evolutionary ecology with imperfect detection: capture–recapture structural equation modeling. Ecology, **93**(2), 248–255.

Currie, I. D., M. Durbán, and P. H. C. Eilers (2004). Smoothing and forecasting mortality rates. Statistical Modelling, **4**, 279–298.

De Moivre, A. (1725). Annuities Upon Lives: Or, The Valuation of Annuities Upon Any Number of Lives. NewsBank Readex.

Desprez, M., McMahon, C. R., Hindell, M. A., Harcourt, R., and Gimenez, O. (2013). Known unknowns in an imperfect world: incorporating uncertainty in recruitment estimates using multi-event capture–recapture models. Ecology and Evolution, **3**, 4658–4668.

Fernández-Chacón, A., Genovart, M., Álvarez, D. et al. (2015). Neighbouring populations, opposite dynamics: influence of body size and environmental variation on the demography of stream-resident brown trout (*Salmo trutta*). Oecologia, **78**, 379–389.

Genovart, M., Pradel, R., Oro, D. (2012). Exploiting uncertain ecological fieldwork data with multi-event capture–recapture modelling: an example with bird sex assignment. Journal of Animal Ecology, **81**, 970–977.

Gimenez, O., Rossi, V., Choquet, R., et al. (2007). State-space modelling of data on marked individuals. Ecological Modelling, **206**(3–4), 431–438.

Gimenez, O., Cam, E., and Gaillard, J. M. (2018a). Individual heterogeneity and capture–recapture models: what, why and how? Oikos, **127**(5), 664–686.

Gimenez, O., Lebreton, J. D., Choquet, R., and Pradel, R. (2018b). R2ucare: an r package to perform goodness-of-fit tests for capture–recapture models. Methods in Ecology and Evolution, **9**(7), 1749–1754.

Gompertz, B. (1825). XXIV. On the nature of the function expressive of the law of human mortality, and on a new mode of determining the value of life contingencies. In a letter to Francis Baily, Esq. FRS. Philosophical Transactions of the Royal Society of London, (115), 513–583.

Grosbois, V. and Tavecchia, G. (2003). Modeling dispersal with capture–recapture data: disentangling decisions of leaving and settlement. Ecology **84**, 1225–1236.

Grosbois, V., Gimenez, O., Gaillard, J.-M., et al. (2008). Assessing the impact of climate variation on survival in vertebrate populations. Biological Reviews, **83**, 357–399.

Hamel, S., Gaillard, J. M., Douhard, M., Festa-Bianchet, M., Pelletier, F., and Yoccoz, N. G. (2018). Quantifying individual heterogeneity and its influence on life-history trajectories: different methods for different questions and contexts. Oikos, **127**(5), 687–704.

Heligman, L., and Pollard, J. H. (1980). The age pattern of mortality. Journal of the Institute of Actuaries, **107**(1), 49–80.

Hooten, M. B., and Hobbs, N. T. (2015). A guide to Bayesian model selection for ecologists. Ecological Monographs, **85**(1), 3–28.

Jenouvrier, S., Tavecchia, G., Thibault, J., Choquet, R., and Bretagnolle, V. (2008). Recruitment processes in long-lived species with delayed maturity: estimating key demographic parameters. Oikos, **117**, 620–628.

Jolly, G. M. (1965). Explicit estimates from capture–recapture data with both death and immigration-stochastic model. Biometrika, **52**(1/2), 225–247.

Jönsson, K. I., Harms-Ringdahl, M., and Torudd, J. (2005). Radiation tolerance in the eutardigrade *Richtersius coronifer*. International Journal of Radiation Biology, **81**(9), 649–656.

Kaplan, E. L., and Meier, P. (1958). Nonparametric estimation from incomplete observations. Journal of the American Statistical Association, **53**(282), 457–481.

Kendall, W. L. and Nichols, J. D. (2002). Estimating state-transition probabilities for unobservable states using capture–recapture/resighting data. Ecology, **83**, 3276–3284.

Kéry, M. and Gregg, K. B. (2004). Demographic analysis of dormancy and survival in the terrestrial orchid *Cypripedium reginae*. Journal of Ecology, **92**, 686–695.

Kéry, M., and Schaub, M. (2011). Bayesian Population Analysis Using WinBUGS: A Hierarchical Perspective. Academic Press, Cambridge, MA.

King, R., Morgan, B., Gimenez, O., and Brooks, S. (2009). Bayesian Analysis for Population Ecology. Chapman and Hall/CRC Press, Boca Raton.

Kirkwood B. R., and Sterne J. A. C. (2003). Essential Medical Statistics, 2nd edition. Blackwell, Oxford.

Klein, J. P., and Moeschberger, M. L. (2006). Survival analysis: techniques for censored and truncated data. Springer Science + Business Media, Berlin.

Klein, J. P., Van Houwelingen, H. C., Ibrahim, J. G. and Scheike, T. H. (eds). (2016). Handbook of Survival Analysis. CRC Press, Boca Raton.

Laake, J. L., Johnson, D. S., and Conn, P. B. (2013). marked: an R package for maximum likelihood and Markov Chain Monte Carlo analysis of capture–recapture data. Methods in Ecology and Evolution, **4**(9), 885–890.

Lagrange, P., Pradel, R., Bélisle, M., and Gimenez, O. (2014). Estimating dispersal among numerous sites using capture–recapture data. Ecology, **95**, 2316–2323.

Langrock, R., and King, R. (2013). Maximum likelihood estimation of mark–recapture–recovery models in the presence of continuous covariates. Annals of Applied Statistics, **7**(3), 1709–1732.

Lebreton, J.-D, Burnham, K. P., Clobert, J., and Anderson, D. R. (1992). Modeling survival and testing biological hypotheses using marked animals—a unified approach with case-studies. Ecological Monographs, **62**, 67–118.

Lebreton, J.-D., Almeras, T., and Pradel, R. (1999). Competing events, mixtures of information and multistrata recapture models. Bird Study, 46,S39–S46.

Lee, R. D., and Carter, L. R. (1992). Modeling and forecasting US mortality. Journal of the American Statistical Association, **87**(419), 659–671.

Lele, S. R., and Dennis, B. (2009). Bayesian methods for hierarchical models: are ecologists making a Faustian bargain? Ecological Applications, 19(3), 581–584.

Loprinzi, C. L., Laurie, J. A., Wieand, H. S., et al. (1994). Prospective evaluation of prognostic variables from patient-completed questionnaires. North Central Cancer Treatment Group. Journal of Clinical Oncology. 12(3), 601–607.

Lyons, J. E., Kendall, W. L., Royle, J. A., Converse, S. J., Andres, B. A., and Buchanan, J. B. (2016). Population size and stopover duration estimation using mark–resight data and Bayesian analysis of a superpopulation model. Biometrics, 72(1), 262–271.

Makeham, W. M. (1860). On the law of mortality and the construction of annuity tables. Journal of the Institute of Actuaries, 8(6), 301–310.

McCarthy, M.A. and Masters, P. I. P. (2005). Profiting from prior information in Bayesian analyses of ecological data. Journal of Applied Ecology, 42(6), 1012–1019.

McCrea, R. S., and Morgan, B. J. (2014). Analysis of Capture–Recapture Data. Chapman and Hall/CRC Press, Boca Raton.

Metcalf, C. J. E., and Pavard, S. (2007). Why evolutionary biologists should be demographers. Trends in Ecology and Evolution, 22(4), 205–212.

Miller Jr, R. G. (2011). Survival Analysis (Vol. 66). John Wiley & Sons, New York.

Nichols, J. D. and Kendall, W. L. (1995). The use of multi-state capture–recapture models to address questions in evolutionary ecology. Journal of Applied Statistics, 22, 835–846.

Ono, F., Mori, Y., Takarabe, F. A, et al. (2016) Effect of ultra-high pressure on small animals, tardigrades and *Artemia*. Cogent Physics, 3(1).

Papaïx, J., Cubaynes, S., Buoro, M., Charmantier, A., Perret, P., and Gimenez, O. (2010). Combining capture–recapture data and pedigree information to assess heritability of demographic parameters in the wild. Journal of Evolutionary Biology, 23(10), 2176–2184.

Perks, W. (1932). On some experiments in the graduation of mortality statistics. Journal of the Institute of Actuaries, 63(1), 12–57.

Peron, G., Crochet, P.A., Choquet, R., Pradel, R., Lebreton, J.-D. and Gimenez, O. (2010). Capture–recapture models with heterogeneity to study survival senescence in the wild. Oikos, 119, 524–532.

Peto, R., Pike, M. C., Armitage, P., et al. (1977). Design and analysis of randomized clinical trials requiring prolonged observation of each patient. II. Analysis and examples. British Journal of Cancer, 35(1), 1–39.

Plummer, M. (2003). JAGS: a program for analysis of Bayesian graphical models using Gibbs sampling. Proceedings of the 3rd International Workshop on Distributed Statistical Computing, 124(125.10).

Plummer, M., Stukalov, A., Denwood, M., and Plummer, M. M. (2018). Package 'rjags'. update, 16, 1.

Pollock, K., Nichols, J., Brownie, C., and Hines, J. (1990). Statistical inference for capture–recapture experiments. Wildlife Monographs, 107, 3–97.

Pradel, R. (1993). Flexibility in survival analysis from recapture data: handling trap-dependence. In: J.-D. Lebreton and P. M. North (eds), Marked Individuals in the Study of Bird Population (pp. 29–37). Birkhäuser, Basel.

Pradel, R. (2005). Multi-event: an extension of multi-state capture–recapture models to uncertain states. Biometrics, 61, 442–447.

Pradel, R. and Lebreton, J.-D. (1999). Comparison of different approaches to the study of local recruitment of breeders. Bird Study, 46, 74–81.

Pradel, R. and Sanz-Aguilar, A. (2012). Modeling trap-awareness and related phenomena in capture–recapture studies. PLOS One, 7, e32666.

Pradel, R., Hines, J. E., Lebreton, J.-D., and Nichols, J. D. (1997). Capture–recapture survival models taking account of transients. Biometrics, 53, 60–72.

Rouan, L., Choquet, R., and Pradel, R. (2009). A general framework for modeling memory in capture–recapture data. Journal of Agricultural Biological and Environmental Statistics, 14, 338–355.

Royle, J. A. (2008). Modeling individual effects in the Cormack–Jolly–Seber model: a state–space formulation. Biometrics, 64(2), 364–370.

Salguero-Gómez, R., Siewert, W., Casper, B.B. and Tielbörger, K. (2012). A demographic approach to study effects of climate change in desert plants. Philosophical Transactions of the Royal Society B: Biological Sciences, 367, 3100–3114.

Santidrián Tomillo, P., Robinson, N., et al. (2017). High and variable mortality of leatherback turtles reveal possible anthropogenic impacts. Ecology, 98, 2170–2179.

Sanz-Aguilar, A., Tavecchia, G., Mínguez, E., et al. (2010). Recapture processes and biological inference in monitoring burrowing nesting seabirds. Journal of Ornithology, 151, 133–146.

Sanz-Aguilar, A., Igual, J.M., Oro, D., Genovart, M., and Tavecchia, G. (2016). Estimating recruitment and survival in partially monitored populations. Journal of Applied Ecology, 53, 73–82.

Schaub, M. and Pradel, R. 2004. Assessing the relative importance of different sources of mortality from recoveries of marked animals. Ecology, 85, 930–938.

Schaub, M., Gimenez, O., Schmidt, B. R., and Pradel, R. (2004). Estimating survival and temporary emigration in the multi-state capture–recapture framework. Ecology, 85(8), 2107–2113.

Schwarz, C. J., and Arnason, A. N. (1996). A general methodology for the analysis of capture–recapture experiments in open populations. Biometrics, 52, 860–873.

Schwarz, C. J., Schweigert, J. F., and Arnason, A. N. (1993). Estimating migration rates using tag-recovery data. Biometrics, 49, 177–193.

Seber, G. A. (1965). A note on the multiple-recapture census. Biometrika, **52**(1/2), 249–259.

Service, P. M. (2000). Heterogeneity in individual mortality risk and its importance for evolutionary studies of senescence. The American Naturalist, **156**(1), 1–13.

Shefferson, R. P. and Tali, K. (2007). Dormancy is associated with decreased adult survival in the burnt orchid, *Neotinea ustulata*. Journal of Ecology, **95**, 217–225.

Shefferson, R. P., Sandercock, B. K., Proper, J., and Beissinger, S. R. (2001). Estimating dormancy and survival of a rare herbaceous perennial using mark–recapture models. Ecology, **82**, 145–156.

Shefferson, R. P., Proper, J., Beissinger, S. R., and Simms, E. L. (2003). Life history trade-offs in a rare orchid: the costs of flowering, dormancy, and sprouting. Ecology, **84**, 1199–1206.

Shefferson, R. P., Kull, T., Hutchings, M. J., et al. (2018). Drivers of vegetative dormancy across herbaceous perennial plant species. Ecology Letters, **21**, 724–733.

Siler, W. (1983). Parameters of mortality in human populations with widely varying life spans. Statistics in Medicine, **2**(3), 373–380.

Slade, N. A., Alexander, H. M., and Kettle, W. D. (2003). Estimation of population size and probabilities of survival and detection in Mead's milkweed. Ecology, **84**, 791–797.

Spiegelhalter, D. J., Best, N. G., Carlin, B. P., and van der Linde, A. (2002). Bayesian measures of model complexity and fit (with discussion). Journal of the Royal Statistical Society B, **64**, 583–639.

Stearns, S. C. (1992). The Evolution of Life Histories (No. 575 S81). Oxford University Press, London.

Tavecchia, G., Pradel, R., Boy, V., Johnson, A. R., and Cezilly, F. (2001). Sex- and age-related variation in survival and cost of first reproduction in greater flamingos. Ecology, **82**, 165–174.

Tavecchia, G., Adrover, J., Navarro, A. M., and Pradel, R. (2012). Modelling mortality causes in longitudinal data in the presence of tag loss: application to raptor poisoning and electrocution. Journal of Applied Ecology, **49**, 297–305.

Tavecchia, G., Sanz-Aguilar, A., and Cannell, B. (2016). Modelling survival and breeding dispersal to unobservable nest sites. Wildlife Research, **43**, 411–417.

Thatcher, R., Kannisto, V., and Vaupel, J. W. (1998). The force of mortality at ages 80 to 120. Monographs on Population Aging, 5. Odense University Press, Odense.

University of California, Berkeley (USA), and Max Planck Institute for Demographic Research (Germany). (2019). Human Mortality Database. www.mortality.org or www.humanmortality.de (accessed 2019).

Vaupel, J. W. (1997). Trajectories of Mortality at Advanced Ages. Between Zeus and the Salmon: The Biodemography of Longevity, 17–37. National Academies Press, Washington, DC.

White, G. C. and Burnham, K. P. (1999). Program MARK: survival estimation from populations of marked animals. Bird Study, **46**, S120–S139.

Wienke, A. (2010). Frailty Models in Survival Analysis. Chapman and Hall/CRC Press, Boca Raton, FL.

Williams, B. K., Nichols, J. D., and Conroy, M. J. (2002). Analysis and Management of Animal Populations. Academic Press, Cambridge, MA.

Worthington, H., King, R., and Buckland, S. T. (2015). Analysing mark–recapture–recovery data in the presence of missing covariate data via multiple imputation. Journal of Agricultural, Biological, and Environmental Statistics, **20**(1), 28–46.

CHAPTER 14

Efficient use of demographic data: integrated population models

Marlène Gamelon, Stefan J. G. Vriend, Marcel E. Visser, Caspar A. Hallmann, Suzanne T. E. Lommen, and Eelke Jongejans

14.1 Introduction: why use integrated population models?

To understand the dynamics of wild populations, various types of demographic data can be collected in the field, including population counts, productivity surveys, capture–mark–recapture (CMR), mark–recovery (MR) data, and telemetry data (see Chapters 4 and 5). As seen in previous chapters, it is possible to gain an understanding of population sizes from population counts (see Chapter 7). Likewise, from CMR data, survival probabilities may be estimated (see Chapter 13). Sometimes, all these demographic data are collected in the same population. Considered separately, each of these data sets provides useful information on population size and vital rates such as survival, fecundity, or immigration/emigration, but inference is limited when component data sets are compared. Indeed, population sizes fluctuate from year to year due to all of these processes together: the loss and gain of individuals through survival, fecundity, immigration, and emigration. Consequently, these data sources share common demographic information about the studied population.

Integrated population models (IPMs) (Besbeas et al. 2002) use different types of demographic data by jointly analysing them. For instance, census data and CMR data can be analysed jointly. This integrated approach has several advantages. First, parameter estimates such as population sizes or survival rates, shared among different data sets, are more precise and accurate. Second, imperfect detection and observation error inherently associated with data sampled in the field are accounted for. Third, some parameter values can be estimated due

to the integrated approach, which would otherwise not be possible due to missing data. Fourth, changes in population structure and interactions at the population (e.g. population counts) and at the individual level (e.g. CMR and fecundity data) are modelled (Plard et al. 2019a). Fifth, separating the true pattern with observation error and reconciling across data types allows more robust matching of structural demography with observed patterns. Thus, resulting population models are more likely to follow the observed temporal fluctuations in population size than when such models are based on vital-rate data only.

The use of IPMs began in the 1980s in fisheries (see Maunder and Punt 2013 for a review). Twenty years later, IPMs were developed in terrestrial systems (see Besbeas et al. 2002 for a study on birds), and from 2002 onwards the number of studies applying IPMs has markedly increased in the literature (see Zipkin and Saunders 2018 for a review). For a long time, most of the studies using IPMs have focused on the statistical developments of methods. For instance, Abadi et al. (2010a) explored to what extent parameter estimates are more accurate and precise under varying sample sizes. But in recent years, IPMs have become an important tool in ecological studies to obtain more precise and accurate estimates of vital rates (see e.g. Lee et al. 2015 for a case study on the Svalbard reindeer *Rangifer tarandus platyrhynchus*) or to estimate vital rates that are difficult to estimate with classical approaches such as capture–mark–recapture (CMR) models (see Abadi et al. 2010b; van Oosten et al. 2015 for an estimation of immigration rate). IPMs also appear to be useful tools in retrospective analyses to study transient dynamics

Marlène Gamelon et al., *Efficient use of demographic data: integrated population models*. In: *Demographic Methods Across the Tree of Life*.
Edited by Roberto Salguero-Gómez and Marlène Gamelon, Oxford University Press.
© Oxford University Press (2021). DOI: 10.1093/oso/9780198838609.003.0014

(see Chapter 11) and make inference on the demographic and structural drivers of realised population growth rate (Koons et al. 2016), as well as in prospective analyses (Oppel et al. 2014). IPMs offer many other possibilities by, for instance, including spatially explicit CMR data (Chandler and Clark 2014) or two sexes (Tenan et al. 2016).

Because of all the possibilities offered by IPMs, they have also started to be widely used for conservation purposes (see Zipkin and Saunders 2018 for a review). In declining populations, the data collected are generally scarce due to low sample sizes and because researchers are reluctant to catch and mark individuals in vulnerable populations. In such cases, IPMs are useful tools; by making efficient use of all data available, key vital rates that drive the dynamics of declining populations are identified and appropriate conservation actions can be advised (see e.g. Lieury et al. 2015, 2016). IPMs also are relevant tools in a management context, where they provide important insights on the dynamics of exploited populations and thus help managers to conduct habitat manipulations and develop relevant harvest strategies (Arnold et al. 2018).

More generally, IPMs are appropriate tools to test hypotheses in life history theory, conservation, and eco-evolutionary biology (Plard et al. 2019a), as the submodels are amenable to common regression structures. Recent methodological advances, for instance the introduction of Markov Chain Monte Carlo (MCMC) methods in ecology, have highly contributed to the expansion of this approach that can be implemented in freely available software (e.g. JAGS, STAN, NIMBLE, or BUGS). Moreover, even if implementing IPMs can be complex, example codes can easily be found in the pivotal Kéry and Schaub (2012) book and in many published studies that provide the annotated scripts (see e.g. Gamelon et al. 2016, 2019).

14.2 How to build an IPM?

The three main steps required to build an IPM are: (1) define the population model that best represents the ecology of the species and links the vital rates to population sizes; (2) define the likelihoods for each data set to estimate the vital rates (e.g. survival, fecundity); and (3) define the joint likelihood of all data sets combined. Here, we provide a brief overview of these three steps. Note that we will not mention any goodness-of-fit tests in this chapter as we are still at early days for goodness-of-fit tests for IPMs (Besbeas and Morgan 2014). This is currently the main limitation associated with the use of the IPMs.

14.2.1 Population model

The first step to build an IPM is to define a population model that links the vital rates (e.g. survival, fecundity) to population sizes. This is basically a (st)age-structured matrix population model that allows (st)age-specific population sizes at year $t+1$ to be estimated from (st)age-specific population sizes at year t and vital rates (see Chapter 9 for an introduction to matrix population models). Showing these transitions between (st)ages with a life cycle graph based on the life history of the studied species may help to construct the population model.

14.2.2 Individual likelihoods

The second step in building an IPM is to define the likelihood functions for each data set (e.g. CMR data, census data) separately, which describe the probability of an observed outcome (i.e., the data) conditional on particular parameter values. For instance, for population count data, state-space models (de Valpine and Hastings 2002) are often used. They consist of a process model describing how the true population size and structure change while accounting for observation error over years (Besbeas et al. 2002). The process model corresponds to the population model described in section 14.2.1. The observation model describes the link between the population counts and the true population sizes. For CMR data, among the variety of likelihoods available for survival analyses and depending on data structure and research question(s), Cormack–Jolly–Seber models (Lebreton et al. 1992) can be used, and a multinomial likelihood links the data to survival and recapture probabilities. For fecundity data, a simple Poisson regression model is generally used.

14.2.3 Joint likelihood

The third and last step to build an IPM is to define the joint likelihood, that is, the likelihood of the integrated model. Assuming independence among the data sets, the likelihood of the IPM is given by the product of the likelihoods of the different data sets. But often, the same individuals are found in different data sets, and so the assumption of independence might be violated. While the data types (e.g. CMR data) might determine the extent of the independence issue, a simulation study of IPM performance has shown that the violation of the assumption of independence does not have a strong impact on the performance of the estimators and highlighted a gain in precision and accuracy when data sets such as population counts, fecundity, and CMR data are analysed simultaneously into an IPM (Abadi et al. 2010a). Note, however, that in the case of JAGS/BUGS/NIMBLE/STAN there is no need to state the likelihood product explicitly.

14.3 Case studies

In this section, we illustrate how IPMs can be used to address major questions in demography in both animals and plants. The blue tit is our first case study. This is a short-lived bird species, abundant in European gardens and woodlands as year-round residents. From a unique long-term monitoring of a Dutch blue tit population, we show how different data sources can be analysed simultaneously and how such an integrated analysis might help us to obtain a comprehensive picture of its dynamics. While most of the IPMs published so far have been built for animal species, our second case study on common ragweed shows that IPMs may also be applicable and very useful in the plant kingdom.

14.3.1 Blue tit

Main questions tackled

Different questions can be addressed in a Dutch blue tit *Cyanistes caeruleus* population while using an IPM. We provide below a (nonexhaustive) list of questions:

- What are the 'true' annual population sizes?
- Do the 'true' annual population sizes differ from the annual population counts?
- What are the annual age-specific numbers of females/survival rates/recruitment rates/recapture rates?
- How does beech crop size influence age-specific survival rates/recruitment rates?

Data collection

The data analysed come from a long-term study of a blue tit population at Hoge Veluwe National Park in the Netherlands (52°02′N, 5°51′E). The blue tit is a small short-lived sedentary passerine bird species. This cavity-nester nests in boxes, making it possible to monitor the entire breeding population as long as more nest boxes than required are provided. Only a few females in the study area bred in natural cavities, partly because 50 out of 450 nest boxes have a small entrance hole only accessible to blue tits. More nest boxes than required were provided to ensure that the availability of nest boxes did not

influence population density. As the study area is surrounded by a matrix of potential suitable habitats for blue tits, the population is open to immigration and emigration. Females usually start reproducing at age 1 (i.e. in the second calendar year of life). The number of recruits, that is, the number of females produced that survive and breed in the nest boxes the next year, is a common measure of breeding success. Three types of demographic data were collected in this population between 1990 and 2017: (1) CMR data, (2) population counts, and (3) fecundity data.

Beech mast is an important food resource for blue tits, especially during winter when other resources are scarce. It is also indicative of seed production of other tree species (Perdeck et al. 2000). The beech crop index (BCI), measured as the net weight of all nuts per square metre, was recorded from 1990 to 2017.

CMR data

Each year, nest boxes were visited at least once per week during the breeding season (April to June), and all young were ringed. The ringed adult females (previously caught and marked at the nest as young) were identified, and the unringed adult females were given a ring, allowing future identifications. Unringed adult females were assumed to have immigrated into the population in the current year. It was possible to exactly age most of these immigrants. Overall, CMR data were available for 708 breeding females of known age (locally fledged females plus females that fledged elsewhere but immigrated to the study area to breed).

Population counts

The second type of demographic data available was an estimate of the total number of counted breeding females each year t (C_t). As females start to reproduce at age 1 with a high breeding rate, breeding population size is a good proxy for the total number of females of age 1 or older. The population count C_t at year t corresponds to the sum of the number of breeding females of known and unknown age at year t and an estimate of the number of breeding females not caught (and thus not identified nor aged) because they have deserted their clutches early in the breeding attempt at year t. The number of females of the latter type is the number of clutches found without the mother being identified.

Fecundity data

The last type of data was age-class-specific counts of breeding females and the age-class-specific contribution to recruitment. The observed number of female fledglings in year t that successfully became a first-year breeding female in year $t+1$ corresponds to the recruitment for year t (J_t). This recruitment could be broken down by the age class of the adult female: first-year breeder, second-year, and so on. This provided estimates of the number of recruits by adult females of age class i in year t ($J_{i,t}$). We also recorded the total number of breeding females of each age class i in year t ($R_{i,t}$). In total, 89 daughters of known age adult females were locally recruited during the study period. Note that in most of the examples available in the literature, fecundity and first-year survival are structural factors, whereas in the current example they are combined in the recruitment rate.

The integrated population model

We aimed to estimate the annual age-specific survival and recruitment rates as well as the true annual age-specific numbers of females. Population counts contain several sources of observation error. There were females for which the age class was unknown (some immigrants). Moreover, not all females were recaptured, resulting in errors and uncertainty in the number of females in the different age classes. This also leads to errors and uncertainty in the estimates of survival and recruitment. In addition, there was a possibility of double counts: for instance, if one female produced two clutches but was only identified in one of them because she had deserted one of the clutches. Lastly, some clutches could have been missed because females bred in natural cavities within the population instead of breeding in nest boxes. To account for these issues, we jointly analysed CMR data of known-age females, population counts (C_t), and fecundity data ($J_{i,t}$ and $R_{i,t}$) using an IPM.

Population model

Based on the life history of blue tits, we considered a prebreeding age-structured female model with four age classes: age class 1 corresponded to the first year of breeding, age class 2 to the second year, age class 3 to the third year, and age class 4 to older breeding females. A very low proportion of individuals were in the oldest age class (about 13%). We assumed that reproduction started when females were 1 year old. The total number of breeding females of age class i in year t was denoted $N_{i,t}$ and defined as $N_{i,t} = n_{i,t} + A_{i,t} + I_{i,t}$, where $n_{i,t}$ was the number of local females in each age class i in year t, $A_{i,t}$ was the number of known age

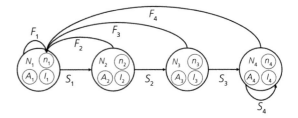

Figure 14.1 Life cycle of Hoge Veluwe blue tit population. Each arrow describes an annual transition from age i to age $i + 1$. N_i is the age-specific number of breeding females, which is the sum of the number of local breeding females (n_i), the number of immigrant females of known age (A_i), and the number of immigrant females of unknown age (I_i); S_i is the age-specific survival; and F_i is the age-specific recruitment.

immigrants of age class i (added as a deterministic number), and $I_{i,t}$ was the number of immigrants of age class i but for which it was impossible to give an exact age.

To account for demographic stochasticity, especially important in small populations (Lande et al. 2003), we used Poisson and binomial processes to describe the number of local breeding females in each age class ($n_{i,t}$): $n_{1,t+1} \sim \text{Poisson}(N_{1,t} \times F_{1,t} + N_{2,t} \times F_{2,t} + N_{3,t} \times F_{3,t} + N_{4,t} \times F_{4,t})$, $n_{2,t+1} \sim \text{Binomial}(N_{1,t}, S_{1,t})$, $n_{3,t+1} \sim \text{Binomial}(N_{2,t}, S_{2,t})$ and $n_{4,t+1} \sim \text{Binomial}(N_{3,t}, S_{3,t}) + \text{Binomial}(N_{4,t}, S_{4,t})$, where F_i and S_i were the age-specific recruitment and survival rates, respectively. For the immigrant females $I_{i,t}$ for which it was impossible to give an exact age, we considered that each year t, a random proportion of them are females of age class i. Modelling the number of immigrants as a function of population size would have led to biased estimates (Schaub and Fletcher 2015). This also allowed considering uncertainty on the age of these immigrant females without making any explicit assumption about their exact age. Females were classified as immigrants only in their year of arrival into the study population and then joined the local population. The true number of breeding females in the population at year t ($N_{tot,t}$) was then $\Sigma N_{i,t}$ (Figure 14.1).

Likelihood functions separately for each data set

The likelihood functions were specified for the different data sets as follows. For CMR data of breeding females of known age, we used the Cormack–Jolly–Seber model (Lebreton et al. 1992) which allowed estimation of annual survival between age class i and $i+1$ ($S_{i,t}$) and recapture (p_t) probabilities. For fecundity data, we assumed that the observed number of female recruits per age class i ($J_{i,t}$) was Poisson distributed with $J_{i,t} \sim \text{Poisson}(R_{i,t} \times F_{i,t})$, where $R_{i,t}$ was the observed number of breeding females of age class i and $F_{i,t}$ was the recruitment rate of females of age class i at year t. $F_{i,t}$ was the term we were estimating and was the contribution of adult females of age class i to recruitment into the age class 1 breeding class next year. Note that emigration (which mostly occurs within the first year of age) and mortality in the first year of life were included in the recruitment estimate.

For the population count data, we used a state-space model (de Valpine and Hastings 2002) which consisted of a process model (i.e. the population model described in the previous section) and an observation model. This latter described the link between the population counts C_t and the true number of females in the population ($N_{tot,t}$). We assumed that $C_t \sim \text{Normal}(N_{tot,t}, \sigma^2_c)$ truncated to positive values, which incorporated observation error of the counts. The observation model accounted for both count errors (females unobserved or double-counted) and lack of fit of the state equations to the true dynamics of the population (Schaub and Abadi 2011).

Joint likelihoods

Assuming independence among the data sets, the likelihood function of the IPM was the product of the likelihoods specified for each of the three different data sets, that is, population counts, fecundity data, and CMR data (Figure 14.2). Our aim was to estimate annual vital rates, which were thought to originate from a random process with a common mean and constant temporal variance (see e.g. Schaub et al. 2012, 2013;

Gamelon et al. 2016 for a similar approach). We used (1) recruitment specified on the log scale $\log(F_{i,t}) = \mu_{F\,i} + \varepsilon_{F\,i,t}$, where $\mu_{F\,i}$ was the mean contribution of adult females of age class i to recruitment next year and $\varepsilon_{F\,i,t}$ was the age-specific temporal residual; (2) the logit link for survival so that $\text{logit}(S_{i,t}) = \mu_{S\,i} + \varepsilon_{S\,i,\cdot}$. ε was the matrix including the temporal residuals of the eight vital rates (i.e. one temporal residual for recruitment rate $\varepsilon_{F\,i,t}$ and one temporal residual for survival rate $\varepsilon_{S\,i,t}$ per age class i). We used a multivariate normal (MVN) distribution $\varepsilon \sim \text{MVN}(0, \Sigma)$, where Σ was the variance–covariance matrix allowing correlated variability among vital rates; and (3) recapture probability, assumed to be age-independent, modelled with random time variation as well as $\log(p_t) = \mu_p + \varepsilon_{P\,t}$.

The IPM was fit within a Bayesian framework, and noninformative priors were specified for all the parameters. MCMC simulation was used for parameter estimation. To assess convergence, we ran four independent chains with different starting values for 200,000 MCMC iterations, with a burn-in of 150,000 iterations, thinned every 100th observation, resulting in 2,000 posterior samples. We used the Brooks and Gelman diagnostic $\hat{R}$ to assess the convergence of the simulations and used the rule $\hat{R} < 1.02$ to determine whether convergence was reached (Brooks and Gelman 1998). The analyses were implemented using JAGS (Plummer 2003) called from R (R Development Core Team 2017) with package R2jags (Su and Yajima 2012).

As mentioned earlier, IPMs are appropriate tools for hypothesis testing on all vital rates. As an example, outside the IPM, we then explored the effect of beech crop resources BCI on age-specific survival and recruitment

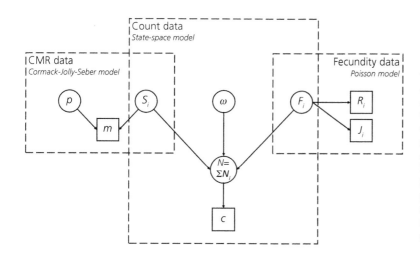

Figure 14.2 Directed acyclic graph of the IPM. Squares represent the data, circles represent the parameters to be estimated, and dashed lines represent the individual submodels. Arrows represent dependencies. Three types of data are collected: CMR data (m), count data (C), and fecundity data (number of daughters locally recruited (J) and number of mothers (R) in each age class i). Estimated parameters are the recapture probability p, age-specific survival S_i, age-specific recruitment F_i, immigration rate ω, and age-specific numbers N_i.

rates. We modelled the effect of BCI on survival and recruitment as follows:

$$\text{logit}\,(S_{i,t}) = \alpha_{Si} + \beta_{Si} \times BCI_t + \text{res}_{Si,t},$$

$$\log\,(F_{i,t}) = \alpha_{Fi} + \beta_{Fi} \times BCI_t + \text{res}_{Fi,t},$$

where α were the intercepts, res the residuals of the regressions, and β_i the age-specific effect of BCI on survival and recruitment.

The data and the JAGS code for the analyses are freely available.

Results

The estimated total number of breeding females (i.e. the 'true' annual population sizes) correlates well with the population counts (Pearson correlation coefficient r = 0.782 [95% CRI 0.720, 0.834]), and fluctuates between 25.91 (95% CRI 18.00, 35.00) and 122.69 (95% CRI 106.98, 140.00) over the study period (Figure 14.3a). Although the annual estimates of the

population growth rate were variable, the mean population growth rate was 1.024 (95% CRI 1.010, 1.036), indicating an overall increasing population trend. The population is dominated by recruits (age class 1), which show high temporal variation in their numbers (Figure 14.3b). The remaining age classes (2 to 4) are approximately equally represented and show less temporal variation (Figure 14.3b).

Annual survival estimates vary among ages and show high temporal variation (Figure 14.3c). Mean survival probability increased with age: from 0.277 (95% CRI 0.234, 0.327) in age class 1 to 0.432 (95% CRI 0.348, 0.520) in age class 2, 0.440 (95% CRI 0.311, 0.573) in age class 3, and 0.504 (95% CRI 0.282, 0.675) in age class 4. Annual recruitment estimates also show variation among ages and through time (Figure 14.3d). Mean recruitment rate is the lowest for age class 1 (0.130; 95% CRI 0.097, 0.169), higher for age class 2 (0.222; 95% CRI 0.151, 0.311) and age class 3 (0.209; 95% CRI 0.089, 0.395), and the highest for age class 4

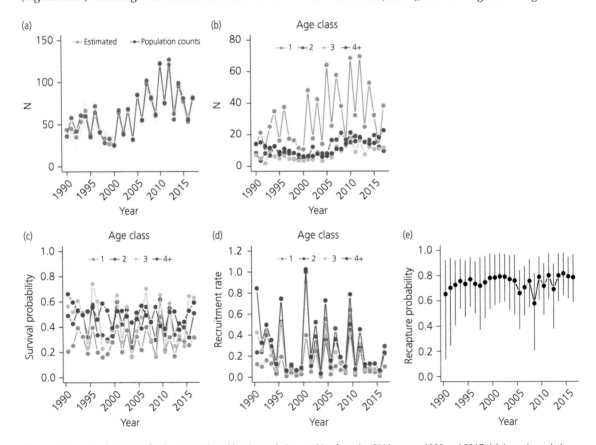

Figure 14.3 Annual estimates for the Hoge Veluwe blue tit population resulting from the IPM between 1990 and 2017. (a) Annual population counts (blue dots) and annual posterior means and associated 95% CRI of the estimated total number of breeding females (grey area).
(b) Posterior means of the annual age-specific numbers of breeding females. (c) Posterior means of the annual age-specific survival probabilities.
(d) Posterior means of the annual age-specific recruitment rates. (e) Posterior means and associated 95% CRI of the annual recapture probabilities.

(0.315; 95% CRI 0.137, 0.569). Annual estimates of the recapture probability were approximately constant throughout the study period (Figure 14.3e), with a mean recapture probability of 0.745 (95% CRI 0.647, 0.828).

We generally find a positive effect of BCI on survival and recruitment rates, and this effect is age-specific for survival. In particular, β_S is estimated to be 0.028 (95% CRI 0.007, 0.049) for age class 1, 0.039 (95% 0.018, 0.061) for age class 2, 0.048 (95% CRI −0.011, 0.109) for age class 3, and 0.002 (95% CRI −0.029, 0.031) for age class 4. For recruitment rates, we find that β_F equals 0.127 (95% CRI 0.065, 0.189) for age class 1, 0.104 (95% CRI 0.049, 0.158) for age class 2, 0.112 (95% CRI 0.052, 0.170) for age class 3, and 0.107 (95% CRI 0.053, 0.161) for age class 4.

14.3.2 Common ragweed

Data collection

The second case study is the estimation of seed bank density, germination, and survival rates of the invasive plant *Ambrosia artemisiifolia*, also known as common ragweed. Common ragweed is an annual plant that flowers in early autumn and overwinters as seed in the soil seed bank. Seeds in the soil seed bank can stay alive for decades. The species is problematic, as high densities of the plant can reduce agricultural crop yield and the pollen are allergenic to humans (Essl et al. 2015). To study the spatial and temporal variation in population dynamics, Lommen et al. (2018b, 2018c) designed a demographic field study that involved multiple European countries. Here we use data from one field season (2014) in Magnago, Italy, as an example.

Two population censuses (after the majority of the seedlings has established: mid-June census, and at seed set at the end of the summer: mid-September census) are performed in 13 square 0.25 m² plots. The total number of plants is counted at each census, but only a random subset of the young plants is marked to record their fate at the end of summer. Additionally, soil samples are taken adjacent to half of the plots at the end of summer (before seed rain) to monitor seed bank density. The major challenge with this field approach is that in order to estimate seedling establishment rates (an important aspect in the population dynamics of common ragweed), seed densities at the start of the growing season need to be estimated. Our proposed model here integrates seasonal plant data and seed bank data before seed rain at the end of summer to provide simultaneous estimates of germination rates, seedling survival rate,

and spring seed bank densities. We explain each of the collected data sets below.

Seedling data

On 12 June 2014, the number of seedlings was recorded in 13 square 0.25 m² plots. The timing of this first census was chosen such that the peak of germination had already taken place, but later germination did occur. Per plot, 4–15 seedlings were marked with aluminium tags to follow their survival and maturation till seed set.

Plant data

On 16 September 2014, the number of plants per plot (including both individuals that were already present in the first census and ones that had germinated since the first census) was determined. Whether individually marked plants had survived was also recorded.

Seed bank data

At the time of the second census, 16 September 2014, soil samples were taken to estimate the number of seeds per square metre (see protocol outlined in Lommen et al. 2018a). As the soil samples were taken before seed rain, the soil samples did not contain fresh seeds but only 1-year-old or older seeds.

Main questions tackled

While Integrated Population Models are rarely used in plant demography, mechanistic models fitted to multiple data sets are very useful for estimating vital rates that have not been estimated directly. Here we aim to use an IPM, and the data introduced above, to answer the following questions:

- What is the seed density (number per square metre) in the soil in spring (i.e. just before germination starts)?
- What is the per-seed establishment rate by the time of the first census?
- What is the per-seed establishment rate between the two censuses?
- What is the probability that a seedling survives up to flowering (i.e. the second census)?

Estimation of these vital rates is key to the construction of periodic population models, and this estimation can only be properly done by combining all three data sets. The IPM could be extended by incorporating plant growth and size-dependent seed production (and specifically the influx of fresh seeds into the soil seed bank), which would also allow the

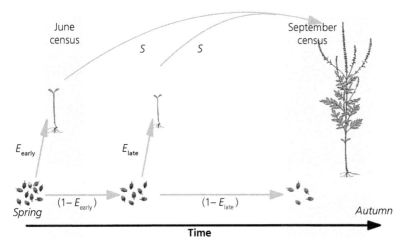

Figure 14.4 Schematic of the population dynamics of common ragweed during the growing season.

estimation of seed survival over winter. However, to keep things tractable, we here focus on the processes that happen during the growing season.

The integrated population model

In order to simultaneously estimate seedling establishment rates, seedling survival, and spring seed densities, we constructed a simple model of the dynamics of a common ragweed population over summer. As shown in Figure 14.4, our model keeps track of the number of seeds, seedlings, and plants over the course of the growing season. In spring the entire population is composed of seeds in the soil, but by mid-June a proportion of those seeds have germinated and successfully established themselves as seedlings: $L_{June} = 0.25 \times D_{Spring} \times E_{early}$, where E_{early} is the probability of seedling establishment per seed, D_{Spring} the density of seeds in the soil (per square metre), and L_{June} the number of seedlings per 0.25 m^2 plot during the first census. A similar process is modelled for late seedling establishment (even though seedlings that established after mid-June were not recorded until the final census in September): $L_{late} = 0.25 \times D_{Spring} \times (1-E_{early}) \times E_{late}$. After establishment, both early and late seedlings have a certain probability of surviving and being counted as plants in September: $P_{Sep} = (L_{early}+L_{late}) \times S$, in which S is the seedling survival rate and P_{Sep} the number of plants per plot in September.

Likelihood functions

We wrote a mechanistic JAGS model (R code provided) based on the population model shown in Figure 14.4 in order to estimate the unknown parameters D_{Spring}, E_{early}, E_{late}, and S using the combined information in the soil, seedling, and plant data sets. We assumed that the observed number of viable seeds in the soil samples in September (i.e. before influx of fresh seeds) follows a binomial distribution, with $D_{Sep} \sim$ Binomial($D_{soilSampleSpring}$, $(1-E_{early}) \times (1-E_{late})$). In this function, $D_{soilSampleSpring}$ is the unmeasured number of seeds in spring in a soil area (1/62.5 m^2) equal to that of the soil samples collected in September. For the realisation of $D_{soilSampleSpring}$ we assume a Poisson distribution: $D_{soilSampleSpring} \sim$ Poisson(D_{Spring}/62.5).

For the observed seedling densities in June (L_{June}) we assumed a binomial process: $L_{June} \sim$ Binomial(0.25 × D_{Spring}, E_{early}). The function for the third data set, the number of plants per plot in September, is more complex because three different groups of plants could have contributed to the total number of plants in the plots: $P_{Sep} \sim$ Poisson($P_{early,marked}$ + $P_{early,unmarked}$ + $P_{late,unmarked}$). Plants of these three groups differed in whether they were marked in June or not and whether unmarked plants had germinated early or late. The latter distinction cannot be made when observing the plants in September, stressing once more the usefulness of an IPM. For each of the three contributions to P_{Sep} we assumed a binomial process: $P_{early,marked} \sim$ Binomial($L_{June,marked}$, S); $P_{early,unmarked} \sim$ Binomial($L_{June,unmarked}$, S); and $P_{late,unmarked} \sim$ Binomial(D_{Spring}/4 – L_{June}, $E_{late} \times S$). This combination of a Poisson process with three underlying binomial processes gives the optimisation algorithm a lot of flexibility to find the most likely set of parameter values given these three data sets.

Please note that we made a few simplifying assumptions in this modelling exercise. For instance, we assumed that seeds do not die during the growing season. Soil samples were taken close to the demographic plots (which were operated for 3 years), so we assumed that seed densities are homogeneous over such short distances (but see Detto et al. 2019 for cautionary

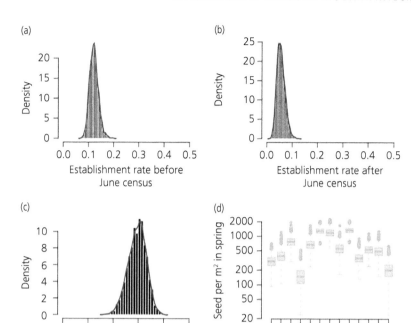

Figure 14.5 Posterior distributions of the estimated parameters of the *Ambrosia artemisiifolia* IPM. (a) E_{early}, the seedling establishment rate before the June census; (b) E_{late}, the seedling establishment rate after the June census; (c) S, the seedling-to-plant survival rate; and (d) D_{Spring}, the seed density in the soil in spring (i.e. before germination).

advice on such matters). Also, we fitted the same survival rate to seedlings prior and after the June census (i.e. no phenological dependent survival rates). The IPM was fitted in a very similar way as the blue tit model above (see technical details in the provided R code).

Results

The posterior distributions of the fitted model parameters are narrow, suggesting that the IPM has converged on a set of parameter values that is very likely, given the observations. This notion is confirmed by looking at the stability and similarity of the three separate chains that were run. Early seedling establishment, E_{early}, was estimated at 0.122 (95% CRI 0.091, 0.160), which was higher than the probability of late establishment, E_{late}: 0.055 (95% CRI 0.028, 0.091). Subsequent seedling survival, S, was estimated at 0.795 (95% CRI 0.720, 0.859). The plot-specific estimates of soil seed densities in spring varied between 159 and 1,367, with a median of 546 seeds per m². The average (among the plots) standard deviation was 151, indicating that although variable, also the seed densities were estimable (Figure 14.5).

Discussion of the plant example

These results can be used to parameterise population projection models (see Chapter 9) or to parametrise integrated integral projection models. This

new generation of IPMs includes both an integral projection model (Chapter 10) and an integrated population model (Plard et al. 2019b). Here, the integrated approach was especially useful because in the September plant count it was impossible to tell individuals apart that had established before or after the June census, and because soil seed density estimates were only available for 5 of the 13 plots and not for spring. When comparing these E_{early} and E_{late} estimates with those of other populations or years, one has to realise that the timing of the first census can have a large effect on ratio of these two seedling establishment probabilities. Now, E_{early} was larger than E_{late}, meaning that the moment of the census was correctly chosen according to protocol: after the germination peak.

Acknowledgements

We are grateful to all those who have collected data on blue tits *Cyanistes caeruleus* at Hoge Veluwe as well as the board for their permission to work in their forest. We are also grateful to all those who have collected data on *Ambrosia artemisiifolia* in Magnago, Italy. We acknowledge financial support from the EU COST Action FA1203 'Sustainable management of *Ambrosia artemisiifolia* in Europe (SMARTER)' and from the Research Council of Norway through its Centres of Excellence funding scheme (project number 223257).

References

Abadi, F., O. Gimenez, R. Arlettaz, and M. Schaub. 2010a. An assessment of integrated population models: bias, accuracy, and violation of the assumption of independence. Ecology 91:7–14.

Abadi, F., O. Gimenez, B. Ullrich, R. Arlettaz, and M. Schaub. 2010b. Estimation of immigration rate using integrated population models. Journal of Applied Ecology 47: 393–400.

Arnold, T. W., R. G. Clark, D. N. Koons, and M. Schaub. 2018. Integrated population models facilitate ecological understanding and improved management decisions. Journal of Wildlife Management 82:266–274.

Besbeas, P., and B. J. T. Morgan. 2014. Goodness-of-fit of integrated population models using calibrated simulation. Methods in Ecology and Evolution 5:1373–1382.

Besbeas, P., S. N. Freeman, B. J. T. Morgan, and E. A. Catchpole. 2002. Integrating mark–recapture–recovery and census data to estimate animal abundance and demographic parameters. Biometrics 58: 540–547.

Brooks, S. P., and A. Gelman. 1998. General methods for monitoring convergence of iterative simulations. Journal of Computational and Graphical Statistics 7: 434–455.

Chandler, R. B., and J. D. Clark. 2014. Spatially explicit integrated population models. Methods in Ecology and Evolution 5:1351–1360.

De Valpine, P., and A. Hastings. 2002. Fitting population models incorporating process noise and observation error. Ecological Monographs 72:57–76.

Detto, M., M. D. Visser, S. J. Wright, and S. W. Pacala. 2019. Bias in the detection of negative density dependence in plant communities. Ecology Letters 22:1923–1939.

Essl, F., K. Biró, D. Brandes, et al. 2015. Biological flora of the British Isles: *Ambrosia artemisiifolia*. Journal of Ecology 103:1069–1098.

Gamelon, M., V. Grøtan, S. Engen, E. Bjørkvoll, M. E. Visser, and B. Sæther. 2016. Density dependence in an age-structured population of great tits: identifying the critical age classes. Ecology 97:2479–2490.

Gamelon, M., S. J. G. Vriend, S. Engen, et al. 2019. Accounting for interspecific competition and age structure in demographic analyses of density dependence improves predictions of fluctuations in population size. Ecology Letters 22:797–806.

Kéry, M., and M. Schaub. 2012. Bayesian Population Analysis using WinBUGS: A hierarchical perspective. Academic Press, Boston.

Koons, D. N., D. T. Iles, M. Schaub, and H. Caswell. 2016. A life-history perspective on the demographic drivers of structured population dynamics in changing environments. Ecology Letters 19:1023–1031.

Lande, R., S. Engen, and B.-E. Sæther. 2003. Stochastic Population Dynamics in Ecology and Conservation. Oxford University Press, Oxford.

Lebreton, J.-D., K. P. Burnham, J. Clobert, and D. R. Anderson. 1992. Modeling survival and testing biological hypotheses using marked animals: a unified approach with case studies. Ecological Monographs 62:67–118.

Lee, A. M., E. M. Bjørkvoll, B. B. Hansen, et al. 2015. An integrated population model for a long-lived ungulate: more efficient data use with Bayesian methods. Oikos 124:806–816.

Lieury, N., M. Gallardo, C. Ponchon, A. Besnard, and A. Millon. 2015. Relative contribution of local demography and immigration in the recovery of a geographically-isolated population of the endangered Egyptian vulture. Biological Conservation 191:349–356.

Lieury, N., A. Besnard, C. Ponchon, A. Ravayrol, and A. Millon. 2016. Geographically isolated but demographically connected: immigration supports efficient conservation actions in the recovery of a range-margin population of the Bonelli's eagle in France. Biological Conservation 195:272–278.

Lommen, S., C. A. Hallmann, B. Chauvel, et al. 2018a. Field survey of the population dynamics of common ragweed (*Ambrosia artemisiifolia*). Protocols, doi: dx.doi. org/10.17504/protocols.io.mmyc47w.

Lommen, S. T. E., C. A. Hallmann, E. Jongejans, et al. 2018b. Explaining variability in the production of seed and allergenic pollen by invasive *Ambrosia artemisiifolia* across Europe. Biological Invasions 20:1475–1491.

Lommen, S. T. E., E. Jongejans, M. Leitsch-Vitalos, et al. 2018c. Time to cut: population models reveal how to mow invasive common ragweed cost-effectively. NeoBiota 39:53–78.

Maunder, M. N., and A. E. Punt. 2013. A review of integrated analysis in fisheries stock assessment. Fisheries Research 142:61–74.

Oppel, S., G. Hilton, N. Ratcliffe, et al. 2014. Assessing population viability while accounting for demographic and environmental uncertainty. Ecology 95:1809–1818.

Perdeck, A. C., M. E. Visser, and J. H. van Balen. 2000. Great tit *Parus major* survival and the beech-crop cycle. Ardea 88:99–106.

Plard, F., R. Fay, M. Kéry, A. Cohas, and M. Schaub. 2019a. Integrated population models: powerful methods to embed individual processes in population dynamics models. Ecology 100:e02715.

Plard, F., D. Turek, M. U. Grüebler, and M. Schaub. 2019b. IPM2: toward better understanding and forecasting of population dynamics. Ecological Monographs 89:e01364.

Plummer, M. 2003. JAGS: a program for analysis of Bayesian graphical models using Gibbs sampling. Pp. 20–22 *in* Proceedings of the 3rd International Workshop on Distributed Statistical Computing. Hornik K, Leisch F, Zeileis A (eds), Vienna, Austria.

R Development Core Team. 2017. R: a language and environment for statistical computing. https://www. R-project.org/.

Schaub, M., and F. Abadi. 2011. Integrated population models: a novel analysis framework for deeper insights into population dynamics. Journal of Ornithology 152: 227–237.

Schaub, M., and D. Fletcher. 2015. Estimating immigration using a Bayesian integrated population model: choice of parametrization and priors. Environmental and Ecological Statistics 22:535–549.

Schaub, M., T. S. Reichlin, F. Abadi, M. Kéry, L. Jenni, and R. Arlettaz. 2012. The demographic drivers of local population dynamics in two rare migratory birds. Oecologia 168:97–108.

Schaub, M., H. Jakober, and W. Stauber. 2013. Strong contribution of immigration to local population regulation: evidence from a migratory passerine. Ecology 94:1828–1838.

Su, Y., and M. Yajima. 2012. R2jags: a package for running jags from R. R package version 0.03–08.

Tenan, S., A. Iemma, N. Bragalanti, et al. 2016. Evaluating mortality rates with a novel integrated framework for nonmonogamous species. Conservation Biology 30: 1307–1319.

van Oosten, H. H., C. van Turnhout, C. A. Hallmann, et al. 2015. Site-specific dynamics in remnant populations of northern wheatears *Oenanthe oenanthe* in the Netherlands. Ibis 157:91–102.

Zipkin, E. F., and S. P. Saunders. 2018. Synthesizing multiple data types for biological conservation using integrated population models. Biological Conservation 217: 240–250.

PART 3

Applications

CHAPTER 15

Spatial demography

Guillaume Péron

15.1 Introduction

Eco-evolutionary processes are rooted in the way fitness varies across space and the way individuals implement dispersal strategies in response to that variation. As climate and land use changes redistribute the abiotic drivers that influence demography, the importance of spatial aspects in eco-evolutionary dynamics is increasingly obvious (Thuiller et al. 2013). In particular, the balance between extinctions and colonisations contributes to range shifts in metapopulations (Hanski and Gyllenberg 1997; Gandon and Michalakis 1999; McCauley et al. 2013). In the broader community context, the balance between local adaptation (Bohonak 1999; McRae et al. 2008; Burton et al. 2010) and dispersal/movement also creates ways for species and individuals to coexist by implementing different space use tactics (Leibold et al. 2004; Wolf et al. 2007; Péron et al. 2019).

This chapter reviews the analysis of demographic data for inference about spatial variation in fitness components and about dispersal strategies (Table 15.1). It also includes a section about prospective analysis, that is, the study of model properties for predictive inference (section 15.4: the feedback between space use and fitness).

15.2 Spatial variation in demographic rates

The *intrinsic population growth rate*, the balance between local births and local deaths, documents whether a place acts as a *population source* that contributes to the overall increase in population (Pulliam 1988) or a *population sink* likely to attract more immigrants than it emits emigrants (Novaro et al. 2005). Mapping out the spatial variation in intrinsic population growth rate is thus key to understanding the spatial functioning of populations, pinpoint the areas deserving of conservationists' attention, infer whether dispersing individuals are using the population density or breeding performance of their conspecifics to decide where to settle, and determine gene flow from source–sink dynamics. I will first review the generic nonparametric statistical tools to estimate spatial variation in any variable (section 15.2.1) and provide a nonexhaustive review of recent nonparametric analyses pertaining to spatial demography (section 15.2.2). Parametric approaches will only be briefly discussed (section 15.2.3) because they have been reviewed elsewhere and extensively (e.g., Dormann et al. 2007; Beale et al. 2010).

15.2.1 Statistical tools for nonparametric inference about spatial variation

The objective in this section is to provide an overview of often complex statistical techniques and point the reader in the right direction, hopefully.

First, at the data exploration stage, users may want to use Moran's index of autocorrelation to determine whether there is spatial structure in their data (Moran 1950; Cressie 1993). Moran's index detects the presence of a systematic similarity between data points that are close together compared to data points that are further apart. In practice, one may use the function Moran from the raster package for R if the data are in raster format and assuming a linear decay in autocorrelation strength with Euclidian distance; or the function moran from the spdep package to accommodate less regular samples (Bivand et al. 2013).

I will now review the three main options for the analysis of spatial variation in demographic data. Note

Guillaume Péron, *Spatial demography*. In: *Demographic Methods Across the Tree of Life*.
Edited by Roberto Salguero-Gómez and Marlène Gamelon, Oxford University Press.
© Oxford University Press (2021). DOI: 10.1093/oso/9780198838609.003.0015

Table 15.1 List of the spatial demography methods reviewed in this chapter. Note that this chapter does not consider movement tracking data or genetic data, only demographic data in the sense of information about vital rates and population abundance.

Spatial variation in demographic rates

Method name	Objectives	Section
Splines	Continuous spatial variation in a demographic rate, nonparametric (no environmental predictor needed)	15.2.1
Autoregressive models	Continuous variation, nonparametric	15.2.1
Random forest algorithm	Continuous variation, nonparametric	15.2.1
Linear models	Continuous or discrete variation, parametric (environmental predictor needed)	15.2.3
Multisite capture–recapture model	Discrete spatial variation in survival probability (i.e. analysis of the differences between predefined population units)	15.3.2
Integrated metapopulation model	Same as above but ability to estimate fecundity too	15.3.2

Dispersal fluxes

Method name	Objectives	Section
Integrated population model	The net flux of individuals into or outside of a focal population	15.3.1
Multisite capture–recapture model	The rates at which individuals disperse between discrete population units	15.3.2
Integrated metapopulation model	Same as above but increased statistical power and spatial extent	15.3.2
Diffusion equation	Landscape resistance to population homogenisation in a continuously distributed population	15.3.3

that these methods can be applied directly to spatially explicit sets of fecundity data or population survey data, but, to apply them to capture–recapture data for inference about survival and lifespan (cf. Chapter 13), they must first be incorporated into the appropriate state-space model that deals with imperfect detection of individuals (Gimenez et al. 2006; Péron et al. 2011).

Random forest algorithms

Also known as hierarchical clustering algorithms, this class of method belongs to the data mining and machine learning movement. It owes its name to its building blocks which are decision trees (Breiman 2001). A decision tree is a sequence of multiple-choice questions based on clustering criteria or *features* (e.g. 'Is the object red, blue, or green?', 'Is it a square or a circle?'). The *training dataset* is grouped into *bags* according to these features (e.g. bag 1 = 'green squares'). The value of the dependent variable in each bag is computed from the training data (e.g. 'bag 1 was good to eat'). The decision tree is fitted to the training data using traditional statistics (e.g. least square criterion or receiver operating characteristic curve). The interest then lies in the predictive power of the tree, that is, its ability to bin

new objects into bags without *a priori* knowledge of the dependent variable and only knowing the features of the object (e.g. 'determine palatability based on colour and shape'). The main issue with this type of approach is overfitting, for example, when the tree uses more features than supported by the training data and identifies small-sample artefacts rather than meaningful trends. To prevent overfitting without losing information, a *random forest* is a group of decision trees each based on a reduced subset of features. The prediction is the consensus between all the trees in the forest. For the random forest to perform better than a single decision tree, we simply need some of the features to be meaningful, that is, related to the predicted variable, and that the predictions made by individual trees have low correlations with each other (Breiman 2001). The latter condition is usually enforced by training each tree in the forest with a different subset of the data and by selecting the features of each tree at random irrespective of their predictive power.

When applied to spatial demography, the key strength of the random forest framework is the ability to input ancillary information in the form of spatial covariates. If the covariates are meaningful and well chosen, the algorithm makes it possible to downscale or disaggregate the demographic data at the resolution of these ancillary covariates (Stevens et al. 2015). However, the random forest algorithm also performs adequately if provided with nothing more than the geographical coordinates of the samples, in which case it will look for clusters of adjacent samples with similar demographic properties (Brickhill et al. 2015; Rushing et al. 2016). In practice, the random forest algorithm for spatial data is implemented in, for example, the `ranger` package for R (Wright and Ziegler 2017).

Splines

This section is going to be more technical than the rest of the chapter. The principle of this family of techniques is to approximate the spatial variation in a variable Y by a sum of K unimodal functions with desirable mathematical properties, which are termed the *splines* (Ruppert et al. 2003). The main feature of the splines is their location, that is the position of their *knot* κ_k, which is typically decided using a data-driven space-filling algorithm (Nychka and Saltzman 1998). The analytical shape of the spline is less influencial. Hereafter, I will consider radial splines denoted $D(\kappa_k, \cdot)$, where D is the distance operator. The dependent variable Y at location (x, y), is then modelled through the link function ℓ, as $\ell(Y)(x, y) = \beta_0 + \beta_1 x + \beta_2 y + \sum_{k=1}^{K} b_k \cdot D(\kappa_k, (x, y))$. The b parameters represent the weights associated to each of the K splines. The β parameters represent clinal variation. To fit that model, we need to minimise the least-square criterion $\mathcal{C}(b, \beta) = \|\ell(Y) - X \cdot \beta - Z \cdot b\|^2$, where Y contains the N observations of the focal variable, X is a $N \times 3$ matrix containing the coordinates at the sampling locations preceded by a 1 for the intercept, and Z is a $N \times K$ matrix, with $Z_{i,k}$ the value of spline k at sampling location i.

The main issue with this basic implementation of the spline model is, like in the decision tree case, overfitting. To avoid giving too much weight to idiosyncratic patterns occurring at the sampled locations, one typically incorporates a *penalisation term* designed to force the sum of the squared bs to stay below a threshold (Ruppert et al. 2003, p. 65). The criterion to be minimised then becomes of the form $\mathcal{C}(b, \beta, \lambda) = \|\ell(Y) - X \cdot \beta - Z \cdot b\|^2 + \lambda^2 \cdot b^T \cdot \Omega \cdot b$. Following Ruppert et al. (2003, p. 73), we use for Ω the $K \times K$ distance matrix with $\Omega_{k,l} = D(\kappa_k, \kappa_l)$. λ is the *smoothing parameter* controlling how much the idiosyncrasies will be smoothed out, to be estimated directly from the data themselves. λ is akin to the bandwidth parameter in kernel density estimators or the penalisation coefficient in lasso regularised regressions. To estimate the parameters, following Ruppert et al. (2003, p. 108ff), we exploit

the fact that the criterion $\mathcal{C}(b, \beta, \lambda)$ is the same as that of a linear mixed model with predictor X, fixed-effect regression coefficient β, design matrix $\tilde{Z} = Z \cdot \Omega^{-1/2}$, Gaussian random effects $u = \Omega^{1/2} \cdot b$, and individually and independently distributed Gaussian error term ε such that $\mathrm{Var}(\varepsilon) = \lambda^2 \mathrm{Var}(u)$. This means that we can fit the model using any software designed to fit linear mixed effect models, as long as the software interface allows us to specify a custom design matrix. Implementations using Bayesian software are available (Crainiceanu et al. 2005), including within capture–recapture models that account for imperfect detection of individuals (Gimenez et al. 2006; Péron et al. 2011).

Autoregressive models

This section is also going to be more technical than the rest of the chapter. Note, however, that autoregressive models are readily available in R packages `spdep` and `nlme` (Pinheiro et al. 2013) and are extensively used in ecology (Dormann et al. 2007). The underlying assumption of this class of model is that the value of the target variable Y at location (x, y) can be predicted from the value at nearby location $(x+dx, y+dy)$ and inversely. There are two nonexclusive options to implement that principle (Cressie 1993).

- In a *spatial lag* model, the observations are directly regressed against each other: $\ell(Y)(x, y) = \mu(x, y) + \rho W \ell(Y) + \varepsilon(x, y)$, where μ is a mean term that usually depends on spatially explicit linear predictors, ρ is the autocorrelation coefficient, W is a weight matrix typically corresponding to the Euclidian distances between samples, and ε contains individually and independently distributed (IID) error terms.

- In a *spatial error* model, the spatial autocorrelation affects the error term only: $\ell(Y)(x, y) = \mu(x, y) + e(x, y)$ with $e(x, y) = \lambda W e + \varepsilon(x, y)$. This type of model is typically used when the sampling noise is spatially autocorrelated. If this autocorrelation in the sampling error was not accounted for, the estimation of μ, that is, the ecological process of how the dependent variable Y depends on environmental covariates at location (x, y), would be biased, sometimes severely (Beale et al. 2010). Importantly, however, spatial error models work under the assumption that only the sampling error is autocorrelated. If it is not the case, that is, there is a biological process of spatial autocorrelation, using a spatial error model will assign too much spatial variance to the error and not enough variance to the mean.

In terms of R implementation, one may use the functions `lagsarlm` for the spatial lag model and `errorsarlm` for the spatial error model, both from package `spdep`. The spatial lag and spatial error models can also be merged together to obtain the simultaneous autoregressive models (SAR) that features both autocorrelation structures. There is also the conditional autoregressive model (CAR) or autologistic model, a variant of the spatial lag model. CAR is often preferred in ecological applications because of its flexibility and ability to accommodate missing data (Yackulic et al. 2012; Péron et al. 2016). CAR is obtained by designing the weight matrix according to user-specified lists of 'neighbours' for each sample, for example, the adjacent cells in a raster. We specify W as the adjacency matrix with a 1 if two samples are neighbours and 0 otherwise, and in addition, D is the normalisation matrix, a diagonal matrix with $D_{ii} = \sum_j W_{ij}$. Then the CAR model is a Gaussian random field (Gelfand and Vounatsou 2003), of mean μ and variance $\sigma^2 (D - \rho W)^{-1}$ (Cressie 1993). Following Besag (1974), there is an analytical solution under some constraints. In practice, implementations like the function `spautolm` from the `spdep` package rely on numerical optimisation to offer more flexibility regarding what goes in μ and how the W matrix is defined, at the cost of increased risk of numerical errors and optimisation failures. Many ecological applications furthermore include the CAR structure within a more complex hierarchical structure, for example, a partially observed colonisation/extinction process (Yackulic et al. 2012; Péron et al. 2016).

15.2.2 Recent implementations in spatial demography

Spatially explicit matrix population models

By 'spatially explicit matrix population model' I mean an array of grid cell-specific matrix population models (cf. Chapter 9), parameterised using the nonparametric estimation techniques outlined in section 15.2.1. They describe the spatial variation in intrinsic population growth rate but do not account for dispersal. Actually, I will only review the studies of spatial variation in survival rate, not in intrinsic population growth rate *per se*. This is because I could not locate any study of the spatial variation in fecundity that used a nonparametric technique. However, see data sets of passerine clutch size and nest fate (Baillie 1990; Brickhill et al. 2015; Eglington et al. 2015), acorn production in oak (*Quercus* sp.) forests (Koenig and Knops 2013; Touzot et al. 2018), and egg or foetus mass in harvested individuals from exploited species like commercial fish (Bell et al. 1992; Kraus et al. 2000; Stige et al. 2017) and game (Karns 2014; Gamelon et al. 2018).

Even survival studies are quite rare to the best of my knowledge. Saracco et al. (2010) used autoregressive models to generate maps of survival probability for the wood thrush (*Hylocichla mustelina*), a migratory passerine. The way survival increased in the northern and especially northwesternmost sections of the range is congruent with the observed increase in abundance in the northwestern part of the range (Rushing et al. 2016) and aligns with predictions from climate change. The particularity of the Saracco et al. study is that they accommodated transience at the bird ringing locations, that is, an excess of individuals captured only once (Pradel et al. 1997)—an important nuisance parameter to accommodate when individuals are not site-faithful.

Péron et al. (2011) used the spline method to delineate population sinks as areas where the predicted overwinter survival probability of Eurasian woodcocks (*Scolopax rusticola*) fell below the population renewal rate, itself a function of spatially invariant fecundity and summer survival rates. Their results confirm the additive nature of hunting mortality (Péron 2013) and imply that the dispersal of juveniles into population sinks is critical to the persistence of the species in the sinks.

Campbell et al. (2018) also used the autoregressive method to perform a spatially explicit population viability analysis of the Sonoran desert tortoise (*Gopherus morafkai*) based on spatially explicit survival rates. In this desert-adapted species, their results provide a demographic explanation for the 'abundant centre hypothesis' (Péron and Altwegg 2015b), that is, demographic performance was best at the core of the species' range and worst near the edge of the species' range. Interestingly, a decrease in the age at maturity near the edge of the range failed to compensate for the decreased survival rates.

To my knowledge, there is no study to date using the random forest method to estimate spatial variation in survival rates. This is because the cross-validation part of the random forest algorithm is not straightforwardly compatible with the state-space modelling framework required to estimate survival from capture–recapture data with imperfect detection.

Modelling the population growth rate from count data

When the vital rates themselves are not available, and if detailed time-specific and spatially-explicit population count data are available, these counts may be interpolated both spatially and temporally to compute the spatial variation in the population growth rate directly (Renner et al. 2013; Rushing et al. 2016). However, in most cases, data sets have proven too sparse and authors either separated the spatial and temporal components or they pooled the data at a coarse ecoregion level. In the latter case, random forest algorithms appear well suited to delineate these ecoregions based on the demographic data themselves (Brickhill et al. 2015; Rushing et al. 2016). In any case, as reviewed further in section 15.3, in vertebrates it is extremely rare to collect exact population counts at a fine spatial resolution and over large geographical areas. What is typically available instead is crowd-sourced, rasterised presence/absence or minimum count data (e.g. Péron and Altwegg 2015a). Using the link between the overall abundance in the landscape and the probability that any given location is occupied, or alternatively the link between the local abundance and the probability that at least one individual is detected (Royle and Nichols 2003), we may use these data as a proxy for the local abundance (Williams et al. 2017) and compute the spatial variation in the population growth rate that way.

15.2.3 Discussion

Empirical discrepancies between the three nonparametric regression methods

Empirically, the outputs of the three nonparametric methods are often significantly different (e.g. 'RF' vs. 'MARS' in Figure 6 of Renner et al. 2013). This is in part because of differences in the amount of smoothing. As illustrated by Kie (2013) in the context of kernel density estimators, the smoothing parameter is of paramount importance (see also 'GAM' vs. 'MARS' in Figure 6 Renner et al. 2013; and 'KDE' vs. 'KDEr' and vs. 'AKDE' in Figure 1 of Péron 2019b). In other words, the differences in output at least in part stem from the different amounts of smoothing. If the noise/signal ratio is large, that is, if nonspatial variance exceeds spatial variance in the data, the random forest is the most likely to overfit the data, and researchers should therefore prefer either the spline or the autoregressive method. But by contrast, if there are discontinuities or locally sharp gradients in the focal variable, the random forest method is the most relevant because other methods will smooth out these discontinuities.

What about fully parametric approaches?

Instead of a nonparametric, descriptive approach, researchers are often keen on using spatial covariates to predict spatial demography and infer the underlying ecological mechanisms (Beale et al. 2010; Germain et al. 2018). This is especially the case when extrapolating mortality factors from telemetry data which have a

natural spatial component to them (Schwartz et al. 2010; Basille et al. 2013; Péron et al. 2017). However, researchers should be acutely aware that, especially in our data-rich times, some climate and vegetation covariates may adequately predict the observed spatial variation in demography, but without capturing any significant biological mechanism, in other words, by coincidence (Journé et al. 2019). Such correlations are potentially great for interpolating gaps in coverage, but they should not be overinterpreted or used for predictions and extrapolations. For example, within a population of red-billed chough (*Pyrrhocorax pyrrhocorax*), the between-site variation in survival probability was caused by the different natal origins of the individuals in the different sites, not by the environmental attributes of the sites (Reid et al. 2006). In an archipelago population of great tits (*Parus major*), the spatial variation in demography was driven by an evolutionary 'island syndrome', not directly by the variation in climate between islands and the continent (Postma and van Noordwijk 2005). To avoid the pitfall of overinterpreting correlations, one may define broad ecoregions and compute spatial variation in demographic rate at this resolution. Alternatively, all the nonparametric approaches described in section 15.2.1 can be made into semiparametric tools that combine a regression against spatial covariates and a nonparametric spatial structure in the error term of the regression (Gimenez et al. 2006).

On spatial capture–recaptures

In the context of spatial demography, beware that the phrase 'spatial capture–recaptures' (Borchers and Efford 2008; Royle et al. 2013) usually refers to a type of model that is indeed spatially explicit but does not typically yield an estimate of spatial variation in survival. Spatial capture–recapture models are designed to improve the estimation of the population size by accommodating the decrease in individual capture probability with the distance between the home range centroid of the animal and the capture locations. They thereby provide a better fit to the capture–recapture data. They can in some configurations inform spatial variation in population density (Gardner et al. 2010), but in most cases the spatial variation in survival probability remains out of bounds for this type of model and data.

15.3 Estimating dispersal fluxes

Dispersal is defined as movement that has the potential to lead to gene flow (Clobert et al. 2001). Movements within individual home ranges and seasonal migrations are not considered dispersal. In many taxa, adults are site-faithful either by constraint (plants, sessile animals) or due to philopatric behaviour (Greenwood 1980), and thereby gene flow mostly occurs through *natal dispersal*, that is, the movements of immatures or propagules (Greenwood 1980).

Importantly, especially with a vertebrate focus, dispersal is the result of a series of decisions—from the decision to leave the current location to the choice of the new location among the locations that were explored (Grosbois and Tavecchia 2003). These different decisions may use different environmental cues, and different individuals may make different decisions based on the same cues. These decisions, summed over time and across individuals, yield the dispersal kernel: the probability that an individual at location (x_1, y_1) at time t_1 will disperse to location (x_2, y_2) at time t_2. One also refers to 'dispersal fluxes' when the population is spatially structured (the number of individuals dispersing from one site to another site, per unit of time) or to 'dispersal rates' (the number of individuals dispersing from one site to another, per unit of time and per individual in the site of origin) (Table 15.1).

Dispersal may be recorded directly by documenting the changes in individual locations from one breeding attempt to the next or from the place of birth to the first breeding attempt (Doligez et al. 2002; Ducros et al. 2019). But this type of data remains rare and therefore I will mostly review less direct methodologies.

15.3.1 Integrated population models

In this method, the target variable is the net flux of individuals that exit or enter a given study region over a given period of time, which is the difference between the number of emigrants and the number of immigrants. The net flux is estimated as the difference between the overall population growth (ΔN) and the intrinsic population growth ($b - d$, where b is the number of births and d is the number of deaths). In practice, the best approach is to jointly analyse capture–recapture data, fecundity survey data, and population survey data in an integrated population model (Abadi et al. 2010), as described in Chapter 14. Because it remains extremely challenging to collect these three types of data together and at the same time, examples are still rare in the literature (Millon et al. 2019). Importantly, however, several less-data-hungry approaches have been proposed for special cases. For example, Nichols and Pollock (1990) described how to use a 'robust design' of the capture–recapture protocol for voles (*Microtus pennsylvanicus*) that relies on the

precise ageing of the juveniles upon capture to separate local recruits from immigrants during the post-dispersal capture events of the same year. Peery et al. (2006) combined the observed recruitment rate of locally born chicks of marbled murrelets *Brachyramphus marmoratus*, with Pradel's (1996) estimate of anteriority within the breeding adults, in order to compute the immigration rate as the difference between local and total recruitment. Note that both of the latter approaches only quantify immigration; emigration is confounded with mortality in these two approaches. Potential users should furthermore carefully consider the model assumptions and the fieldwork requirements that the above authors listed.

15.3.2 Multisite capture–recapture models and integrated metapopulation models

In these methods, individuals are monitored over discrete population units or 'sites', such as seabird colonies or forest fragments. When an individual is recorded to move from one site to another, this represents a recorded dispersal event. The data are processed through a multisite capture–recapture model to allow for missing observations of some individuals in some years (Lebreton et al. 2009), as described in Chapter 13, yielding an estimate of dispersal rates. To further study the dispersal decision process, the dispersal rates may be decomposed into the probability of site-fidelity and the site selection conditioned on the lack of site-fidelity (Grosbois and Tavecchia 2003). The model may feature a 'memory effect' of the location of birth (Péron et al. 2010b) or of the previous dispersal decision (Lagrange et al. 2014).

In addition, to accommodate the fact that researchers in the field may not be able to monitor all the sites in a metapopulation, a 'ghost' population unit may be incorporated in the model, representing unmonitored population units where an individual may temporarily reside and therefore temporarily escape detection (Schaub et al. 2004). Statistical inference in this case may be facilitated if the unmonitored population units that form the ghost state can be surveyed for population size. These data may be jointly analysed with the capture–recapture data in an 'integrated metapopulation model' (Péron et al. 2010a). The key benefit of the integrated approach is improved accuracy of parameter estimates and the ability to extrapolate the dispersal information from a subset of sites, where marked individuals are monitored, to a larger ensemble of sites where only population count data are collected (Péron et al. 2010a). Lastly, when the interest lies only in the decision to leave or stay in the current site (Schmidt 2004), the dispersal information may be simplified into a binary variable—site-faithful or site-unfaithful—thereby discarding the information about the site identities (Lagrange et al. 2014).

As a final note the dispersal rates in these models do not account for permanent emigration out of the survey area, which is confounded with death-but may be age-specific which researchers should be aware of when studying age-specific dispersal.

In practice, multisite capture–recapture models are best implemented using dedicated software as described in Chapter 13. Integrated metapopulation models are best implemented in the Bayesian framework (e.g. Péron et al. 2012 and references therein), although initial implementations were in the frequentist framework using the Kalman filter to compute the likelihood of the population count data (Péron et al. 2010a and references therein).

15.3.3 Dispersal as a diffusion process along resistance surfaces

The main drawback of the two previous methods is that they require clear boundaries delineating population units. In most cases, these do not exist. The population is instead distributed in a continuous manner along ecological gradients. The dispersal process is then more akin to a diffusion process than to a discrete state-switching process (Ovaskainen et al. 2008; Jongejans et al. 2011; Foltête and Giraudoux 2012; Williams et al. 2017). Nevertheless, dispersal may still be spatially structured by landscape features (McRae et al. 2008; Spear et al. 2010; Berthier et al. 2013; Hanks and Hooten 2013) or by the social cues that the individuals use to decide where to move, such as conspecific population density and conspecific breeding success (Doligez et al. 2002; Péron et al. 2010b).

Traditionally, this type of ecological landscape structure is documented either by *landscape genetics*, whereby the genetic structure of the population is compared to a landcover map, mostly to infer the barriers to dispersal but also the dispersal corridors that promoted genetic homogenisation (McRae et al. 2008; Spear et al. 2010; Hanks and Hooten 2013), or by *movement data analysis*, whereby the tracked movements of focal individuals are extrapolated to infer the barriers to movement and the corridors that would facilitate the movements of potential dispersers (Ovaskainen 2004; Panzacchi et al. 2016; Scharf et al. 2018; Zeller et al. 2016; Wang 2020). The result is a *resistance surface*.

The resistance surface may also be estimated from demographic data. The spatial and temporal variation in the rasterised population abundance N may be modelled as a *partial differential equation* or *diffusion equation* (Skellam 1951; Okubo and Levin 2001; Ovaskainen et al. 2008; Foltête and Giraudoux 2012; see also Neubert and Caswell 2000; Jongejans et al. 2011). That framework allows an elegant partition of the local population growth rate $\frac{dN}{dt}$ into the intrinsic population growth on the one hand and a diffusion term representing the movements of individuals from or into neighbouring locations on the other hand:

$$\frac{dN}{dt} = \overbrace{R \cdot N \cdot \left(1 - \frac{N}{K}\right)}^{intrinsic} + \overbrace{\nabla^2(D \cdot N)}^{diffusion}$$

The diffusion parameters in the matrix in D directly quantify the resistance surface. R and K represent the population growth rates and carrying capacities, and ∇^2 is the Laplace operator that computes the sum of the second partial derivatives relative to geographical coordinates. In practice the diffusion term is approximated by its rasterised formulation, with adjacency matrix W (as explained in section 15.2.1) and raster grid resolution h:

$$\nabla^2(D \cdot N)_i \approx \sum_{j|W_{ij}=1} \frac{(D \cdot N)_j - (D \cdot N)_i}{h^2}$$

The rasterisation transforms the diffusion equation into a set of ordinary differential equations, making it solvable within any numerical solver that can manage large systems of equations. However, such a dispersal kernel does not feature any preferential dispersal direction. In wind-dispersed plants (Jongejans et al. 2011), and in the theoretical situation where individuals would have been selected by climate change to systematically disperse in a particular direction, the contributions to the sum over j in the above equation could be weighted according to the dispersal direction.

Another major caveat, which is not always recognised, is that the diffusion equation as written above assumes that movements are most likely between adjacent grid cells. The framework does not really accommodate long-distance dispersal. Depending on the behaviour of the species at stake, this model may therefore grossly misrepresent the biological process. This may explain some apparent lack of fit of that model to the patterns that occur along the colonisation front of species that routinely perform long-distance dispersal (Louvrier et al. 2019). Alternative formulations of the diffusion term in the above equation therefore appear warranted. Avenues of research in this direction include habitat- rather than distance-based adjacency matrices, but also maybe the Riesz fractional derivatives, as used to represent the infamous (for movement ecologists) 'Lévy flight' (Mandelbrot 1982; Sims et al. 2008; Çelik and Duman 2012).

Eventually, the main issue with this approach is going to be the data requirements, but a new era seems to have opened in that respect. Recently, Williams et al. (2017) proposed to leverage the link between the abundance of a species and the probability that it is detected during any one visit to a location of occurrence (Royle and Nichols 2003). This makes it possible to fit the above model to presence/absence data, which are much easier to obtain nowadays by crowd-sourcing than population survey data which would require an intense professional field effort. The parameters of the dynamic occupancy model (colonisation rates, extinction rates, and detection probabilities; Yackulic et al. 2012; Péron et al. 2016) are expressed as functions of R, K, and D, allowing the estimation of the latter. Recent examples include the description of the colonisation of a newly ice-free bay by sea otters (*Enhydra lutris*) (Williams et al. 2019) and the recolonisation of south-eastern France by wolves (*Canis lupus*) (Louvrier et al. 2019). The main limiting assumption of that presence–absence framework is that the population must remain well below carrying capacity. In other words, the framework applies to rare species or species that are currently expanding. When the population enters the density-dependent regime, most of the study area is occupied, the detection of at least one individual does not depend on abundance anymore (Royle and Nichols 2003), and the intrinsic growth rate is mostly influenced by K rather than R. This makes the model unidentifiable, that is, the parameters cannot be separately estimated in this biological configuration. This is called *weak identifiability*.

15.4 The feedback between space use and fitness

In the previous sections, I described how to estimate the link between the geographic location of an individual, its demographic performance, and its dispersal kernel. Now I will outline how to use this information to parameterise an integral projection model (IPM) to forecast the future distribution of the population based on its spatial demography.

15.4.1 Principles of integral projection models as pertaining to spatial demography

In IPM parlance (cf. Chapter 10), natal dispersal corresponds to 'inheritance' and adult or breeding dispersal corresponds to 'growth'. Besides that, spatial IPMs are not conceptually very different from unidimensional IPMs designed to study the dynamics of quantitative traits like body mass (Lewis et al. 2006; Jongejans et al. 2011; Merow et al. 2014). The basic principle is to define a kernel function k representing the probability that an individual at location (x_1, y_1) at time t_1 will yield an individual at location (x_2, y_2) at time t_2. The kernel function combines all the information about how the individual survived, where it moved, whether it reproduced, and where the offspring moved. Further defining the *function to function operator* $K(\nu)(x_2, y_2) = \iint_\Omega \nu(x_1, y_1) k(x_1, y_1, x_2, y_2) dx_1 dy_1$, where ν represents a distribution function, the IPM equation becomes a single line equation $n(t_2) = K(n(t_1))$, where $n(t)$ denotes the population distribution at time t. In the following I will outline how to analyse the properties of K, but first I will review a lingering question in spatial ecology, that is, the fundamental difference between contingency and agency.

15.4.2 Space use or resource selection?

One of the critical decisions that eventual applications of the above principles will have to make is whether to model space use *per se*—for example, the focal variable is the geographical coordinates of the individual (Lewis et al. 2006; Jongejans et al. 2011)—or whether to model a behavioural trait that governs space use—for example, resource selection coefficients, personality traits, or 'movement syndromes'. 'Movement syndromes' refer to covariations between movement rates and between movement rates and other behavioural traits, for instance explorative or competitive personality (Wolf et al. 2007; Spiegel et al. 2017). Variability in these hard-wired syndromes between species and individuals is known to foster coexistence (Vanak et al. 2013; Péron et al. 2019) and population viability in fragmented landscapes (Zheng et al. 2009; Zera 2017).

In the first case (space use), the model describes the interaction between geographical coordinates and fitness. For geographical coordinates, one may use the centroid of an animal's home range. One may also replace the coordinates with the proportion of a given landcover in the home range (Péron 2019b) or the environmental covariates where a plant is growing (Merow et al. 2014), thereby swapping the geographic coordinates for the ecological coordinates. In this type of application, the IPM will become a relatively phenological model, forecasting the population distribution if current reaction norms are maintained into the future (Neubert and Caswell 2000; Lewis et al.

(a)

Seed production: 26

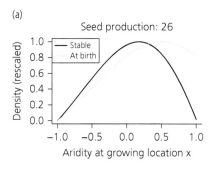

(b)

Seed production: 53

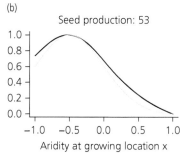

Figure 15.1 Simulation inspired by Rebarber et al. (2012) of a perennial plant population with no seed bank and random seed rain, faced with two scenarios of aridity. (a) a drier scenario. Seeds are more likely to fall in an arid spot (cf. grey curve). Compared to the distribution among seedlings (grey curve), the stable population distribution is shifted towards wetter locations (black curve), mostly because the seedlings that grow in the driest locations fail to survive. This means that over time the population is expected to shift its range centroid towards wetter locations, at a pace that can be estimated, but that there will always be a few individuals growing in arid locations because that is where they fell as a seed. (b) a wetter scenario. The stable population distribution is barely different from the expectation from the seed rain. This means that the population range is not expected to change much over time. Note also among the model outputs the forecast of the average population fitness and reproductive rates. In detail, the model was specified by the survival function $s(x) = logit^{-1}(2 - 2x)$, the fecundity function $c(x) = exp(2 - 2x)$, and the density-dependent recruitment $g(C) = C^{-0.33}$.

2006; Merow et al. 2014). However, that approach will describe but not explain the spatial variation in demography.

In the second case (resource selection/movement syndrome), one first estimates behavioural traits and in a second step links these traits to spatial demography. But in this case, researchers may need to incorporate plastic reaction norms describing how the behavioural trait changes depending on the context. For example, resource selection coefficients depend on what is available for selection, often in a nonlinear way (Forester et al. 2009). A typical example is roe deer (*Capreolus capreolus*), which today are commonly found in open agricultural landscapes but will revert to selecting forest edges if given the opportunity. The realised resource selection coefficients may therefore need to be interpreted with caution and may perhaps best be replaced by alternative behavioural metrics that capture the plasticity of resource selection by each individual (Bonnot et al. 2018).

Importantly, the two approaches are not mutually exclusive. They are only opposed here to reflect on the underlying philosophy and challenges in terms of data acquisition. As a final note, the interpretation of spatial IPMs in terms of evolutionary dynamics is not always straightforward. Similar to the way that offspring may inherit the large or small body mass of their mother because they feed on the same resource, not because they share the same genotype, the inheritance parameter in spatial IPMs combines true heritability and nongenetic inheritance. The nongenetic

inheritance component corresponds to physical barriers that constrain the distance between offspring and parents.

15.4.3 Formal analysis of IPMs

IPMs, once parameterised with field data, can be formally analysed for inference about equilibrium distribution, spatially explicit population growth, sensitivity of the population growth rate to variation in model parameters, individual variation, transitory dynamics (Ellner and Rees 2006), and the speed of any shift in the population geographic distribution, under some additional conditions (Lewis et al. 2006). As an illustration, I will focus on a well-worked example that pertains to monocarpic perennial plants (Rose et al. 2005; Ellner and Rees 2006; Rebarber et al. 2012). The keys to this model are that (1) the distribution of the focal trait among propagules is independent from the trait value of their parents; and (2) density dependence occurs among seedlings/propagules only, meaning that the recruitment rate declines with the number of candidates to recruitment, not the number of adults. There is usually a stable distribution of the trait in such a system (Ellner and Rees 2006; Rebarber et al. 2012) which can be computed as a function of the trait-specific survival function s, the trait distribution at birth b, the trait-specific fecundity function c, and the density-dependent recruitment term g (Figure 15.1; Appendix 1; please go to www.oup.com/companion/SalgueroGamelonDM).

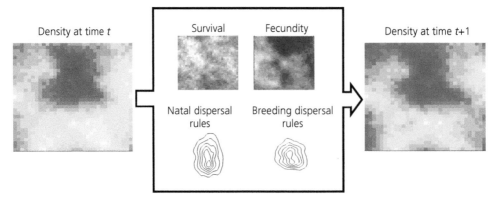

Figure 15.2 Principles of the simulation approach to analyse complex IPMs. The simulated animals are drawn from the density distribution at time *t*, made to perform and move according to the IPM kernel function, yielding the density distribution at time *t*+1. The process may be iterated a large number of times to obtain the asymptotic behaviours of the system. The kernel function may incorporate more complex rules than depicted, for example, dependency on nonstationary environmental covariates and interactions between the model individuals. Importantly, the kernel functions may depend on heritable individual traits that govern the spatial behaviour, rather than directly on the geographical coordinates. For example, data permitting, the trait under study could be the preferential direction of the natal dispersal kernel, and the model would provide inference about the selection pressure on dispersal direction.

Now, to make this model a model of spatial demography, the trait may, for example, represent the aridity at the specific location where the plant is growing. Then, *b* represents the relative frequency of arid and wet places in the landscape where seeds are randomly dispersed, by wind, for example; that is, *b* represents the environmental constraints. The stable distribution represents the eventual population distribution along the aridity gradient, as caused by the demographic performance of the individuals (Figure 15.1). There is, however, no evolution involved: at each generation the seedlings are redistributed at random along the aridity gradient (using the grey curves in Figure 15.1). As a side note, the complex mathematics underlying the formal analysis of density-dependent IPMs (Rebarber et al. 2012) render the framework somewhat costly to tailor to specific situations, which might restrict the type of study systems for which it is relevant and the type of evolutionary questions that can be tackled with it.

15.4.4 Individual-based simulation approach

As highlighted in section 15.4.3, the formal analysis of IPMs may not offer enough flexibility in practice to accommodate the complexity of ecological processes and to easily tailor the IPMs to specific case studies. In the context of spatial demography, a more flexible alternative is individual-based modelling (Chapter 12). The simulation focuses on individual movement decisions and their consequences in terms of survival and

reproduction. The simulation parameters are informed by analyses from section 15.2 about spatially explicit survival and fecundity and section 15.3 about dispersal kernels. The IPM properties are computed as emergent properties of the individual-based simulation (e.g. Wiens et al. 2017). The movement model does not need to be more complex than a step selection model (Signer et al. 2017; Péron 2019a). However, the interest of the framework lies in its flexibility, for example, to incorporate interactions between individuals, to take into account the cost of movement (Péron 2019b), to add heritable individual covariates that influence dispersal decisions, and to outfit the individuals with the ability to gather information about the spatial variation in fitness (Doligez et al. 2002) (Figure 15.2).

Technical bottlenecks that currently hamper the more widespread empirical implementation of spatial IPMs to vertebrates include the need to track the movements and demographic performance of a large number of individuals with known parentage and over long enough periods of time. This is because the repeatability, reaction norms, and age specificity of their space use patterns and fitness components need to be estimated. Despite the democratisation of geotracking technologies, these conditions remain hard to meet.

15.5 Conclusions

Although spatial demography is firmly located at the core of eco-evolutionary thinking, it does not appear to have given rise to a rich empirical literature yet, even if many methods are now readily available. The reasons include on the one hand the prohibitive data requirements and on the other hand the relative lack of communication between method-heavy 'quantitative ecology' projects and concept-heavy theoretical initiatives. This overall makes it challenging to adopt a comprehensive empirical outlook on spatial demography. The increasing availability of crowd-sourced species distribution data and the democratisation of geotracking technologies represent exciting opportunities. The fact that spatial demography bridges some currently isolated subfields like movement ecology and evolutionary biodemography may also open fruitful interfaces.

References

Abadi, F., O. Gimenez, B. Ullrich, R. Arlettaz, and M. Schaub. 2010. Estimation of immigration rate using integrated population models. Journal of Applied Ecology **47**:393–400.

Baillie, S. R. 1990. Integrated population monitoring of breeding birds in Britain and Ireland. Ibis **132**:151–166.

Basille, M., B. Van Moorter, I. Herfindal, et al. 2013. Selecting habitat to survive: the impact of road density on survival in a large carnivore. PLOS One **8**:e65493.

Beale, C. M., J. J. Lennon, J. M. Yearsley, M. J. Brewer, and D. A. Elston. 2010. Regression analysis of spatial data. Ecology Letters **13**:246–264.

Bell, J. D., J. M. Lyle, C. M. Bulman, K. J. Graham, G. M. Newton, and D. C. Smith. 1992. Spatial variation in reproduction, and occurrence of non-reproductive adults, in orange roughy, *Hoplostethus atlanticus* Collett (Trachichthyidae), from south-eastern Australia. Journal of Fish Biology **40**:107–122.

Berthier, K., S. Piry, J.-F. Cosson, et al. 2013. Dispersal, landscape and travelling waves in cyclic vole populations. Ecology letters **17**:53–64.

Besag, J. 1974. Spatial interaction and the statistical analysis of lattice systems. Journal of the Royal Statistical Society: Series B **36**:192–225.

Bivand, R. S., E. Pebesma, and V. Gómez-Rubio. 2013. Applied Spatial Data Analysis with R, 2nd edition. Springer, New York.

Bohonak, A. J. 1999. Dispersal, Gene Flow, and Population Structure. University of Chicago Press.

Bonnot, N. C., M. Goulard, A. J. M. Hewison, et al. 2018. Boldness-mediated habitat use tactics and reproductive success in a wild large herbivore. Animal Behaviour **145**:107–115.

Borchers, D. L. and M. G. Efford. 2008. Spatially explicit maximum likelihood methods for capture–recapture studies. Biometrics **64**:377–385.

Breiman, L. 2001. Random forests. Machine Learning **45**:5–32.

Brickhill, D., P. G. H. Evans, and J. M. Reid. 2015. Spatiotemporal variation in European starling reproductive success at multiple small spatial scales. Ecology and Evolution **5**:3364–3377.

Burton, O. J., B. L. Phillips, and J. M. J. Travis. 2010. Tradeoffs and the evolution of life-histories during range expansion. Ecology Letters **13**:1210–1220.

Campbell, S. P., E. R. Zylstra, C. R. Darst, R. C. Averill-Murray, and R. J. Steidl. 2018. A spatially explicit hierarchical model to characterize population viability. Ecological Applications **28**:2055–2065.

Çelik, C. and M. Duman. 2012. Crank–Nicolson method for the fractional diffusion equation with the Riesz fractional derivative. Journal of Computational Physics **231**:1743–1750.

Clobert, J., E. Danchin, A. A. Dhondt, and J. D. Nichols. 2001. Dispersal. Oxford University Press, New York.

Crainiceanu, C. M., D. Ruppert, and M. P. Wand. 2005. Bayesian analysis for penalized spline regression using WinBUGS. Journal of Statistical Software **14**: 1–24.

Cressie, N. A. C. 1993. Statistics for Spatial Data. Wiley, New York.

Doligez, B., É. Danchin, and J. Clobert. 2002. Public information and breeding habitat selection in a wild bird population. Science **297**:1168–1170.

Dormann, C., J. McPherson, M. Araújo, et al. 2007. Methods to account for spatial autocorrelation in the analysis of species distributional data: a review. Ecography **30**:609–628.

Ducros, D., N. Morellet, R. Patin, et al. 2019. Beyond dispersal versus philopatry? Alternative behavioural tactics of juvenile roe deer in a heterogeneous landscape. Oikos **129**(1):81–92.

Eglington, S. M., R. Julliard, G. Gargallo, et al. 2015. Latitudinal gradients in the productivity of European migrant warblers have not shifted northwards during a period of climate change. Global Ecology and Biogeography **24**: 427–436.

Ellner, S. P. and M. Rees. 2006. Integral projection models for species with complex demography. American Naturalist **167**:410–428.

Foltête, J.-C. and P. Giraudoux. 2012. A graph-based approach to investigating the influence of the landscape on population spread processes. Ecological Indicators **18**:684–692.

Forester, J. D., H. K. Im, and P. J. Rathouz. 2009. Accounting for animal movement in estimation of resource selection functions: sampling and data analysis. Ecology **90**:3554–3565.

Gamelon, M., T. Gayet, E. Baubet, et al. 2018. Does multiple paternity explain phenotypic variation among offspring in wild boar? Behavioral Ecology **29**: 904–909.

Gandon, S. and Y. Michalakis. 1999. Evolutionarily stable dispersal rate in a metapopulation with extinctions and kin competition. Journal of Theoretical Biology **199**:275–290.

Gardner, B., J. Reppucci, M. Lucherini, and J. A. Royle. 2010. Spatially explicit inference for open populations: estimating demographic parameters from camera-trap studies. Ecology **91**:3376–3383.

Gelfand, A. E. and P. Vounatsou. 2003. Proper multivariate conditional autoregressive models for spatial data analysis. Biostatistics **4**:11–15.

Germain, R. R., R. Schuster, C. E. Tarwater, W. M. Hochachka, and P. Arcese. 2018. Adult survival and reproductive rate are linked to habitat preference in territorial, year-round resident song sparrows *Melospiza melodia*. Ibis **160**:568–581.

Gimenez, O., C. Crainiceanu, C. Barbraud, S. Jenouvrier, and B. J. T. Morgan. 2006. Semiparametric regression in capture-recapture modeling. Biometrics **62**: 691–698.

Greenwood, P. J. 1980. Mating systems, philopatry and dispersal in birds and mammals. Animal Behaviour **28**:1140–1162.

Grosbois, V. and G. Tavecchia. 2003. Modeling dispersal with capture–recapture data: disentangling decisions of leaving and settlement. Ecology **84**:1225–1236.

Hanks, E. M. and M. B. Hooten. 2013. Circuit theory and model-based inference for landscape connectivity. Journal of the American Statistical Association **108**:22–33.

Hanski, I. and M. Gyllenberg. 1997. Uniting two general patterns in the distribution of species. Science **275**:397–400.

Jongejans, E., K. Shea, O. Skarpaas, D. Kelly, and S. P. Ellner. 2011. Importance of individual and environmental variation for invasive species spread: a spatial integral projection model. Ecology **92**:86–97.

Journé, V., J. Barnagaud, C. Bernard, P. Crochet, and X. Morin. 2019. Correlative climatic niche models predict real and virtual species distributions equally well. Ecology **101**:e02912.

Karns, G. R. 2014. Spatiotemporal Breeding Strategies within a High Density, Male-Skewed White-Tailed Deer Population. Auburn University.

Kie, J. G. 2013. A rule-based ad hoc method for selecting a bandwidth in kernel home-range analyses. Animal Biotelemetry **1**:13.

Koenig, W. D. and J. M. H. Knops. 2013. Large-scale spatial synchrony and cross-synchrony in acorn production by two California oaks. Ecology **94**:83–93.

Kraus, G., A. Muller, K. Trella, and F. W. Kouster. 2000. Fecundity of Baltic cod: temporal and spatial variation. Journal of Fish Biology **56**:1327–1341.

Lagrange, P., R. Pradel, M. Bélisle, and O. Gimenez. 2014. Estimating dispersal among numerous sites using capture–recapture data. Ecology **95**:2316–2323.

Lebreton, J. -D., J. D. Nichols, R. J. Barker, R. Pradel, and J. A. Spendelow. 2009. Modeling Individual Animal Histories with Multistate Capture–Recapture Models. Advances in Ecological Research **41**:87–173.

Leibold, M. A., M. Holyoak, N. Mouquet, et al. 2004. The metacommunity concept: a framework for multi-scale community ecology. Ecology Letters **7**:601–613.

Lewis, M. A., M. G. Neubert, H. Caswell, J. S. Clark, and K. Shea. 2006. A guide to calculating discrete-time invasion rates from data. Pages 169–192 in M. W. Cadotte, S. M. McMahon, and T. Fukami, editors. Conceptual Ecology and Invasion Biology: Reciprocal Approaches to Nature. Springer, Dordrecht, NL.

Louvrier, J., J. Papaïx, C. Duchamp, and O. Gimenez. 2019. A mechanistic–statistical species distribution model to explain and forecast wolf (*Canis lupus*) colonization in South-Eastern France. https://arxiv.org/abs/1912.09676.

Mandelbrot, B. B. 1982. The Fractal Geometry of Nature. W. H. Freeman and Co., New York.

McCauley, S. J., C. J. Davis, E. E. Werner, and M. S. Robeson. 2013. Dispersal, niche breadth and population extinction: colonization ratios predict range size in North American dragonflies. Journal of Animal Ecology **83**: 858–865.

McRae, B., B. Dickson, T. Keitt, and V. Shah. 2008. Using circuit theory to model connectivity in ecology, evolution, and conservation. Ecology **89**:2712–2724.

Merow, C., A. M. Latimer, A. M. Wilson, S. M. McMahon, A. G. Rebelo, and J. A. Silander. 2014. On using integral projection models to generate demographically driven predictions of species' distributions: development and validation using sparse data. Ecography **37**:1167–1183.

Millon, A., X. Lambin, S. Devillard, and M. Schaub. 2019. Quantifying the contribution of immigration to population dynamics: a review of methods, evidence and perspectives in birds and mammals. Biological Reviews **94**:2049–2067.

Moran, P. A. P. 1950. Notes on continuous stochastic phenomena. Biometrika **37**:17–23.

Neubert, M. G. and H. Caswell. 2000. Demography and dispersal: calculation and sensitivity analysis of invasion speed for structured populations. Ecology **81**: 1613–1628.

Nichols, J. D. and K. H. Pollock. 1990. Estimation of recruitment from immigration versus in situ reproduction using Pollock's robust design. Ecology **71**:21–26.

Novaro, A., M. Funes, and R. Walker. 2005. An empirical test of source–sink dynamics induced by hunting. Journal of Applied Ecology **42**:910–920.

Nychka, D. and N. Saltzman. 1998. Design of Air-Quality Monitoring Networks. Pages 51–76. Springer Nature, Cham.

Okubo, A. and S. A. Levin, editors. 2001. Diffusion and Ecological Problems: Modern Perspectives. Springer, New York.

Ovaskainen, O. 2004. Habitat-specific movement parameters estimated using mark–recapture data and a diffusion model. Ecology **85**:242–257.

Ovaskainen, O., H. Rekola, E. Meyke, and E. Arjas. 2008. Bayesian methods for analyzing movements in heterogeneous landscapes from mark–recapture data. Ecology **89**:542–554.

Panzacchi, M., B. Van Moorter, O. Strand, et al. 2016. Predicting the continuum between corridors and barriers to animal movements using step selection functions and randomized shortest paths. Journal of Animal Ecology **85**:32–42.

Peery, M. Z., B. H. Becker, and S. R. Beissinger. 2006. Combining demographic and count-based approaches to identify source–sink dynamics of a threatened seabird. Ecological Applications **16**:1516–1528.

Péron, G. 2013. Compensation and additivity of anthropogenic mortality: life-history effects and review of methods. Journal of Animal Ecology **82**:408–417.

Péron, G. 2019a. Modified home range kernel density estimators that take environmental interactions into account. Movement Ecology **7**:16.

Péron, G. 2019b. The time frame of home-range studies: from function to utilization. Biological Reviews **94**: 1974–1982.

Péron, G. and R. Altwegg. 2015a. Twenty-five years of change in southern African passerine diversity: nonclimatic factors of change. Global Change Biology **21**: 3347–3355.

Péron, G. and R. Altwegg. 2015b. The abundant centre syndrome and species distributions: insights from closely related species pairs in southern Africa. Global Ecology and Biogeography **24**:215–225.

Péron, G., P.-A. Crochet, P. F. Doherty Jr, and J.-D. Lebreton. 2010a. Studying dispersal at the landscape scale: efficient combination of population surveys and capture–recapture data. Ecology **91**:3365–3375.

Péron, G., J.-D. Lebreton, and P.-A. Crochet. 2010b. Breeding dispersal in black-headed gull: the value of familiarity in a contrasted environment. Journal of Animal Ecology **79**:317–326.

Péron, G., Y. Ferrand, F. Gossmann, C. Bastat, M. Guenezan, and O. Gimenez. 2011. Nonparametric spatial regression of survival probability: visualization of population sinks in Eurasian woodcock. Ecology **92**:1672–1679.

Péron, G., C. A. Nicolai, and D. N. Koons. 2012. Demographic response to perturbations: the role of compensatory density dependence in a North American duck under variable harvest regulations and changing habitat. Journal of Animal Ecology **81**:960–969.

Péron, G., R. Altwegg, G. A. Jamie, and C. N. Spottiswoode. 2016. Coupled range dynamics of brood parasites and their hosts responding to climate and vegetation changes. Journal of Animal Ecology **85**: 1191–1199.

Péron, G., C. H. Fleming, O. Duriez, et al. 2017. The energy landscape predicts flight height and wind turbine collision hazard in three species of large soaring raptor. Journal of Applied Ecology **54**:1895–1906.

Péron, G., C. Bonenfant, R. Gagnon, and C. T. Mabika. 2019. The two oxpecker species reveal the role of movement rates and foraging intensity in species coexistence. Biology Letters **15**:20190548.

Pinheiro, J., D. Bates, S. DebRoy, D. Sarkar, and the R Development Core Team. 2013. nlme: linear and nonlinear mixed effects models. R package version 3.1-111. http://cran.r-project.org/package=nlme.

Postma, E. and A. J. Van Noordwijk. 2005. Gene flow maintains a large genetic difference in clutch size at a small spatial scale. Nature **433**:65–68.

Pradel, R. 1996. Utilization of capture–mark–recapture for the study of recruitment and population growth rate. Biometrics **52**:703.

Pradel, R., J. E. Hines, J. -D. Lebreton, and J. D. Nichols. 1997. Capture–recapture survival models taking account of transients. Biometrics **53**:60–72.

Pulliam, H. R. 1988. Sources, sinks, and population regulation. The American Naturalist **132**:652–661.

Rebarber, R., B. Tenhumberg, and S. Townley. 2012. Global asymptotic stability of density dependent integral population projection models. Theoretical Population Biology **81**:81–87.

Reid, J. M., E. M. Bignal, S. Bignal, D. I. McCracken, and P. Monaghan. 2006. Spatial variation in demography and population growth rate: the importance of natal location. Journal of Animal Ecology **75**:1201–1211.

Renner, M., J. K. Parrish, J. F. Piatt, K. J. Kuletz, A. E. Edwards, and G. L. Hunt. 2013. Modeled distribution and abundance of a pelagic seabird reveal trends in relation to fisheries. Marine Ecology Progress Series **484**:259–277.

Rose, K. E., S. M. Louda, and M. Rees. 2005. Demographic and evolutionary impacts of native and invasive insect herbivores on *Cirsium canescens*. Ecology **86**:453–465.

Royle, J. A., and J. D. Nichols. 2003. Estimating abundance from repeated presence–absence data or point counts. Ecology **84**:777–790.

Royle, J. A., R. B. Chandler, K. D. Gazenski, and T. A. Graves. 2013. Spatial capture–recapture models for jointly estimating population density and landscape connectivity. Ecology **94**:287–294.

Ruppert, D., M. P. Wand, and R. J. Carroll. 2003. Semiparametric Regression. Cambridge University Press, Cambridge.

Rushing, C. S., T. B. Ryder, A. L. Scarpignato, J. F. Saracco, and P. P. Marra. 2016. Using demographic attributes from long-term monitoring data to delineate natural population structure. Journal of Applied Ecology **53**:491–500.

Saracco, J. F., J. A. Royle, D. F. DeSante, and B. Gardner. 2010. Modeling spatial variation in avian survival and residency probabilities. Ecology **91**:1885–1891.

Scharf, A. K., J. L. Belant, D. E. Beyer, M. Wikelski, and K. Safi. 2018. Habitat suitability does not capture the essence of animal-defined corridors. Movement Ecology **6**:18.

Schaub, M., O. Gimenez, B. R. Schmidt, and R. Pradel. 2004. Estimating survival and temporary emigration in the multistate capture–recapture framework. Ecology **85**:2107–2113.

Schmidt, K. A. 2004. Site fidelity in temporally correlated environments enhances population persistence. Ecology Letters **7**:176–184.

Schwartz, C. C., M. A. Haroldson, and G. C. White. 2010. Hazards affecting grizzly bear survival in the Greater Yellowstone ecosystem. Journal of Wildlife Management **74**:654–667.

Signer, J., J. Fieberg, and T. Avgar. 2017. Estimating utilization distributions from fitted step-selection functions. Ecosphere **8**:e01771.

Sims, D. W., E. J. Southall, N. E. Humphries, et al. 2008. Scaling laws of marine predator search behaviour. Nature **451**:1098–1102.

Skellam, J. G. 1951. Random dispersal in theoretical populations. Biometrika **38**:196–218.

Spear, S. F., N. Balkenhol, M.-J. Fortin, B. H. McRae, and K. Scribner. 2010. Use of resistance surfaces for landscape genetic studies: considerations for parameterization and analysis. Molecular Ecology **19**:3576–3591.

Spiegel, O., S. T. Leu, C. M. Bull, and A. Sih. 2017. What's your move? Movement as a link between personality and spatial dynamics in animal populations. Ecology Letters **20**:3–18.

Stevens, F. R., A. E. Gaughan, C. Linard, and A. J. Tatem. 2015. Disaggregating census data for population mapping using random forests with remotely-sensed and ancillary data. PLOS One **10**:e0107042.

Stige, L. C., N. A. Yaragina, Ø. Langangen, B. Bogstad, N. Chr Stenseth, and G. Ottersen. 2017. Effect of a fish stock's demographic structure on offspring survival & sensitivity to climate. Proceedings of the National Academy of Sciences of the United States of America **114**: 1347–1352.

Thuiller, W., T. Münkemüller, S. Lavergne, et al. 2013. A road map for integrating eco-evolutionary processes into biodiversity models. Ecology Letters **16**(Suppl 1): 94–105.

Touzot, L., M. C. Bel-Venner, M. Gamelon, et al. 2018. The ground plot counting method: a valid and reliable assessment tool for quantifying seed production in temperate oak forests? Forest Ecology and Management **430**: 143–149.

Vanak, A. T., D. Fortin, M. Thaker, et al. 2013. Moving to stay in place: behavioral mechanisms for coexistence of African large carnivores. Ecology **94**:2619–2631.

Wang, I. J. 2020. Topographic path analysis for modelling dispersal and functional connectivity: calculating topographic distances in the topoDistance R package. Methods in Ecology and Evolution **11**:265–272.

Wiens, J. D., N. H. Schumaker, R. D. Inman, T. C. Esque, K. M. Longshore, and K. E. Nussear. 2017. Spatial demographic models to inform conservation planning of golden eagles in renewable energy landscapes. Journal of Raptor Research **51**:234–257.

Williams, P. J., M. B. Hooten, J. N. Womble, G. G. Esslinger, M. R. Bower, and T. J. Hefley. 2017. An integrated data model to estimate spatiotemporal occupancy, abundance, and colonization dynamics. Ecology **98**:328–336.

Williams, P. J., M. B. Hooten, G. G. Esslinger, J. N. Womble, J. L. Bodkin, and M. R. Bower. 2019. The rise of an apex predator following deglaciation. Diversity and Distributions **25**:895–908.

Wolf, M., G. S. Van Doorn, O. Leimar, and F. J. Weissing. 2007. Life-history trade-offs favour the evolution of animal personalities. Nature **447**:581–584.

Wright, M. N. and A. Ziegler. 2017. ranger: a fast implementation of random forests for high dimensional data in C++ and R. Journal of Statistical Software 77 (2017).

Yackulic, C. B., J. Reid, R. Davis, J. E. Hines, J. D. Nichols, and E. Forsman. 2012. Neighborhood and habitat effects on vital rates: expansion of the barred owl in the Oregon Coast Ranges. Ecology **93**:1953–1966.

Zeller, K. A., K. McGarigal, S. A. Cushman, P. Beier, T. W. Vickers, and W. M. Boyce. 2016. Using step and path selection functions for estimating resistance to movement: pumas as a case study. Landscape Ecology **31**: 1319–1335.

Zera, A. J. 2017. The biochemical basis of life history adaptation: gryllus studies lead the way. Pages 229–243 *in* H. Noji, W. Horch, T. Mito, A. Popadićhideyo, and O. Sumihare, editors. The Cricket as a Model Organism. Springer, Tokyo.

Zheng, C. Z., O. Ovaskainen, and I. Hanski. 2009. Modelling single nucleotide effects in phosphoglucose isomerase on dispersal in the Glanville fritillary butterfly: coupling of ecological and evolutionary dynamics. Philosophical Transactions of the Royal Society B: Biological Sciences **364**:1519–1532.

CHAPTER 16

Evolutionary Demography

Shripad Tuljapurkar and Wenyun Zuo

16.1 Introduction

Evolutionary demography has grown rapidly in recent years, as witness this volume, and now covers most topics at the intersection of demography and evolution. Particular impetus has come from work on senescence in humans and then in other species, aided by the rapid growth of databases such as the HMD (University of California, Berkeley (USA) and Max Planck Institute for Demographic Research (Germany) 2019), HFD (Max Planck Institute for Demographic Research (Germany) and Vienna Institute of Demography (Austria) 2019), COMPADRE (Salguero-Gómez et al. 2015), and COMADRE (Salguero-Gómez et al. 2016). The biological topics of life history evolution and evolution in population with complex life cycles have benefitted from and contributed to this newer and broader focus on evolutionary biodemography. The goals of this chapter are to provide a critical summary of the central ideas and methods. We emphasise theoretical methods, starting with the main ideas that have attracted attention in the field, the assumptions behind these, and efforts to relax those assumptions, and provide a short account of some new directions. Only a few empirical findings are discussed. The reader will find more extensive accounts of empirical work elsewhere in this volume. We are deliberately not historically complete or comprehensive, but give references that the reader can use as points of departure and discovery.

We begin with the classic work of Peter Medawar and William Hamilton but in somewhat modern form. Their ideas and results are basic in evolutionary demography, and we discuss briefly the connections, applications, assumptions, and limitations. Then we discuss extensions to variable (random) environments and the large body of theory around that topic. Next we discuss how these theoretical methods are related to analyses and theories of post-reproductive life, via the general concept of "borrowing fitness". Finally, we discuss nonlinear models of mutation and selection and density-dependent models.

16.2 The early theory

Medawar (1957) and Hamilton (1966) are foundational to the theory senescence and well worth a read. Both start with an age-structured population in demographic equilibrium, whose individuals have a survival rate p_x and a fertility rate m_x that both depend on age x. At demographic equilibrium the population changes at a constant rate. Usually only one sex is counted, typically females, although extensions are possible and are discussed briefly later. Ages can be discrete or continuous: here, we use discrete age and time (the discrete time interval used may be a day or a month, etc.). Newborns counted at the beginning of a discrete time interval are labelled as being in age class 1 (although some prefer age 0), meaning that they are between exact age 0 and any age up to 1 when counted. For brevity, we often use "age" interchangeably with "age class." In discrete time, the growth rate of the population, called λ, is geometric per interval of time. It is often convenient to use the equivalent exponential rate $r = \log \lambda$. Then r solves the famous Lotka equation,

$$1 = \sum_{\alpha}^{\beta} e^{-rx} l(x) m(x), \qquad (16.1)$$

and using the age-specific survival probabilities

$$p_x = \text{survival rate from age } x \text{ to age } x + 1,$$

$$l_1 = 1, \; l_x = p_1 p_2 \ldots p_{x-1}, \; x \geq 2.$$

Shripad Tuljapurkar and Wenyun Zuo, *Evolutionary Demography*. In: *Demographic Methods Across the Tree of Life*. Edited by Roberto Salguero-Gómez and Marlène Gamelon, Oxford University Press.

In this equation the symbols α and β are the first and last ages of reproduction respectively. The approach we have used here is suited to humans, but details may differ for other species (Caswell 2001).

The Lotka equation makes clear that r (or the equivalent one-period growth rate $\lambda = e^r$) is a function of α, β, p_x, and m_x—and these parameters make up a life history. This equation allows one to analyse the relationship between life history components, but of course the equation, by itself, cannot tell what the values of these quantities can or should be. Rather, this is a "stone tablet" equation. To use it, we must be given (as if on a tablet) a range of life histories, usually defined by life history parameters, and for each parameter set within this range the Lotka equation yields corresponding values of λ. How then should we use the Lotka equation to understand evolution?

The first task is to identify fitness, and Hamilton (and presumably Medawar) assumed that fitness is given by λ in eqn. (16.1). This idea goes back to Fisher (1930) and the foundations of evolutionary theory. One can now work "locally", as Hamilton did, and compute the changes in the fitness λ produced by small changes in the life history. If the small changes are assumed to be due to mutations, the changes in fitness can be interpreted as selection acting on the mutations, and we can think about local evolutionary change of the life history. In our view, Medawar's arguments are essentially similar. This local approach has been extensively developed and expounded in ecological and evolutionary theory (Charlesworth 1994; Caswell 2001). Similar arguments have been used to study random environments and other factors.

The local question that Hamilton asked is: What is the force of selection against a mutation that causes a small change in survival or fertility at age x? The answer lies in the partial derivatives of λ, also known as sensitivities:

$$s_{px} = (\partial\lambda/\partial p_x) \text{ and } s_{mx} = (\partial\lambda/\partial m_x). \quad (16.2)$$

Thus, if a change in only fertility m_x at age x is caused by a mutation acting only at that age, we can interpret s_{mx} as the force of selection acting on that mutation. Of course the forces of selection in eqn. (16.2) are negative when the mutations decrease fertility or survival, as expected of the vast majority of mutations that are likely to be deleterious. If mutations reducing fertility at age x occur at a constant rate ν, and if the effect of mutations on fertility is simply additive, the equilibrium frequency of mutations maintained by balancing the arrival rate against the corresponding force of selection is

$$q^* = \left(\frac{\nu}{|s_{mx}|} \right). \quad (16.3)$$

Key assumptions here are that mutations occur at a single autosomal locus in a large population (so drift is not an issue) and that mutations have an additive effect. Generalisations of eqn. (16.3) to consider effects on multiple ages and correlations across ages are discussed by Charlesworth (2001).

Important aspects and implications of these equations are as follows:

(a) The sensitivities above have been explored in detail, as also have the corresponding elasticities that are defined as follows: given s_{px}, the corresponding elasticity is

$$e_{px} = (p_x/\lambda)\, s_{px} = (\partial\log\lambda/\partial\log p_x). \quad (16.4)$$

An example of these is given for red deer (*Cervus elephas*, Benton et al. (1995) for data) in Figure 16.1. Clearly we can define sensitivities and elasticities for changes in any kind of life cycle, for example, one defined by stage rather than age, or one that requires age+stage. This approach has been used to gain many ecological and evolutionary insights (e.g. cases in this book). See Caswell et al. (2018) for recent developments and extensions, as well as references to earlier work. We should note (Caswell 2001) that selection pressures are directly related to sensitivities, not elasticities.

(b) It should be obvious that the elasticities are zero at ages where the vital rates are currently zero. This is always true when fertility or survival are set at zero by the biology. For example, fertility is always zero at ages x past the maximum potential reproductive age, such as age at menopause for humans. Figure 16.2 shows the difference between sensitivities and elasticities to fertility and survival for a modern human population. The age limit of reproduction is critical to any work that attempts to apply Hamilton-type theories on senescence to species with post-reproductive life. In particular note that the Lotka eqn. (16.1) shows that λ only depends on reproduction and survival between the ages of first and last reproduction. Some previous work ignores this limitation or tries to circumvent this point in unconvincing ways. For example, the otherwise elegant analysis in Abrams and Ludwig (1995) is only relevant when there is no upper limit to the age of reproduction.

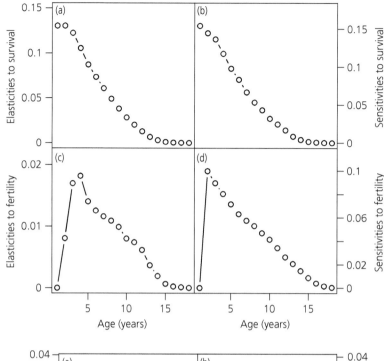

Figure 16.1 The sensitivities and elasticities to survival and fertility for Red deer (*Cervus elephas*). (Benton et al. (1995) for data).

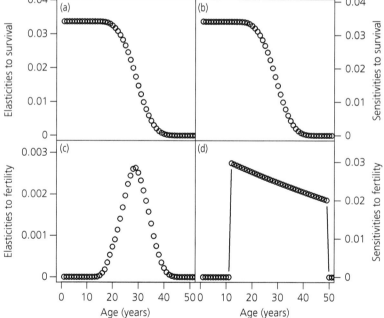

Figure 16.2 The sensitivities and elasticities to survival and fertility for Japan at 2009. (Data provided by van Daalen and Caswell (2017).)

(c) The reader will note that all sensitivities decrease with increasing age (after the age of first repro- duction), a general feature of age-structured life histories (Caswell 2001). In consequence, the equilibrium frequency of mutations implied by

eqn. (16.3) increases at old age. And once repro- duction stops, mutational frequencies should increase towards 1 (i.e. fixation). In such cases, some deleterious mutations will always get in, and more of them get in at late age, so that mutations should "chip" away at late-age reproduction. Thus the Hamilton theory by itself must surely lead to an evolutionarily unstable age of last reproduction. This begs the question, why are we all not like salmon (Tuljapurkar 1997)? Wachter et al. (2013) call this limit, more strikingly, a "wall of death."

(d) The preceding point also poses a fatal obstacle to Williams's celebrated view that we should think about ageing as a result of antagonistic pleiotropy. This view argues that vital rates at early ages are high because late-age vital rates are low, and that such trade-offs are genetic, hence the antagonism (Williams 1996). But such antagonism can only strengthen the instability of life histories (with a defined age of last reproduction), as discussed above. Certainly cases of antagonistic pleiotropy can be found – but antagonistic pleiotropy cannot provide a general explanation for senescence.

(e) One direction of improvement of the theory is to analyse mutations that affect a range of ages, so that the selection pressure does not depend only on one focal late age. In any model using an age range, mutations are selected against over the entire age-range as in Charlesworth (2009). This is, of course, an example of correlated effects of deleterious mutations across several ages, which is an example of positive rather than antagonistic pleiotropy.

(f) Another and even more important direction recognises that existing mutations must decrease individual fitness. As a result, the growth rate λ of a population depends on the distribution of mutations over individuals in the population, which of course requires a nonlinear theory, which has been developed by Evans et al. (2013). In a nonlinear theory, mutations can affect several ages, the effect of mutations on fitness is not additive, and different individuals may carry different numbers of mutations. See the reference for more details.

(g) The great increase in human life span in the past century has mostly occurred at older ages, that is, ages past menopause. This change cannot be explained by the application of traditional Hamilton theory, or even some improved versions of it. A different approach is needed and is discussed below.

(h) The Hamilton theory assumes the population is in demographic equilibrium with constant vital rates. In cases where density is assumed to maintain an equilibrium population, Caswell (2001) shows how a density-independent analysis is relevant to density-dependent models at equilibrium. More generally, the Hamilton theory does not apply directly in variable environments, and we discuss stochastic extensions of the theory below.

One can also work more "in the large," asking whether trade-offs and constraints may be used to define a region of fitness space in which some set of life histories is more fit than alternatives. These constraints may be energetic, as in arguments that metabolic rate reflects the "pace of life." The production rate scales with body mass in the same way as metabolic rate, so that, per unit mass, large sized organisms have slower rates of production than small sized ones (Brown and Sibly 2006). On the other hand, the variations in lifestyles and environmental conditions also affect the evolution of body size and production rates, thus directly affecting energy consumption and allocation (Sibly and Brown 2007). Constraints may also be biomechanical. Biomechanical designs of organisms are critical to performance of organisms which affects their fitness, such as the effect of fish morphology on swimming performance (Langerhans and Reznick 2009; Higham et al. 2016) and plant traits on support and anchorage (Niklas 1992, 1994; Read and Stokes 2006). Finally, factors such as energetics, biomechanics, and genetic correlations also act at the population level and can be based on representations of the ecological dynamics that affect life histories, as in the arguments about "r-K" selection and the fast- or slow-paced lifestyle (Stearns 1992). The latter are discussed later in this chapter.

16.3 Randomly varying environments

Unpredictable environmental changes are so ubiquitous that it makes sense to replace the Lotka eqn. (16.1) by equations that include variability. In the presence of temporal variation, a life history necessarily depends on several factors – the average vital rates, variances and covariances of vital rates at any one time, and serial correlations between times now and in the future (or past). The latter quantities tell us how environments vary and how vital rates respond to that variation.

The relevant theory originated with Furstenberg and Kesten (1960) and Cohen (1976), and later developments and other references are discussed in Tuljapurkar (1990). The theory has since been extensively explained, developed, and applied. Let $\mathbf{n}_t$ denote the vector of population numbers at any time t. Then the population between times $(t-1)$ and t is governed by a random vital rate matrix $\mathbf{A}_t$, and the population vector at time t is given by

$$\mathbf{n}_t = \mathbf{A}_t \mathbf{n}_{t-1}. \tag{16.5}$$

In this interval, the total population sizes change from N_{t-1} at $(t-1)$ to N_t with (time-varying) growth rate

$$\lambda_t = \frac{N_t}{N_{t-1}}. \tag{16.6}$$

At each time, the environmental state determines the projection matrix that applies; for example, if the environment is in state x at time t, that state determines that the projection matrix is $\mathbf{B}_x$. The long-run population growth rate converges to a (deterministic) limit. Over time, environments, or equivalently matrices, occur in a random sequence that we denote by the symbol ω. Over such a sequence, the cumulative growth rate from $t=0$ to $t=T$ is the product of single-period growth rates,

$$\Lambda_T(\omega) = \lambda_T \lambda_{T-1} \ldots, \lambda_1. \tag{16.7}$$

The theory ensures that every random sequence (except sequences of zero probability) has the same stochastic growth rate

$$a = \lim_{T\to\infty} \frac{1}{T} \log \Lambda_T. \tag{16.8}$$

Existence of this limit occurs under wide conditions (Tuljapurkar 1990). In brief, the requirements are weak demographic ergodicity and mixing of the stochastic process (a finite number of matrices is not required). The limit in eqn. (16.8) clearly equals an average over sequences,

$$a = \lim_{T\to\infty} \frac{1}{T} \mathcal{E}[\log \Lambda_T(\omega)]. \tag{16.9}$$

Fitness is now the stochastic growth rate a defined by eqn. (16.9), from which one can compute a growth rate $\lambda_S = e^a$.

Now separate the average projection matrix and the random bit by writing

$$\mathbf{A}_t = \mathbf{B} + \epsilon \mathbf{H}_t, \tag{16.10}$$

$$\mathcal{E}(\mathbf{H}_t) = 0, \tag{16.11}$$

where $\epsilon > 0$ is a scale factor that determines the magnitude of the random effects, and $\mathcal{E}$ indicates an expectation (average) over random environments. The projection matrix $\mathbf{B}$ contains average vital rates and has a dominant eigenvalue λ. This theory applies to age+stage models, so in our discussion we take the projection matrix $\mathbf{A}_t$ to have elements $A_t(i,j)$ in row i and column j. The average matrix $\mathbf{B}$ has sensitivities $s_{i,j}$

with respect to changes in the average $b(i,j)$ of the (i,j) element.

Assume the random matrix $\mathbf{H}_t$ has stationary variance-covariance at all lags (e.g., when the environments are stationary and ergodic). At each time t, so at time lag zero, the variance-covariance matrix has elements

$$\mathcal{E}\{H_1(i,j) H_1(k,l)\}, \tag{16.12}$$

and at times separated by a time lag $M > 0$, the serial variance-covariance matrix has elements

$$\mathcal{E}\{H_{M+1}(i,j) H_1(k,l)\}. \tag{16.13}$$

When the scaling factor ϵ is small, Tuljapurkar (1982) showed that the stochastic growth rate is given by the general small-noise expansion

$$a \approx \log \lambda - W_1 + W_2, \tag{16.14}$$

where both W_1 and W_2 are $O(\epsilon^2)$. The term W_1 is due to interannual variability (i.e., variability within each time interval) whereas W_2 is due to autocorrelation between distinct times. The contribution W_1 is

$$W_1 = \frac{1}{2\lambda^2} \sum_{ij,kl} s_{ij}\, s_{k,l}\, \mathrm{Cov}\{H_1(i,j) H_1(k,l)\}. \tag{16.15}$$

The effects of serial correlations and the nature of W_2 are discussed by Tuljapurkar and Haridas (2006) and more generally by Tuljapurkar et al. (2009).

Important implications of these equations are as follows:

(a) A basic question: Does the relative fitness of a life history depend on the amount of environmentally driven variability in vital rates? The answer is a decided yes. We illustrate with a simple two-stage life history with delays possible at juvenile and/or adult stages (Tuljapurkar and Wiener 2000). These can be simply represented using the projection matrix

$$\mathbf{A}_t = \begin{pmatrix} p & R_t(1-q) \\ (1-p) & q \end{pmatrix}. \tag{16.16}$$

The extent of delay at each stage is measured by parameters (p,q). Thus if $(p,q)=(0,0)$ then all individuals live exactly two time-steps, one as an immature and one as an adult producing R_t offspring. If $p>0$, then immature may remain in that state for several time-steps, and if $q>0$ then adults can "spread" their mean production of R_t offspring over several time-steps. Note that these delay parameters are always fixed here. Stochasticity enters via the reproductive rate R_t which varies

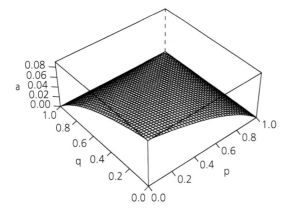

Figure 16.3 Fitness surface for constant environment. *a*: Stochastic growth rate (see eqn. 16.14). *p*: Extent of delay at juvenile stage (see eqn. 16.16). *q*: Extent of delay at adult stage (see eqn. 16.16).

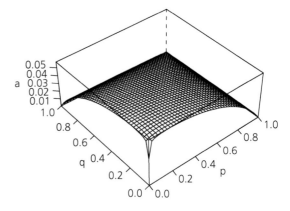

Figure 16.4 Fitness surface for variable environment. *a*: Stochastic growth rate (see eqn. 16.14). *p*: Extent of delay at juvenile stage (see eqn. 16.16). *q*: Extent of delay at adult stage (see eqn. 16.16).

around an average $\overline{R}$ so that

$$R_t = \overline{R} + H_t. \tag{16.17}$$

Now suppose there is a constant environment, in which case the growth rate is the logarithm of the dominant eigenvalue λ of the average projection matrix. Either kind of delay slows down the life history, so $\log\lambda$ decreases as either p or q increases, as shown in Figure 16.3. Therefore in a constant environment we do not expect delayed life cycles to evolve. But in a variable environment the projection matrix (eqn. 16.16) is randomly varying. The stochastic growth rate a in the presence of variability, computed by simulation, shows (Figure 16.4) that a small amount of variability causes the growth rate to increase sharply as the amount of delay (of either kind) increases, and then a falls when delay is long. Clearly this is an argument for the evolution of life cycle delays and would not hold in a constant environment. We note also that, in this model, serial autocorrelation of reproduction R_t can produce a positive feedback effect on fitness and thus on evolution. As seen in Figure 16.5, autocorrelation leads to a greater increase of fitness when just one type of delay increases. Evolution is then expected to lead to life histories in which one or the other kind of delay predominates. This feedback effect is seen more clearly in a contour plot, Figure 16.6.

(b) Fitness a in a variable environment depends on all aspects of how the vital rates vary in response to environments. So fitness depends not just on average vital rates but also on all of the variance-

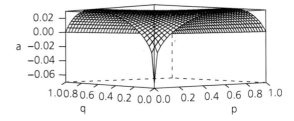

Figure 16.5 Fitness surface for variable environment with a serial autocorrelation. *a*: Stochastic growth rate (see eqn. 16.14). *p*: Extent of delay at juvenile stage (see eqn. 16.16). *q*: Extent of delay at adult stage (see eqn. 16.16).

covariance elements in eqns. 16.12 and 16.13. As a first step, we can simply look at changes in average rates and compute elasticities such as

$$\frac{\partial a}{\partial \log b(i,j)}.$$

Caswell (2001) discusses how the small-noise approach can be used to understand these. But more generally, there are many elasticities of a, as explained with examples by Tuljapurkar et al. (2003). In that paper, we point out that a general decomposition of any change in vital rates can be analysed by combining the effects on fitness when one changes just the mean or just the variance of vital rates. This decomposition is applied to life history evolution in Haridas and Tuljapurkar (2005).

(c) More generally we can examine how a changes with respect to the the variation in the environment and/or the vital rates. A given environmental pattern leads to environmental sequences, and we can change the vital rates associated to each

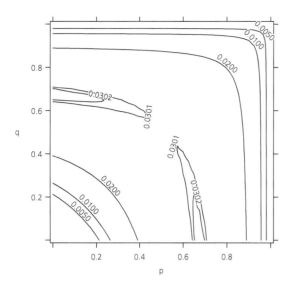

Figure 16.6 Contours for variable environment with a serial autocorrelation. The contours show the stochastic growth rate a (see eqn. 16.14). p: Extent of delay at juvenile stage (see eqn. 16.16). q: Extent of delay at adult stage (see eqn. 16.16).

environmental state – which affects average rates or the deviations in eqn. (16.10). Alternatively, we can maintain the association between vital rates and environments, but change the rules governing environmental sequences. As a result we have two kinds of sensitivities: derivatives of a with respect to vital rates and derivatives of a with respect to environmental pattern. The paper by Steinsaltz et al. (2011) provides fast numerical methods of computing a and both kinds of derivatives.

(d) The importance of autocorelation in understanding evolutionary demography was analysed in detail in an important paper by Orzack (1993). New analyses in Paniw et al. (2018) present analyses of the qualitative and quantitative effects of serial correlation. More general perspectives on the importance of environmental variability in life history evolution are given by Tuljapurkar et al. (2009).

(e) Empirical studies of serial correlation are unfortunately still rare. However, eqn. (16.14) provides useful ways of analysing data to separate the contributions of variability and correlation, as shown by Davison et al. (2010). The small-noise method can also be used in combination with randomization to understand whether a data set evidences serial autocorrelation (Jongejans et al. 2010).

(f) Finally we note the robust conclusion that environmental variability allows for the coexistence of very different life histories that would have unequal fitness in deterministic environments (Orzack 1985; Tuljapurkar and Wiener 2000).

16.4 Post-reproductive life

Post-reproductive life is fairly common among female mammals (Cohen 2004; Ellis et al. 2018). In humans, the end of female reproduction is marked by menopause at age around 50 years. There has been far greater downward secular change in the age at menarche (Frisch and Revelle 1971; Prentice et al. 2010) than in the age at menopause over the past century. Yet this period has seen a dramatic increase, essentially a doubling, in human lifespan (Tuljapurkar and Boe 1998). Gurven and Kaplan (2007) even present data suggesting that early humans lived well past the age of menopause. In recent humans, extended life is unlikely (probably impossible) to be the result of genetic change (Zuo et al. 2018), which means that (1) post-reproductive life is a very plastic trait, and (2) in evolutionary terms, selection must have made this plasticity possible by preventing the accumulation of mutations that would make rapid increases impossible. Even a short human life after menopause needs evolutionary explanation, as more broadly discussed by Levitis et al. (2013).

We address here one general class of evolutionary arguments about post-reproductive life: that older individuals "borrow fitness" from younger females who are reproductively capable. Such borrowing occurs when old individuals (who cannot reproduce) take actions that increase the fitness of younger individuals (who can reproduce). As a result, selection in favour of reproduction by younger females will act indirectly to maintain the longer life of older individuals. One version of this is provided by the celebrated grandmother hypothesis (Hawkes et al. 1998), a second is the older-male hypothesis (Tuljapurkar et al. 2007), and the most general is the transfer hypothesis (Lee 2003). Lee's analysis is framed in the context of economic transfers but applies to transfers of any kind including the provision of care, the sharing of knowledge or information, and so on, and is the most general formulation (Tuljapurkar 2019). Useful matrix models of such models have been analysed by Pavard and Branger (2012).

It is important to note that an analysis of one set of historical data by Moorad and Walling (2017) does not support these hypotheses. However, Moorad and Walling use quantitative genetic models to translate the evolutionary hypotheses into hypotheses about correlations in the data. It may be useful to reexamine the relevant arguments and look at other data sets.

However, some new evolutionary hypotheses may be needed (Cohen 2017)!

The field of evolutionary demography has received considerable impetus from work on human senescence and human health in old age. The existence of extended human lives is, as we have seen, a challenge for evolutionary theory. But for evolutionary arguments to bear on human health, beyond the usual "just-so" stories, is even more of a challenge. Metabolic theory has been of value in many ecological settings (as illustrated by Brown and Sibly (2006); Sibly and Brown (2007)), and a different kind of metabolic theory has been important in human health and healthy aging (Steptoe et al. 2015; Camandola and Mattson 2017). The time is ripe for a synthesis of these streams of thought and may open a useful wedge into the effects of evolution on health. There is of course a large literature on evolutionary medicine, some of which is reviewed by Ewald (2018), but we do not believe this is useful or relevant in the context of the evolution of late-age mortality.

16.5 Nonlinear models for mutation–selection

A distinct and important issue is the correct representation of mutation–selection balance. This is essentially the same issue raised by Kimura and Maruyama (1966) who considered the distribution of mutations in a population. Since mutations affect individuals, any population will contain individuals who carry different numbers of mutations (0, 1, 2, . . . mutations). As Kimura and Maruyama (1966) pointed out, we can expect that selection will result in an equilibrium distribution of mutations. In the simplest situation, there will be a random number M of mutations in any individual, and the probability distribution $p(m) = \Pr\{M = m\}$ will be maintained by a mutation–selection balance. This is quite different from the simple analysis that leads to eqn. (16.3) as the number of mutations per individual. In the present setting, individuals will carry different numbers of mutations, and individual and average fitness will depend on the relationship between mutations and individual-level vital rates.

The general setting and suitable tools are described by Evans et al. (2013). The mathematical treatment here is beyond that usually found in biology, but the book begins with a useful informal description that motivates and explains the theory. A demographic implementation of this nonlinear theory, along with a variety of pleiotropic effects of mutation, has been applied by Wachter et al. (2013). This theory does

provide examples of unstable life histories that, presumably, are not held together by positive pleiotropy. This is a surprise because earlier additive theories for mutation accumulation assumed that positive pleiotropy was a way of transferring fitness across the life cycle. The demographic analysis may seem technically formidable, but the examples in the paper (Wachter et al. 2013) show that numerical implementation should be relatively straightforward. These nonlinear analyses are important for many questions in the evolutionary demography of complex life cycles, and extensions and developments should be valuable.

16.6 Density-dependent models

Finally, we consider density dependence canonically exemplified for one species by the logistic model (Verhulst 1838), is widely used in population and community ecology and describes the population by

$$N_{t+1} - N_t = rN_t \left(1 - \frac{N_t}{K} \right) \qquad (16.18)$$

where N is the population size, t is time, K is the carrying capacity (also known as equilibrium density), and r is the per capita growth rate (also known as the intrinsic rate of increase). In continuous time, the left side of eqn. (16.18) is a derivative; here we ignore differences between continuous and discrete time versions (that may be very large; see e.g. Royama (1992)). In eqn. (16.18), a population starting at $N_0 < K$ grows until it reaches carrying capacity K, which is the maximum number of individuals the environment can support. The growth rate r is the difference between the per capita birth rate and per capita death rate.

MacArthur (1962) and MacArthur and Wilson (1967) proposed the theory of r- and K-selection, the first of its kind addressing density-dependent life history evolution. The fitness in the case of r-selection is just r. In harsh environments (such as extreme seasonality, fires, floods, or storms), populations decline because of increased density-independent mortality, then re-expand. In this scenario, selection would favour individuals that have high productivity as measured by r. Alternatively, the fitness in the case of K-selection is the carrying capacity, K. In stable environments, populations stay close to carrying capacity, so selection would favour individuals with high efficiency of resource utilisation. Later work by Roughgarden (1971) united these two theories of evolution (r- and K-selection) with application to evolution in environments of intermediate harshness. Charlesworth (1971) generalised this by assuming that fitness is a strictly decreasing function of population size. He showed that K_i (the carrying capacity of genotype i) determines whether a stable polymorphism exists and r_i (the per capita growth rate of genotype i) controls the composition of the polymorphism. Moreover, he demonstrated that the amount of genetic load in the population affected the mean survival of genotypes within a generation, producing an effectively density-independent force of selection on both the r and K. In recent years, this form of selection theory has been criticised as too simplistic (Nosil et al. 2018).

Intraspecific dynamics may not be enough to understand or predict life history evolution. Interspecific interactions such as competition and predation may have to be included to get close to reality. When a pair of species (e.g., competitors, or a predator–prey pair) or a multispecies system (such as a food web) is considered, intraspecific interactions may lead to predictions inconsistent with single species models (Wootton 1994; Abrams 2007, 2019). A relevant read on the evolution of predator-prey interactions is Abrams (2000).

A standard model for "consumer-resource" interactions is the Lotka–Volterra. Incorporating a density-dependent resource (MacArthur 1970, 1972) leads to the model

$$N_{t+1} - N_t = rN_t \left(1 - \frac{N_t}{K} \right) - cN_t P_t \qquad (16.19)$$

$$P_{t+1} - P_t = P_t(bcN_t - d). \qquad (16.20)$$

Here N and P are population sizes of resource and consumer respectively, t is time, r and K describe the resource, c is the capture rate by an average consumer, b is the conversion efficiency of captured resource into new consumers, and d is the per capita death rate of consumers. The consumer species has linear functional and numerical responses, and the resource has logistic growth in the absence of the consumer.

Common modifications of eqns. 16.19 and 16.20 are:

(a) replace a linear functional response (the capture rate by an average consumer) with nonlinear versions (Holling 1959; Abrams 1980, 1990, 2009a; Abrams and Ginzburg 2000);
(b) replace a linear numerical response (the relationship between the amount eaten and the per capita consumer reproductive rate) with a nonlinear version (Abrams 2009a);
(c) impose a negative effect of consumer density on the numerical or functional responses (Abrams 1993);

(d) impose a consumer foraging time effect on the functional responses (Abrams 1982);
(e) incorporate nonlinear density-dependence growth of the resource population (Abrams 2001, 2009a).

Such models have been applied to interactions among different numbers of consumers and/or resources (Abrams 1977, 1980, 1998, 2009b).

Any life history is directly or indirectly affected by traits such as body mass (Dieckmann et al. 1995; Brown and Sibly 2006) and locomotion (Langerhans and Reznick 2009). Traits that suit environments and thus increase individual fitness are called adaptive traits; in the opposite case we have maladaptive traits. At the individual level, selection presumably favours adaptive traits. Surprisingly, adaptive traits do not always increase population growth or numbers, and vice versa. These counter-intuitive situations have been both termed as "adaptive decline" (Abrams 2019; Brady et al. 2019b). When and how often does adaptive decline happen? In eqn. (16.19), the individual fitness of a consumer is $(bcN_t - d)$, where b, c, and d are as previously defined, but now the aim is to consider their evolution. When the consumer population is at equilibrium, the equilibrium population size of resource is

$$N^* = \frac{d}{bc}. \qquad (16.21)$$

Since natural selection should always favour increasing individual fitness, adaptive evolution at consumer individual level should increase b and c and/or decrease d whenever possible. But any such change reduces the equilibrium resource population (eqn. (16.21)). Substituting eqn. (16.21) into eqn. (16.20) yields the equilibrium population size of the consumer

$$P^* = \frac{r}{c} \left(1 - \frac{d}{Kbc} \right), \qquad (16.22)$$

which indicates that adaptive evolution on b, c, and d may affect P^* in either direction. However density-independence resource growth, where $(1/K) \approx 0$ in eqn. (16.19), guarantees adaptive decline (see eqn. (16.22)). Abrams (2019) points out that the special case may actually be common, which implies frequent adaptive decline. This may explain why maladaptation is often observed (Brady et al. 2019a, 2019b). For more complex consumer–resource systems, such as food webs, these changes have yet to be explored.

In this way, models of life history evolution built on density independence may be usefully extended to incorporate the effects of certain kinds of density independence.

References

Abrams, P.A. (1977). "Density-Independent Mortality and Interspecific Competition: A Test of Pianka's Niche Overlap Hypothesis". *The American Naturalist* **111**.979, pp. 539–552.

Abrams, P.A. (1980). "Consumer Functional Response and Competition in Consumer Resource Systems". *Theoretical Population Biology* **17**.1, pp. 80–102.

Abrams, P.A. (1982). "Functional Responses of Optimal Foragers". *The American Naturalist* **120**.3, pp. 382–390.

Abrams, P.A. (1990). "The Effects of Adaptive Behavior on the Type-2 Functional Response". *Ecology* **71**.3, pp. 877–885.

Abrams, P.A. (1993). "Why Predation Rate Should Not be Proportional to Predator Density". *Ecology* **74**.3, pp. 726–733.

Abrams, P.A. (1998). "High Competition with Low Similarity and Low Competition with High Similarity: Exploitative and Apparent Competition in Consumer-Resource Systems". *The American Naturalist* **152**.1, pp. 114–128.

Abrams, P.A. (2000). "The Evolution of Predator-Prey Interactions: Theory and Evidence". *Annual Review of Ecology and Systematics* **31**.1, pp. 79–105.

Abrams, P.A. (2001). "The Effect of Density-Independent Mortality on the Coexistence of Exploitative Competitors for Renewing Resources". *The American Naturalist* **158**.5. PMID: 18707301, pp. 459–470.

Abrams, P.A. (2007). "Defining and Measuring the Impact of Dynamic Traits on Interspecific Interactions". *Ecology* **88**.10, pp. 2555–2562.

Abrams, P.A. (2009a). "Determining the Functional Form of Density Dependence: Deductive Approaches for Consumer Resource Systems Having a Single Resource". *The American Naturalist* **174**.3, pp. 321–330.

Abrams, P.A. (2009b). "The Implications of Using Multiple Resources for Consumer Density Dependence". *Evolutionary Ecology Research* **11**, pp. 517–540.

Abrams, P.A. (2019). "How Does the Evolution of Universal Ecological Traits Affect Population Size? Lessons from Simple Models". *The American Naturalist* **193**.6. PMID: 31094600, pp. 814–829.

Abrams, P.A. and Lev R. Ginzburg (2000). "The Nature of Predation: Prey Dependent, Ratio Dependent or Neither?" *Trends in Ecology & Evolution* **15**.8, pp. 337–341.

Abrams, P.A and Donald Ludwig (1995). "Optimality Theory, Gompertz'law and the Disposable Soma Theory of Senescence". *Evolution* **49**.6, pp. 1055–1066.

Benton, T. G., A. Grant, and T. H. Clutton-Brock (1995). "Does Environmental Stochasticity Matter? Analysis of Red Deer Life-Histories on Rum". *Evolutionary Ecology* **9**.6, pp. 559–574.

Brady, S.P. et al. (2019a). "Causes of Maladaptation". *Evolutionary Applications* **12**.7, pp. 1229–1242.

Brady, S.P. et al. (2019b). "Understanding Maladaptation by Uniting Ecological and Evolutionary Perspectives". *The American Naturalist* **194**.4. PMID: 31490718, pp. 495–515.

Brown, J.H. and R.M. Sibly (2006). "Inaugural Article: Life-History Evolution Under a Production Constraint". *Proceedings of the National Academy of Sciences* **103**.47, p. 17595.

Camandola, S. and Mark P Mattson (2017)." Brain Metabolism in Health, Aging, and Neurodegeneration". *The EMBO Journal* **36**.11, pp. 1474–1492.

Caswell, H. (2001). *Matrix Population Models: Construction, Analysis, and Interpretation*. 2nd edition. Sunderland, MA: Sinauer Associates.

Caswell, H. et al. (2018). "Age × Stage-Classified Demographic Analysis: A Comprehensive Approach". *Ecological Monographs* **88**.4, pp. 560–584.

Charlesworth, B. (1971). "Selection in Density-Regulated Populations". *Ecology* **52**.3, pp. 469–474.

Charlesworth, B. (1994). *Evolution in Age-Structured Populations*. Cambridge: Cambridge University Press.

Charlesworth, B. (2001). "Patterns of Age-Specific Means and Genetic Variances of Mortality Rates Predicted by the Mutation-Accumulation Theory of Ageing". *Journal of Theoretical Biology* **210**.1, pp. 47–65.

Charlesworth, B. (2009). "Effective Population Size and Patterns of Molecular Evolution and Variation". *Nature Reviews Genetics* **10**.3, pp. 195–205.

Cohen, A.A. (2004). "Female Post-Reproductive Lifespan: A General Mammalian Trait". *Biological Reviews* **79**.4, pp. 733–750.

Cohen, A.A. (2017). "The Mystery of Life Beyond Menopause". *Nature Ecology and Evolution* **1**.11, p. 1604.

Cohen, J.E. (1976). "Ergodicity of Age Structure in Populations with Markovian Vital Rates, I: Countable States". *Journal of the American Statistical Association* **71**.354, pp. 335–339.

Davison, R. et al. (2010). "Advances in Plant Demography Using Matrix Models". *Journal of Ecology* **98**.2, pp. 255–267.

Dieckmann, U., Paul Marrow, and Richard Law (1995). "Evolutionary Cycling in Predator-Prey Interactions: Population Dynamics and the Red Queen". *Journal of Theoretical Biology* **176**.1, pp. 91–102.

Ellis, S. et al. (2018). "Postreproductive Lifespans are Rare in Mammals". *Ecology and Evolution* **8**.5, pp. 2482–2494.

Evans, S.N., David Steinsaltz, and Kenneth W Wachter (2013). A Mutation–Selection Model with Recombination for General Genotypes. Vol. 222. Providence, RI: American Mathematical Society.

Ewald, P.W. (2018). Darwinian Medicine: Evolutionary Approaches to Disease. In: H. Callan (ed.) *The International Encyclopedia of Anthropology*, pp. 1–9. John Wiley & Sons, Hoboken.

Fisher, R. A. (1930). *The Genetical Theory of Natural Selection*. Oxford: Clarendon Press.

Frisch, R.E. and Roger Revelle (1971). "Height and Weight at Menarche and a Hypothesis of Menarche". *Archives of Disease in Childhood* **46**.249, pp. 695–701.

Furstenberg, H. and H. Kesten (1960). "Products of Random Matrices". *The Annals of Mathematical Statistics* **31**.2, pp. 457–469.

Gurven, M. and Hillard Kaplan (2007). "Longevity Among Hunter-Gatherers: A Cross-Cultural Examination". *Population and Development Review* **33**.2, pp. 321–365.

Hamilton, WD (1966). "The Moulding of Senescence by Natural Selection". *Journal of Theoretical Biology* **12**.1, pp. 12–45.

Haridas, C. V. and S. Tuljapurkar (2005). "Elasticities in Variable Environments: Properties and Implications". *The American Naturalist*. **166**.4, pp. 481–95.

Hawkes, K. et al. (1998). "Grandmothering, Menopause, and the Evolution of Human Life Histories". *Proceedings of the National Academy of Sciences* **95**.3, pp. 1336–1339.

Higham, T.E. et al. (2016). "Speciation Through the Lens of Biomechanics: Locomotion, Prey Capture and Reproductive Isolation". *Proceedings of the Royal Society B: Biological Sciences* **283**.1838, p. 20161294.

Holling, C. S. (1959). "The Components of Predation as Revealed by a Study of Small-Mammal Predation of the European Pine Sawfly". *The Canadian Entomologist* **91**.5, 293–320.

Jongejans, E., Hans De Kroon, S. Tuljapurkar, and Katriona Shea (2010). "Plant Populations Track Rather than Buffer Climate Fluctuations". *Ecology Letters* **13**.6, pp. 736–743.

Kimura, M. and T. Maruyama (1966). "The Mutational Load with Epistatic Gene Interactions in Fitness". *Genetics* **54**.6, pp. 1337–1351.

Langerhans, R. B. and David N. Reznick (2009). Ecology and Evolution of Swimming Performance in Fishes: Predicting Evolution with Biomechanics. *Fish Locomotion: An Etho-Ecological Perspective*. Ed. by P. Domenici and B.G. Kapoor. Boca Raton, FL: Taylor & Francis Group. Chap. 7, pp. 200–248.

Lee, R.D. (2003). "Rethinking the Evolutionary Theory of Aging: Transfers, not Births, Shape Senescence in Social Species". *Proceedings of the National Academy of Sciences* **100**.16, pp. 9637–9642.

Levitis, D.A., Oskar Burger, and Laurie Bingaman Lackey (2013). "The Human Post-Fertile Lifespan in Comparative Evolutionary Context". *Evolutionary Anthropology: Issues, News, and Reviews* **22**.2, pp. 66–79.

MacArthur, R.H. (1962). "Some Generalised Theorems of Natural Selection". *Proceedings of the National Academy of Sciences* **48**.11, pp. 1893–1897.

MacArthur, R.H. (1970). "Species Packing and Competitive Equilibrium for Many Species". *Theoretical Population Biology* **1**.1, pp. 1–11.

MacArthur, R.H. (1972). *Geographical Ecology*. Princeton, NJ: Princeton University Press.

MacArthur, R.H. and E. O. Wilson (1967). *The Theory of Island Biogeography*. Princeton, NJ: Princeton University Press.

Max Planck Institute for Demographic Research (Germany) and Vienna In stitute of Demography (Austria) (2019). *Human Fertility Database*. Available at www.humanfertility.org.

Medawar, P.B. (1957). An Unsolved Problem of Biology. In: P.B. Medawar (ed.) *The Uniqueness of the Individual*. London: Routledge.

Moorad, J.A. and Craig A Walling (2017). "Measuring Selection for Genes that Promote Long Life in a Historical Human Population". *Nature Ecology & Evolution* **1**.11, p. 1773.

Niklas, K.J. (1992). *Plant Biomechanics: An Engineering Approach to Plant Form and Function*. Chicago, IL: University of Chicago Press.

Niklas, K.J. (1994). "Morphological Evolution Through Complex Domains of Fitness". *Proceedings of the National Academy of Sciences* **91**.15, pp. 6772–6779.

Nosil, P. et al. (2018). "Natural Selection and the Predictability of Evolution in Timema Stick Insects". *Science* **359**.6377, pp. 765–770.

Orzack, S. H. (1985). "Population Dynamics in Variable Environments. V. The Genetics of Homeostasis Revisited". *The American Naturalist*, pp. 550–572.

Orzack, S. H. (1993). "Life History Evolution and Population Dynamics in Variable Environments: Some Insights from Stochastic Demography". *Lecture Notes in Biomathematics* **98**, pp. 63–104.

Paniw, M., Arpat Ozgul, and Roberto Salguero-Gomez (2018). "Interactive Life-History Traits Predict Sensitivity of Plants and Animals to Temporal Autocorrelation". *Ecology Letters* **21**.2, pp. 275–286.

Pavard, S. and Frédéric Branger (2012). "Effect of Maternal and Grand Maternal Care on Population Dynamics and Human Life-History Evolution: A Matrix Projection Model". *Theoretical Population Biology* **82**.4, pp. 364–376.

Prentice, S. et al. (2010). "Evidence for a Downward Secular Trend in Age of Menarche in a Rural Gambian Population". *Annals of Human Biology* **37**.5, pp. 717–721.

Read, J. and Alexia Stokes (2006). "Plant Biomechanics in an Ecological Context". *American Journal of Botany* **93**.10, pp. 1546–1565.

Roughgarden, J. (1971). "Density-Dependent Natural Selection". *Ecology* **52**.3, pp. 453–468.

Royama, T. (1992). Analytic Population Dynamics. London: Chapman and Hall.

Salguero-Gómez, R. et al. (2015). "The compadre Plant Matrix Database: An Open Online Repository for Plant Demography". *Journal of Ecology* **103**.1, pp. 202–218.

Salguero-Gómez, R. et al. (2016). "COMADRE: A Global Data Base of Animal Demography". *Journal of Animal Ecology* **85**.2, pp. 371–384.

Sibly, R. M. and James H. Brown (2007). "Effects of Body Size and Lifestyle on Evolution of Mammal Life Histories". *Proceedings of the National Academy of Sciences* **104**.45, pp. 17707–17712.

Stearns, S. C. (1992). *The Evolution of Life Histories*. Oxford: Oxford University Press.

Steinsaltz, D., S. Tuljapurkar, and C. Horvitz (2011). "Derivatives of the Stochastic Growth Rate". *Theoretical Population Biology* **80**.1, pp. 1–15.

Steptoe, A., Angus Deaton, and Arthur A Stone (2015). "Subjective Wellbeing, Health, and Ageing". *The Lancet* **385**.9968, pp. 640–648.

Tuljapurkar, S (1982). "Population Dynamics in Variable Environments. III. Evolutionary Dynamics of r-Selection". *Theoretical Population Biology* **21**.1, pp. 141–165.

Tuljapurkar, S (1990). *Population Dynamics in Variable Environments*. New York: Springer.

Tuljapurkar, S. (1997). The Evolution of Senescence. In: K.W. Wachter and C.E. Finch (eds) *Between Zeus and the Salmon—The Biodemography of Longevity*, pp. 65–77. Washington, DC: National Academy Press.

Tuljapurkar, S. (2019). The Challenges of Evolutionary Biodemography and the Example of Menopause. In: O. Burger, R. Lee, and R. Sear (eds) *Human Evolutionary Demography*, pp. 319–334. Cambridge: Open Book Publishers.

Tuljapurkar, S. and Carl Boe (1998). "Mortality Change and Forecasting: How Much and How Little do we Know?" *North American Actuarial Journal* **2**.4, pp. 13–47.

Tuljapurkar, S. and C. V. Haridas (2006). "Temporal Autocorrelation and Stochastic Population Growth". *Ecology Letters* **9**.3, pp. 327–337.

Tuljapurkar, SD, Cedric O Puleston, and Michael D Gurven (2007). "Why Men Matter: Mating Patterns Drive Evolution of Human Lifespan". *PLOS One* **2**.8, e785.

Tuljapurkar, S. and P. Wiener (2000). "Escape in Time: Stay Young or Age Gracefully?" *Ecological Modelling* **133**.1–2, pp. 143–159.

Tuljapurkar, S., C. Horvitz, and J Pascarella (2003). "The Many Growth Rates and Elasticities of Populations in Random Environments". *The American Naturalist* **162**.4, pp. 489–502.

Tuljapurkar, S., J-M. Gaillard, and T. Coulson (2009). "From Stochastic Environments to Life Histories and Back". *Philosophical Transactions of the Royal Society of London. Series B, Biological Sciences* **364**.1523, pp. 1499–1509.

University of California, Berkeley (USA) and Max Planck Institute for Demographic Research (Germany) (2019). *Human Mortality Database*. Available at www.mortality.org or www.humanmortality.de.

van Daalen, S.F. and Hal Caswell (2017). "Lifetime Reproductive Output: Individual Stochasticity, Variance, and Sensitivity Analysis". *Theoretical Ecology* **10**.3, pp. 355–374.

Verhulst, Pierre-François (1838). "Notice sur la loi que la Population Suit dans son Accroissement". *Correspondance Mathématique et Physique.* **10**, pp. 113–126.

Wachter, K.W., Steven N Evans, and David Steinsaltz (2013). "The Age-Specific Force of Natural Selection and Biodemographic Walls of Death". *Proceedings of the National Academy of Sciences* **110**.25, pp. 10141–10146.

Williams, G.C. (1996). *Adaptation and Natural Selection: A Critique of Some Current Evolutionary Thought*. Princeton, NJ: Princeton University Press.

Wootton, J. T. (1994). "The Nature and Consequences of Indirect Effects in Ecological Communities". *Annual Review of Ecology and Systematics* **25**, pp. 443–466.

Zuo, W. et al. (2018). "Advancing Front of Old-Age Human Survival". *Proceedings of the National Academy of Sciences* **115**.44, pp. 11209–11214.

CHAPTER 17

Reproductive value and analyses of population dynamics of age-structured populations

Bernt-Erik Sæther and Steinar Engen

17.1 Introduction

Many populations show large temporal variation in age structure, often related to fluctuations in the environment (Coulson et al. 2001; Planque et al. 2010; Vlam et al. 2017) or the effects of density regulation (Lande et al. 2006). In particular, exploited populations tend to show large changes in age distributions caused by variation in the rate of harvesting (Toresen and Østvedt 2000; Rouyer et al. 2011; Pukk et al. 2013). These fluctuations generate variation in population size, which often introduce time-lags in the population dynamics (Lande et al. 2002). This makes it difficult to identify the underlying factors causing variation in population growth. Thus, the effects of fluctuations in age distribution are important to include in analyses of how populations are expected to respond to variation in the environment (Perry et al. 2010).

The purpose of this chapter is to summarise the use of an approach originally suggested by Fisher (1930) based on the concept of the total reproductive value, that is, the sum of all individual reproductive values, to study the dynamics of populations. Because fluctuations in total reproductive value approximates well to temporal variation in population size even for species with complex life histories (Engen et al. 2007), this variable provides an approach to account for the effects of stochastic fluctuations in the age distribution on the population dynamics. Another advantage of this approach is that the use of this concept enables decomposition of the long-run growth rate (Lande et al. 2003) into stochastic components caused by age-specific variation in demographic and environmental

stochasticity. We will illustrate the practical use of this approach by analyses of the dynamics of a small island population of moose (*Alces alces*). Finally, we will show that these methods can be particularly useful in viability analyses of small populations of vulnerable or endangered species by estimating extinction probabilities of Eurasian lynx (*Lynx lynx*) in Scandinavia.

17.2 Stochastic theory

In populations with no age structure but overlapping generations, the contribution W from a female to the next generation is her number of female offspring surviving to the next census (usually at time steps of one year) plus one when mother survives herself. During a time step the population then changes by $\Delta N = \sum W - N$, where N is the population size. Writing Z for a vector of variables expressing the environment that the population experience, the variance of the contributions can then according to Engen et al. (1998) be decomposed into stochastic components as

$$\mathrm{var}(W) = \mathrm{Evar}(W \mid Z) + \mathrm{varE}(W \mid Z) = \sigma_d^2 + \sigma_e^2,$$

where $\sigma_d^2 = \mathrm{Evar}(W \mid Z)$ is the variance of individual contributions within years averaged over time, called the demographic variance, while $\sigma_e^2 = \mathrm{varE}(W \mid Z)$ is the temporal variance of the mean individual contribution within years, called the environmental variance. The expected change in population size is then

$$\mathrm{E}(\Delta N \mid N) = NEW - N = (\lambda - 1)N,$$

Bernt-Erik Sæther and Steinar Engen, *Reproductive value and analyses of population dynamics of age-structured populations*.
In: *Demographic Methods Across the Tree of Life*. Edited by Roberto Salguero-Gómez and Marlène Gamelon, Oxford University Press.
© Oxford University Press (2021). DOI: 10.1093/oso/9780198838609.003.0017

where $\lambda = EW$ and the variance of change in population size is $var(\Delta N|N) = \sigma_d^2 N + \sigma_e^2 N^2$ (Engen et al. 1998).

We then consider the corresponding stochastic theory for an age-structured model with projection matrix L fluctuating among years with no temporal autocorrelations so that the population size next year is $\mathbf{n} + \Delta \mathbf{n} = \mathbf{Ln}$. Here $\mathbf{n}$ is the column vector $(n_1, n_2 \ldots n_k)^T$, where n_i is the number of individuals in age class i (Caswell 2001). Under a prebreeding census scenario, the first row of L is the mean fecundities in the age classes including first-year survival, and the subdiagonal defines the fractions surviving to the next year.

Let the temporal expectation of the stochastic projection matrix be $l = EL$ and let λ now denote the real dominant eigenvalue of l. The associated right and left eigenvectors $\mathbf{u}$ (column vector) and v (row vector) are then defined by $\mathbf{lu} = \lambda\mathbf{u}$ and $\mathbf{vl} = \lambda\mathbf{v}$, respectively. In mathematical literature these vectors are usually scaled to have unit length, while in demography it is more appropriate to require $\sum u_i = 1$ and $\mathbf{vu} = \sum u_i v_i = 1$ (Engen et al. 2009b), where $\mathbf{u}$ is the stable age distribution in the deterministic model with projection matrix l. The components of the vector v are called the reproductive values of the different age classes (Caswell 1978), a concept originating from Fisher (1930) when dealing with age-structured models in continuous age and time. The age-specific distribution of reproductive values scaled so that $\sum u_i v_i = 1$ is termed the *Fisherian stable age distribution* (Engen et al. 2011) and expresses how the total reproductive value V is distributed among age classes if the population is at the stable age distribution $\mathbf{u}$. Sæther et al. (2013) empirically showed that in several species of birds and mammals the Fisherian stable age distribution is generally characterised by a large contribution of younger age classes to V, with values declining with age. The total reproductive value $V = \mathbf{vn} = \sum n_i v_i$ equals the population size $N = \sum n_i$ if the population is exactly at its stable age distribution, $n_i = Nu_i$. In the average environment (no stochasticity) $V + \Delta V = \mathbf{v}(\mathbf{n} + \Delta \mathbf{n}) = \mathbf{vln} = \lambda\mathbf{vn} = \lambda V$ (Fisher 1930; Crow and Kimura 1970; Charlesworth 1994; Lion 2018), so that the total reproductive value has a constant multiplicative growth rate exactly equal to λ.

Engen et al. (2005), (2007) showed that in a fluctuating environment the age distribution will fluctuate around $\mathbf{u}$, whereas the total reproductive value V fluctuates around N with no temporal autocorrelation to the first-order approximation. This means that the noise in the process fluctuates independently of years, which also has been found in several populations of birds and mammals (Engen et al. 2013; Sæther et al.

2013). As in the case of no age structure, the variance of the yearly changes in total reproductive value can be decomposed into additive environmental and demographic components, thus:

$$var(\Delta V|V) \approx \sigma_e^2 N^2 + \sigma_d^2 N \approx \sigma_e^2 V^2 + \sigma_d^2 V.$$

To analyse population growth and extinction, one commonly works on the logarithmic scale (Lewontin and Cohen 1969; Cohen 1977, 1979; Tuljapurkar 1982; Lande et al. 2003; Engen et al. 2009b), which shows that

$$var(\Delta \ln V|V) \approx \frac{\sigma_e^2}{\lambda^2} + \frac{\sigma_d^2}{\lambda^2} V^{-1}$$

and hence the expected growth on log scale, to the same order of approximation, is

$$E(\Delta \ln V|V) \approx r - \frac{\sigma_e^2}{2\lambda^2} + \frac{\sigma_d^2}{2\lambda^2} V^{-1}$$

where $r = \ln \lambda$. When working on the log-scale the factors λ^{-2} are often included in the definitions of demographic and environmental variances (Lande et al. 2003).

Under small and moderate stochastic fluctuations in age structure the process V can be approximated by a diffusion with infinitesimal mean and variance $\mu(v) = \lambda v$ and $\nu(v) = \sigma_e^2 v^2 + \sigma_d^2 v$, respectively. When the population size is large enough to ignore the demographic noise, then $\ln V$ is approximately a Brownian motion with drift parameter $r - \sigma_e^2/2$, which is often called the long-run growth rate, and variance σ_e^2.

Many species subject to population viability analyses have complex life histories. Such complexities make it difficult to provide reliable population trajectories and hence obtain estimates of the risk of extinction (Fieberg and Ellner 2001). A useful metric in such analyses may be the probability of ultimate extinction, that is, the probability the population will reach a certain lower threshold (e.g. 0 or 1 individual) (Sæther and Engen 2003). Using the diffusion approximation one can show using the general formula for the first passage time to 1 (or infinite) (Karlin and Taylor 1981) that the probability of ultimate extinction for a process initiated at a total reproductive value of V_0 and extinction barrier at 1 is approximately

$$\left(\frac{\sigma_d^2 + \sigma_e^2}{\sigma_d^2 + \sigma_e^2 V_0} \right)^{2r/\sigma_e^2 - 1}$$

Thus, the probability of ultimate extinction shows a strong increase with increasing σ_d^2 (Sæther and Engen 2019).

17.3 The concept of individual reproductive value

The statistical analysis based on individual data on survival and reproduction at different ages is significantly simplified by introducing the concept of individual reproductive value, defined as each individual's contribution to the total reproductive value of the population the next year (Engen et al. 2009b). Consider first individuals in a given age class i with reproductive value v_i. One of these individual's contributions to the total reproductive value next year is

$$W_i = J_i v_{i+1} + B_i v_i,$$

where J_i is an indicator variable for survival (equals 1 for survival and otherwise 0) and B_i is the number of offspring. W_i are stochastic quantities assumed to have the same distribution for all n_i individuals in age class i. Then $EW_i = \lambda v_i$ because the total reproductive value of any subpopulations grows with expected rate λ. The mean over all individual reproductive values is accordingly $E\bar{W}_i = \lambda \bar{v}_i = \lambda \sum u_i v_i = \lambda$. Thus, the individual reproductive value can be decomposed into additive components $J_i v_{i+1}$ and $B_i v_1$ due to individual variation in survival and fecundity schedules, respectively (Box 17.1).

In contrast to the general decrease with age in the age-specific components of the total reproductive value found in several populations of birds and mammals, there were much larger interspecific differences in the pattern of age-specific variation in individual reproductive value (Sæther et al. 2013). In six of the populations included in the study of Sæther et al. (2013) maximum individual reproductive value was achieved at some intermediate age (Figure 17.1). Consequently, those age classes contributing most to the total reproductive value of the population may not necessarily be those consisting of individuals with large contributions, because the former is more strongly dependent on the stable age distribution of

---

Box 17.1 Decomposing individual reproductive value into components due to reproduction and survival

First, let us ignore the environmental stochasticity. In this case, the variance of reproductive value next year is $\sigma_{di}^2 = \text{var}(W_i) = s_{di}v_{i+1}^2 + f_{bi}v_1^2 + 2c_{di}v_{i+1}v_1$, where $c_{di} = \text{cov}[J_i, B_i]$ and p_i and $s_{di} = p_i(1 - p_i)$ are the mean and variance of J_i, respectively, while f_{di} is the variance of B_i. The demographic variance of the contribution to the next year by this age class is then $\sum n_i \text{var}(W_i)$. Using the approximation $n_i \approx Nu_i$, the total demographic variance is

$$\text{var}(V + \Delta V | V) \approx N \sum_{i=1}^{k} u_i [s_{di}v_{i+1}^2 + f_{di}v_1^2 + 2c_{di}v_{i+1}v_1]$$

$$= N \sum u_i \sigma_{di}^2 = N\sigma_d^2$$

where

$$\sigma_d^2 = \sum_{i=1}^{k} u_i [s_{di}v_{i+1}^2 + f_{di}v_1^2 + 2c_{di}v_{i+1}v_1] = \sum u_i \sigma_{di}^2$$

Here $\sigma_{di}^2 = \text{var}(W_i)$ is the within-year variance of individual reproductive value among individuals of age i.

Second, let us consider the environmental variance by assuming a very large population, where the demographic noise can be ignored. In this case we have $n + \Delta n = L(Z)n$, where the projection matrix $L(Z)$ is now determined primarily by the temporal fluctuations in the environment Z. Then we get $V + \Delta V = v(n + \Delta n) = vLn =$

$\sum_{ij} L_{ij} n_j v_i$. Under small or moderate fluctuations in the age distribution so that n_j can be approximated by Nu_j, the variance in the change of the total reproductive value is

$$\text{var}(V + \Delta V | V) \approx N^2 \sum_{ij} v_i v_k u_j u_l \text{cov}(L_{ij}, L_{kl}) = N^2 \sigma_e^2$$

so that the environmental variance is

$$\sigma_e^2 = \sum_{ij} v_i v_k u_j u_l \text{cov}(L_{ij}, L_{kl}).$$

Defining the survival probabilities $s_{eij} = \text{cov}(L_{i+1,i}, L_{j,j+1})$ and fecundities $f_{eij} = \text{cov}(L_{1i}, L_{1j})$, and the covariance $c_{eij} = \text{cov}(L_{i+1,i}, L_{1,j})$, we obtain the decomposition of the environmental variance into components from reproduction and survival, thus

$$\sigma_e^2 = \sum_{ij} u_i u_j [s_{eij}v_{i+1}v_{j+1} + f_{eij}v_1^2 + c_{eij}v_{i+1}v_1]$$

$$= \sum_{ij} u_i u_j \tau_{eij}.$$

In analogy with the model without any age structure we can now add the demographic and environmental variance terms, provided that the demographic components are replaced by their temporal mean values under environmental fluctuations.

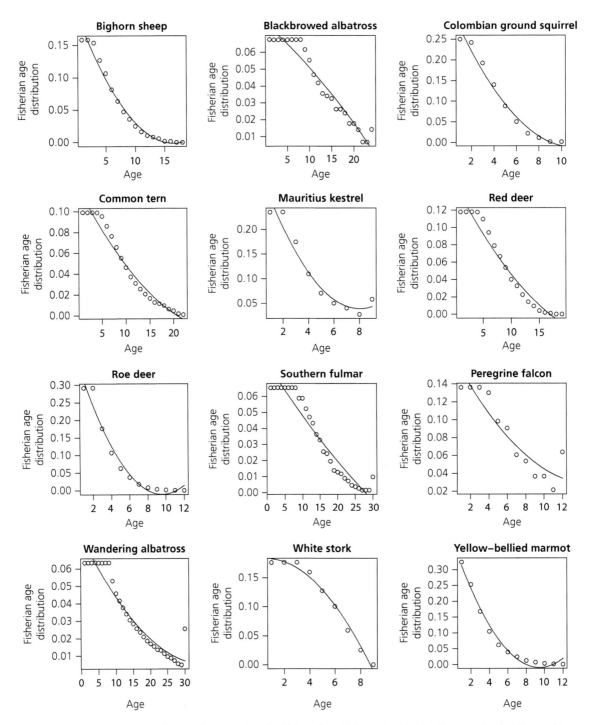

Figure 17.1 The Fisherian age distribution in five mammal species (bighorn sheep (*Ovis canadensis*), Columbian ground squirrel (*Spermophilus columbianus*), red deer (*Cervus elaphus*), roe deer (*Capreolus capreolus*), and yellow-bellied marmot (*Marmota flaviventris*)) and seven bird species (blackbrowed albatross (*Thalassarche melanophrys*), common tern (*Sterna hirundo*), Mauritius kestrel (*Falco punctatus*), peregrine falcon (*Falco peregrinus*), southern fulmar (*Fulmarus glaicialoides*), wandering albatross (*Diomedea exulans*), and white stork (*Ciconia ciconia*). This age distribution refers to the reproductive value in age class *i*, scaled so that $\sum u_i = 1$ and $vu = \sum u_i v_i = 1$, where *u* is the stable age distribution for the deterministic model. This shows the relative contribution of the different age classes to the total reproductive value of the population (see Sæther et al. 2013 for further details).

the population. Similarly, a curvilinear relationship between reproductive value and age has been found in four passerines (Moller and De Lope 1999; Keller et al. 2008; Brown and Roth 2009; Bouwhuis et al. 2011) as well as in the sparrow-hawk (*Accipiter nisus*) (Newton and Rothery 1997), but the reproductive value increased with age in female red-billed chough (*Pyrrhocorax pyrrhocorax*) (Trask et al. 2017).

The distribution of reproductive values throughout the life cycle provides important information about how different stages contribute to the population growth (Engen et al. 2009b). In an analyses of birds and mammals, Sæther et al. (2013) showed that the larger variation among species in the contribution to the total reproductive value of the population and in the demographic variance was found in the younger age classes (relative to generation time). Furthermore, the contribution of earlier stages within the life cycle to stochastic population growth rate was greater in long-lived than in short-lived species. In contrast, environmental stochasticity at different life history stages had similar effects on population fluctuations per generation. Similarly, the distribution of reproductive values throughout the life span of plants differed dependent on life form (Mbeau-Ache and Franco 2013). The results of these analyses indicated that the attributes related to pace were constrained by duration of life, generating large differences among life forms in the age-specific distribution of reproductive values.

17.4 Temporal autocovariance in the environmental noise

Until now we have assumed that the yearly realised Leslie matrices are temporally uncorrelated. In age-structured models, as in models with no age structure, environmental autocorrelation affects the long-run growth rate (Tuljapurkar 1982; Caswell 2001; Doak et al. 2005); Tuljapurkar and Haridas 2006; Morris et al. 2011) through its effect on the environmental noise. The impact over a time interval is proportional to the length t, whereas the standard deviation is proportional to the square root of t. This term is usually negative and may be smaller or larger in magnitude than σ_e^2, dependent on the mean projection matrix as well as the strength of temporal autocorrelations. In this more general model, the long-run growth rate takes the form

$$s = r - \frac{1}{2}\lambda^{-2}\sigma_e^2 + \lambda^{-1}\eta$$

where $\sigma_e^2 = E(\mathbf{v}\varepsilon_t\mathbf{u})^2 = \sum v_i u_j v_k u_m \mathrm{cov}[L_{ijt}, L_{kmt}]$ is the environmental variance, $\varepsilon_t = \mathbf{L}_t - E\mathbf{L}$, and the last term

$\lambda^{-1}\eta$ is generated by the temporal autocorrelations. Engen et al. (2013) showed that

$$\eta = \sum_j \mathrm{cov}(\bar{W}_j, n_j/V)$$

where $\bar{W}$ is the mean individual reproductive value of individuals in age class j at a given year, n_j is the number of individuals in this class, and V is the total reproductive value of the population. When observations of these variables are available over several years, η can be estimated by the cross-product sum over years divided by one less than the number of years.

For further details on how to further decompose this into contributions from survival and fecundity see Engen et al. (2013).

17.5 Case studies

17.5.1 Moose on the island of Vega

Moose (*Alces alces*) is a sexual dimorphic species that is subject to intense harvesting in most parts of Scandinavia (Lavsund et al. 2003). Its life history is characterised by strong age dependency in fecundity, with high twinning rates among prime-aged females (Sæther and Haagenrud 1983). In addition, several aspects of individual variation in reproductive success of females are dependent on the body mass of females early in life (Sæther and Haagenrud 1985; Lee et al. 2017; Markussen et al. 2018). In 1985, three moose colonised the island of Vega in Northern Norway. In 1992, we used this opportunity of having access to a newly established population to initiate a long-term study to examine the ecological and evolutionary consequences of harvest. This involved radio-collaring of almost all individuals on the island, which facilitated construction of the pedigree using genetic parentage assignment (Markussen et al. 2019). The environmental conditions for moose on the island are favourable, resulting in large body masses (Solberg et al. 2008). The age and sex compositions of the population have in periods been altered through directional harvest of particular categories of animals (Sæther et al. 2003, 2004).

The reproductive value of the moose showed strong age-specific variation in both sexes, with a maximum at age 2 years in females and 1 year later in bulls (Figure 17.2). This difference was caused by particularly high mating success among prime-aged bulls (Markussen et al. 2019). The population growth estimated by means of the total reproductive value of

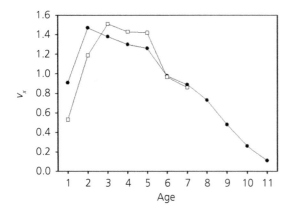

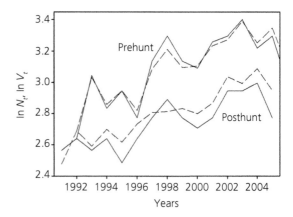

Figure 17.2 Age-specific variation in the reproductive value v_x of female (open squares) and male (black circles) moose (*Alces alces*) on the island of Vega in northern Norway (Kvalnes et al. 2016). This demonstrates that in both sexes the greatest contribution to the growth of this population occurs from age classes around age at maturity, which then decreases with increasing age.

Figure 17.3 Annual variation in (log-) population size (solid lines) of female moose on the island of Vega just before (Prehunt) and after (Posthunt) the hunting season. The dashed lines indicate the (log-) total reproductive value of the population (Sæther et al. 2007a). This shows that the total reproductive value of the population describes the population fluctuations quite well (see also Engen et al. 2007).

the population closely resembles the actual number of females in the population both before and after the hunting season (Figure 17.3). This shows that V actually can describe the population fluctuations well even in populations that have undergone large variation in age composition through directed harvest towards certain age classes.

17.5.2 A practical application: estimating risk of extinction in harvested Scandinavian lynx populations

Most large carnivores are subject to human exploitation. For example, in Scandinavia, Eurasian lynx (*Lynx lynx*) is regularly harvested to reduce the conflict level associated with depredation on free-ranging livestock (e.g. domestic sheep (*Ovis aries*)) and semi-domestic reindeer (*Rangifer tarandus*). In addition, the lynx is subject to substantial rates of illegal killings (Andrén et al. 2006). Here we will demonstrate that analyses based on the total reproduction of the population can be a useful tool to predict the consequences of such human-induced demographic changes for the probability of extinction.

Such analyses are difficult because a general characteristic of the life history of large carnivores is that they are long lived once they have reached sexual maturity. For instance, the natural annual survival rate of adult female lynx in Fennoscandia was 92.3% (Andrén et al. 2006). Harvest that induces changes in age structures will therefore easily result in time lags

in the population dynamics (Caswell 2001; Lande et al. 2003), which can erroneously be interpreted as environmental stochasticity and result in biased predictions of future population fluctuations (Holmes 2001).

We will introduce the population prediction interval (PPI) (Engen et al. 2001; Sæther and Engen 2002) as a useful concept in developing reliable population projections. The PPI is defined as the stochastic interval that includes a given population size with probability $(1 - \alpha)$, where α is the probability that the variable we want to predict is not contained in the stochastic interval. For example, we can model a population of initial size of 20 females for years. After 50 years, the distribution of predicted population sizes is graphed using quantiles, showing the proportion of population size estimates in, for example, 10 percentile groups. The width (e.g. the difference between the 5% and 95% percentiles) indicates the precision of the predictions and the probability that the population size will be below a critical threshold (often 0 or 1 individual). The advantage of this approach is that changes in expected population size, population variability due to demographic and environmental stochasticity, and uncertainties in parameter estimates all affect the width of the PPI. Large uncertainties in parameter estimates as well as strong stochastic influences on the population dynamics will typically result in wide PPI. Furthermore, the width of the PPI will increase through time until some stationary distribution is reached (see examples in Sæther et al. 2002a, 2002b, 2007b). Finally, the PPI can be used to assess the quality of our predictions

(Sæther et al. 2009, 2019). If our model makes correct predictions of future population fluctuations, then the quantiles of simulated population sizes at a given time step will be uniformly distributed across the quantiles of the PPI.

An important advantage of this approach is that the PPI relates directly to the International Union for Conservation of Nature (IUCN)'s criteria E for quantitative classification of population extinction risk (IUCN 2012). For instance, a population is defined as vulnerable, that is, it has a 10% probability of extinction within 100 years, if the lower 10% quantile of PPI decreases below 1 before 100 years. The PPI is found by parametric bootstrapping from the sampling distributions of the parameter estimates, which is now commonly used in population viability analyses (Holmes 2004; Holmes et al. 2007). This approach has also previously been used to explore management strategies for wolverines and brown bears in Scandinavia (Sæther et al. 1998, 2005; Tufto et al. 1999).

Sæther et al. (2010) analysed temporal variation in the total reproductive value of two Scandinavian populations of lynx (Hedmark in Norway and Sarek in Sweden) for which long-term individual-based demographic data were available (Andrén et al. 2006, 2002) to estimate the population growth rate as well as the influence of demographic and environmental stochasticities on the population dynamics. The estimate of the population growth rate was significantly higher and more certain in Sarek ($\lambda = 1.09$; 95% confidence interval = 1.00–1.17) than in Hedmark ($\lambda = 1.06$; 95% confidence interval = 0.93–1.19). The demographic variance was similar in the two populations ($\sigma_d^2 = 0.312$ and $\sigma_d^2 = 0.309$ in the northern and southern population, respectively), whereas the environmental variance σ_e^2 was not significantly different from 0 in any of the two populations. These estimates derived from the analyses of reproductive values in the two populations (Boxes 17.1 and 17.2) can then be used to construct PPIs of future population fluctuations to suggest sustainable

---

Box 17.2 Estimation

Estimation of variance components requires that individual data on survival and fecundity are available for all age classes over several years. As in models with no age structure, estimates of demographic variance components can be rather certain even if the observation period is only a few years, provided that observations are available from a rather large number of individuals. Estimates of environmental components, however, are in general rather uncertain even if data are available over long periods of times (Sæther et al. 2008). We use a single subscript i for age, writing J_i, B_i, and W_i for stochastic variables expressing age-specific survival and reproduction that are independent within years, and i and j when two age classes are considered jointly, such as in the covariance of contributions of individuals to the next generation $\text{cov}(W_i, W_j)$. In a given year t data from n_{it} individuals numbered by m are available from age class i. The complete data set consists therefore of records of survival and reproduction, (J_{imt}, B_{imt}), for ages $i = 1, 2, \ldots k$, in years $t = 1, 2, \ldots T$, and individuals $m = 1, 2, \ldots n_{it}$. When considering two different individuals in an age class in a given year we use subscript q as well as m for numbering individuals. Similarly, when required we use y as well as t for time.

The exact relationship between the population vector at two subsequent years is given by the Leslie matrix L with elements that are mean survivals $\bar{J}$ at the subdiagonal and mean fecundities $\bar{B}$ at the first row (Engen et al. 2005).

Assuming no demographic covariances (Engen et al. 1998), the demographic stochastic components are independent between individuals so that the demographic contribution to $\text{cov}(L_{ij}, L_{kl})$ are zero for $j \neq l$ and thus there are only environmental contributions, that is, $\text{cov}(L_{i+1,i}, L_{j+1,j}) = s_{eij}$, $\text{cov}(L_{1,i}, L_{1,j}) = f_{eij}$, and $\text{cov}(L_{i+1,i}, L_{1,j}) = c_{eij}$. Within a column ($i = j$) there may be demographic components due to nonzero individual covariances between survival and fecundity. The components s_{di}/N_i, f_{di}/N_i, and c_{di}/N_i must therefore be added, giving, for example $\text{cov}(L_{i+1,i}, L_{i+1,i}) = s_{eii} + s_{di}/N_i$.

The elements of the expected projection matrix $l = EL$ can be estimated by mean values of all records of (J_{imt}, B_{imt}). From this matrix we can compute the corresponding estimate of the multiplicative growth rate λ, the stable age distribution u, and reproductive values v. To find the uncertainties it is recommended to simulate bootstrap replicates of the data, resampling the recordings of (J_{imt}, B_{imt}) with replacement within each year.

Estimating demographic variance components

An important advantage of this approach is that all individual reproductive values $W_{imt} = J_{imt} v_{i+1} + B_{imt} v_1$ can

continued

Box 17.2 *Continued*

---

be observed, provided that estimates of u and v are available. Using the definition $\sigma_{di}^2 = \mathrm{Evar}(W_i|\mathbf{Z})$ we then first estimate $\sigma_{di}^2(\mathbf{Z}) = \mathrm{var}(W_i|\mathbf{Z})$ for each year (with different but unknown values of Z) by the simple sum of squares

$$\hat{\sigma}_{di}^2(\mathbf{Z}) = \frac{1}{n_{it} - 1} \sum_{m=1}^{n_{it}} (W_{imt} - \bar{W}_{it})^2$$

where $\bar{W}_{it} = n_{it}^{-1} \sum_{m=1}^{n_{it}} W_{imt}$. The weighted mean of the above estimates over years with at least two records from individuals in the age class with weights $n_{ti} - 1$ then yields an estimate $\hat{\sigma}_{di}^2$ of the contribution from age class i, while the total demographic variance is estimated by $\hat{\sigma}_d^2 = \sum u_i \hat{\sigma}_{di}^2$.

Bootstrap replicates should be produced as for the Leslie matrix, although replicates for the sum of squares for year t and age class i should be multiplied by $n_{it}/(n_{it} - 1)$ to make them unbiased (Engen et al. 2009b).

The age-specific components generated by survival and fecundity, s_{di} and f_{di}, can similarly be estimated by replacing W_{imt} by J_{imt} and B_{imt}, respectively, while the estimate for the covariance c_{di} is the corresponding sum of cross products.

Estimating environmental stochastic components

Assuming that all individual demographic components are independent for a given environment Z and applying the decomposition

$$\mathrm{cov}(W_{imt}, W_{jqt}) = \mathrm{Ecov}[W_{imt}, W_{jqt}|\mathbf{Z}]$$
$$+ \mathrm{cov}[\mathrm{E}(W_{imt}|\mathbf{Z}), \mathrm{E}(W_{jqt}|\mathbf{Z})],$$

it appears that

$$\mathrm{cov}(W_{imt}, W_{jqt}) = \mathrm{cov}[\mathrm{E}(W_{imt}|\mathbf{Z}), \mathrm{E}(W_{jqt}|\mathbf{Z})] = \tau_{eij}$$

where $m \neq q$ when i and j refer to the same age class $(i = j)$. Engen et al. (2009b) showed that $\frac{1}{2}(W_{imt} - W_{im'y})(W_{jqt} - W_{jq't})$ has expectation τ_{eij}, so that an unbiased estimate of τ_{eij} is obtained as the mean value of these products over all combinations of $t \neq y$ and all appropriate combinations of the second subscript. This step can easily be computed even if the number of terms is large. Engen et al. (2009b) also showed how the survival- and fecundity-specific variance components as well as the contribution from the environmental covariance between survival and fecundity can be estimated in the same way.

---

harvest strategies to ensure the populations remain viable.

More specifically, Sæther et al. (2010) analysed the effects of adopting three different harvest strategies: (1) proportional harvesting (Lande et al. 1995, 1997); (2) threshold harvesting (Lande et al. 1995, 1997); and (3) proportional threshold harvesting (Engen et al. 1997; Tufto et al. 1999). For all three harvest strategies the harvested females were assumed to be chosen at random, since the hunters typically are unable to visually differentiate between the age and sex of lynx in a hunting situation. Proportional harvesting implies that a certain fraction p of the females is harvested each year. Under threshold harvesting the population $N_1 + N_2$ is harvested down to a threshold c if $N_1 + N_2 > c$; otherwise no hunting occurs. Proportional threshold harvesting has in particular been shown to be a good strategy when the population estimates $\hat{N}_1$ and $\hat{N}_2$ are uncertain (Engen et al. 1997; Lillegård et al. 2005), in which case the harvest is $(\hat{N}_1 + \hat{N}_2)a$ when $N_1 + N_2 > c$, and otherwise there is no harvest.

The sustainability of the current harvest strategies of lynx in Norway was explored by simulating future population fluctuations based on bootstrap replicates

of the demographic parameters estimated for the two populations using the reproductive values (Box 17.2). In the county of Hedmark in southern Norway the minimum (politically determined) threshold for the population size was 10 family groups. The proportion of adult females that had kittens indicated an adult population size of 19 adult ($\geq$ 2-year-old) females (Andrén et al. 2002). In addition, based on the stationary age distribution, eight 1-year-old females were also assumed to be present. As expected from large uncertainties in the parameter estimates and the large demographic stochasticity, the population prediction intervals soon became wide (Figure 17.3). However, the simulations showed that if the population size was harvested down to the minimum number of 10 reproductive adult females, the lower 10% quantile was below 1 already after 34 years. Similarly, this population will have a 50% probability of extinction after 273 years.

We then repeated this exercise for the population in the county of Nordland in northern Norway, where we assumed similar demography as in the lynx population in Sarek in northern Sweden. Here, a threshold harvest strategy with no harvest when less than 10 family groups were recorded will result in stationary

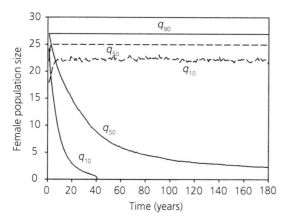

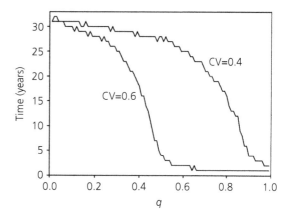

Figure 17.4 Prediction of population fluctuations of lynx (*Lynx lynx*) in the Norwegian counties of Hedmark (solid lines) and Nordland (dashed lines) assuming threshold harvesting. Values of q_{10} q_{50}, and q_{90} denote the 10%, 50%, and 90%, respectively, quantiles of the population prediction interval (PPI). For the northern population, demographic characteristics were assumed to be similar to those estimated for the Sarek population in northern Sweden. In both areas the threshold is set at 10 reproductive females (i.e. females recorded during winter with kittens). The simulations were done using the population counts in 2007 as initial population size and assuming that harvest occurred according to the proportion of each age class in the population. For further details see Sæther et al. (2010).

Figure 17.6 The time it takes for the lower 10% quantile of the population prediction interval (PPI) q_{10} to reach one female in relation to the proportion of animals above the threshold $c = 23$ that is removed in Hedmark for different uncertainties in the population estimates (CV). We assume an initial size of the populations of eight 1-year-old and nineteen adult ($\geq$ 2-year-old) females. No harvest occurs below the harvest threshold.

relatively low values of the threshold (Figure 17.5). In contrast, the population in Hedmark was likely to go extinct even if a large threshold is chosen (Figure 17.5). However, assuming a proportional harvesting strategy, extinction is likely irrespective of the demographic characteristics of the population (Sæther et al. 2010). In addition, large observation errors in population estimates must require more careful harvesting to ensure population viability. In such cases, Engen et al. (1997) proposed that proportional threshold harvesting should be applied. For the lynx in Hedmark, uncertainties in the population estimates strongly affected the estimates of time to extinction (Figure 17.6). For a coefficient of variation (CV) in the population estimate of 40%, removing more than 60% of the population above the threshold of 27 female lynx will lead to rapid increase in the risk of extinction.

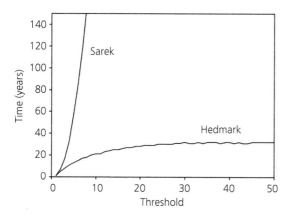

Figure 17.5 The time it takes for the lower 10% quantile of the population prediction interval (PPI) q_{10} to reach one female in relation to the harvest threshold of females ($\geq$ 1 year old) c for the two populations in Hedmark and Sarek. We assume the same initial sizes of the populations (eight 1-year-old and nineteen adult ($\geq$ 2-year-old) females). No harvest occurs below the harvest, and all females above the threshold are immediately harvested ($p = 1.0$).

fluctuations between 15 and 23 1-year-old females (Figure 17.4). In general, a threshold harvest strategy of the population in the county of Nordland with demographic characteristics similar to the Sarek population will ensure viable populations even for

17.6 Conclusions

In this chapter we have shown that characterising the population dynamics through analyses of temporal variation in the total reproductive value of the population has several advantages: (1) In a fluctuating environment the total reproductive value V will fluctuate around N with no temporal autocorrelation and will have a constant multiplicative growth rate exactly equal to λ. (2) The variance of yearly changes in the total reproductive value can be decomposed into additive components caused by environmental and demographic stochasticity. (3) The Fisherian age distribution scaled so that $\sum u_i v_i = 1$ expresses how the total reproductive value V is distributed among age classes. (4) The dynamics of the total reproductive value of the population provides information about important characteristics of the population dynamics, such as the effects of temporal autocorrelations in the environmental noise and the probability of ultimate extinction. (5) The distribution of reproductive values throughout the life cycle show how different stages contribute to the population growth, which then can be decomposed into age-specific contributions from fecundity and survival as well as their covariance (Box 17.1). Thus, the application of Fisher's (1930) concept of the total reproductive value of the population provides an important tool in analyses of the population dynamics especially of species with a complex life history in which fluctuations in age distributions are likely to inflate estimation of key parameters. Our analyses of the lynx population in Scandinavia illustrate that this approach can be particularly useful in analyses of the viability of populations of vulnerable species because it enables us to estimate the stochastic effects on the dynamics as well as the uncertainty in predictions (Figure 17.4). This requires, however, that the reproductive success and survival are known for an unbiased sample of individuals from the different age classes.

Reproductive value has since the concept was introduced by Fisher (1930) been commonly used in analyses of selection in age-structured populations. For example, Williams (1966) decomposed reproductive value into components due to current reproduction and residual reproductive value to derive the optimal reproductive effort. This approach provided the foundation for the maximisation principle derived by Schaffer (1974) and Caswell (1982), who showed that in age-structured populations a life history maximising fitness is equivalent to one that maximises the reproductive value of each age class. Similarly, after the pioneering work of Hamilton (1966), reproductive values have been widely applied in analyses of selection on age-specific variation in several life history traits covering both animals and plants (e.g. Silvertown et al. 2001; Brommer et al. 2007; Bouwhuis et al. 2011; Caswell and Salguero-Gomez 2013; Mbeau-Ache and Franco 2013; Zhang et al. 2015). Application of the concept individual reproductive value that includes stochastic components facilitates further decomposition of the contributions to the total reproductive value due to age-specific variation in survival and fecundity as well as their covariance (Box 17.1). This extends

the possibilities for analyses of selection and evolution in age-structured populations (Crow 1979; Engen et al. 2009a, 2010, 2011, 2012; Barton and Etheridge 2011) and provides an approach to link evolutionary changes to variation in the environment even in organisms with quite complex life cycles (Kvalnes et al. 2016).

A complication in analyses of population dynamics based on the total reproductive value of the population is introduced by density regulation. Lande et al. (2006) estimated the overall strength of density dependence as well as the total stochasticity (including the demographic stochasticity) acting on the population dynamics of red deer (*Cervus elaphus*) using fluctuations in density-dependent reproductive values allowing for density dependence within and between age groups. This approach was based on time series of fluctuations in the complete age structure and showed that the matrix of density-dependent coefficient describing how fluctuations in the size of different age classes affected the population growth rate despite high census accuracy became subject to substantial sampling error. Such complexities in the structure of density-dependent effects, which occur even in species with quite simple life histories (Gamelon et al. 2016), make it extremely difficult to approximate the dynamics of density-regulated populations by temporal variation in reproductive values.

Acknowledgements

This study was funded by the Research Council of Norway (SFF-III 223,257/F50). We are grateful to John Fryxell for comments on a previous version of this paper.

References

Andrén, H., Linnell, J. D. C., Liberg, O., et al. 2002. Estimating total lynx *Lynx lynx* population size from censuses of family groups. *Wildlife Biology*, **8**, 299–306.

Andrén, H., Linnell, J. D. C., Liberg, O., et al. 2006. Survival rates and causes of mortality in Eurasian lynx (*Lynx lynx*) in multi-use landscapes. *Biological Conservation*, **131**, 23–32.

Barton, N. H. and Etheridge, A. M. 2011. The relation between reproductive value and genetic contribution. *Genetics*, **188**, 953–973.

Bouwhuis, S., Choquet, R., Sheldon, B. C., and Verhulst, S. 2011. The forms and fitness cost of senescence: age-specific recapture, survival, reproduction, and reproductive value in a wild bird population. *The American Naturalist*, **179**, E15–E27.

Brommer, J. E., Wilson, A. J., and Gustafsson, L. 2007. Exploring the genetics of aging in a wild passerine bird. *The American Naturalist*, **170**, 643–650.

Brown, W. P. and Roth, R. R. 2009. Age-specific reproduction and survival of individually marked wood thrushes, *Hylocichla mustelina*. *Ecology*, **90**, 218–229.

Caswell, H. 1978. A general formula for the sensitivity of population growth rate to changes in life history parameters. *Theoretical Population Biology*, **14**, 215–230.

Caswell, H. 1982. Optimal life histories and the maximization of reproductive value: a general theorem for complex life-cycles. *Ecology*, **63**, 1218–1222.

Caswell, H. 2001. *Matrix Population Models*, 2nd edition. Sunderland, MA, Sinauer.

Caswell, H. and Salguero-Gomez, R. 2013. Age, stage and senescence in plants. *Journal of Ecology*, **101**, 585–595.

Charlesworth, B. 1994. *Evolution in Age—Structured Populations*, 2nd edition. Cambridge, Cambridge University Press.

Cohen, J. 1977. Ergodicity of age structure in populations with Markovian vital rates, III: finite-state moments and growth rate; an illustration. *Advances in Applied Probability*, **9**, 462–475.

Cohen, J. E. 1979. Ergodic theorems in demography. *Bulletin of the American Mathematical Society*, **1**, 275–295.

Coulson, T., Catchpole, E. A., Albon, S. D., et al. 2001. Age, sex, density, winter weather, and population crashes in Soay sheep. *Science*, **292**, 1528–1531.

Crow, J. F. 1979. Gene frequency and fitness change in an age-structured population. *Annals of Human Genetics*, **42**, 355–370.

Crow, J. F. and Kimura, M. 1970. *An Introduction to Population Genetics Theory*. New York, Harper & Row.

Doak, D. F., Morris, W. F., Pfister, C., Kendall, B. E., and Bruna, E. M. 2005. Correctly estimating how environmental stochasticity influences fitness and population growth. *The American Naturalist*, **166**, E14–E21.

Engen, S., Lande, R., and Sæther, B.-E. 1997. Harvesting strategies for fluctuating populations based on uncertain population estimates. *Journal of Theoretical Biology*, **186**, 201–212.

Engen, S., Bakke, Ø., and Islam, A. 1998. Demographic and environmental stochasticity—concepts and definitions. *Biometrics*, **54**, 840–846.

Engen, S., Sæther, B.-E. and Møller, A. P. 2001. Stochastic population dynamics and time to extinction of a declining population of barn swallows. *Journal of Animal Ecology*, **70**, 789–797.

Engen, S., Lande, R., Sæther, B.-E., and Weimerskirch, H. 2005. Extinction in relation to demographic and environmental stochasticity in age-structured models. *Mathematical Biosciences*, **195**, 210–227.

Engen, S., Lande, R., Sæther, B.-E., and Festa-bianchet, M. 2007. Using reproductive value to estimate key parameters in density-independent age-structured populations. *Journal of Theoretical Biology* **244**, 208–317.

Engen, S., Lande, R., and Sæther, B.-E. 2009a. Reproductive value and fluctuating selection in an age-structured population. *Genetics*, **183**, 629–637.

Engen, S., Lande, R., Sæther, B.-E., and Dobson, F. S. 2009b. Reproductive value and the stochastic demography of age-structured populations. *The American Naturalist*, **174**, 795–804.

Engen, S., Lande, R., Sæther, B.-E., and Gienapp, P. 2010. Estimating the ratio of effective to actual size of an age-structured population from individual demographic data. *Journal of Evolutionary Biology*, **23**, 1148–1158.

Engen, S., Lande, R., and Sæther, B.-E. 2011. Evolution of a plastic quantitative trait in age-structured population in a fluctuating environment. *Evolution*, **65**, 2893–2906.

Engen, S., Saether, B.-E., Kvalnes, T., and Jensen, H. 2012. Estimating fluctuating selection in age-structured populations. *Journal of Evolutionary Biology*, **25**, 1487–1499.

Engen, S., Sæther, B.-E., Armitage, K. B., et al. 2013. Estimating the effect of temporally autocorrelated environments on the demography of density-independent age-structured populations. *Methods in Ecology and Evolution*, **4**, 573–584.

Fieberg, J. and Ellner, S. P. 2001. Stochastic matrix models for conservation and management: a comparative review of methods. *Ecology Letters*, **4**, 244–266.

Fisher, R. A. 1930. *The Genetical Theory of Natural Selection*. Oxford, Oxford at the Clarendon Press.

Gamelon, M., Grøtan, V., Engen, S., Bjørkvoll, E., Visser, M. E., and Sæther, B.-E. 2016. Density-dependence in an age-structured population of great tits: identifying the critical age-class. *Ecology*, **97**, 2479–2490.

Hamilton, W. D. 1966. The moulding of senescence by natural selection. *The Journal of Theoretical Biology* **12**:12–45.

Holmes, E. E. 2001. Estimating risks in declining populations with poor data. *Proceedings of the National Academy of Sciences*, **98**, 5072–5077.

Holmes, E. E. 2004. Beyond theory to application and evaluation: diffusion approximations for population viability analysis. *Ecological Applications*, **14**, 1272–1293.

Holmes, E. E., Sabo, J. L., Viscido, S. V., and Fagan, W. F. 2007. A statistical approach to quasi-extinction forecasting. *Ecology Letters*, **10**, 1182–1198.

IUCN 2012. *IUCN Red List Categories and Criteria. Version 3.1*, 2nd edition. Gland, Switzerland, IUCN.

Karlin, S. and Taylor, H. M. 1981. *A Second Course in Stochastic Processes*. New York, Academic Press.

Keller, L. F., Reid, J. M., and Arcese, P. 2008. Testing evolutionary models of senescence in a natural population: age and inbreeding effects on fitness components in song sparrows. *Proceedings of the Royal Society London B*, **275**, 597–604.

Kvalnes, T., Sæther, B.-E., Haanes, H., Roed, K. H., Engen, S., and Solberg, E. J. 2016. Harvest-induced phenotypic selection in an island population of moose, *Alces alces*. *Evolution*, **70**, 1486–1500.

Lande, R., Engen, S., and Sæther, B.-E. 1995. Optimal harvesting of fluctuating populations with a risk of extinction. *The American Naturalist*, **145**, 728–745.

Lande, R., Sæther, B.-E., and Engen, S. 1997. Threshold harvesting for sustainability of fluctuating resources. *Ecology*, **78**, 1341–1350.

Lande, R., Sæther, B.-E., Engen, S., Filli, F., Matthysen, E., and Weimerskirch, H. 2002. Estimating density dependence from population time series using demographic theory and life-history data. *The American Naturalist*, **159**, 321–332.

Lande, R., Engen, S., and Sæther, B.-E. 2003. *Stochastic Population Dynamics in Ecology and Conservation*. Oxford, Oxford University Press.

Lande, R., Engen, S., Sæther, B.-E., and Coulson, T. 2006. Estimating density dependence from time series of population age structure. *The American Naturalist*, **168**, 76–87.

Lavsund, S., Nygren, T., and Solberg, E. J. 2003. Status of moose populations and challenges to moose management in Fennoscandia. *Alces*, 39, 109–130.

Lee, A. M., Sæther, B.-E., Markussen, S. S., and Engen, S. 2017. Modelling time to population extinction when individual reproduction is autocorrelated. *Ecology Letters*, **20**, 1385–1394.

Lewontin, R. C. and Cohen, D. 1969. On population growth in a randomly varying environment. *Proceedings of the National Academy of Sciences*, **62**, 1056–1060.

Lillegård, M., Engen, S., Sæther, B.-E., and Toresen, R. 2005. Harvesting strategies for Norwegian spring-spawning herring. *Oikos*, 110, 567–577.

Lion, S. 2018. Class structure, demography, and selection: reproductive-value weighting in nonequilibrium, polymorphic populations. *The American Naturalist*, **191**, 620–637.

Markussen, S. S., Loison, A., Herfindal, I., et al. 2018. Fitness correlates of age at primiparity in a hunted moose population. *Oecologia*, **186**, 447–458.

Markussen, S. S., Herfindal, I., Loison, A., et al. 2019. Determinants of age at first reproduction and lifetime breeding success revealed by full paternity assignment in a male ungulate. *Oikos*, **128**, 328–337.

Mbeau-ache, C. and Franco, M. 2013. The time distribution of reproductive value measures the pace of life. *Journal of Ecology*, **101**, 1273–1280.

Moller, A. P. and De Lope, F. 1999. Senescence in a short-lived migratory bird: age-dependent morphology, migration, reproduction and parasitism. *Journal of Animal Ecology*, **68**, 163–171.

Morris, W. F., Altmann, J., Brockman, D. K., et al. 2011. Low demographic variability in wild primate populations: fitness impacts of variation, covariation, and serial correlation in vital rates. *The American Naturalist*, **177**, E14–E28.

Newton, I. and Rothery, P. 1997. Senescence and reproductive value in sparrowhawks. *Ecology*, **78**, 1000–1008.

Perry, R. I., Cury, P., Brander, K., Jennings, S., Mollmann, C., and Planque, B. 2010. Sensitivity of marine systems to climate and fishing: concepts, issues and management responses. *Journal of Marine Systems*, **79**, 427–435.

Planque, B., Fromentin, J. M., Cury, P., et al. 2010. How does fishing alter marine populations and ecosystems sensitivity to climate? *Journal of Marine Systems*, **79**, 403–417.

Pukk, L., Kuparinen, A., Jarv, L., Gross, R., and Vasemagi, A. 2013. Genetic and life-history changes associated with fisheries-induced population collapse. *Evolutionary Applications*, **6**, 749–760.

Rouyer, T., Ottersen, G., Durant, J. M., et al. 2011. Shifting dynamic forces in fish stock fluctuations triggered by age truncation? *Global Change Biology*, **17**, 3046–3057.

Sæther, B.-E. and Engen, S. 2002. Including uncertainties in population viability analysis using population prediction intervals. *In*: Beissinger, S. R. and Mccullough, D. R. (eds.) *Population Viability Analysis*, pp. 191–212. Chicago: University of Chicago Press.

Sæther, B.-E. and Engen, S. 2003. Routes to extinction. *In*: Blackburn, T. M. and Gaston, K. J. (eds.) *Macroecology. Concepts and Consequences*, pp. 218–236. Oxford: Blackwell.

Sæther, B.-E. and Engen, S. 2019. Towards a predictive conservation biology: the devil is in the behaviour. *Philosophical Transactions of the Royal Society B London* **374** (1781).

Sæther, B.-E. and Haagenrud, H. 1983. Life history of the moose (*Alces alces*): fecundity rates in relation to age and carcass weight. *Journal of Mammalogy*, **64**, 226–232.

Sæther, B.-E. and Haagenrud, H. 1985. Life history of the moose (*Alces alces*): relationship between growth and reproduction. *Holarctic Ecology*, **8**, 100–106.

Sæther, B.-E., Engen, S., Swenson, J. E., Bakke, Ø., and Sandegren, F. 1998. Assessing the viability of Scandinavian brown bear, *Ursus arctos*, populations: the effects of uncertain parameter estimates. *Oikos*, **83**, 403–416.

Sæther, B.-E., Engen, S., Filli, F., Aanes, R., Schröder, W., and Andersen, R. 2002a. Stochastic population dynamics of an introduced Swiss population of the ibex. *Ecology*, **83**, 3457–3465.

Sæther, B.-E., Engen, S., Lande, R., Visser, M., and Both, C. 2002b. Density dependence and stochastic variation in a newly established population of a small songbird. *Oikos*, **99**, 331–337.

Sæther, B.-E., Solberg, E. J., and Heim, M. 2003. Effects of altering sex ratio structure on the demography of an isolated moose population. *Journal of Wildlife Management*, **67**, 455–466.

Sæther, B.-E., Solberg, E. J., Heim, M., Stacy, J. E., Jakobsen, K. S., and Olstad, R. 2004. Offspring sex ratio in moose *Alces alces* in relation to paternal age: an experiment. *Wildlife Biology*, **10**, 51–57.

Sæther, B.-E., Engen, S., Persson, J., Brøseth, H., Landa, A., and Willebrand, T. 2005. Management strategies for the wolverine in Scandinavia. *Journal of Wildlife Management*, **69**, 1001–1014.

Sæther, B.-E., Engen, S., Solberg, E. J., and Heim, M. 2007a. Estimating the growth of a newly established moose population using reproductive value. *Ecography*, **30**, 417–421.

Sæther, B.-E., Lillegård, M., Grøtan, V., Filli, F., and Engen, S. 2007b. Predicting fluctuations of re-introduced ibex populations: the importance of density-dependence, environmental stochasticity and uncertain population estimates. *Journal of Animal Ecology*, **76**, 326–336.

Sæther, B.-E., Grøtan, V., Engen, S., Noble, D. G., and Freckleton, R. P. 2009. Critical paramters for predicting populations of some British passerines. *Journal of Animal Ecology*, **78**, 1063–1075.

Sæther, B.-E., Engen, S., Odden, J., Linnell, J. D. C., Grotan, V., and Andren, H. 2010. Sustainable harvest strategies for age-structured Eurasian lynx populations: the use of reproductive value. *Biological Conservation*, **143**, 1970–1979.

Sæther, B.-E., Coulson, T., Grøtan, V., et al. 2013. How life history influences population dynamics in fluctuating environments. *The American Naturalist*, **182**, 743–759.

Sæther, B.-E., Engen, S., Gamelon, M., and Grøtan, V. 2019. Predicting the effects of climate change on bird population dynamics. *In:* Dunn, P. O. and Møller, A. P. (eds.) *Effects of Climate Change on Birds*, 2nd edition. Oxford: Oxford University Press.

Schaffer, W. M. 1974. Selection for optimal life histories: effects of age structure. *Ecology*, **55**, 291–303.

Silvertown, J., Franco, M., and Perez-ishiwara, R. 2001. Evolution of senescence in iteroparous perennial plants. *Evolutionary Ecology Research*, **3**, 393–412.

Solberg, E. J., Garel, M., Heim, M., Grøtan, V., and Sæther, B.-E. 2008. Lack of compensatory body growth in a high performance moose *Alces alces* population. *Oecologia*, **158**, 485–498.

Toresen, R. and Østvedt, O. J. 2000. Variation in abundance of Norwegian spring-spawning herring (*Clupea harengus*, Clupeidae) throughout the 20th century and the influence of climatic fluctuations. *Fish and Fisheries*, **1**, 231–256.

Trask, A. E., Bignal, E. M., Mccracken, D. I., Piertney, S. B., and Reid, J. M. 2017. Estimating demographic contributions to effective population size in an age-structured wild population experiencing environmental and demographic stochasticity. *Journal of Animal Ecology*, **86**, 1082–1093.

Tufto, J., Sæther, B.-E., Engen, S., Swenson, J. E., and Sandegren, F. 1999. Harvesting strategies for conserving minimum viable populations based on World Conservation Union criteria: brown bears in Norway. *Proceedings of the Royal Society London B, Biological Sciences*, **266**, 961–967.

Tuljapurkar, S. D. 1982. Population dynamics in variable environments. II. Correlated environments, sensitivity analysis and dynamics. *Theorethical Population Biology*, **21**, 114–140.

Tuljapurkar, S. and Haridas, C. V. 2006. Temporal autocorrelation and stochastic population growth. *Ecology Letters*, **9**, 327–337.

Vlam, M., Van der Sleen, P., Groenendijk, P., and Zuidema, P. A. 2017. Tree age distributions reveal large-scale disturbance-recovery cycles in three tropical forests. *Frontiers in Plant Science*, **7**.

Williams, G. C. 1966. Natural selection: costs of reproduction and a refinement of Lack's principle. *The American Naturalist*, **100**, 687–690.

Zhang, H., Rebke, M., Becker, P. H., and Bouwhuis, S. 2015. Fitness prospects: effects of age, sex and recruitment age on reproductive value in a long-lived seabird. *Journal of Animal Ecology*, **84**, 199–207.

CHAPTER 18

Applying comparative methods to different databases: lessons from demographic analyses across mammal species

Jean-Michel Gaillard, Victor Ronget, Jean-François Lemaître, Christophe Bonenfant, Guillaume Péron, Pol Capdevila, Marlène Gamelon, and Roberto Salguero-Gómez

18.1 Introduction

The diversity of life histories displayed by species along the Tree of Life has fascinated researchers for millennia. For several centuries, the *Natural History of Animals* written by Aristotle (350 BCE) was the only work reporting a comprehensive evidence of this diversity. Much later, Georges-Louis Leclerc de Buffon proposed his *Histoire Naturelle* (1749–1767), adding a tremendous amount of new observations on life histories and unique synthesis, including the first human life tables in contrasting environments (i.e. cities vs. countryside). This very early work was essentially descriptive, and proper comparative demographic analyses across species searching for a structure and organisation of life histories appeared much later. For instance, one of the first comparative demographic analyses was proposed by Mitchell (1911), although it was not rooted in the grounds of life history evolution. The first question formulated in an evolutionary framework was posed by Lamont Cole (1954) about the optimal timing of first reproduction, which led him to launch the paradox of the rarity of semelparous species in the Tree of Life despite their evolutionary advantage over iteroparous species.

The modern and still current era of comparative analyses of life histories emerged with the seminal work by Stearns (1976, 1977), which generated a burst of comparative studies in the 1980s and 1990s (see Gaillard et al. 2016 for a review). These life history analyses generally included vital rates (e.g. survival and fecundity) among a set of diverse life history traits describing the life cycle of organisms, and covered a wide range of organisms (e.g. several groups: Fenchel 1974; Blueweiss et al. 1978; insects: Blackburn 1991; plants: Silvertown et al. 1993; homothermous vertebrates: Gaillard et al. 1989). Overall, these comparative studies enabled the identification of covariation among nondemographic life history traits (e.g. Russell 1982; Bennett and Harvey 1987), vital rates (e.g. Gaillard et al. 1989; Silverton et al. 1993), or both types (e.g. Promislow et al. 1990) across species. For a focal taxonomic group (order, e.g. Swihart 1984, or class, e.g. Stearns 1983), species-specific life history traits were collected from literature surveys. The association between traits was investigated through bivariate correlations first, and then multivariate covariations among traits were examined. It became rapidly evident that allometric rules (i.e. linear scaling of a life history trait [e.g. life expectancy] as a function of size [e.g. body mass] on a log-log scale; Huxley 1932) strongly constrain life history variation among species (Peters 1983; Calder 1984) and should be accounted for to identify reliably covariation patterns among life history traits. Following Felsenstein's work (1985), it was also soon realised that phylogenetic relatedness—that is, the propensity of closely related species to share

Jean-Michel Gaillard et al., *Applying comparative methods to different databases: lessons from demographic analyses across mammal species.*
In: *Demographic Methods Across the Tree of Life.* Edited by Roberto Salguero-Gómez and Marlène Gamelon, Oxford University Press.

more similar trait values than distantly related ones, when everything else is kept constant—should also be accounted for in comparative analyses (Harvey and Pagel 1991; Garamszegi 2014; Chapter 1).

The comparative studies of life histories across species published in the last decades of the twentieth century covered a broad range of taxonomic groups (e.g. from insects to mammals among animals). Taken together, these studies consistently showed that the first axis of life history variation across species corresponds to a time-scale ranking species along a slow–fast continuum of life histories, opposing species with a slow pace of life (i.e. with long biological times [i.e. any life history trait measured in units of time such as developmental time, age at first reproduction, or life span] and low frequencies (i.e. the inverse of a biological time, such as heart beats per minute or annual fecundity) to species with a fast pace of life (i.e. with opposite characteristics). Offspring number (e.g. litter size) and mass (e.g. birth mass), which are often included in comparative analyses of life histories (e.g. Promislow et al. 1990; Bielby et al. 2007), correspond to dimensionless number and volume, respectively. Therefore, while these traits can be linked to the slow–fast continuum of life histories, they cannot be interpreted as defining this continuum. It is noteworthy that this continuum is retrieved both with and without accounting for allometric effects on traits, although, obviously, the rank of species along this continuum can differ markedly between analyses performed with and without correction for body size (e.g. Gaillard et al. 1989). Another continuum involving developmental (e.g. Stearns 1983) or reproductive (e.g. Gaillard et al. 1989) traits is often reported as a second structuring axis of life history variation across species (see also Salguero-Gómez et al. 2016a for a study on plants).

While most studies performed to date have investigated the structure of life history variation (reviewed in Gaillard et al. 2016), only a few have examined the ecological correlates (*sensu* Harvey and Clutton-Brock 1985) of specific life history covariations. Firm evidence that ecological factors or traits describing the lifestyle (in terms of, e.g., habitat, diet, mating system, or degree of sociality; *sensu* Dobson 2007) influence the covariation pattern among life history traits is so far limited to the role played by hibernation (Turbill et al. 2011) and by the habitat, with aerial or subterranean species being slower than species living in other habitats (Healy et al. 2019) and species living in tropical environments being slower than species living in temperate areas (Wiersma et al. 2007). However, while living in aquatic habitats has selected for obvious morphological changes in mammalian

evolution (Eisenberg 1981), whether aquatic mammals display different demographic patterns compared to terrestrial mammals has remained little investigated, although habitat is a template (*sensu* Southwood 1977) that influences size, life history, and behaviour of organisms. We will thus test whether habitat types (aquatic vs. terrestrial) influence allometric relationships of demographic metrics and/or the contribution of dimensionless demographic metrics to major axes, which shape the variation in demographic trajectories.

In this chapter, we will illustrate the common approach currently used to identify covariation among life history traits by performing a comparative demographic analysis across mammalian species. More specifically, we illustrate the principle and the power of the comparative demographic approach by focusing on the shape of demographic trajectories (*sensu* Ronget and Gaillard 2020) across mammalian species in relation to their occurrence in terrestrial vs. aquatic habitats. The choice of this question was motivated by two reasons.

First, while whether allometric constraints should be accounted for when performing comparative analyses has been often discussed (e.g. Jeschke and Kokko 2009), covariation among dimensionless metrics, which solves this problem, has not been yet investigated, although recent comparative studies of actuarial senescence included shape in their analyses (see e.g. Baudisch 2011; Ronget and Gaillard 2020). Performing analysis of dimensionless metrics gives us the opportunity to match the criteria of a dimensional analysis (*sensu* Stahl 1962). Dimensional analysis, which tracks the dimensions (e.g. length, volume, or time) and the measurement units when comparing among traits, is key to comparative analyses because life history traits do not share a common dimension. For instance, life expectancy is measured in units of time, skeletal size is measured in units of length, and body mass is measured in units of volume. As such, biologists should be careful not to compare apples to oranges when performing comparative analyses encompassing very diverse traits. For instance, mixing biological times and dimensionless numbers with a same analysis often leads to identifying two independent axes of variation simply because of differences in units (as in, e.g., Bielby et al. 2007, where biological times structured the first axis and litter size the second one in most taxonomic-based analyses of mammals reported in that work). Charnov (1993) popularised this type of analysis by encouraging the use of dimensionless numbers that can be compared directly among species, without worrying about allometric effects.

Second, while we expect allometric constraints to differ between terrestrial (with high constraints on locomotion) and aquatic (with less constraints on locomotion) habitats, this has been mostly overlooked in previous life history analyses (but see Capdevila et al. 2020 for a recent comparative analysis of differences between terrestrial and aquatic species of plants and animals). With these issues at the forefront, we designed the current study with the following three key considerations, which highlight common pitfalls in comparative demographic research.

First, as the deposition of data in widely available repositories is rapidly becoming common in biological research, we have to identify the most suitable primary data to address our question. To illustrate the potential influence of the source of species-specific data retrieved, we will compare the results of the same analyses replicated on two different reference data sets. The first one was based on the detailed fully age-specific lx and mx series (see Chapter 8) estimated for a single population (i.e. lx and mx series came from the same population for a given species) (Malddaba database;[1] see Lemaître et al. 2020a, 2020b for recent applications using age-specific mortality and reproductive data, respectively). The second one was based on st(age)-structured population matrix models (see Chapter 9) that included most often an age for which vital rates were assumed to be constant, independent of age (i.e. in most cases adult survival and adult fecundity is assumed constant over ages) for a given species (Comadre database; see Salguero-Gómez et al. 2016b for an overview of the database). The two databases have very different purposes. Malddaba only includes age-specific traits (i.e. survival, reproduction, body mass) of mammals and currently contains ca. 200 species with at least partial age-specific demographic information. On the other hand, Comadre encompasses all major taxa and includes at least partial demographic information for thousands of species. We thus expected different outcomes for our analyses, depending on the database used.

Second, from both databases, we estimated biological times and dimensionless metrics. We performed an allometric analysis to test whether allometric expectations were met. Indeed, we expected biological times (i.e. metrics expressed in units of time) to be tightly related to body mass across species, contrary to

dimensionless metrics. We also tested whether habitat types (i.e. terrestrial vs. aquatic) and the source of the demographic data (i.e. Malddaba vs. Comadre) influenced the relationship between body mass and each demographic metric. We expected the relationships between demographic metrics and body mass to be stronger in Malddaba than in Comadre data sets due to the inclusion of full age-dependence in vital rates in Malddaba (see e.g. Festa-Bianchet et al. 2003). We also expected stronger allometric relationships in species living in terrestrial habitats because the biophysical constraints of a large body size in terms of locomotion, and thereby of resource acquisition, are much weaker in aquatic than in terrestrial environments (see Gaillard et al. 1989 for a discussion).

Lastly, we performed phylogenetically controlled statistical analyses to determine the covariation that structured interspecific variation in the shape of demographic trajectories. We thus used dimensionless metrics that can be interpreted in terms of shape of demographic trajectories (see Ronget and Gaillard 2020). Comparing the analysis performed using data sets extracted from Malddaba and Comadre allowed us to assess the role of age dependence in vital rates for shaping variation in demographic trajectories across species.

18.2 Methods

18.2.1 Building databases and calculating demographic metrics

We first looked for age-specific demographic data (i.e. lx and mx series) for all mammalian species included in Malddaba. As generally done in comparative analyses of life history traits (but see Lemaître et al. 2020a), only the mean trait value for each species was considered. For this analysis, we selected each species in which lx series (i.e. percentage of individuals remaining alive at age x) and mx series (i.e. average number of female offspring born per female of age x) were reported for the same population. We retrieved 66 species (54 species with terrestrial habitats and 12 with aquatic habitats) where this required information was available. When species-specific data were available for more than one population, we retained the population for which the data were of highest quality (i.e. larger sample size, longer monitoring, more reliable estimation methods of vital rates). Next, for all these species, we extracted female adult body mass (in grams) from Malddaba, and we calculated the five following demographic metrics using the lx and mx series:

[1] The database Malddaba (built and managed by JMG, VR, and JFL) is currently under progress but will be released online and freely available for the scientific community in the coming years. In the meantime, any query regarding the database should be sent to JMG, VR, or JFL.

Generation time (T): We calculated *T* as recommended in Staerk et al. (2019) using the following formula:

$$T = \frac{\sum x\lambda^{-x}lxmx}{\sum \lambda^{-x}lxmx}$$

which corresponds to *Tb* discounted for observed population growth (see Table 1 in Staerk et al. 2019, p. 3). Assuming the population has reached a stable age distribution and overlapping generations, *Tb* corresponds to the mean weighted age of mothers in a population. The observed population growth (λ) was calculated from a population projection model (see Chapter 9) built from the full *lx* and *mx* schedules using the R package popbio (Stubben et al. 2007). *T* provides a measure of the ranking of a given species along the slow–fast continuum of life histories (Gaillard et al. 2005). *T* has a dimension of time (in years).

Age at first reproduction (AFR): AFR was defined as the earliest age at which at least 5% of females in a population give birth to newborn. AFR has a dimension of time (in years).

Mortality entropy from birth (H_{birth}): We calculated H_{birth} as the Keyfitz entropy (Keyfitz 1985) using the following formula:

$$H_{birth} = -\frac{\sum lx \ln(lx)}{\sum lx}$$

H_{birth} is often used to assess the shape of the mortality curve over the lifetime (Chapter 8). A value of H_{birth} lower than 1 is typically interpreted as an increase of mortality with age, and a value of H_{birth} higher than 1 is typically interpreted as a decrease of mortality with increasing age. However, identical values of H_{birth} can correspond to rather different forms of mortality curves (see e.g. Goldman and Lord 1986). We will interpret more generally H_{birth} as a metric measuring the lifetime lost due to age-specific mortality relative to the life expectancy at birth (Goldman and Lord 1986). H_{birth} is highly sensitive to differences between juvenile and adult mortality. It generally decreases with survival increase at older ages (Fernandez and Beltran-Sanchez 2015) and can be interpreted as a measure of the diversity of ages at death. H_{birth} is a positive nondimensional number.

Mean number of recruits per lifetime (φ): We calculated φ as the weighted (with the proportional asymptotic representation of the age class in the population as weight) mean number of recruits produced by females over their life span using the following formula:

$$\varphi = \sum w\, mx$$

with *w* the asymptotic distribution of the age class in the population as weight. The metric φ measures the mean lifetime reproductive output and has a dimension of time (years⁻¹).

Number of genealogies (S): We calculated *S* as the Demetrius's entropy (Demetrius 1979) calculated using the following formula:

$$S = -\sum \lambda^{-x}lxmx * \ln\left(\lambda^{-x}lxmx\right)$$

This trait corresponds to the number of genealogies compatible with a given population growth rate and can be interpreted as a measure of the diversity of reproductive trajectories within the focal population. The metric *S* is a dimensionless number.

For the species retrieved from the Comadre database, body mass and all the metrics described above were directly obtained from a recent comparative analysis that derived them from matrix population models (see R code and Supplementary Online Materials available in Capdevila et al. 2020). In this case, the information was available for 70 mammalian species (55 species with a terrestrial habitat and 15 with an aquatic habitat).

18.2.2 Statistical analyses accounting for phylogenetic inertia

We obtained two data sets (Malddaba and Comadre) with species-specific information for five demographic metrics, one measure of habitat type, and adult body mass. We first analysed the allometric relationships using phylogenetic regressions linking demographic metrics and body mass, as well as assessing the potential influence of habitat type on these relationships. Demographic traits and adult body mass were all log-transformed (Pélabon et al. 2018). We fitted five different models: (1) a model including interactive effects of adult body mass and habitat type on a given demographic trait only (i.e. different allometries between habitat types); (2) a model including additive effects of adult body mass and habitat type on a given demographic trait (i.e. same allometry in both habitat types with a constant difference in a given demographic trait at a given adult body mass); (3) a model only including the influence of adult body mass on a given demographic trait (i.e. same allometry and same trait value at a given adult body mass in both habitat types); (4) a model only including the influence of habitat type on a given demographic trait (i.e. no allometry and a constant difference in a given demographic trait between habitats); and (5) a model including no influence of adult body mass or habitat type on a given demographic trait (i.e. intercept-only model corresponding to a same average value for a given demographic trait at any size in any habitat type). Importantly, to avoid nonindependence of traits among species caused by the phylogenetic relatedness (see Chapter 1), we used phylogenetically corrected analyses. We thus fitted phylogenetic generalised least square (PGLS) models (Freckleton et al. 2002) and used a consensus phylogenetic tree (Bininda-Emonds et al. 2007, 2008). Although more recent phylogenetic trees have been published (see e.g. Upham et al. 2019), we keep using the most common tree considered in comparative analyses of mammalian life histories (e.g. Lemaître et al. 2020a). We estimated the strength of the phylogenetic signal for each model by using Pagel's λ (Pagel 1999) that varies from 0 (no phylogenetic signal detected) to 1 (perfect phylogenetic signal). Although this signal is highly informative, it is important to keep in mind that it requires some strong assumptions to be fulfilled, such as Brownian motion for trait evolution and the same rate of trait change across the whole tree. It is noteworthy that Pagel's λ should not be confused with population growth (also called λ). We performed model selection based on AIC (Burnham and Anderson 2002). We retained the model with the lowest AIC, so long as the difference in AIC scores was more than 2 points. If the difference in AIC scores between candidate models was less than 2, we retained the model with the least parameters based on the criterion of parsimony.

We next identified the axes of variation that structure the shape of demographic trajectories from the variation observed in dimensionless demographic metrics across species with a phylogenetic principal component analysis (PPCA) (Revell 2012). We entered four dimensionless demographic metrics as PPCA loadings, including the two measures of entropy defined above and available in both data sets (Keyfitz's entropy H_{birth} and Demetrius's entropy S, log-transformed) and two other measures corresponding to biological constructs based on different demographic metrics defined earlier. We used the $\log_e$ transformation in all cases. We defined a measure of dimensionless productivity (*Prod*) as the product between Log(T) (generation time in years) and Log(ϕ) (number of recruits per lifetime in years^{-1}), and a measure of dimensionless timing of first reproduction (*TFR*) as the ratio between Log(*AFR*) (age at first reproduction in years) and Log(T) (generation time in years). Although the use of ratios in life history analyses has been criticised for leading to spurious or biased relationships (e.g. Atchley et al. 1976), they

are valid and generate nonspurious results when the resulting metrics are biological constructs (rather than mathematical constructs) based on highly correlated traits (Smith 1999), as proposed in the present analysis. We assessed the number of principal components (i.e. defining the structuring axes) to retain based on the broken-stick method (Jackson 1993) and tested the potential influence of habitat type and of Log(T) (as a measure of the ranking of a given species along the slow–fast continuum of life histories; Gaillard et al. 2005) on this/these component(s). We fitted models including both additive and interactive effects of generation time (Log(T)) and habitat type and, again, used AIC for model selection.

The script to perform the statistical analyses in R is provided in the Appendix, together with the data sets and the phylogenetic trees (please go to www.oup.com/companion/SalgueroGamelonDM).

18.3 Results

Both Malddaba and Comadre data sets encompassed a wide range of taxonomic groups of mammals (Figure 18.1). The two databases shared 22 common species.

18.3.1 Allometric analyses

As expected, we detected allometric relationships for both demographic metrics with a dimension of time (i.e. generation time, T, and age at first reproduction, *AFR*) across mammalian species, from both Malddaba (Table 18.1a) and Comadre data sets (Table 18.1b). However, the strength of these relationships was consistently higher when using Malddaba data, with larger allometric coefficients and larger proportion of variance explained (i.e. larger R^2) than allometric relationships assessed from Comadre data (Table 18.2a vs. Table 18.2b). As a result, because allometric relationships are strongly influenced by phylogeny in mammals (e.g. Riek and Geiser 2013), the phylogenetic signal was stronger in analyses performed using Malddaba than using Comadre (Table 18.2), indicating a higher explanatory power. For the third demographic trait with a dimension of time, the number of recruits per lifetime ϕ, we found that its response to mass changes differed between databases. A quite strong positive relationship was found when using the Malddaba data (Table 18.2a), whereas the number of recruits per lifetime ϕ was independent of adult body mass changes when using Comadre (Table 18.2b). The analyses of dimensionless

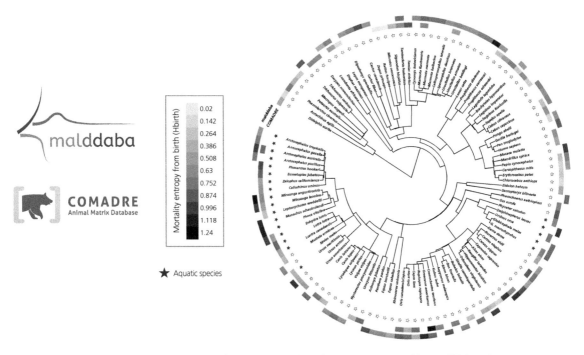

Figure 18.1 Phylogenetic tree of mammalian species for which demographic information was extracted from Malddaba and COMADRE databases to estimate the demographic metrics under study. The habitat type (terrestrial vs. aquatic) of each species is reported. As an illustration, database-specific values of one demographic metric (H_{birth}) are displayed.

Table 18.1 Model selection of PGLS models fitted to assess allometric relationships of a given demographic trait estimated for mammalian species from Malddaba (a) and Comadre (b) databases. The table displays AIC of each model fitted, and the selected model occurs in bold.

(a)

Model	np	T	H_{birth}	ϕ	S	AFR
Log(*BM*) × *Hab*	4	47.523	6.745	130.984	−74.887	78.930
Log(*BM*) + *Hab*	3	46.237	5.578	129.924	−76.407	77.145
Log(*BM*)	2	**45.110**	10.572	**128.005**	**−77.219**	**75.412**
Hab	2	76.931	**3.585**	143.878	−49.671	102.765
1	1	77.474	9.792	142.672	−48.528	102.330

(b)

Model	np	T	H_{birth}	ϕ	S	AFR
Log(*BM*) × *Hab*	4	167.203	−8.522	−166.040	69.277	131.338
Log(*BM*) + *Hab*	3	165.770	−9.888	−165.714	67.373	129.718
Log(*BM*)	2	**163.798**	**−11.590**	**−167.593**	66.265	**127.720**
Hab	2	171.805	−7.565	−153.902	65.412	137.081
1	1	170.999	−8.141	−152.633	**64.686**	136.832

T: generation time (time); H_{birth} (dimensionless): mortality entropy from birth; ϕ (time^{-1}): mean number of recruits produced per female during its lifetime; *S* (dimensionless): Demetrius's entropy; *AFR* (time): age at first reproduction; np: number of parameters of the model; BM: adult body mass; Hab: habitat type.

Table 18.2 Parameter estimates of the PGLS model retained to assess allometric relationships of a given demographic trait estimated for mammalian species from Malddaba (a) and Comadre (b) databases. For each demographic trait, the table displays the selected model; the parameters (intercept, allometric slope, and habitat effect), with standard error in parentheses; the estimate of the phylogenetic signal and its 95% confidence interval; and the explanatory power of the selected model (measured by R^2, the proportion of variance explained).

(a)

Demographic Trait	Model selected	Intercept	Slope	Effect of Hab	λ	R^2
T	Log(BM)	0.463 (0.414)	**0.158 (0.023)**	NA	0.891 [0.733 – 0.970]	0.435
H_{birth}	Hab	−0.309 (0.033)	NA	**0.028 (0.078)**	0.000 [0.000 – 0.188]	0.117
ϕ	Log(BM)	0.329 (0.689)	−0.181 (0.040)	NA	0.818 [0.582 – 0.947]	0.242
S	Log(BM)	0.315 (0.163)	**0.057 (0.009)**	NA	0.887 [0.738 – 0.962]	0.393
AFR	Log(BM)	−0.790 (0.441)	**0.165 (0.026)**	NA	0.783 [0.550 – 0.921]	0.381

(b)

Demographic Trait	Model selected	Intercept	Slope	Effect of Hab	λ	R^2
T	Log(BM)	**1.448 (0.319)**	**0.096 (0.031)**	NA	0.000 [0.000 – 0.334]	0.123
H_{birth}	Log(BM)	**0.204 (0.091)**	**0.025 (0.009)**	NA	0.000 [0.000 – 0.488]	0.107
ϕ	Log(BM)	**0.205 (0.030)**	−0.013 (0.003)	NA	0.000 [0.000 – 0.471]	0.221
S	1	**0.413 (0.046)**	NA	NA	0.000 [0.000 – 0.215]	0.000
AFR	Log(BM)	0.804 (0.247)	**0.082 (0.024)**	NA	0.000 [0.000 – 0.251]	0.147

λ: phylogenetic signal.

demographic traits (i.e. Demetrius's entropy S and Keyfitz's entropy H_{birth}) showed contrasting outcomes between the two databases. Keyfitz's entropy H_{birth} did not depend on adult body mass in the analyses using Malddaba (Table 18.2a), whereas a positive but weak association between Keyfitz's entropy H_{birth} and adult body mass was identified in the analyses using Comadre (Table 18.2b). Demetrius's entropy S was positively associated with adult body mass for data from Malddaba, whereas S did not depend on adult body mass from Comadre analyses. From the two databases we did not find any support for an influence of living in water or on land on mammalian demographic metrics either directly or through a differential response to adult body mass changes (Table 18.1). Only Keyfitz's entropy H_{birth} displayed an average difference between terrestrial and aquatic species from Malddaba data, with aquatic mammals having higher mortality entropy at birth than their terrestrial counterparts.

18.3.2 Assessing axes of variation among dimensionless demographic metrics

From the PPCA performed using Malddaba we uncovered two major axes of variation among dimensionless demographic metrics (Figure 18.2a). The first axis (PPC1) accounted for 42.94% of the total demographic variance among the four metrics (Table 18.3) and was mostly structured by the trade-off between dimensionless timing of first reproduction and dimensionless productivity and, to a much lesser extent, by the trade-off between diversity of mortality trajectories and diversity of reproductive trajectories (Figure 18.3). This structuring axis thus included a continuum opposing species with a relatively late age at first reproduction (correlation between PPC1 and dimensionless timing of first reproduction TFR of 0.740), a high diversity of reproductive trajectories (correlation between PPC1 and Demetrius's entropy S of 0.577), a low relative number of recruits (correlation between PPC1 and dimensionless productivity $Prod$ of −0.893), and to a lesser extent a low diversity of ages at death (correlation between PPC1 and Keyfitz's entropy H_{birth} of −0.200) to species presenting the opposite set of metrics. The second axis (PPC2) accounted for 31.34% of the total demographic variance among the four traits (Table 18.3) and was mostly structured by a continuum of diversity of trajectories in terms of both mortality and reproduction. This structuring axis corresponded to a continuum opposing species with a high diversity of ages at death (correlation between PPC2 and Keyfitz's entropy H_{birth} of 0.855), a high diversity of

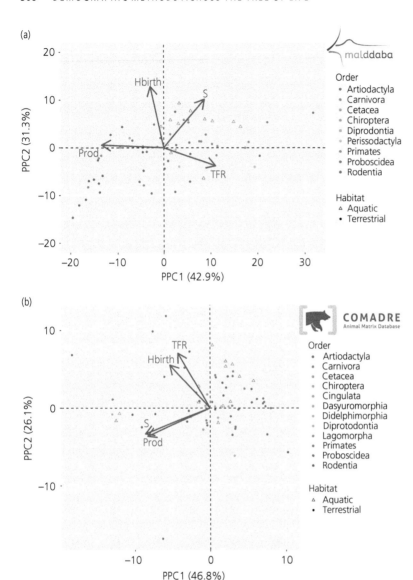

Figure 18.2 Graphical displays of the first factorial plan (PPC2 vs. PPC1) obtained from the multivariate analysis of dimensionless demographic traits performed on Malddaba (a) and on COMADRE (b) databases. Each data point corresponds to a mammalian species, and the nonindependence among species data points was corrected for by using a phylogenetically controlled PCA. The demographic metrics (lH_{birth}: log-transformed mortality entropy from birth; lS: log-transformed Demetrius's entropy; *Prod*: dimensionless productivity; *TFR*: dimensionless timing of first reproduction) are displayed in red.

reproductive trajectories (correlation between PPC2 and Demetrius's entropy S of 0.676), and to a lesser extent a relatively early age at first reproduction (correlation between PPC2 and dimensionless timing of first reproduction *TFR* of −0.253) to species with opposite metrics. The dimensionless productivity contributed very little to PPC2 (correlation between PPC2 and dimensionless productivity *Prod* of 0.036). The two remaining axes accounted for only one-quarter of the demographic variance across species when pooled and cannot be interpreted in biological terms (Table 18.3). Interestingly, these two independent axes structuring the variation among dimensionless

demographic metrics were both closely related to generation time T, which measures the ranking of species along the slow–fast continuum (Figure 18.3). Thus, PPC1 was strongly correlated positively with generation time T, which explained most observed variation in PPC1 (R^2 of 0.958). Likewise, although mathematically independent of PPC1, PPC2 was also positively correlated with generation time T, which explained almost half the variation observed in PPC2 (R^2 of 0.427).

We analysed the influence of habitat type on PPCs, on generation time T, and on the covariation between structuring axes and generation time T. On average,

Table 18.3 Proportion of variance explained by the four phylogenetically corrected principal components (PPC1–PPC4) obtained from PPCA performed on four dimensionless demographic metrics measured from data extracted from Malddaba and Comadre databases. Only the first two PPCs (in bold) are interpreted in terms of structuring axes of demographic variation due to their relevance in passing the broken-stick rule (Jackson 1993).

Database	PPC1	PPC2	PPC3	PPC4
Malddaba	**0.4294**	**0.3134**	0.1700	0.0871
Comadre	**0.4680**	**0.2614**	0.1986	0.0720

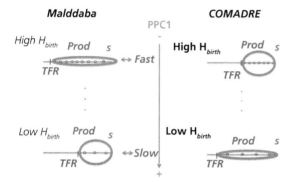

Figure 18.3 Biological interpretation of the main structuring factor shaping the variation among dimensionless demographic metrics in the examined mammalian species according to the database used. When analysing data from Malddaba, mammalian species are distributed along a main axis opposing species with a relatively early age at first reproduction (low TFR, brown trait), producing a relatively large number of recruits (high Prod, open red circles) and displaying a low diversity of reproductive trajectories for a given population growth rate (low S, green ellipses) and, to a much lesser extent, a high diversity of mortality trajectories from birth (H_{birth}, in black) to species with opposing characteristics. The size of the acronym indicates the contribution of the metric to the axis. This axis is closely correlated to the ranking of mammalian species along the continuum opposing species with a fast pace of life (Fast, in red) to species with a slow pace of life (Slow, in red), with generation time (T) as a metric. On the other hand, when analysing data from Comadre, the main axis of covariation among dimensionless demographic metrics opposes species with a relatively late age at first reproduction (high TFR, brown trait), producing a relatively large number of recruits (high Prod, open red circles) and displaying both a high diversity of reproductive trajectories for a given population growth rate (low S, green ellipses) and a high diversity of mortality trajectories from birth (H_{birth}, in black) to species with opposing characteristics. This axis provides a ranking of mammalian species independent of their ranking along the slow–fast continuum of life histories obtained when using generation time (T) as a metric. Note that in both cases, the intervals over which these covariations are assessed (flat blue lines) are constant and dimensionless. These intervals can be interpreted biologically as time-independent life cycles that are identical across all examined mammalian species.

aquatic species had higher values of PPC1 (intercept difference of 6.731 ± 3.448, P = 0.055) and higher values of PPC2 (intercept difference of 5.557 ± 1.614, P = 0.001), likely because they displayed a slower pace of life on average (i.e. generation time T longer by 0.483 ± 0.203 [log-scale], P = 0.020) than terrestrial species. We did not detect any evidence of an influence for habitat type on the slope between PPCs and generation time T (a slope difference of −1.010 ± 2.149, P = 0.640 and 4.540 ± 8.881, P = 0.611 with terrestrial habitats as a reference for PPC1 and PPC2, respectively). On the other hand, aquatic species had lower scores for PPC1 (i.e. towards early and high reproduction with a low diversity of reproductive trajectories) and higher scores for PPC2 (i.e. towards high diversity of both reproductive trajectories and ages at death) than terrestrial species for a given generation time T (intercept difference of −1.313 ± 0.750, P = 0.085 and 3.192 ± 1.340, P = 0.020 with terrestrial habitat as a reference for PPC1 and PPC2 (Figure 18.4), respectively).

The PPCA using Comadre also yielded two major axes of variation among dimensionless demographic metrics (Figure 18.2b). The first axis (PPC1) accounted for 46.80% of the total demographic variance among the four traits (Table 18.3). This structuring axis corresponded to a continuum opposing species with a relatively early age at first reproduction (correlation between PPC1 and dimensionless timing of first reproduction TFR of −0.417), a low diversity of reproductive trajectories (correlation between PPC1 and Demetrius's entropy S of −0.859), a low relative number of recruits (correlation between PPC1 and dimensionless productivity Prod of −0.828), and a low diversity of ages at death (correlation between PPC1 and Keyfitz's entropy H_{birth} of −0.525) to species with opposite metrics. The second axis (PPC2) accounted for 26.14% of the total demographic variance among the four metrics (Table 18.3). This structuring axis corresponded to a continuum opposing species with a high diversity of ages at death (correlation between PPC2 and Keyfitz's entropy H_{birth} of 0.553), a low diversity of reproductive trajectories (correlation between PPC2 and Demetrius's entropy S of −0.332), a relatively late age at first reproduction (correlation between PPC2 and dimensionless timing of first reproduction TFR of 0.706), and a low relative number of recruits (correlation between PPC2 and dimensionless productivity Prod of −0.362) to species with opposite metrics. The two other axes also accounted for slightly more than one-quarter of the demographic variance across species when pooled and cannot be interpreted in biological terms (Table 18.3). While the partition of

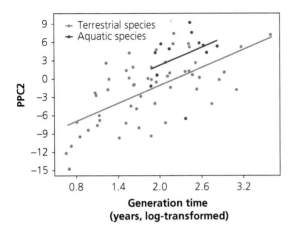

Figure 18.4 Relationship between generation time (T, on a log scale) and PPC2 across mammalian species from Malddaba. For a given generation time, aquatic mammalian species have higher PPC2 scores, which correspond to higher diversity of reproductive and mortality trajectories and to a lesser extent an earlier age at first reproduction.

demographic variance among axes of variation was similar (Table 18.3), the covariation among demographic traits differed depending on which database was analysed. Contrary to what we reported from analyses performed on Malddaba, these two independent axes structuring the variation among dimensionless demographic metrics were not related to the ranking of species on the slow–fast continuum. Thus, neither PPC1 (slope of -0.434 ± 0.836, P = 0.605) nor PPC2 (slope of -0.309 ± 0.592, P = 0.603) were associated with generation time T, and the two negative trends only accounted for a negligible proportion of observed variation in PPC1 and PPC2 (R^2 of ~0.004 in both cases). We did not detect any evidence of an influence of habitat types on either of the two examined PPCs (intercept difference of -0.589 ± 1.650, P = 0.722 and 1.596 ± 1.153, P = 0.171 with terrestrial habitat as a reference for PPC1 and PPC2, respectively), and generation time T did not differ between aquatic and terrestrial species either (intercept difference of 0.256 ± 0.237, P = 0.283).

18.4 Conclusions

Our comparative demographic analyses provide four major findings. First, the demographic structures including allometric relationships of biological times and the key axes of variation in demographic metrics are retrieved in both Malddaba and Comadre databases. This suggests that detailed information on age-specific vital rates (Malddaba database) has little effect on these patterns. Second, the strength of allometric relationships and the demographic covariation identified from the multivariate PPCAs differ between Malddaba and Comadre. These different outcomes cannot be explained either by a difference in explanatory power because the number of species was slightly larger in analyses performed on Comadre or by a different representation of mammalian groups between data sets, which both equally covered the main mammalian orders (see Figure 18.1). This finding supports the hypothesis that Malddaba, which includes full age dependence in lx and mx schedules, provides an accurate description of the age-based demographic trajectories across mammals. Third, both axes of demographic variation assessed from dimensionless traits are closely associated with the slow–fast continuum, supporting the significance of this time-scale to shape demographic tactics across mammals (Figure 18.3). Fourth, terrestrial and aquatic mammals differ only slightly in terms of demographic covariation among their dimensionless demographic metrics. However, in support of a recent comparative analysis of the demographic influence of habitat type performed across the Tree of Life (Capdevila et al. 2020), aquatic species reproduce earlier and produce a larger number of recruits than terrestrial species relative to their position on the slow–fast continuum.

Our findings clearly support the view that demographic metrics with a dimension of time are hypo-allometrically (i.e. with an allometric coefficient less than 1) linked with body mass across species (Peters 1983; Calder 1984), independently of the database used. On the other hand, dimensionless metrics displayed different patterns in relation to the database used, which demonstrate that dimensionless metrics are not necessarily independent of adult body size and can display a monotonous relationship with body mass. Using Malddaba, we found clear evidence that S, Demetrius's entropy, is positively associated with adult body size across mammals. Demetrius (1979) defined his dimensionless metric as the number of genealogies in a population expected for a given population growth. This high diversity of reproductive trajectories in slow-living species compared to fast-living ones might be the result of strong selection pressure against mortality patterns (illustrated by the low diversity of Keyfitz's entropy H_{birth} in slow-living compared to fast-living species we report), leading reproductive trajectories to respond to species-specific lifestyle and ecology. While there is only one way to have a slow pace of life (i.e. by avoiding risk to not jeopardise survivorship; Péron et al. 2019), the diversity of reproductive trajectories allows individuals in slow-living species to fine-tune state-dependent adjustments to the ecological context. We can expect this number to increase with the duration of the life cycle, which itself increases with adult body size.

The most conflicting results between Malddaba and Comadre occurred for H_{birth}, Keyfitz's entropy of mortality from birth. This is not surprising because Keyfitz's entropy H_{birth} is highly sensitive to age dependence in mortality during the lifetime (Goldman and Lord 1986). For a substantial proportion of species, Keyfitz's entropy H_{birth} in Comadre was calculated without including potential actuarial senescence. On the contrary, Keyfitz's entropy H_{birth} in Malddaba was calculated from the full age dependence of mortality reported, including the mortality increase with age caused by actuarial senescence (i.e. the increase in mortality rate with increasing age). As actuarial senescence is the rule rather than the exception in mammals (Nussey et al. 2013), we recommend including age dependence in adult mortality to consider Keyfitz's entropy H_{birth} in comparative demographic analyses focused on a given taxon (e.g. order or class). However, at the broad level of the Tree of Life, this problem is likely negligible because of the huge diversity of mortality curves (see e.g. Jones et al. 2014), whereby mammals can be simply characterised by a decreased mortality between the juvenile and adult stage and a low mortality during the adult stage.

The multivariate analysis of covariations among dimensionless demographic metrics identified the same structuring of the shape of demographic trajectories, with a main axis of variation accounting for more than 40% of the overall variation and the two first components accounting for about three-quarters of the overall variation, irrespective of the database analysed. However, the nature of this axis of variation differed between analyses performed using Malddaba and Comadre, indicating that the occurrence, the strength, and the shape of age dependence in survival and reproduction have a strong effect on the shape of demographic trajectories, at least across mammalian species. This finding should not come as a surprise, as previous studies have pointed out the need to include full age dependence when analysing demographic patterns. For instance, Eberhardt (1985) was one of the first to highlight the importance of considering age dependence in survival within the

adult stage to get an accurate picture of the demography of a given population. More recently, Colchero et al. (2019) showed how including age dependence in vital rates during the adult stage markedly modulates the estimates of stochastic population growth, which is crucial in the current context of climate change.

When including detailed age dependence in vital rates, there was another key finding. The ranking of mammalian species along the main axis of variation among dimensionless demographic metrics perfectly matched the species ranking along the slow–fast continuum. Thus, when standardising all mammalian species to the same dimensionless life cycle duration starting at a value of 0 and ending at a value of 1, they ranked along an axis of demographic covariations among dimensionless metrics from early, high, and weakly diverse reproductive trajectories with (at a much lesser extent) a high diversity of ages at death to late, low, and highly diverse reproductive trajectories with (at a much lesser extent) a low diversity of ages at death.

The identical nature of these two markedly different continua demonstrates that the speed of the life cycle and the shape of the life cycle are strongly associated, suggesting strong constraints in demographic variation across species. As pointed out by Péron et al. (2019), the constraint of adopting a risk-adverse life history strategy to be long-lived and display a slow pace of life strongly reduces variation in ages at death. Likewise, the low demographic impact of a given change in reproduction for slow-living species (Hamilton 1966) allows them to display a large array of reproductive trajectories, often shaped by environmental stochasticity (Tuljapurkar et al. 2009). This similarity of biological meaning between different continuums is parallel to the similarity of the nature of variation among demographic traits between analyses correcting or not for allometric effects of body size (Stearns 1983; Gaillard et al. 1989) and indicates that time is key to shape not only the timing of demographic events but also the shape of variation in the occurrence in these events along the life cycle.

Finally, our comparison between databases provides a kind of proof of concept. Some evolutionary ecologists not used to dimensional analyses (see Stahl 1962) and the appropriate use of ratios (see Smith 1999) might be tempted to see some circularity in our findings using the Malddaba database. After all, two of our dimensionless traits included generation time explicitly in their definition, one in a product (*Prod*) and the other in a ratio (*TFR*). At first glance, it may seem trivial that a compound variable XxZ should be negatively correlated with a compound Y/Z and thus strongly structure the first PCA axis when included with two other mathematically independent traits, and lead the PCA axis to be strongly related to Z. However, the replication of the analyses to the Comadre data set empirically proves that it is not the case, mostly because, as well explained by Smith (1999) when X and Y are each highly correlated with Z (which was the case here because all metrics included in *Prod* and *TRF* were biological times that are strongly associated across species; see e.g. Calder 1984), the correlation between the ratios is not at all spurious. Indeed, the first PCA axis from the Comadre analysis based on the same metrics revealed a positive association between *Prod* and *TFR* instead of the negative association that a sceptical evolutionary ecologist would have expected.

18.5 Prespectives

The inability to retrieve these demographic outcomes when using Comadre data is highly informative but does not preclude analyses from this database. As the devil is in the detail, reducing age dependence to an opposition between juvenile and adult stages might be more appropriate for broad-scale analyses at the level of the Tree of Life, the clear objective of this database (see e.g. Paniw et al. 2018; Healy et al. 2019). First, including detailed st(age)-specific demographic information on species ranging from invertebrates to vertebrates will tremendously reduce the number of species analysed as our current knowledge of demographic traits across the Tree of Life is very scarce (Conde et al. 2019). Then, even if available, including detailed demographic information across a broad range of species would likely prevent us from identifying clear structures in demographic variation because variation within taxon will likely blur the big picture of across-taxon variation. Therefore, instead of opposing databases, our analysis should be interpreted as illustrating the need to carefully choose the database to analyse for answering a specific question about life history variation. This chapter illustrates the philosophy of this book: different types of questions can be tackled with different data sources and different (and appropriate!) methodologies. Thus, to assess fine-scaled demographic variation, a specific database including the key factor of variation (e.g. age in mammals) such as Malddaba is required. However, when looking for the big picture across the Tree of Life, a general database including simplified descriptions of species-specific life cycles for a large number of diverse species such as Comadre allows suitable comparisons.

References

Atchley, W. R., Gaskins, C. T., & Anderson, D. (1976). Statistical properties of ratios. I. Empirical results. *Systematic Zoology*, **25**:137–148.

Baudisch, A. (2011). The pace and shape of ageing. *Methods in Ecology and Evolution*, **2**:375–382.

Bennett, P. M. & Harvey, P. H. (1987). Active and resting metabolism in birds: allometry, phylogeny and ecology. *Journal of Zoology*, **213**:327–344.

Bielby, J., Mace, G. M., Bininda-Emonds, O. R., et al. (2007). The fast-slow continuum in mammalian life history: an empirical reevaluation. *The American Naturalist*, **169**:748–757.

Bininda-Emonds, O. R., Cardillo, M., Jones, K. E., et al. (2007). The delayed rise of present-day mammals. *Nature*, **446**:507–512.

Bininda-Emonds, O. R., Cardillo, M., Jones, K. E., et al. (2008). Erratum: the delayed rise of present-day mammals. *Nature*, **456**:274-274.

Blackburn, T. M. (1991). Evidence for a fast–slow continuum of life-history traits among parasitoid Hymenoptera. *Functional Ecology*, **5**:65–74.

Blueweiss, L., Fox, H., Kudzma, V., Nakashima, D., Peters, R., & Sams, S. (1978). Relationships between body size and some life history parameters. *Oecologia*, **37**: 257–272.

Burnham, K. P. & Anderson, D. R. (2002). *A practical information-theoretic approach. Model selection and multimodel inference*, 2nd edition. Springer, New York, 2.

Calder, W. A. (1984). *Size, function, and life history*. Harvard University Press, Cambridge.

Capdevila, P., Beger, M., Blomberg, S. P., Hereu, B., Linares, C., & Salguero-Gómez, R. (2020). Longevity, body dimension and reproductive mode drive differences in aquatic versus terrestrial life-history strategies. *Functional Ecology*, **34**:1613–1625.

Charnov, E. L. (1993). *Life history invariants: some explorations of symmetry in evolutionary ecology*. Oxford University Press, Oxford.

Colchero, F., Jones, O. R., Conde, D. A., et al. (2019). The diversity of population responses to environmental change. *Ecology Letters*, **22**:342–353.

Cole, L. C. (1954). The population consequences of life history phenomena. *Quarterly Review of Biology*, **29**: 103–137.

Conde, D. A., Staerk, J., Colchero, F., et al. (2019). Data gaps and opportunities for comparative and conservation biology. *Proceedings of the National Academy of Sciences*, **116**:9658–9664.

Demetrius, L. (1979). Relations between demographic parameters. *Demography*, **16**:329–338.

Dobson, F. S. (2007). A lifestyle view of life-history evolution. *Proceedings of the National Academy of Sciences*, **104**: 17565–17566.

Eberhardt, L. L. (1985). Assessing the dynamics of wild populations. *Journal of Wildlife Management*, **40**:997–1012.

Eisenberg, J. F. (1981). *The mammalian radiations: an analysis of trends in evolution, adaptation, and behaviour*. Chicago University Press, Chicago.

Felsenstein, J. (1985). Phylogenies and the comparative method. *The American Naturalist*, **125**:1–15.

Fenchel, T. (1974). Intrinsic rate of natural increase: the relationship with body size. *Oecologia*, **14**:317–326.

Fernandez, O. E. & Beltran-Sanchez, H. (2015). The entropy of the life table: a reappraisal. *Theoretical Population Biology*, **104**:26–45.

Festa-Bianchet, M., Gaillard, J. M., & Côté, S. D. (2003). Variable age structure and apparent density dependence in survival of adult ungulates. *Journal of Animal Ecology*, **72**:640–649.

Freckleton, R. P., Harvey, P. H., & Pagel, M. (2002). Phylogenetic analysis and comparative data: a test and review of evidence. *The American Naturalist*, **160**:712–726.

Gaillard, J. M., Yoccoz, N. G., Lebreton, J. D., et al. (2005). Generation time: a reliable metric to measure life-history variation among mammalian populations. *The American Naturalist*, **166**:119–123.

Gaillard, J. M., Pontier, D., Allainé, D., et al. (1989). An analysis of demographic tactics in birds and mammals. *Oikos*, **56**:59–76.

Gaillard J. M, Lemaître, J. F., Berger, V., et al. (2016). Life histories, axes of variation. In R. M. Kliman, ed. *Encyclopedia of evolutionary biology*. vol. 2. Academic Press, Oxford. Pp. 312–323.

Garamszegi, L. Z. (ed.). (2014). *Modern phylogenetic comparative methods and their application in evolutionary biology: concepts and practice*. Springer, Berlin.

Goldman, N. & Lord, G. (1986). A new look at entropy and the life table. *Demography*, **23**:275–282.

Hamilton, W. D. (1966). The moulding of senescence by natural selection. *Journal of Theoretical Biology*, **12**:12–45.

Harvey, P. H. & Clutton-Brock, T. H. (1985). Life history variation in primates. *Evolution*, **39**:559–581.

Harvey, P. H. & Pagel, M. D. (1991). *The comparative method in evolutionary biology*. Oxford University Press, Oxford.

Healy, K., Ezard, T. H., Jones, O. R., Salguero-Gómez, R., & Buckley, Y. M. (2019). Animal life history is shaped by the pace of life and the distribution of age-specific mortality and reproduction. *Nature Ecology & Evolution*, **3**:1217–1224.

Huxley, J. S. (1932). *Problems of relative growth*. Methuen and Co., London.

Jackson, D. A. (1993). Stopping rules in principal components analysis: a comparison of heuristical and statistical approaches. *Ecology*, **74**:2204–2214.

Jeschke, J. M. & Kokko, H. (2009). The roles of body size and phylogeny in fast and slow life histories. *Evolutionary Ecology*, **23**:867–878.

Jones, O. R., Scheuerlein, A., Salguero-Gómez, R., et al. (2014). Diversity of ageing across the tree of life. *Nature*, **505**:169–173.

Keyfitz, N. (1985). *Applied mathematical demography*, 2nd edition. Springer, New York.

Lemaître, J. F., Ronget, V., Tidière, M., et al. (2020a). Sex differences in adult lifespan and aging rates of mortality across wild mammals. *Proceedings of the National Academy of Sciences*, **117**:8546–8553.

Lemaître, J. F., Ronget, V., & Gaillard, J. M. (2020b). Female reproductive senescence across mammals: a high diversity of patterns modulated by life history and mating traits. *Mechanisms of Ageing and Development*, **192**:111377.

Mitchell, P. C. (1911). On longevity and relative viability in mammals and birds; with a note on the theory of longevity. *Proceedings of the Zoological Society of London*, **81**:425–548.

Nussey, D. H., Froy, H., Lemaitre, J. F., Gaillard, J. M., & Austad, S. N. (2013). Senescence in natural populations of animals: widespread evidence and its implications for bio-gerontology. *Ageing Research Reviews*, **12**:214–225.

Pagel, M. (1999). Inferring the historical patterns of biological evolution. *Nature*, **401**:877–884.

Paniw, M., Ozgul, A., & Salguero-Gómez, R. (2018). Interactive life-history traits predict sensitivity of plants and animals to temporal autocorrelation. *Ecology Letters*, **21**:275–286.

Pélabon, C., Tidière, M., Lemaître, J. F., & Gaillard, J. M. (2018). Modelling allometry: statistical and biological considerations—a reply to Packard. *Biological Journal of the Linnean Society*, **125**:664–671.

Péron, G., Lemaître, J. F., Ronget, V., Tidière, M., & Gaillard, J. M. (2019). Variation in actuarial senescence does not reflect life span variation across mammals. *PLOS Biology*, **17**:e3000432.

Peters, R. H. (1983). *The ecological implications of body size*. Cambridge University Press, Cambridge.

Promislow, D. E. & Harvey, P. H. (1990). Living fast and dying young: a comparative analysis of life-history variation among mammals. *Journal of Zoology*, **220**: 417–437. "

Revell, L. J. (2012). phytools: an R package for phylogenetic comparative biology (and other things). *Methods in Ecology and Evolution*, **3**:217–223.

Riek, A. & Geiser, F. (2013). Allometry of thermal variables in mammals: consequences of body size and phylogeny. *Biological Reviews*, **88**:564–572.

Ronget, V. & Gaillard, J. M. (2020). Assessing ageing patterns for comparative analyses of mortality curves: going beyond the use of maximum longevity. *Functional Ecology*, **34**:65–75.

Russell, E. M. (1982). Patterns of parental care and parental investment in marsupials. *Biological Reviews*, **57**:423–486.

Salguero-Gómez, R., Jones, O. R., Jongejans, E., et al. (2016a). Fast–slow continuum and reproductive strategies structure plant life-history variation worldwide. *Proceedings of the National Academy of Sciences*, **113**:230–235.

Salguero-Gómez, R., Jones, O. R., Archer, C. R., et al. (2016b). Comadre: a global data base of animal demography. *Journal of Animal Ecology*, **85**:371–384.

Silvertown, J., Franco, M., Pisanty, I., & Mendoza, A. N. A. (1993). Comparative plant demography—relative importance of life-cycle components to the finite rate of increase in woody and herbaceous perennials. *Journal of Ecology*, **81**:465–476.

Smith, R. J. (1999). Statistics of sexual size dimorphism. *Journal of Human Evolution*, **36**:423–458.

Southwood, T. R. (1977). Habitat, the templet for ecological strategies? *Journal of Animal Ecology*, **46**:337–365.

Staerk, J., Conde, D. A., Ronget, V., Lemaitre, J. F., Gaillard, J. M., & Colchero, F. (2019). Performance of generation time approximations for extinction risk assessments. *Journal of Applied Ecology*, **56**:1436–1446.

Stahl, W. R. (1962). Similarity and dimensional methods in biology. *Science*, **137**:205–212.

Stearns, S. C. (1976). Life-history tactics: a review of the ideas. *Quarterly Review of Biology*, **51**:3–47.

Stearns, S. C. (1977). The evolution of life history traits: a critique of the theory and a review of the data. *Annual Review of Ecology and Systematics*, **8**:145–171.

Stearns, S. C. (1983). The influence of size and phylogeny on patterns of covariation among life-history traits in the mammals. *Oikos*, **41**:173–187.

Stubben, C. & Milligan, B. G. (2007). Estimating and analyzing demographic models using the popbio package in R. *Journal of Statistical Software*, **22**:1–23.

Swihart, R. K. (1984). Body size, breeding season length, and life history tactics of lagomorphs. *Oikos*, **43**: 282–290.

Tuljapurkar, S., Steiner, U. K., & Orzack, S. H. (2009). Dynamic heterogeneity in life histories. *Ecology Letters*, **12**: 93–106.

Turbill, C., Bieber, C., & Ruf, T. (2011). Hibernation is associated with increased survival and the evolution of slow life histories among mammals. *Proceedings of the Royal Society B: Biological Sciences*, **278**:3355–3363.

Upham, N. S., Esselstyn, J. A., & Jetz, W. (2019). Inferring the mammal tree: species-level sets of phylogenies for questions in ecology, evolution, and conservation. *PLOS Biology*, **17**:e3000494

Wiersma, P., Muñoz-Garcia, A., Walker, A., & Williams, J. B. (2007). Tropical birds have a slow pace of life. *Proceedings of the National Academy of Sciences*, **104**:9340–9345.

CHAPTER 19

Adaptive management: making recurrent decisions in the face of uncertainty

James D. Nichols

19.1 Introduction

Many, if not most, decisions in conservation biology and wildlife management require a focus on system dynamics. One or more system state variables is identified as being relevant to programme objectives, and management then focuses on effects of potential management actions on these state variables. These effects of management actions generally influence system state via their effects on system vital rates, for example, the demographic parameters that are the focus of this book. Thus, estimation of vital rates and the functional relationships between management actions and vital rates is a key information need for conservation programmes. These functional relationships form the basis for projecting effects of management on system state and hence for selecting optimal management actions.

This focus of decision-making on ecological state variables and the vital rates that drive their dynamics is very general, extending across taxa. Single-species conservation problems typically focus on population size as the focal state variable, and rates of survival, reproductive recruitment, and movement as the vital rates that drive dynamics. Larger-scale single-species problems sometimes focus on the fraction of a landscape occupied by a species, with vital rates identified as local probabilities of extinction and colonisation. For community studies, species richness may be the focal state variable, with vital rates being local species-level rates of extinction and colonisation. In some land management programmes, the number of sites in each of a number of potential vegetation types may be the focal state variable, with vital

rates the transition probabilities that govern underlying dynamics. Finally, studies of disease dynamics may focus on the fraction of a population infected by the disease, with vital rates the individual transition probabilities among uninfected, infected, and post-infection states, as well as state-specific mortality rates. Some of these vital rates may not be immediately recognisable as demographic parameters, but I argue (see section 19.6) that they are all functions of survival, reproduction, and movement and hence relevant to the central topic of this book.

Decision processes for all of these situations require estimates of system state, the vital rates that drive state dynamics, and the relationships between management actions and these vital rates. Estimates of these quantities must typically deal with nondetection (counts of entities in a spatial sample are typically undercounts, as entities are missed) and sometimes misclassification (e.g. an individual classified as not infected may actually carry the infection). The estimates of system state are used to make state-dependent decisions, to assess progress towards meeting objectives, and to learn about system dynamics (see section 19.2). Estimates of vital rates and their relationships with management actions are used for developing projection models in order to make management decisions.

In this chapter, I first lay out the basic principles for structured decision-making. I then focus on the class of decisions that are recurrent and characterised by uncertainty, and the adaptive management (AM) approach developed to deal with such decisions. I then move to three examples that use AM, outlining

James D. Nichols, *Adaptive management: making recurrent decisions in the face of uncertainty*.
In: *Demographic Methods Across the Tree of Life*. Edited by Roberto Salguero-Gómez and Marlène Gamelon,
Oxford University Press. © Oxford University Press (2021). DOI: 10.1093/oso/9780198838609.003.0019

the basic approach and emphasising the role of estimating vital rates in this approach. I conclude with a discussion of integrated demographic parameters and a plea for more methodological sharing among disciplines.

19.2 Structured decision-making

Humans are faced with numerous decisions every day, some of which have important consequences for our lives. In fact, an individual's life can be viewed as a decision process, as Gandalph reminded Frodo: 'All we have to decide is what to do with the time that is given us' (Tolkien 1954). Given this central role of decisions, it is surprising that most of us spend so little time thinking about useful processes for making them. 'Management' and 'conservation' both entail actions designed to achieve specified objectives, yet it is possible to obtain graduate degrees in wildlife management and conservation biology without taking a single course in decision-making!

Structured decision-making, for the purposes of this chapter, is simply an attempt to formalise a common-sense approach to making decisions. The approach decomposes a decision into component parts, first focusing on each separately and later combining them into a complete decision process (e.g. Clemen and Reilly 2001; Martin et al. 2009). An important advantage of this approach is reduction of confusing discussions among stakeholders who disagree and tend to shift arguments from one component to another. The approach also blunts the natural tendency of stakeholders to focus directly on preferred actions, sometimes without adequately considering true objectives.

The components of structured decision-making are objectives, potential actions, models of system response to actions, monitoring, and a decision algorithm. Objectives are simply clear statements about what stakeholders would like to achieve through management of the focal system. Even when all stakeholders are in general agreement, it can be surprisingly difficult to develop specific objectives. In the more common case, when different groups of stakeholders have competing objectives, the development of compromise objectives can be very difficult. However, objectives are not optional, as the concept of management depends on them. Indeed, objectives drive the entire process and are an important consideration in development of all other decision components. Management objectives might include maintenance of population size of an endangered species above some threshold value, attainment of a threshold value of species diversity, or perhaps maximisation of species harvest, conditional on population size not dropping below a specified threshold level.

Managers must be able to identify a set of potential actions (including 'do nothing') that can move the system in a direction consistent with management objectives. At each decision point, the decision problem is to select which action is most consistent with programme objectives. Each decision thus requires models that predict system responses to the potential actions such that utility of each action can be judged relative to objectives. Some managers are reluctant to use models, viewing them as academic exercises with little grounding in reality. However, models are simply structures that permit predictions about system responses, and without such predictions there is no basis for selection of an appropriate action; hence informed management is not possible. Demographic vital rates are key components of management models, and indeed the key relationships in these models predict the effects of potential actions on vital rates.

Monitoring of system state variables (e.g. population size) and other variables relevant to objectives is important for multiple reasons. One is simply to assess the degree to which management is successful. Another principal reason is that decisions are typically state-dependent. That is, we would expect to make very different decisions depending on whether a population size is above or below some target value based on objectives. A third reason is for learning, as comparison of estimated state variables with model-based predictions provides a basis for discriminating among competing models of system response to management (see section 19.3.3). Finally, monitoring data can be used to obtain updated estimates of model parameters.

The final component of structured decision-making is a decision algorithm. The algorithm is used to select the action most consistent with objectives, based on the current state of the system (estimated via monitoring) and the projected consequences of the different potential actions (based on models). Frequently, the decision algorithm resides in the head of the decision-maker, but use of optimisation methods is more objective, transparent, and defensible.

19.3 Adaptive management (AM)

Adaptive management (e.g. Walters and Hilborn 1976, 1978; Walters 1986; Williams et al. 2007) is a special case of structured decision-making developed to deal with recurrent decision processes characterised by uncertainty. The basic idea underlying AM is that repeated management decisions can be used not only to manage the focal system, but also to reduce uncertainty about system responses to management. Each management decision and resultant action is based on model predictions of system responses, and these predictions can be compared against estimates of actual response, as obtained via monitoring, to learn about the predictive abilities of the model(s). The process of AM includes three main phases: deliberative, iterative, and double-loop learning.

19.3.1 Deliberative phase

The deliberative, or set-up, phase focuses on developing the components of AM. The first two components, objectives and potential actions, typically require input from all relevant stakeholder groups. Many decisions in conservation and natural resource management are about systems that are valued in different ways by different stakeholder groups. For example, decisions commonly involve systems that some groups want to protect from any potential deleterious action, whereas others want to exploit at least some system components. The deliberative phase of AM frequently includes stakeholder meetings such that all relevant groups can have input into development of programme objectives and actions. Such meetings can become contentious when different stakeholders have competing interests, and efforts must be made to develop compromise objectives to which all stakeholders agree. The objectives should ultimately be incorporated into an objective function, an explicit (usually mathematical) statement of objectives. In the case of multiple objectives, an objective function may include different 'weights' for the different components of the function. Another approach to dealing with multiple objectives is to focus on one objective component, subject to a constraint or threshold associated with another component (e.g. Martin et al. 2009). For example, an objective function might entail maximising harvest of a system component, subject to the constraint of that component exceeding some threshold value (e.g. Johnson et al. 1997).

Development of a set of potential actions usually requires stakeholder input and may be difficult as well. For example, some actions (e.g. predator control) that are acceptable to some stakeholders may be unacceptable to others for ethical or other reasons associated with individual values. There are frequently

disagreements about the likely effectiveness of potential actions, with a tendency in wildlife management and conservation biology to overestimate the effects of many potential actions (Nichols 2012). Adaptive management provides the opportunity to eventually resolve such disagreements via learning. Decisions about the set of potential management actions are sometimes as contentious as those involving objectives. In summary, stakeholder meetings can be difficult to direct and navigate but are an integral part of the deliberative phase of AM for most decision problems. The purpose of such meetings is to emerge with clear statements of decision-process objectives and a set of available actions.

Models of system response to management actions, and monitoring designed to estimate system state variables and other variables relevant to objectives, must also be developed. These deliberative phase tasks usually involve a smaller group than the stakeholders who develop objectives and actions. Development of models and monitoring is typically handled by professional managers and associated scientists. A common approach is to establish working groups that include some scientists and managers who are experts for the focal systems and others who are experts in the quantitative aspects of modelling and monitoring. Once developed, models and monitoring plans may be shared with the larger stakeholder groups, and even modified in response to stakeholder concerns, as a means of insuring transparency and stakeholder engagement in the process.

Use of AM is predicated on a desire to reduce uncertainty about system response to management actions. This uncertainty is typically expressed in system models in one of two ways. The approach with which I have most experience actually identifies multiple models that incorporate the range of plausible system behaviours, and hopefully the range of stakeholder beliefs. For example, in the case of resource exploitation, stakeholders with protective conservation perspectives frequently believe in larger effects of exploitation on the focal system, whereas those favouring exploitation often believe in minimal effects. One advantage of AM is the ability to incorporate these competing views of system responses. If competing models are weighted equally in the deliberative phase, as is often wise in the case with competing stakeholder views, then no single model has undue influence in the initial management decision. As learning proceeds in the iterative phase, models with greater predictive abilities exert greater influence in the decision process. If the model set includes at least one model that is a good approximation to reality, then management decisions should improve.

The other way to incorporate uncertainty about system response entails use of a single general model that includes one (or perhaps more) key parameter(s), different values of which can produce very different system responses to actions. In the deliberative phase, it may be decided to begin the process with a value that is intermediate with respect to the different effect hypotheses under consideration. The iterative phase then leads to increasingly better estimates for use in management decisions.

Finally, a decision algorithm must be selected. Recurrent decision processes for state-dependent decisions typically cannot simply consider returns based on the upcoming time-step and hence require *dynamic* optimisation algorithms. For example, consider a simple objective of maximising harvest of a focal species over a 10-year time horizon. Optimisation of harvest for a one-step problem would be to harvest as many individuals as possible. However, in a recurrent problem, such an action may leave few or no individuals to harvest in the next time step. The decision at time t must thus consider not only returns between t and $t+1$, but also all time steps for the remainder of the management programme time horizon. Such decisions require dynamic optimisation algorithms such as stochastic dynamic programming (Bellman 1957; Puterman 1994; Williams et al. 2002).

In summary, the purpose of the deliberative phase is to develop all of the components required for the iterative phase in which AM is actually implemented. Hopefully, this development is conducted in a manner that satisfies all relevant stakeholders and keeps them engaged and supportive of the AM process.

19.3.2 Iterative phase

Armed with the described components of structured decision processes, as developed during the deliberative phase of AM, the iterative phase proceeds as outlined in Figure 19.1. At each decision point, denoted as time t, the decision algorithm is used to select the optimal action based on the other process components, that is, objective function, set of potential actions, system model(s), and estimated state of the system (based on monitoring). That action is taken (1) driving the system to a new state and (2) providing returns (desired changes in variables specified in objectives). The new state of the system is then identified via monitoring and is compared against predictions based on each model, yielding changes in model weights (see below). At the next decision point, $t+1$, a new decision is made using the new model weights and new estimate of system state. The process proceeds in this manner through time.

In this case of multiple discrete models, learning occurs via the evolution of model weights. Specifically, we define the 'information state' as a vector of model weights, $\pi_t (model\ i)$ for model i at time t, that reflect the relative predictive abilities of models in the model set. In the case of M models in the set:

$$\sum_{i=1}^{M} \pi_t\ (model\ i) = 1. \qquad (19.1)$$

Management of Dynamic Resources

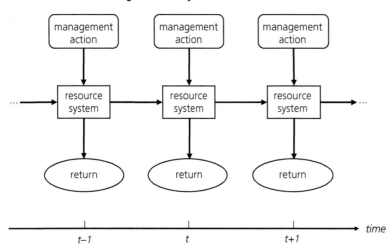

Figure 19.1 Diagrammatic representation of a recurrent decision problem. At each decision point an action is taken based on the objectives, potential actions, models, and current system state. The action induces a return and drives the system to a new state that becomes the basis for the next decision. (Source: Nichols and Williams 2013.)

We are more confident in models with higher weights, viewing them as more likely to represent reasonable abstractions of the modelled natural processes. The weights of the models at a decision point determine their relative influence on the optimal decision. Initial weights (at the beginning of the management programme) can be based on historic information (e.g. one might use AIC weights from a single analysis; Burnham and Anderson 2002) or intuition, or simply set equal $(1/M)$ for each model.

Subsequent model weights evolve with new observations according to:

$$(\pi_{t+1} \ (model \ i) \mid data_{t+1})$$

$$= \frac{\pi_t \ (model \ i) \ \Pr \ (data_{t+1} \mid model \ i)}{\sum_i^M \pi_t \ (model \ i) \ \Pr \ (data_{t+1} \mid model \ i)}, \quad (19.2)$$

where $\Pr(data_{t+1} \mid$ model $i)$ is the probability that the new observations at time $t+1$ would have arisen, given that model i is a good representation of the actual process that generated them. Equation (19.2) is based on Bayes' theorem or rule (Bayes and Price 1763). The updating of model weights is based on the relative confidence in the model that has accumulated through time t, π_t (model i), and the consistency of the new observations with that model, $\Pr(data_{t+1} \mid$ model $i)$. If the model set includes a good approximating model, then the weight for that model should evolve to approach 1, whereas the weights of models that predict more poorly should eventually approach 0 (see Figure 19.2).

The iterative phase thus entails making decisions that are optimal with respect to specified objectives,

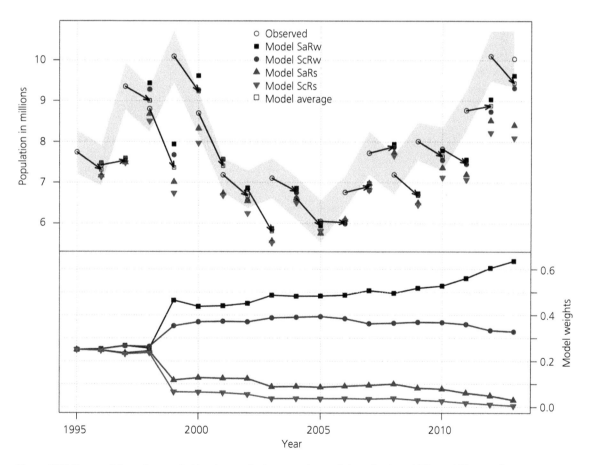

Figure 19.2 Top: population estimates of midcontinent mallards compared to predictions of each model (Sa = additive mortality; Sc = compensatory mortality; Rw = weakly density-dependent reproduction; Rs = strongly density-dependent reproduction) from 1996 to 2013. Grey shading represents 95% confidence intervals for population estimates. Arrows begin at the population estimate of year t and point to the weighted mean annual prediction for year $t+1$. Bottom: changes in model weights from 1995 (equal weights) to 2013. (Source: Nichols et al. 2015.)

while simultaneously reducing uncertainty and learning which model(s) represents the best approximation to the processes governing system responses to management actions. This combination of wise decision-making and simultaneous learning distinguishes adaptive management from other forms of management.

19.3.3 Double-loop learning phase

Learning that occurs during the iterative phase of AM may lead managers to revisit decisions made during the deliberative phase. For example, objectives may evolve, resulting in a need for revision (Johnson et al. 2015). Similarly, if experience with the set of identified actions indicates that none of them is particularly effective, development of new actions will be useful. In addition, model weights may fluctuate nonmonotonically, and/or none of the models in the set may be predicting well. In such cases, it may be useful to revisit the model set, perhaps adding models and even deleting some. Finally, the monitoring may not be providing estimates of sufficient precision to permit wise decisions and rapid learning, leading to a need to revise the current monitoring programme.

Thus, a perception of programme organisers that any of the process components should be revised may lead to a double-loop phase in which one or more elements of the deliberative phase is revisited. These double-loop learning phase steps typically occur on a larger time-scale than the decision points of the iterative phase. So if the iterative phase entails a 1-year time-step, for example, the double-loop steps might occur at irregular 5- or 10-year intervals, as needed.

19.4 Demographic methods

The demographic methods that form the subject matter of this book are integral to the application of AM. Management typically focuses on key state variables of managed systems, as well as other variables associated with management objectives. Dynamics of system state variables are determined by system vital rates, and management actions are usually designed to influence dynamics by affecting these vital rates. AM typically entails estimation of system state variables, the vital rates that drive them, and the relationships between vital rates and both management actions and influential environmental variables. These estimates are then used to populate models of system dynamics in order to project consequences of different management actions. Thus, models for estimation of vital rates and state variables (e.g. Seber 1982; Williams et al. 2002) play an important role in AM. Similarly, projection models for predicting the consequences of management actions for system dynamics and other variables identified in objectives are essential for AM as well, forming the basis for optimal decisions. In sections 19.4.1–19.4.4, I briefly discuss inference for four common state variables and their vital rates.

19.4.1 Population size/abundance

Frequently, the state variable of single-species management is population size, and the associated vital rates are survival, reproduction, and movement in and out of the focal population. Ecologists have developed a variety of methods for estimating population size in the face of unknown detection probabilities. These methods include closed and open capture–recapture models, removal methods, catch-effort methods, change-in-ratio methods, distance sampling, double-observer models, time-at-detection models, and N-mixture models, as well as approaches that combine two or more of these methods (e.g. see Chapters 5, 7, and 14 and reviews in Seber 1982; Williams et al. 2002; Lancia et al. 2005; Nichols et al. 2009).

Methods for inference about survival rates include open capture–recapture models and band recovery models, both of which deal with unknown detection probabilities, and methods based on known deaths for some radio-telemetry studies (e.g. see Chapter 13 and Seber 1982; Lebreton et al. 1992, 2009; Williams et al. 2002). Inference about reproductive rates may focus on an overall reproductive rate (e.g. young per breeding female) or the various components of reproductive rate (e.g. the proportion of reproductive-age animals that breed, nest success, brood survival) (e.g. Seber, 1982; Williams et al. 2002; Skalski et al. 2010). Rates of movement in and out of populations are often estimated using capture–recapture or band recovery methods (e.g. Williams et al. 2002; Lebreton et al. 2009). Virtually all of the methods developed for estimation of vital rates can also be used to draw inferences about the effects of management actions and environmental and other variables on these rates.

Population projection models used to predict consequences of management actions on population size are also well developed. The most commonly used approaches are based on discrete age or stage classes (Chapters 9 and 11 and reviews by Caswell 2001; Tuljapurkar and Caswell 2012), but integral models using continuous variation among individuals in traits influencing vital rates are also available (Chapter 10 and Rees and Ellner 2009). For both of these classes of model, dynamics usually follow first-order Markov processes, providing access to the variety of decision algorithms developed for such problems (e.g. Bellman 1957; Puterman 1994).

19.4.2 Occupancy

Another state variable used in single-species management is occupancy, or the proportion of area occupied by a species. Large areas are often subdivided into patches the size of individual animal territories or home ranges, but sample unit size can be more arbitrary. The state variable is then defined as the proportion of those designated units that is occupied by the focal species, recognising that species may occupy a site yet go undetected. This state variable is thus not equivalent to abundance, although the two are related.

Inference about occupancy in the face of nondetection is possible using a basic approach that has been extended to a variety of useful applications (MacKenzie et al. 2018). Dynamics are driven by the vital rates of unit-level extinction and colonisation. Local extinction and colonisation, and the relationship between these rates and management actions and environmental variables, can be estimated using multiseason occupancy models developed for these purposes. These same multiseason models provide the framework for projecting occupancy dynamics through time (e.g. Yackulic et al. 2014). These dynamic occupancy models are also first-order Markov processes and thus amenable to use with many decision algorithms.

19.4.3 Species richness

Some community-level studies focus on species richness for specific taxonomic groups. Richness is defined as simply the number of species inhabiting an area. Because species are not always detected at a location when present, inference methods must deal with unknown detection probabilities. Richness estimation is often based on capture–recapture thinking, with a variety of specific estimators now available (Williams et al. 2002; Chao et al. 2014). Occupancy modelling can also be used in either of two forms, one dealing with species data from multiple sites and the other dealing with data from only one site (Zipkin et al. 2009; MacKenzie et al. 2018).

Vital rates associated with dynamics of species richness are local rates of species-level extinction and colonisation. When richness is estimated using capture–recapture thinking, these rate parameters can be estimated using a community-level version of Pollock's (1982) robust design (Nichols et al. 1998; Hines et al. 1999). Rate of local colonisation can be defined in multiple ways, depending on whether the members of the species pool of a sample location are known *a priori*. If so, then a reasonable approach is to define colonisation as the probability that a member of the species pool not present at the focal location at time t is present at time $t+1$ (see Nichols et al. 1998; Boulinier et al. 2001). When occupancy approaches are used, the most general models permit estimation of species-specific rates of local extinction and colonisation using multiseason occupancy models (MacKenzie et al. 2018).

A natural way to model dynamics of species richness at a focal site, when the species pool associated with the site is known, is to consider the number of pool species in each of two states, present and not present at the focal site at time t. Rates of local colonisation and extinction are then used to project the system state through time. These vital rates can be modelled as functions of management actions, habitat characteristics, and weather variables in order to produce time-specific projections. The first-order Markovian nature of the system dynamics model (Boulinier et al. 2001) is again preadapted for use with decision algorithms. This same basic approach can also be applied to systems in which the species pool is not known but must be estimated, for example using the data augmentation approach of Royle et al. (2007; see Kery and Royle 2008).

19.4.4 Landcover vegetative states

Demographic performance and population growth of many species are dependent on habitat (e.g. Breininger et al. 2009). When habitat is dynamic, projection of focal species dynamics often requires joint models that include the managed species and associated habitat (e.g. Johnson et al. 2011; Miller et al. 2012). In such situations, habitat can be categorised into discrete states, such that the state variable for a managed area is the fraction of patches or habitat units in each of the different habitat classes. In some situations where ground assessments are made, the state of each habitat unit can be made with no error or need to deal with nondetection or misclassification. When state assignment is based on aerial or satellite imagery, it may be necessary to deal with misclassification, which can be accomplished using methods similar to those developed in multistate capture–recapture modelling (Pradel 2005; Veran et al. 2012).

Habitat state dynamics can be modelled using matrices of state transition probabilities (e.g. Breininger et al. 2010; Johnson et al. 2011). Transition probabilities can be estimated directly using multinomial modelling in the absence of state uncertainty. When state assignment is based on imagery characterised by uncertainty, capture–recapture thinking can be used to estimate transition probabilities using data from multiple seasons (Veran et al. 2012).

The state variable is a vector consisting of the number of patches in each habitat state, for all k habitat classes. The associated transition matrices are of dimension $k \times k$ and are used to project the state variable through time. Transition probabilities may be modelled as functions of management actions and environmental conditions, or different matrices may be constructed for each potential management action. This kind of modelling is Markovian and amenable to use with decision algorithms.

19.5 Case studies

A large part of the motivation for this book is to demonstrate the importance of demographic methods across a variety of taxa and types of questions. The AM approach described in this chapter is similarly general and applicable to a variety of taxa and kinds of management problems. As stated, AM is especially applicable for recurrent management problems characterised by uncertainty, especially uncertainty associated with system responses to management. AM is thus applicable to a very large class of ecological and resource problems. This section includes overviews of three different problems conducive to use of AM, involving vertebrates, plants, and disease organisms. The vertebrate example is by far the most detailed, describing the longest-running formal AM programme in the world. An overview is provided for the vegetation management example, which corresponds to a relatively new AM programme. The disease example is hypothetical and does not correspond to any real-world example that I know of. I simply provide a sketch of what an AM programme for disease management should look like.

19.5.1 Mallard (*Anas platyrhynchos*) duck harvest in North America

The US Fish and Wildlife Service (FWS) is charged with setting annual hunting regulations for migratory birds

harvested in the United States. In the 1960s and early 1970s, divergent views among stakeholders about the effects of hunting on duck populations led to annual arguments about appropriate hunting regulations (Nichols 2000). In an effort to reduce this uncertainty and associated contentious debate, FWS scientist F. A. Johnson led an effort to develop an AM programme for the setting of annual hunting regulations for mid-continent mallard ducks. This programme was formally adopted by FWS in 1995 and has been in use since that time (see Johnson et al. 1993, 1997, 2015; Nichols et al. 1995a, 2007, 2015; Williams et al. 1996; US Fish and Wildlife Service 2018).

The established objective of harvest management for mid-continent mallards is to maximise cumulative harvest over the long term, requiring persistence of a viable population into the future. This objective is constrained to avoid regulations expected to result in a population size below a specified numerical goal of 8,500,000 birds. Based on this constraint, the value of harvest decreases proportionally as the difference between the goal and expected population size increases, for expected population sizes less than the goal. Management actions include four discrete options defined by length of the hunting season and daily bag limits: closed season, restrictive, moderate, and liberal regulations; with season length and daily limits increasing from restrictive to liberal packages (US Fish and Wildlife Service 2018).

The relevance of demographic analyses is clear in the modelling and monitoring components of this AM programme. The basic model projecting the mallard population through time is:

$$N_{t+1} = N_t \left[mS_{t,AM} + \left(1 - m\right) \left(S_{t,AF} + R_t \right. \right.$$
$$\left. \left. \left(S_{t,JF} + \frac{S_{t,JM}\phi_F^{sum}}{\phi_M^{sum}}\right)\right)\right] \tag{19.3}$$

where
N_t = breeding population size in spring of year t,
m = proportion of males in the breeding population,
$S_{t,AM}, S_{t,AF}, S_{t,JF}, S_{t,JM}$ = annual survival rates (year t) for adult males, adult females, juvenile females, and juvenile males, respectively,
R_t = reproductive rate for year t, defined as autumn age ratio (juveniles per adult) of females,
$\phi_F^{sum}/\phi_M^{sum}$ = ratio of female to male summer survival rate.

If we choose to focus on adult females and males, separately, we can write eqn. (19.1) as a matrix projection model:

$$\begin{bmatrix} N_{t+1,AM} \\ N_{t+1,AF} \end{bmatrix} = \begin{bmatrix} S_{AM} & RS_{JM}\phi_F^{sum}/\phi_M^{sum} \\ 0 & S_{AF} + RS_{JF} \end{bmatrix} \begin{bmatrix} N_{t,AM} \\ N_{t,AF} \end{bmatrix}$$

(19.4)

The stakeholder controversy over the effects of hunting focused on the demographic rates of survival and reproduction. The basic model structure of eqns. (19.3) and (19.4) was thus used to create four different models defined by two different survival functions and two different reproductive rate functions (Johnson et al. 1997; US Fish and Wildlife Service 2018). The two survival functions are termed the compensatory and additive mortality hypotheses (Anderson and Burnham 1976; Cooch et al. 2014). Survival rate under the additive mortality hypothesis is written as:

$$S_{t,sex,age} = S_{0,sex,age}^A \left(1 - K_{t,sex,age}\right),$$ (19.5)

whereas under the compensatory mortality hypothesis, survival is:

$$S_{t,sex,age} = \left\{ \begin{array}{l} S_{0,sex,age}^C \ if \ K_{t,sex,age} \leq 1 - S_{0,sex,age}^C \\ 1 - K_{t,sex,age} \ if \ K_{t,sex,age} > 1 - S_{0,sex,age}^C \end{array} \right\},$$

(19.6)

where
$S_{0,sex,age}^H$ = annual sex- and age-specific survival rate in the absence of hunting under hypothesis H (additive, A, or compensatory, C),
$K_{t,sex,age}$ = sex- and age-specific hunting mortality rate for year t.

Hunting mortality always decreases overall survival under the additive mortality hypothesis. However, under the compensatory mortality hypothesis, changes in hunting mortality rates over a specified range cause no such changes in overall survival.

Estimation of annual survival rates, survival rates in the absence of hunting, and hunting mortality rates is based on large-scale banding of ducks on the breeding grounds just prior to the hunting season and reported recoveries of bands during the subsequent hunting seasons (e.g. see Anderson 1975; Brownie et al. 1985). These models allow for the nonreporting of ducks that are shot and retrieved by hunters. The specified relationships between annual survival and both hunting mortality rate and survival in the absence of hunting are estimated directly using either likelihood or Markov Chain Monte Carlo (MCMC) approaches.

Reproductive rate is then modelled as either strongly or weakly density dependent, representing two extremes in the ability of a population to respond to hunting via increased reproduction. The general model defining reproductive rate is:

$$R_t^H = \beta_0^H + \beta_1^H P_t - \beta_2^H N_t,$$ (19.7)

where
R_t^H = reproductive rate under the hypothesis of weak ($H = W$) or strong ($H = S$) density dependence,
P_t = number of ponds (in millions) in prairie Canada breeding   grounds in May of year t,
β_0^H = intercept parameter under hypothesis H,
β_1^H = slope parameter for the relationship between reproductive rate and ponds (in millions) under hypothesis H,
β_2^H = slope parameter for the relationship between reproductive rate and mallard breeding population size (in millions) under hypothesis H.

Reproductive rate for the mallard population is estimated via combined use of banding and recovery data and data from harvest surveys based on hunter questionnaire and parts collection surveys (Martin and Carney 1977). Specifically, reproductive rate (autumn age ratio of females) is estimated as:

$$\hat{R}_t = \frac{\hat{H}_{t,J}/\hat{H}_{t,A}}{\hat{h}_{t,J}/\hat{h}_{t,A}},$$ (19.8)

where
$\hat{H}_{t,age}$ = estimated total harvest for year t and age = J (juvenile) or A (adult), based on the FWS harvest survey (Martin and Carney 1977),
$\hat{h}_{t,age}$ = estimated harvest rate for year t and age = J (juvenile) or A (adult), based on banding and recovery data.

Equation (19.8) can be viewed as a ratio of Lincoln–Petersen estimates (e.g. see Lincoln 1930; Seber 1982) for age-specific autumn abundance. For example, the number of adults present in the autumn of year t can be estimated as the ratio of number of adults in the harvest to the probability that an autumn bird became part of that harvest, that is, $\hat{H}_{t,A}/\hat{h}_{t,A}$.

Because pond numbers are such important determinants of mallard reproductive rate, they are treated as a state variable as well. Specifically, ponds are modelled as:

$$P_{t+1} = 2.2127 + 03420P_t + \varepsilon_t,$$ (19.9)

where ε_t is distributed as normal (0, 1.2567). Using the estimates of reproductive rate and pond numbers, the coefficients of the relationship of eqn. (19.7) were estimated via maximum likelihood. The parameters of the two density-dependent hypotheses were then based on these estimates and their associated variances.

The two different hypotheses about survival rates, and the two about reproductive rates, were combined in pairs to yield four different models intended to capture the uncertainty about mallard population responses to hunting. Let Sa denote the additive mortality hypothesis (eqn. 19.5), Sc denote the compensatory mortality hypothesis (eqn. 19.6), Rs denote strongly density-dependent reproductive rate (eqn. 19.7), and Rw indicate weakly density-dependent reproductive rate. The resulting four population-dynamic models are thus SaRs (additive mortality, strongly density-dependent reproduction), SaRw (additive mortality, weakly density-dependent reproduction), ScRs (compensatory mortality, strongly density-dependent reproduction), and ScRw (compensatory mortality, weakly density-dependent reproduction). This AM programme of harvest management was formally adopted for mid-continent mallards in 1995 and has been ongoing ever since. In 1995, all four models were given equal weights of 0.25.

Multiple FWS monitoring programmes support this harvest management programme for mid-continent mallards. Breeding population size and pond numbers are the two system state variables and are estimated via an extensive aerial survey programme (Smith 1995). A large-scale banding programme is used to estimate survival, harvest, and hunting mortality rates, based on band recovery models (Brownie et al. 1985) and band reporting rates (Nichols et al. 1995b). Finally, total age- and sex-specific harvest is estimated using the harvest survey (Martin and Carney 1977). These various monitoring programmes are summarised by Martin et al. (1979) and Blohm (1989).

The iterative phase of mallard AM entails selecting a regulations package each year in the summer, just after aerial survey monitoring results are available and shortly before the hunting season begins. This selection uses stochastic dynamic programming (e.g. Williams 1996; Lubow 1997) and is based on programme objectives, the four regulations packages, the models and their respective weights, and the estimated state of the system (from monitoring). The optimal regulations are implemented during the subsequent autumn–winter hunting season, driving the system to a new state (estimated the following spring via monitoring) and yielding returns (harvest, also estimated via monitoring). The predictions of each of the four models are compared against the new estimate of population size,

and model weights are modified, increasing for models that predict well and decreasing for those predicting poorly. The next season's regulations are then selected with the new model weights and estimated system state, and the process proceeds.

The AM process developed by FWS for mid-continent mallards has been a major success, producing optimal decisions that are defensible, objective, and transparent. Learning has progressed, with model SaRw achieving twice the weight of ScRw and weights of the other two models approaching 0 (Figure 19.2). These changes in model weights have produced corresponding changes in the optimal state-specific regulations (Johnson 2011).

The central roles of demographic methods in this process are clear. Estimates of survival rates and reproductive rates are key components of the underlying population projection models on which optimal decisions are based. Relationships between these rates and both management actions and environmental variables are key sources of uncertainty that are reduced during the management process. Estimates of the key state variable, population size, are essential for informing state-dependent decisions and for learning about population responses to management.

19.5.2 Management of vegetation states for Florida scrub jays (*Aphelocoma coerulescens*)

The Florida scrub jay is a threatened species under the US Endangered Species Act and generally viewed as a flagship species for scrub and flatwoods habitat found along central Florida's Atlantic coast. Florida scrub jays prefer a transitional stage of vegetative succession in these systems. However, human influences have fragmented preferred habitat and reduced fire frequency, resulting in more habitat in older stages of succession that are not as productive for jays. Traditional habitat management for Florida scrub jays has used prescribed burning to set back succession and provide more habitat suitable for jays.

Florida scrub jay is often categorised into four different habitat states: short (*Sh*: <1.2 m tall; open sand between patches of oak scrub; burned within the last few years), optimal (*Op*: 1.2–1.7 m tall; open sand between patches of medium-height oak scrub), tall mix (*Tm*: short- or medium-height oaks with large patches of tall oak), and tall (*Ta*: no short–medium oaks, all tall with little open sand) (Breininger et al. 2009). Florida scrub jay survival and reproductive rates vary across these habitat states such that birds with territories in the optimal habitat state have higher survival and reproductive rates than birds in the other three habitat types (Breininger et al. 1996, 2009; Breininger and Carter 2003; Breininger and Oddy 2004). Johnson et al. (2011) used these vital rates to compute habitat-specific population growth rates, λ^H, with only $\lambda^{Op} > 1$.

Two AM programmes have been proposed for Florida scrub jay populations in central Florida. The first programme was developed by F. A. Johnson and D. R. Breininger for Merritt Island National Wildlife Refuge. Johnson et al. (2011) carried out the deliberative phase of AM, focusing on maximising the amount of habitat in each management unit that is in the optimal state. D. R. Breininger, M. J. Eaton, and F. A. Johnson have brought together a group of state and private landowners and managers to initiate another AM programme for the species in Brevard County, FL (Eaton et al. 2021). The programmes have somewhat different approaches, but habitat management is central to each programme. Here, I summarise the approach of Johnson et al. (2011), which focuses on maximising the amount of optimal habitat in a management unit over a long time horizon. This approach is consistent with the purpose of this section to focus on vegetation dynamics.

The objective of maximising the number of 10-ha patches (the approximate territory size of Florida scrub jays) of optimal habitat in a management unit requires a set of actions designed to alter vegetation structure. One possible set of three potential actions is: to do nothing (action 1), to implement prescribed burning of a management unit (action 2), and to use mechanical cutting of vegetation, followed by burning (action 3). The decision at each time-step (e.g. each year) for each management unit is then selection of which action to implement.

The basic model for habitat transitions among the four habitat states can be written as:

$$\begin{bmatrix} n_{t+1}^{Sh} \\ n_{t+1}^{Op} \\ n_{t+1}^{Tm} \\ n_{t+1}^{Ta} \end{bmatrix} = \begin{bmatrix} \psi_t^{ShSh} & \psi_t^{OpSh} & \psi_t^{TmSh} & \psi_t^{TaSh} \\ \psi_t^{ShOp} & \psi_t^{OpOp} & \psi_t^{TmOp} & \psi_t^{TaOp} \\ \psi_t^{ShTm} & \psi_t^{OpTm} & \psi_t^{TmTm} & \psi_t^{TaTm} \\ \psi_t^{ShTa} & \psi_t^{OpTa} & \psi_t^{TmTa} & \psi_t^{TaTa} \end{bmatrix} \begin{bmatrix} n_t^{Sh} \\ n_t^{Op} \\ n_t^{Tm} \\ n_t^{Ta} \end{bmatrix},$$

(19.10)

where

n_t^H = number of patches in habitat state H at time t,
$\psi_t^{HH'}$ = probability that a patch in state H at time t is in state H' at time $t+1$.

The key to using eqn. (19.10) in AM is to recognise that each of the management actions leads to a different set of transition probabilities, that is, a different transition matrix. As an example, consider the transition

probability ψ_t^{TmSh}, for a patch going from state tall mix to short. Under the 'do nothing' management action (action 1), this probability is very small, approaching 0. Under the burning action (action 2), it is larger, and under mechanical cutting (action 3), this probability is larger still (i.e. $\psi_t^{TmSh}(1) < \psi_t^{TmSh}(2) < \psi_t^{TmSh}(3)$, where management action is denoted in parentheses).

Estimation of transition probabilities is based on observed patch-level state transitions under the different management actions. Estimation is accomplished via any software that provides maximum likelihood estimates for multinomial model parameters. Breininger et al. (2009, 2010) and Johnson et al. (2011) used multistate capture–recapture models (e.g. Lebreton et al. 2009) for which detection probabilities were set equal to 1 (because all habitat patches were observed and classified correctly to habitat type each year).

Uncertainty about system response to management actions could be acknowledged by developing multiple discrete models reflecting different transition probabilities, $\psi_t^{HH'}$, for at least some habitat pairs, HH', and actions. The approach to learning (reducing uncertainty) taken in the AM programme developed by Eaton et al. (2021) for Brevard County is to view this as an estimation problem, focusing on transition probability estimates, with more observations of state transitions leading to better estimates.

Monitoring in support of this AM problem entails visits to each patch every year in order to assess habitat type. These assessments are conducted on the ground (as opposed to use of remote imagery), hence assumed to be made without error, although approaches to dealing with misclassification error are available (Veran et al. 2012). Monitoring of Florida scrub jay patch occupancy also occurs using modified occupancy models (MacKenzie et al. 2018), but as the focus here is on vegetation dynamics, I will not describe these inference methods.

Equation (19.10) represents a first-order Markov process, so optimal decisions can be obtained using such approaches as stochastic dynamic programming (Bellman 1957; Puterman 1994). Johnson et al. (2011) provide examples of optimal decisions for different action-specific matrices of transition probabilities. The AM programme for Brevard County is in the early stages of implementation and appears to be working well. One important message emerging from the AM programmes for Florida scrub jays is the degree to which we are constrained by available management actions. The nature of succession in scrub oak–flatlands habitat in central Florida, and the effects of the identified management actions on the successional process, are such that maintaining substantial amounts of transitional (in our case, optimal) habitat is extremely difficult, leading to the need to at least consider other possible management actions for influencing Florida scrub jay populations.

19.5.3 Disease control via AM

Although AM has been recommended as a potentially useful approach to epidemiological programmes (e.g. Bogich et al. 2013; Shea et al. 2014; Nichols 2019), I am not aware of formal adoption of AM for disease control. The argument in favour of AM is a simple one. Agencies responsible for disease monitoring and control efforts tend to devote substantial efforts to disease modelling and monitoring, often using rigorous mathematical approaches. Yet the final decisions about control actions such as vaccination strategies are often made by groups of experts studying estimates and model results, integrating them in their heads, and then making a decision. The methodological asymmetry between modelling and estimation on the one hand and decision-making on the other is striking. The recommendation is thus to confront uncertainty in a rigorous manner, using the same rigour to make decisions as is currently used in modelling and estimation. Here I simply provide sketches of potential AM programmes at two different geographic scales in order to emphasise their applicability and discuss the kinds of demographic parameters useful to such endeavours.

First consider an effort within a large city to reduce the number of people infected by a disease such as hepatitis C that is spread through intravenous drug use with dirty or shared needles. A formal statement of objectives might specify a reduction below some threshold number of infections. Potential management actions might include such divergent efforts as public programmes to supply drug users with clean needles or increased law enforcement targeting intravenous drug users and suppliers.

Because objectives focus on numbers of persons infected with the disease, compartmental models (e.g. SIR models; Kermack and McKendrick 1927; Bailey 1975; Anderson and May 1991; Cooch et al. 2012) are a natural choice. Such models categorise individuals of a focal population into disease states (e.g. susceptible [S], infected [I], and recovered or removed [R], depending on the disease). Numbers of individuals in each state are thus the state variables of this modelling. These models can be viewed as stage-structured population models in which stages or states are defined with respect to presence of disease organisms.

Two classes of demographic parameters drive the underlying dynamics of such multistate models, state-specific transition probabilities, and state-specific mortality rates. Estimation of both the state variables and the underlying rate parameters can be accomplished using multiple list data at different points in time. For example, one list (data source) might be individuals diagnosed while seeking treatment in hospital emergency rooms. Another might be diagnoses of individuals tested when entering drug rehabilitation programmes. Such lists are characterised by detection probabilities that are <1 and that vary among lists, leading to the suggestion to use multistate capture–recapture models (Arnason 1973; Brownie et al. 1993; Schwarz et al. 1993; Lebreton et al. 2009) for inference about numbers of individuals in each state and associated transition and mortality rates (see Cooch et al. 2012; Nichols et al. 2017; Nichols 2019).

Substantial uncertainty is likely to surround the effectiveness of potential management actions (e.g. clean needles, law enforcement), leading to multiple models that include large and small effects for each action. Specifically, actions would be expected to affect the probability that an individual in state S at time t will be in state I at time $t+1$, denoted as ψ_t^{SI}. This transition probability or infection rate is a key driver of SIR models (Anderson and May 1991; Cooch et al. 2012), an important source of uncertainty, and a natural target of management actions. Compartmental models are usually written as first-order Markov processes and thus amenable to use of associated decision algorithms.

Now consider the epidemiological problem of managing disease problems such as measles at a country level. The management objective might be to minimise the number of locations (e.g. counties) with infected individuals, and actions might include various components of a nationwide vaccination strategy (specific disease strain(s) targeted, types of efforts to increase number of vaccinations [mandatory for school attendance, kinds and intensities of advertising], possible geographic variation in these components to target predicted hotspots, etc.).

An AM programme at this country-level scale might use as a state variable the number or proportion of counties at which the infection is present, suggesting the use of occupancy modelling (MacKenzie et al. 2018). The occupancy parameter estimates the proportion of sites with at least one infected individual, and the vital rates are probabilities of location-specific colonisation (disease moving into a county) or extinction (loss of disease from a county that contained infected individuals during the prior time-step). The fundamental equation governing occupancy dynamics is:

$$\theta_{t+1} = \theta_t (1 - \varepsilon_t) + (1 - \theta_t) \gamma_t, \qquad (19.11)$$

where

θ_t = proportion of sites occupied (disease present) at time t,

ε_t = probability that a site occupied at time t is not occupied at time $t+1$,

γ_t = probability that a site not occupied at time t is occupied at time $t+1$.

County-specific vaccination actions and effectiveness might be expected to influence colonisation probability, and perhaps local extinction as well. Both colonisation and extinction may also be affected by occupancy (disease presence) in surrounding counties, suggesting the use of autologistic occupancy models (Yackulic et al. 2012; Eaton et al. 2014; MacKenzie et al. 2018). These effects of management actions and neighbourhood disease status on colonisation (disease spread) and extinction are key sources of uncertainty that would be foci for an AM programme. Equation (19.11) is a first-order Markov process and thus amenable to the various decision algorithms developed for such processes.

19.6 Conclusions

Development of demographic models and estimation of model parameters are not inherently useful activities. Instead, they attain their value as parts of larger endeavours, typically either science or management. The focus of this chapter has been on a form of management developed to deal especially with the common problem of uncertainty, particularly uncertainty associated with system responses to management actions.

One theme of this chapter is that a natural approach to modelling the effects of management is to focus on vital rates. It would be possible to develop models in which implementation of a management action at time t would yield a direct change in the state variable(s) at time $t+1$. I have argued that a more natural and mechanistic approach is to model the effects of management actions on the vital rates underlying system dynamics. This perspective is consistent with the theme of this book and its focus on demographic rates and associated modelling.

The fundamental equation of demography focuses on population size and the vital rates that influence it: rates of mortality, reproduction, and movement in and out of the population. I have argued that for some kinds of questions and management problems, different state variables and vital rates should be considered. Here, I claim that these different kinds of parameterisations can simply be viewed as different ways of integrating the fundamental demographic variables and parameters.

Consider occupancy, for example, defined as the proportion of sites occupied by a focal species; that is, the proportion of sites at which abundance is >0. The vital rates of local extinction and colonisation can be written as functions of the fundamental demographic variables and parameters. For example, local extinction, ε_t, for a site must entail death or movement out of the site by all individuals present at t and no movement into the site by individuals from elsewhere. Dynamics of species richness are similarly determined by demographic variables and parameters, this time of multiple species rather than just one. The state variables of occupancy and richness thus focus on one tail of the abundance distribution (abundance = 0 or >0), and associated vital rates are direct functions of rates of death, reproduction, and movement of individual organisms. State dynamics of vegetation can also be viewed as integrating demographic variables and parameters. In this case, the state variables are defined by the abundances of different species. Transition probabilities for moving from one state to another are functions of species-specific rates of death, reproduction, and movement (often at the seed stage). In the case of compartmental disease models, we are modelling disease dynamics by focusing on dynamics of host organisms classified by their disease state. Transition probabilities of hosts from one disease state to another reflect the survival, reproduction, and movement of disease organisms within the host population.

All of the examples in this chapter are thus based on demography. The state variables of occupancy, species richness, and vegetative state can all be written in terms of species abundances. Similarly, the vital rates of local extinction, colonisation, and state transition can all be rewritten as functions of species-specific rates of death, reproduction, and movement. As is always the case for modelling, the choices of state variables and associated vital rates are very much determined by the objectives of the modelling effort (Nichols 2001). The use of integrated variables and parameters is simply an approach to increasing efficiency and reducing costs when modelling objectives can be met with integrated parameters. This general approach is consistent with Levins' (1966) discussion of sufficient parameters, where 'sufficient' is defined relative to model use (see Nichols 2001; Williams et al. 2002).

Despite the importance of demography and associated methods to so many kinds of questions, classes of decision problems, and classes of organisms, substantial methodological variation exists among the different disciplines that investigate demography. Some of this variation is justified. Human demographers, for example, have developed research methods that rely on direct counts of individuals experiencing different fates over time. Scientists studying animal populations, however, have developed methods that account for undercounts (usually), recognising that detection probabilities associated with virtually all animal sampling methods are <1. Similarly, classification mistakes are sometimes made, with species being misidentified or misclassified to various state variables (sex, age, reproductive status). I believe that many of these methods developed to deal with nondetection may be useful to human investigations in epidemiology and the social sciences, especially when dealing with 'hidden' human populations (e.g., homeless, criminals) and new data sources (e.g. health web site queries with symptom lists indicative of specific diseases) (Nichols 2019). I believe the emphasis of this book on the commonalities of demographic processes across taxa is a useful step to the integration of associated methods.

Acknowledgements

Thanks to Marlene Gamelon and Roberto Salguero-Gomez for developing the concept of this book and inviting me to contribute to it. Thanks to Mitch Eaton and an anonymous reviewer for providing useful comments on the initial draft manuscript. My ideas on this topic have been influenced by far more people than I can list here, but some major influences include Scott Boomer, Mike Conroy, Evan Cooch, Paul Fackler, Fred Johnson, Bill Kendall, Mike Runge, and Ken Williams.

References

Anderson, D.R. (1975). *Population Ecology of the Mallard: V. Temporal and Geographic Estimates of Survival, Recovery and Harvest Rates*. USFWS Resource Publication No. 125. US Fish and Wildlife Service, Washington, DC.

Anderson, D.R. and Burnham, K.P. (1976). *Population Ecology of the Mallard: VI. The Effect of Exploitation on Survival*. USFWS Resource Publication No. 128. US Fish and Wildlife Service, Washington, DC.

Anderson, R.M. and May, R.M. (1991). *Infectious Diseases of Humans: Dynamics and Control*. Oxford University Press, Oxford.

Arnason, A.N. (1973). The estimation of population size, migration rates, and survival in a stratified population. *Researches on Population Ecology*, **15**, 1–8.

Bailey, N. T. J. (1975). *The Mathematical Theory of Infectious Diseases*, 2nd edition. Macmillan, London.

Bayes, T. and R. Price. (1763). An essay towards solving a problem in the doctrine of chances. *Philosophical Transactions* (1683–1775), **53**, 370–418.

Bellman, R. (1957). *Dynamic Programming*. Princeton University Press, Princeton, NJ.

Blohm, R.J. (1989). Introduction to harvest: understanding surveys and season setting. *Proceedings of the International Waterfowl Symposium*, **6**, 118–133.

Bogich, T.L., Anthony, S.J., and Nichols, J.D. (2013). Surveillance theory applied to virus detection: a case for targeted discovery. *Future Virology*, **8**, 1201–1206.

Boulinier, T., Nichols, J.D., Hines, J.E. Sauer, J.R., Flather, C.H., and Pollock, K.H. (2001). Forest fragmentation and bird community dynamics: inference at regional scales. *Ecology*, **82**, 1159–1169.

Breininger, D.R. and Carter, G.M. (2003). Territory quality transitions and source-sink dynamics in a Florida scrub-jay population. *Ecological Applications*, **13**, 516–529.

Breininger, D. R. and Oddy, D.M. (2004). Do habitat potential, population density, and fires influence Florida scrub-jay source-sink dynamics? *Ecological Applications*, **14**, 1079–1089.

Breininger, D.R., Larson, V.L., Oddy, D.M., Smith, R.B., and Barkaszi, M.J. (1996). Florida scrub-jay demography in different landscapes. *Auk*, **112**, 617–625.

Breininger, D.R., Nichols, J.D., Carter, G.M., and Oddy, D.M. (2009). Habitat-specific breeder survival of Florida scrub-jays: inferences from multistate models. *Ecology*, **90**, 3180–3189.

Breininger, D.R., Nichols, J.D., Duncan, B.W., et al. (2010). Multistate modeling of habitat dynamics: factors affecting Florida scrub transition probabilities. *Ecology*, **91**, 3354–3364.

Brownie, C., Anderson, D.R., Burnham, K.P., and Robson, D.S. (1985). *Statistical Inference from Band Recovery Data: A Handbook*, 2nd edition. USFWS Resource Publication No. 156. US Fish and Wildlife Service, Washington, DC.

Brownie, C., Hines, J. E., Nichols, J.D., Pollock, K.H., and Hestbeck, J.B. (1993). Capture–recapture studies for multiple strata including non-Markovian transition probabilities. *Biometrics*, **49**, 1173–1187.

Burnham, K.P. and Anderson, D.R. (2002). *Model Selection and Multimodel Inference*, 2nd edition. Springer, New York.

Caswell, H. (2001). *Matrix Population Models*. Sinauer Associates, Sunderland, MA.

Chao, A., Gotelli, N.J., Hsieh, T.C., et al. (2014). Rarefaction and extrapolation with Hill numbers: a framework for sampling and estimation in species diversity studies. *Ecological Monographs*, **84**, 45–67.

Clemen, R.T. and Reilly, T. (2001). *Making Hard Decisions with Decision Tools*. Duxbury Press, Pacific Grove, CA.

Cooch, E.G., Conn, P.B., Ellner, S.P., Dobson, A.P., and Pollock, K.H. (2012). Disease dynamics in wild populations: modeling and estimation: a review. *Journal of Ornithology*, **152**, 485–509.

Cooch, E.G., Guillemain, M., Boomer, G.S., Lebreton, J.-D., and Nichols, J.D. (2014). The effects of harvest on waterfowl populations. *Wildfowl Special Issue*, **4**, 220–276.

Eaton, M.J., Hughes, P.T., Hines, J.E., and Nichols, J.D. (2014). Testing metapopulation concepts: effects of patch characteristics and neighborhood on occupancy dynamics of an endangered lagomorph. *Oikos*, **123**, 662–676.

Eaton, M.J., D.R. Breininger, J.D. Nichols, et al. (2021). Integrated hierarchical models to inform management of transitional habitat and the recovery of a habitat specialist. *Ecosphere*, **12**(1), e03306.

Hines, J.E., Boulinier, T., Nichols, J.D., Sauer, J.R., and Pollock, K.H. (1999). COMDYN: software to study the dynamics of animal communities using a capture–recapture approach. *Bird Study*, **46** (Suppl.), S209–S217.

Johnson, F.A. (2011). Learning and adaptation in the management of waterfowl harvests. *Journal of Environmental Management*, **92**, 1385–1394.

Johnson, F.A., Williams, B.K., Nichols, J.D., et al. (1993). Developing an adaptive management strategy for harvesting waterfowl in North America. *Transactions of the North American Wildlife and Natural Resources Conference*, **58**, 565–583.

Johnson, F.A., Moore, C.T., Kendall, W.L., et al. (1997) Uncertainty and the management of mallard harvests. *Journal of Wildlife Management*, **61**, 202–216.

Johnson, F.A., Breininger, D.R., Duncan, B.W., Nichols, J.D., Runge, M.R., and Williams, B.K. (2011). A Markov decision process for managing habitat for Florida scrub-jays. *Journal of Fisheries and Wildlife Management*, 2, 234–246.

Johnson, F.A., Boomer, G.S., Williams, B.K. Nichols, J.D., and Case, D.J. (2015). Multi-level learning in the adaptive management of waterfowl harvests: 20 years and counting. *Wildlife Society Bulletin*, **39**, 9–19.

Kermack, W.O. and McKendrick, A.G. (1927). A contribution to the mathematical theory of epidemics. *Proceedings of the Royal Society of London, Series A*, **115**, 700–721.

Kery, M. and Royle, J.A. (2008). Hierarchical Bayes estimation of species richness and occupancy in spatially replicated surveys. *Journal of Applied Ecology*, **45**, 589–598.

Lancia, R.A., Kendall, W.L., Pollock, K.H., and Nichols, J.D. (2005). Estimating the number of animals in wildlife populations. In C.E. Braun (ed.) *Research and Management Techniques for Wildlife and Habitats*, pp. 106–153. The Wildlife Society, Bethesda, MD.

Lebreton, J.-D., Burnham, K.P., Clobert, J., and Anderson, D.R. (1992). Modeling survival and testing biological hypotheses using marked animals: a unified approach with case studies. *Ecological Monographs*, **62**, 67–118.

Lebreton, J.-D., Nichols, J.D., Barker, R., Pradel, R., and Spendelow, J. (2009). Modeling individual animal histories with multistate capture–recapture models. *Advances in Ecological Research*, **41**, 87–173.

Levins, R. (1966). The strategy of model building in population biology. *American Scientist*, **54**, 421–431.

Lincoln, F.C. (1930). *Calculating Waterfowl Abundance on the Basis of Banding Returns*. Circular Number 118. US Department of Agriculture, Washington, DC.

Lubow, B. (1997). *Adaptive Stochastic Dynamic Programming (ASDP) Supplement to SDP User's Guide. Version 2.0.* Colorado Cooperative Fish and Wildlife Research Unit, Colorado State University, Fort Collins, CO.

MacKenzie, D.I., Nichols, J.D., Royle, J.A., Pollock, K.H., Bailey, L.A. and Hines, J.E. (2018). *Occupancy Modeling and Estimation*, 2nd edition. Academic Press, San Diego, CA.

Martin, E.M. and Carney, S.M. (1977). *Population Ecology of the Mallard: IV. A Review of Duck Hunting Regulations, Activity and Success, with Special Reference to the Mallard*. USFWS Resource Publication No. 130. US Fish and Wildlife Service, Washington, DC.

Martin, F.W., Pospahala, R.S., and Nichols, J.D. (1979). Assessment and population management of North American migratory birds. In J. Cairns, G.P. Patil, and W.E. Waters (eds.) *Environmental Biomonitoring, Assessment, Prediction, and Management. Certain Case Studies and Related Quantitative Issues. Statistical Ecology Vol. Sll*, pp. 187–239. International Co-operative Publishing House, Fairland, MD.

Martin J., Runge, M.C., Nichols, J.D., Lubow, B.C. and Kendall, W.L. (2009). Structured decision making as a conceptual framework to identify thresholds for conservation and management. *Ecological Applications*, **19**, 1079–1090.

Miller, D.A.W., Brehme, C.S., Hines, J.E., Nichols, J.D. and Fisher, R.N. (2012). Joint estimation of habitat dynamics and species interactions; disturbance reduces co-occurrence of non-native predators with an endangered toad. *Journal of Animal Ecology*, **81**, 1288–1297.

Nichols, J.D. (2000). Evolution of harvest management for North American waterfowl: selective pressures and preadaptations for adaptive harvest management. *Transactions of the North American Wildlife and Natural Resources Conference*, **65**, 65–77.

Nichols, J.D. (2001). Using models in the conduct of science and management of natural resources. In T.M. Shenk and A.B. Franklin (eds). *Modeling in Natural Resource Management: Development, Interpretation and Application*, pp. 11–34. Island Press, Washington, DC.

Nichols, J.D. (2012). Evidence, models, conservation programs and limits to management. *Animal Conservation*, **15**, 331–333.

Nichols, J.D. (2019). Confronting uncertainty: contributions of the wildlife profession to the broader scientific community. *Journal of Wildlife Management*, 83, 519–533.

Nichols, J.D. and Williams, B.K. (2013). Adaptive management. In A.H El-Shaarawi and W.W. Piegorsch (eds).

Encyclopedia of Environmetrics, 2nd edition, p.28. John Wiley, New York.

Nichols, J.D., Boulinier, T., Hines, J.E., Pollock, K.H., and Sauer, J.R. (1998). Estimating rates of local extinction, colonization and turnover in animal communities. *Ecological Applications*, **8**, 1213–1225.

Nichols, J.D., Johnson, F.A., and Williams, B.K. (1995a). Managing North American waterfowl in the face of uncertainty. *Annual Review of Ecology and Systematics*, **26**, 177–199.

Nichols, J.D., Reynolds, R.E. Blohm, R.J., Trost, R.E., Hines, J.E., and Bladen, J.P. (1995b). Geographic variation in band reporting rates of mallards based on reward banding. *Journal of Wildlife Management*, **59**, 697–708

Nichols, J.D., Runge, M.C., Johnson, F.A., and Williams, B.K. (2007). Adaptive harvest management of North American waterfowl populations: a brief history and future prospects. *Journal of Ornithology*, **148** (Suppl. 2), S343–S349.

Nichols, J.D., Thomas, L., and Conn, P.B. (2009). Inferences about landbird abundance from count data: recent advances and future directions. In D.L. Thomson, E.G. Cooch, and M.J. Conroy (eds.) *Modeling Demographic Processes in Marked Populations*, pp. 201–235. Springer, New York.

Nichols, J.D., Johnson, F.A., Williams, B.K., and Boomer, G.S. (2015). On formally integrating science and policy: walking the walk. *Journal of Applied Ecology*, **52**, 539–543.

Nichols, J.D., Hollman, T., and Grand, J.B. (2017). Monitoring for the management of disease risk in animal translocation programmes. *EcoHealth*, **14**, S156–S166.

Pollock, K.H. (1982). A capture–recapture design robust to unequal probability of capture. *Journal of Wildlife Management*, **46**, 752–757.

Pradel, R. (2005). Multievent: an extension of multistate capture–recapture models to uncertain states. *Biometrics*, **61**, 442–447.

Puterman, M.L. (1994). *Markov Decision Processes: Discrete Stochastic Dynamic Programming*. Wiley, New York.

Rees, M. and Ellner, S.P. (2009). Integral projection models for populations in temporally varying environments. *Ecological Monographs*, **79**, 575–594.

Royle, J.A., Dorazio, R.M., and Link, W.A. (2007). Analysis of multinomial models with unknown index using data augmentation. *Journal of Computational and Graphical Statistics*, **16**, 67–85.

Schwarz, C.J., Schweigert, J.F., and Arnason, A.N. (1993). Estimating migration rates using tag-recovery data. *Biometrics*, **49**, 177–193.

Seber, G. A. F. (1982). *The Estimation of Animal Abundance and Related Parameters*. MacMillan, New York.

Shea, K., Tildesley, M.J., Runge, M.C., Fonnesbeck, C.J., and Ferrari, M.J. (2014). Adaptive management and the value of information: learning via intervention in epidemiology. *PLOS Biology*, 12(10), e1001970.

Skalski, J.R., Ryding, K.E., and Millspaugh, J.J. (2010). *Wildlife Demography: Analysis of Sex, Age and Count Data*. Elsevier/Academic Press, New York.

Smith, G.W. (1995). *A Critical Review of the Aerial and Ground Surveys of Breeding Waterfowl in North America*. Biological Scientific Report 5. US Department of the Interior, Washington, DC.

Tolkien, J.R.R. (1954). *The Fellowship of the Ring*. George Allen and Unwin, London.

Tuljapurkar, S. and Caswell, H. (2012). *Structured-Population Models in Marine, Terrestrial, and Freshwater Systems*. Chapman and Hall, New York.

US Fish and Wildlife Service. (2018). *Adaptive Harvest Management: 2019 Hunting Season*. US Department of the Interior, Washington, DC.

Veran, S., Kleiner, K.J., Choquet, R., Collazo, J., and Nichols, J.D. (2012). Modeling habitat dynamics accounting for possible misclassification. *Landscape Ecology*, **27**, 943–956.

Walters, C.J. (1986). *Adaptive Management of Renewable Resources*. MacMillan, New York.

Walters, C.J. and Hilborn, R. (1976). Adaptive control of fishing systems. *Canadian Journal of the Fisheries Research Board of Canada*, **33**, 145–159.

Walters, C.J. and Hilborn, R. (1978). Ecological optimization and adaptive management. *Annual Review of Ecology and Systematics*, **9**, 157–188.

Williams, B.K. (1996). Adaptive optimization and the harvest of biological populations. *Mathematical Biosciences*, **136**, 1–20.

Williams, B.K., Johnson, F.A., and Wilkins, K. (1996). Uncertainty and the adaptive management of waterfowl harvests. *Journal of Wildlife Management*, **60**, 223–232.

Williams, B.K., Nichols, J.D., and Conroy, M.J. (2002). *Analysis and Management of Animal Populations*. Academic Press, San Diego, CA.

Williams, B.K., Szaro, R.C., and Shapiro, C.D. (2007). *Adaptive Management: US Department of the Interior Technical Guide*. Adaptive Management Working Group. US Department of the Interior, Washington, DC.

Yackulic, C.B., Reid, J., Davis, R., Hines, J.E., Nichols, J.D., and Forsman, E. (2012). Neighborhood and habitat effects on vital rates: expansion of the barred owl in the Oregon Coast Ranges. *Ecology*, **93**, 1953–1966.

Yackulic, C.B., Reid, J., Nichols, J.D. Hines, J.E., Davis, R., and Forsman, E. (2014). The roles of competition and habitat in the dynamics of populations and species distributions. *Ecology*, **95**, 265–279.

Zipkin, E.F., DeWan, A., and Royle, J.A. (2009). Impacts of forest fragmentation on species richness: hierarchical approach to community modelling. *Ecological Applications*, **46**, 815–822.

CHAPTER 20

Heritability, polymorphism, and population dynamics: individual-based eco-evolutionary simulations

Anna Kuparinen

20.1 Introduction

Life history traits, such as rate of somatic growth and metrics of reproductive output, vary across individual phenotypes. Consequently, the demographic rates expressed at the population level depend on the phenotypic composition of the populations. For example, if the somatic growth rates of young individuals are higher than those of older individuals, a population dominated by young individuals shows higher biomass growth per unit time. On the other hand, reproductive output of a population might be promoted by the presence of large, experienced individuals that produce viable juveniles. Thus, the trade-off between somatic growth and reproduction at the level of individuals manifests as differences in population demographic rates, mediated by differences in population age structure.

More generally, trade-offs are manifested at the level of individuals, among phenotypic traits, such as life history invariants (Charnov 1993), but they may also translate to population-level differences in basic demographic traits. The key is to understand that both the individual- and population-level interfaces of a biological system affect each other and to identify their interaction pathways, such as age structure. Interactions among individual traits and population dynamics are not limited to ecological dynamics, as phenotypes can evolve on a decadal time-scale or across few generations (Reznick 1997). Such evolution within contemporary time-scales is likely to alter demographic rates and the dynamics of populations (Hendry 2016).

Research that integrated the fundamentals of population ecology and evolutionary biology was long denoted as evolutionary ecology. It was characteristic of evolutionary ecology to study how ecological differences in the environment caused selection and drove evolution of species. Conversely, adaptive differences among populations could cause differences in demographic properties of different populations of the same species. Most commonly, the focus was on the one-way feedbacks from ecology to evolution or vice versa (Hendry 2016).

More recently, increased computational power and numeric simulation approaches have enabled a broadening of evolutionary ecology from the study of one-way feedbacks to iterative dynamics and the equilibrium conditions in which the eco-evolutionary dynamics settle into a stable state or fluctuate about it. Of course, entirely stable states rarely exist in nature due to continuously changing environmental conditions. Eco-evolutionary dynamics should therefore be seen as an ongoing continuum of feedbacks between ecological and evolutionary processes orchestrated by selection arising from ecological conditions and modifying population evolution within genetic, physiological, and spatiotemporal boundaries (Hendry 2016).

20.2 Heritability and selection in the context of population dynamics

Most phenotypic traits such as age and size at reproduction and growth rate are to some extent heritable (Mousseau and Roff 1987), meaning that traits are partially coded by genes. The trait frequencies in a population can, therefore, be modified through selection on the coding genes. This may further lead

Anna Kuparinen, *Heritability, polymorphism, and population dynamics: individual-based eco-evolutionary simulations.*
In: *Demographic Methods Across the Tree of Life.* Edited by Roberto Salguero-Gómez and Marlène Gamelon,
Oxford University Press. © Oxford University Press (2021). DOI: 10.1093/oso/9780198838609.003.0020

to co-evolution of correlated traits even if they were not subjected to selection directly. The extent to which phenotypic variation is associated with genetic variation is typically quantified through heritability, the ratio of genetic variance, and phenotypic variance (Roff 1992).

In theory, one can estimate evolutionary responses to selection by using the Breeder's equation $R = h^2S$, where R is the response to selection, h^2 is the heritability, and S is the selection differential (i.e. the change in the cohort trait value from birth to reproduction). This equation will quantify the expected response to selection in the presence of nonoverlapping generations but also makes several restricting assumptions (Morrissey et al. 2010). For example, coupled with traditional life table analyses, one can then approximate how an evolutionary change in age at maturity alters a population's net reproductive rate and the intrinsic rate of population growth. However, this approach gives at its best only very rough approximations. Firstly, in most populations generations do overlap. Secondly, the selection differential integrates all the selection on the given trait from birth up to reproduction, and, specifically, it cannot be estimated based on a single selective event such as elevated mortality at a certain age or stage. Thirdly, the selection differential will likely be affected by many correlated traits: a trait generally favoured by selection at one stage of life may be disadvantageous at other life stages. The selection differential will jointly integrate the outcome of multiple selective events across the lifetime of an individual (Roff 1992). Capturing the complexity of eco-evolutionary dynamics thus requires extensive long-term experiments or modelling approaches with empirically based parameterisations. This is the case even if eco-evolutionary dynamics were studied only at the level of one population or species. Involving community or ecosystem aspects obviously makes things even more complicated. In the following, this chapter focuses on an individual-based eco-evolutionary simulation approach as a tool to understand trait evolution in the context of population dynamics.

20.3 Generic ways to model the evolution of demographic rates

A straightforward and intuitively appealing way to integrate phenotypic evolution into ecological dynamics is to use individual-based simulation models (Box 20.1; see also Chapter 12). In such models, growth, survival, mating, and reproduction are described at the level of individuals, allowing for full consideration of trade-offs and correlations among life history traits. Moreover, no restricting assumptions need to be made as would often be the case in analytically solvable models. Evolutionary changes in the population simply depend on which individuals survive to maturity and with whom they mate. The juveniles will to some extent reflect the properties of their parents, and the new cohort will reflect the ways in which selection filtered the parental cohorts. The annual number of survivors, individuals that died, and new juveniles will reflect demographic rates of survival, mortality, and reproduction. Box 20.1 and Boxed Figure 20.1 provide a generic illustration of how to construct such a simulation approach in practice.

20.3.1 Parental trait values coupled with known heritability

In the case of quantitative traits, the trait of an offspring can be approximated based on the trait values of its parents coupled with some random variability typically drawn from a normal distribution with a mean of zero and the standard deviation calibrated to yield the desired level of phenotypic versus genetic variability, that is, the desired level of heritability. This approach has been used particularly in so-called ecogenetic simulation models applied to predict how fishing-induced evolution shapes fish phenotypes and the dynamics of the population, particularly its reproductive capacity and recovery ability (Dunlop 2009; Enberg 2009). These modelling approaches gave important insights into broad-scale eco-evolutionary dynamics of natural populations and influenced recommendations regarding fisheries management practices (Jørgensen et al. 2007).

However, from an eco-evolutionary theory perspective, the above approach suffers from some limitations. Assuming a constant heritability might overestimate the true adaptive capacity of the population—namely, in the absence of, for example, mutations and immigration—heritability should decline in the presence of strong, directional selection, as selection removes less-fit phenotypes/genotypes from the population. As heritability declines, trait evolution slows down, and one expects that the population finds a new adaptive optimum at which its resilience and persistence are maximised, given the prevailing conditions (Chevin et al. 2010).

20.3.2 Dummy genes coding the traits in a classic Mendelian manner

An alternative to fixing the heritability was presented by Savolainen et al. (2004), and this approach is based

Box 20.1 The basic concept of individual-based eco-evolutionary simulations.

The simulation tracks through time, from one time-step to the next one (e.g. day or year), and within each time–step a set of demographic and evolutionary processes are modelled at the level of each individual, going through all individuals in the population one by one. These processes are shown in Boxed Figure 20.1 and involve the following:

(1) Mortality: Based on a random number, an individual either survives or dies, and in the case of the latter, the individual is simply removed from the population.
(2) Growth: The surviving individual grows, depending on the growth model used. It can progress along a growth curve or, in a discrete stage-based setting, shift from one life history stage to another. Again, a random number can be involved to make the process stochastic, and density-dependent restrictions can be applied.
(3) Maturity: At some point of its life, an individual starts to reproduce. It may exceed a simple size or age threshold, or its possibility to reproduce may depend on the population or age class/stage density. The following steps apply to mature individuals only.
(4) Number of juveniles: Depending on the individual's body size, age, or condition, it produces a number of juveniles. The individual can have one or several mates within the time step.

(5) Juvenile genotype: The genotype of each individual juvenile depends on its parents. Genotype can be determined based on the average of the parental trait values coupled with some stochasticity or drawn from parental alleles.
(6) Juvenile phenotype: Once juvenile genotype is known, the juvenile is assigned with some environmental variation in the phenotypic expression.
(7) Updated population: New juveniles will be added to the population.

After going through steps 1–7 for each individual and their juveniles, the population will be updated with new juveniles and time shifts to the next time step. The above processes will be repeated for each time-step. Each of the ecological and evolutionary processes can be based on random numbers drawn from appropriate distributions, given the average rate and the desired level of stochasticity. The choice of distributions, the scale of time, forms of density dependence, models for growth, maturity, and fecundity, as well as parameter values are species- and study-system specific.

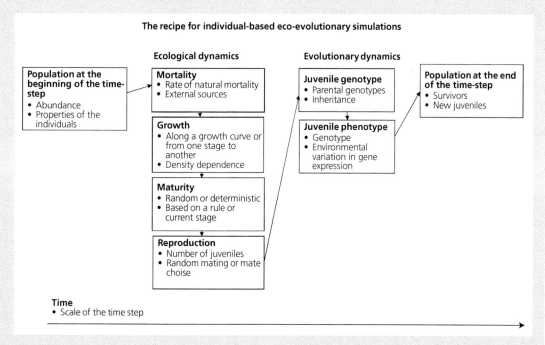

The recipe for individual-based eco-evolutionary simulations

Boxed Figure 20.1 Key processes affecting individual-based eco-evolutionary dynamics of a population. Examples of factors that may affect each process are detailed in the boxes. In principle, simulations track trough time, and each process is simulated for each individual, thus updating the population from the previous to the next time point.

on the idea that most quantitative traits are determined by small, additive effects of a large number of loci (Roff 1992).

In this approach, a trait of interest is determined by a set number of loci, the alleles of which are coded by 0 and 1. Now, the sum across the allele values can be translated into a realistic range of trait values, such that sums close to 0 are associated with the lowest trait value and the highest sums with the highest trait value. Naturally, the direct trait value sum must be coupled with some random variability to allow for environmental variation in phenotypes. This variability can be initially tuned to yield realistic heritabilities in pristine or stable conditions. When disturbances such as harvesting or changes in environmental drivers are then induced to the simulated eco-evolutionary population dynamics, the heritability can change similarly to the other population-level parameters.

It is notable that the number of loci considered in such simulations does not need to reflect the actual number of genes/loci coding the trait. Instead, the modelled loci are simply a representation of the genetic basis of the trait in question. Therefore, the number of modelled loci is mainly just a computational question: There needs to be enough loci such that the trait can evolve fairly smoothly but not too many to slow down the computation speed. Namely, in this approach, genes of an offspring depend on the genes of the parents and are passed via the principles of classic Mendelian heritance. This involves going through each locus one by one and randomly drawing the juvenile genes from the two alleles of both the parents (Figure 20.1). While this may sound laborious, the obvious advantage is that the inherent stochasticity associated with passing genes to offspring is accounted for in the model.

The flexibility that this approach offers is that the trait in question does not necessarily need to be coded by many loci or by their additive effects. The relative influences of the genes can be easily varied, and the number of coding genes can be low or high (i.e. the model allows different genetic architectures of traits). This will be demonstrated in the third case study (section 20.4.3). Similarly, one can set a rate of mutation for all or some loci, if this seems relevant for the time-frame of the study or the system in question.

20.4 Case studies

The objective of the present book is to present demographic methods that can be applied across different types of study organisms. In the context of eco-evolutionary dynamics, this applies not only to animals and plants but also to traits that are coded in very different ways and, thus, can evolve very differently, with associated effects of co-evolving traits and ecological dynamics. To this end, all the chosen case studies utilise the approach detailed in section 20.3.2 and demonstrate its functioning for different types of organisms and genetic architectures.

20.4.1 Boreal forest adaptation to warming climate

Global warming has increased temperatures in boreal areas, and this warming is likely to continue in the future. In practice, forest trees living close to or at their species ranges will experience improved growth conditions not only due to increased temperature during the growing season but also as a result of the prolonged growth season. Namely, transplantation experiments have demonstrated that in the northernmost forests a tree's individual fitness will increase as those trees are reared at slightly more southern latitudes than where their parental population originated from (Savolainen et al. 2007). This observation suggests that, as opposed to the core of the species range, where individuals are typically adapted to their local conditions and suffer from transplantation northwards or southwards, at the

Each individual has X loci with two alleles in each, coded as 0 or 1

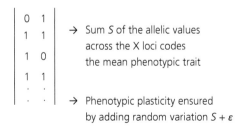

→ Sum S of the allelic values across the X loci codes the mean phenotypic trait

→ Phenotypic plasticity ensured by adding random variation $S + \varepsilon$

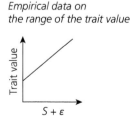

Figure 20.1 Schematic illustration of the model, where a phenotypic trait is coded by X loci and two alleles in each.

northern edge of the species range local adaptation does not occur, or it breaks down. This can be explained by gene flow from more southern populations that is not balanced by gene flow from the north, as the species does not exist beyond its northern range boundary. Thus, northernmost forests are likely to experience some level of fitness benefit from more 'southern-type' conditions (Savolainen et al. 2007).

One obstacle in this reasoning is that in boreal forest trees the growth cessation in the autumn is to some extent genetically controlled (Savolainen et al. 2004). As the first frosts occur, a tree that has not ceased its growth and developed winter hardening will be harmed, and the younger the tree, the higher the potential for death. On the other hand, too early growth cessation is a disadvantage in competition among better growing conspecifics. Purely ecological forest growth models do not take into account this evolutionary aspect of forest growth but assume that forests will be able to utilise their full growth capacity given the prevailing conditions.

This mismatch makes it interesting to explore how much genetic control of growth cessation might affect tree population dynamics, compared to an assumption that growth cessation is able to track the changing climate without any delay. Answering this question requires coupling the phenotypic trait evolution to forest tree population dynamics.

In northern Fennoscandia, Scots pine (*Pinus sylvestris*) and silver birch (*Betula pendula*) are the dominant forest trees. Following the approach detailed in section 20.3.2, Kuparinen et al. (2010) assumed that the timing of growth cessation is coded by ten loci with additive effects. The loci sum was coupled with phenotypic variability, to yield heritabilities similar to those estimated in Scots pine common garden experiments (Notivol et al. 2007). For birch the phenotypic variability was assumed the same, in the absence of similar estimates as for Scots pine.

Scots pine and silver birch differ substantially in their maturation times and pollen and seed dispersal abilities. These ecological differences between the species affect their population renewal rates and, thus, potentially modify the rate at which evolution towards later growth cessation could take place (Kuparinen et al. 2010. To investigate this, we constructed an eco-evolutionary simulation model following the principles described in Box 20.1. The model had two distinct developmental life history stages: seedling pool stage and the established adult stage.

The eco-evolutionary dynamics of ten populations along an 800-km cline were simulated. These

simulations looked into how the populations adapted to the predicted change in the growth season duration across 100 years: from 170 to 220 days in the south end of the cline and from 120 to 170 days in the north end of the cline (Savolainen et al. 2004). In between, the change was linearly related to the location along the cline, and the shift in the growth season duration also occurred linearly across the 100 years.

Genetic variability and gene flow from southern populations (in the forms of pollen and seeds) as well as phenotypic variability served as sources for such seedlings, whose phenotypic growth season length was longer than the local adaptive average. These seedlings were able to outcompete seedlings with shorter growth season, but if the phenotypic growth season of a seedling exceeded that at the location at the given year, it was considered killed by frost. Old established trees were assumed more cold tolerant than seedlings. Therefore, selection acting at the seedling stage favoured longer growth season but had also a knife-edge threshold. The final step in the evolution process was the establishment of a seedling into an adult reproducing tree; this required that an established tree died first, which occurred at the annual probability of 2–4%. Once a previously established adult died, ten seedling-stage individuals were randomly drawn from the population. Among these, the seedling with the latest growth cessation possible, given the prevailing climate at the current time point, was allowed to grow into an established adult stage.

The simulated population adaptation to increasing growth season length was much slower than the environmental change; that is, evolution lagged behind the changing environment (Kuparinen et al. 2010) (Figure 20.2). Thus, forests were not able to utilise right away their improved growth conditions, and forest growth predictions that omitted this evolutionary limitation would therefore be vastly overestimated. The slow adaptation was not due to the lack of genetic variability but the slow rate at which the seedlings could establish themselves. As the second author, Professor Outi Savolainen, interpreted the results: 'The pattern is similar to academic life: young scientists can't get positions because those are held by old established professors until they retire' (O. Savolainen, personal communication).

20.4.2 Co-evolution of correlated traits in Atlantic cod

The previous example was simple in the sense that it only looked at one evolving trait and how that

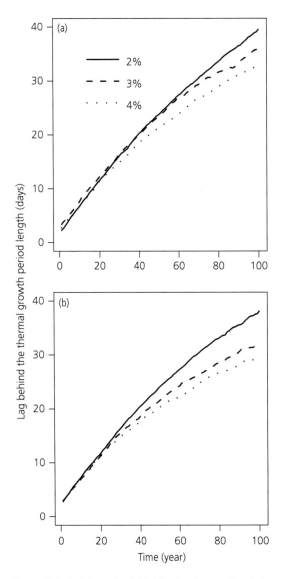

Lag behind the thermal growth period length (days)

Figure 20.2 Evolutionary lag behind the changing growth period length for (a) Scots pine and (b) silver birch. The lag between the mean genotypic growth period length and the climatic optimum (thermal growth period length) is shown by the y-axis. The lag is given as a function of time during which the environment changes gradually (x-axis). Results for different rates of adult tree mortality are indicated by different line types. (Reproduced from Kuparinen et al. (2010) by permission of Elsevier.)

by simply adding the number of loci and having certain loci to code each trait. However, this has major points of consideration: (1) computation time can increase substantially; and (2) one needs to make assumptions about the correlations among the traits (or lack of correlation). If strong well-established correlations exist, those can be utilised without the need to increase the number of loci. For example, let's assume we have three traits, A, B, and C, of interest and we know how they are correlated with each other. We can then just simulate the loci coding trait A and then determine the value of B and C based on the value of A. Thus, the traits co-evolve according to their correlation structure, but only the loci coding trait A is simulated.

An alternative way, of course, would be to allow overlap among the sets of loci coding each trait, but this requires tuning the model such that the emerging correlations match with the empirically detected ones. Computation time would also increase as the number of simulated loci increases. Nonetheless, such an approach is interesting and worth testing in the future.

In ectotherms, many key life history traits are correlated, and these correlations are often known as life history invariants (Charnov 1993). In fishes, these correlations are closely involved with the so-called von Bertalanffy growth model (Figure 20.3), which tells at a given age t the length of the fish $L(t)$. The model is based on three parameters: the asymptotic body length

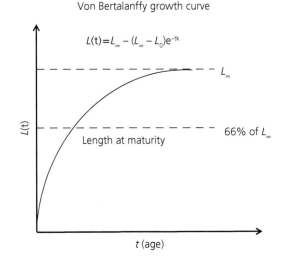

Von Bertalanffy growth curve

$$L(t) = L_\infty - (L_\infty - L_0)e^{-tk}$$

L_∞

66% of L_∞

Length at maturity

$L(t)$

t (age)

Figure 20.3 Von Bertalanffy growth model and its life history invariants (von Bertalanffy 1938; Charnov 1993). $L(t)$: length at age t; L_∞: asymptotic body length; L_0: length at age $t=0$; k: the rate at which an individual reaches its L_∞.

affected the projected population dynamics. However, in reality, fitness is often the product of a few important life history traits. The method presented in section 20.3.2 and utilised throughout the three presented applications (sections 20.4.1–20.4.3) can be straightforwardly extended to several evolving traits

L_∞, the rate k at which $L(t)$ approaches L_∞, and the average length at t = 0. The two key invariants related to L_∞ are as follows: (1) k and L_∞ are strongly and negatively correlated; and (2) length at maturity is about 66% of L_∞ (Charnov 1993; Jensen 1997).

By utilising these correlations, the evolution of fish growth trajectory and fish maturation can be modelled by simulating the loci that code L_∞ and then predicting the age and size at maturation and the growth parameter k based on L_∞. While the length at maturity (and thus the age at maturity) can be derived straightforwardly from the value of L_∞, predicting k requires knowledge about the empirical relationship between L_∞ and k. In fishes this information can be obtained by reading lengths-at-age of the fish from scales or otoliths. Von Bertalanffy growth curves can then be fitted to individual growth trajectories and k modelled as a function of L_∞. If the relationship is particularly skewed, this can be fixed by taking a log-transformation of k, as shown in Figure 20.4. Now the linear model can be fitted to describe the association of L_∞ and k. For each value of L_∞, the value of k can be predicted using the fitted model and coupled with random variation, the magnitude of which should match the residual variation in the model fit.

The above described modelling approach was utilised to predict eco-evolutionary responses of Atlantic cod (*Gadus morhua* L.) (Figure 20.5) to selective fishing (Kuparinen and Hutchings 2012; Kuparinen et al. 2014). As theory, empirical observations, and experiments demonstrate (e.g. Law 2007), increased

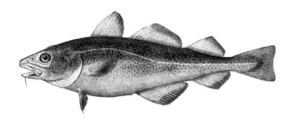

Figure 20.5 Atlantic cod (*Gadus morhua* L.). (Source: Wikipedia.)

mortality due to fishing is likely to select for earlier maturity at smaller body size. These life history changes, in turn, can affect fish population dynamics, more specifically population productivity, resilience to disturbances, and the ability to recover from low abundances.

Atlantic cod is in many ways an iconic species to study fishing impacts, owing to several population collapses in the 1980s and 1990s in the Atlantic Ocean and the Baltic Sea, followed by cod fishing closures in many management units. Despite dramatic reductions in cod fishing pressure, in many areas such as the coasts of Newfoundland and the Gulf of St Lawrence in Canada, cod has shown very little or no signs of recovery. While many reasons such as environmental conditions, ecosystem regime shifts, and changes in predator–prey dynamics can hinder recovery, the dramatic reductions in cod size and age at maturity suggest that some life history evolution might have occurred, and this may interact with the ecological dynamics of cod populations. Namely, based on demographic population parameters alone, fisheries scientists have for long predicted using their stock-assessment models that cod should recovery rapidly after fishing pressure relaxes.

Indeed, simulated eco-evolutionary dynamics of cod populations exposed to fishing suggest that cod L_∞ might evolve to substantially smaller values within just a few decades and under moderate fishing pressure (Kuparinen and Hutchings 2012). Not only the average age and size of the reproducing individuals is reduced but also the net reproductive rate of the population R_0 and per-capita population growth rate r are affected, particularly at low population abundances (Kuparinen et al. 2014). Similarly, after the life history changes the populations could not rebuild their biomasses back to prefishing levels but only to biomasses that were 10–20% lower. Thus, life history evolution can also change the population carrying capacity, a fundamental ecological parameter regulating population dynamics (Kuparinen and Hutchings 2012).

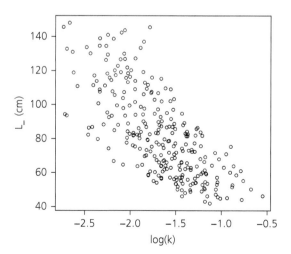

Figure 20.4 Association between L_∞ and k parameters of the von Bertalanffy growth models fitted to lengths-at-age read from 258 Atlantic cod otoliths collected by Hardie (2007).

20.4.3 Evolution of a sexually polymorphic age at maturity in Atlantic salmon

The previous two examples were based on the assumption that quantitative traits, such as the length of the growth season in trees (section 20.4.1) and asymptotic body length (L_∞) in fishes (section 20.4.2), are coded by a large number of loci with small, additive effects (Roff 1992). While this might be the case for many organisms and traits, more recent genomic discoveries have demonstrated that the genetic architecture of a phenotypic trait can be more diverse.

Atlantic salmon (*Salmo salar* L.) (Figure 20.6) is an anadromous fish species inhabiting rivers in the northern hemisphere. Many Atlantic salmon populations have declined to vulnerable levels or even gone extinct (WWF 2001). Due to their critical state, successful reproduction is extremely important for the future of many populations. One key trait involved in salmon reproduction is the age at which the individual returns to its natal river to spawn. The longer the salmon postpones maturity, the lower its odds of reproducing before it dies. On the other hand, larger older spawners produce much more and possibly better quality juveniles, so postponing might also be beneficial. This largely depends on prereproductive mortality and uncertainty about it. Consequently, the question of what triggers maturity in Atlantic salmon has puzzled scientists for decades.

Recently, genomic analyses have discovered that salmon age at maturity is strongly controlled by one single locus with sexually dimorphic expression; this one locus explains about 40% of variation in salmon age at maturity (Barson et al. 2015). In practice, homozygotes for one allele tend to mature young and homozygotes for the other allele tend to mature old, irrespective of the sex. But the age at maturity of the heterozygotes depends on the sex: heterozygote

females tend to mature late, similarly to the homozygotes for the one allele, but heterozygous males are more similar to the early maturing homozygotes for the other allele (Table 20.1). This genetic architecture makes the age at maturity in Atlantic salmon a polymorphic trait, with phenotypes being either early or late reproducers. However, this polymorphism is partly masked by sex and can therefore be difficult to distinguish by eye from raw data. In fact, more specifically one would say that the trait is sexually dimorphic among heterozygotes.

Flexibility of the modelling approach described in section 20.3.2 becomes particularly convenient when modelling a specific genetic trait architecture such as the one discovered for Atlantic salmon age at maturity. Instead of an arbitrary number of loci, salmon age at maturity can now be described by a single locus with two alleles. The trait expression is then sex-dependent. But in individual-based simulations, tracking a trait like sex is simple. And, as usual, there is some environmental variance about the alleles coding the trait. In the case of Atlantic salmon maturation, gene expression is defined in a probabilistic manner: different genotypes are associated with the probabilities for postponing maturity or, inversely, becoming mature at the given age. Thus, for an individual at a given age *t* with a given genotype and sex, one can simply draw a binomial random number to determine whether the individual postpones maturity or becomes mature at age *t*. Simulated in this manner, there will be some variability about the expression of the same genotypes across a number of individuals.

Compared to the traditional assumption that quantitative traits are coded by a large number of genes with small additive effect, the genetic coding of Atlantic salmon age at maturity is quite different. Firstly, there are much fewer genotypes; and, secondly, the sexual dimorphism of the heterozygotes is likely to promote genetic diversity in a population (Barson et al. 2015). This is because even if selection strongly favoured early or late maturity, heterozygote males or females carry alleles promoting the opposite maturity schedule. Therefore, it becomes very interesting to investigate how evolution driven by directional selection differs between the two scenarios for trait control: (1) one locus with sexually dimorphic expression; and (2) multiple loci coding the trait additively without sex-dependence.

The above question was addressed by systematic simulations of the evolution of Atlantic salmon age at maturity in the presence of fishing targeting old (large) individuals (Kuparinen and Hutchings 2017). These simulations revealed several interesting aspects

Figure 20.6 Atlantic salmon (*Salmo salar* L.). (Source: Wikipedia.)

Table 20.1 Probabilities of postponing maturity beyond threshold ages for Vgll3TOP genotype × sex, as estimated by Barson et al. (2015). These probabilities were utilised in salmon fishing simulations, in the model where the age at maturity was coded by one locus having two alleles (0 and 1). (The table is reproduced from Kuparinen and Hutchings (2017).)

Sex		Homozygote (11)	Heterozygote (10 or 01)	Homozygote (00)
Female	2 SW → 3 SW	0.754	0.949	0.983
	1 SW → 2 SW	0.101	0.404	0.665
Male	2 SW → 3 SW	0.266	0.277	0.835
	1 SW → 2 SW	0.058	0.061	0.467

about the interaction of the genetic trait architecture, the nature of trait evolution, and its feedbacks to population dynamics.

Among the simulated populations, single-locus control did not lead to large changes in average age at maturity but vastly increased the variance among the simulated populations (Figure 20.7). The pattern was opposite to multi-locus control, as in this scenario fishing led to an expected decrease in age at maturity but also reduced variability among replicated populations. In brief, multi-locus control led to the expected and predictable pattern, whereas single-locus control made the evolutionary response to selection highly uncertain. Uncertainty was further manifested within populations, as temporal variability in age at maturity was much higher in the presence of single-locus control compared to multi-locus control (Kuparinen and Hutchings 2017). Thus, one can conclude that sexually dimorphic single-locus control of age at maturity causes trait evolution to be disruptive and divergent. The former of these terms means increased variability among identical populations, while the latter describes increased temporal variability within each population. Further insights into the relative roles of single-locus control and sexual dimorphism were recently provided by Oomen al. (2020), as the same simulation design was repeated in the absence of dimorphism: the increased drift that underlies divergent and disruptive elements of the trait evolution was virtually completely subject to the single-locus control, as compared to multi-locus control.

From practical conservation and fisheries management perspectives, the nature of this trait evolution has several consequences. Firstly, an Atlantic salmon population might experience fishing-induced evolution that cannot be inferred from phenotypic data. As shown in Figure 20.7, fishing can either slightly increase or decrease the age at maturity or leave it relatively unchanged. This results from the inherent stochasticity among the evolutionary trajectories. One simply cannot predict how a trajectory develops.

This is opposite to the multi-locus scenario without sexual dimorphism, where all evolutionary trajectories are very similar and, thus, indicative of the general outcome.

The second important aspect is that Atlantic salmon populations that seem to respond differently to the same fishing pressure and selectivity need not differ in any of their other genetic or ecological aspects. The simulated populations in Figure 20.7 are identical, but the single-locus sexually dimorphic control of age at maturity drives them completely randomly in different directions when exposed to fishing. This increased uncertainty within and among populations in terms of evolutionary dynamics also reflects their ecological dynamics: Both in the absence and presence of selective fishing, the per-capita population growth rate r varies more if age at maturity is under single-locus sexually dimorphic control. This metric (r) is the major correlate of population recovery ability and, inversely, the risk of extinction (Dulvy et al. 2004; Mace et al. 2008). Thus, instability in r may increase the uncertainty of population persistence and recovery ability, which can be particularly problematic if the population is exposed to external environmental stochasticity in addition to its internal inherent stochasticity.

In summary, the case of Atlantic salmon age at maturity suggests that population dynamics and demography can be intimately linked with the genetic architecture of key life history traits and that this linkage and its ecological consequences are far from intuitive.

20.5 Conclusions and future directions

The above three examples (sections 20.4.1–20.4.3) illustrate why it is important to consider evolutionary processes as a part of demographic analyses. This is because evolution can change key demographic parameters and population dynamics, and these, in turn, can feed back to selection driving evolution. In brief, ecology and evolution are just two sides of

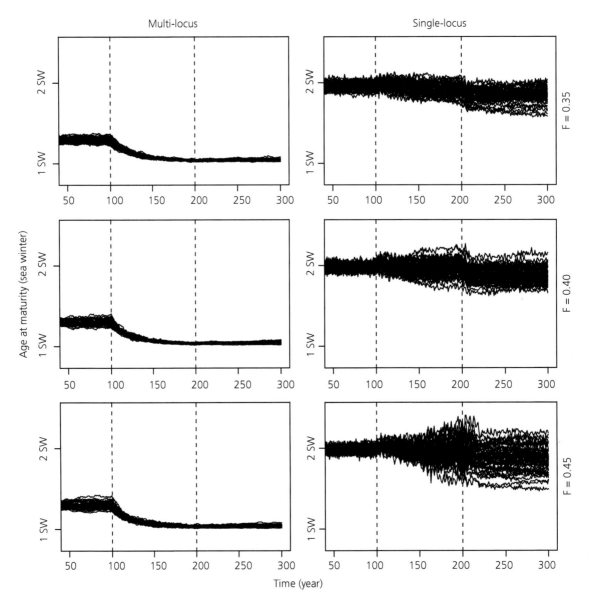

Figure 20.7 Development of age at maturity in Atlantic salmon under two scenarios for the genetic architecture of the trait (multi- and single-locus). Alternative fishing mortalities (F) are indicated on the right of each pair of graphs; the beginning and end of the fishing period are indicated by dashed vertical lines. Each individual line represents one replicated simulation line (*n* = 50). (Reproduced from Kuparinen and Hutchings (2017). Reproduced with permission from Royal Society Publishing.)

the same coin—the eco-evolutionary dynamics (see Chapter 16).

The main objective of this book is to present demographic methods that are applicable across the Tree of Life, and the present chapter did this in the context of eco-evolutionary trait dynamics. More specifically, the present chapter provided a recipe (Boxed Figure 20.1) to integrate the evolution of a phenotypic trait into

population dynamics and further demonstrated why this is important from the perspectives of ecological dynamics, conservation, and the management of natural populations.

As described in section 20.2, the method applied here is by no means the only way to couple evolution with individual-based ecological dynamics, but it has some advantages, being both flexible and allowing for heritability to change in response to selection. Likewise, the method has been applied to plants, animals, and different modes of inheritance. Even though the method has here been applied to the evolution of fitness-related traits, it can also be used to describe neutral genetic diversity and drift, as well as how those are affected by ecological dynamics and directional selection (Kuparinen et al. 2016).

Rigorous simulation-based analyses of eco-evolutionary population dynamics are still a very new field in biological research. Nonetheless, the need for such analyses becomes increasingly clear as new case studies and empirical observations demonstrate the interplay between contemporary evolution and ecological dynamics. At the same time, the need for biological forecasts is increasing: projections for how human impacts and climate change affect natural populations, particularly their resilience, recovery potential, and macroecological dynamics, require mechanistic eco-evolutionary modelling tools (Chapter 19). Although exploitation, climate change, and species range shifts are anticipated to drive rapid evolutionary changes in life histories, evidence of such change remains provisional (e.g. Gienapp et al. 2008; Audzijonyte et al. 2016). Thus, evolution seems to be an integral component of current and future ecological dynamics of species.

As illustrated by the Atlantic salmon case study (section 20.4.3), our understanding of the drivers of eco-evolutionary population dynamics will likely drastically change as new genomic discoveries accumulate (e.g. Barson et al. 2015; Kuparinen and Hutchings 2017). To date, we know little about the genomic structures coding multiple correlated traits. Nonetheless, as those are discovered, we will be in a much better position to predict the boundaries of phenotypic adaptation and how phenotypes might evolve under human-induced selection (see Chapter 1). In practice, much tighter collaboration among genomicists, evolutionary biologists, and ecologists will be needed to synthesise and integrate research findings into an eco-evolutionary framework.

On a final point, one aspect often lacking from these analyses is an ecosystem perspective. Basic demographic parameters such as natural mortality are often a product of predator–prey interactions. Likewise, fluctuations in species abundances often relate to fluctuations in prey availability (Olsen et al. 2004) and environmental stochasticity mediated by the food web (Kuparinen et al. 2019), or are generated by phenotypic changes in the focal or other species in the community (Wood et al. 2018). While ecological networks and their functioning have been studied theoretically

already for decades, practical applications to conservation, management, and eco-evolutionary predictions remain rare.

The range of demographic methods available to analyse and predict species dynamics is already wide (Chapter 5), but there are two clear directions for future development: (1) Efficient integration of genomic knowledge is necessary to better understand the nature and boundaries of phenotypic evolution as well as eco-evolutionary feedbacks to population dynamics. (2) The consideration of species dynamics as a 'dimension' of multispecies dynamics in an ecosystem is likely to illuminate mechanisms underlying fundamental demographic parameters and population dynamical properties. Capturing complex biological dynamics from genes to communities and ecosystems is undoubtedly challenging, but the potential gains are also tremendous: discoveries of the mechanisms behind the patterns, identification of the key pathways that regulate the eco-evolutionary system dynamics, and improved ability to predict the destiny species face with environmental changes and anthropogenic disturbances.

References

Audzijonyte, A., Fulton, B., Haddon, M., et al. (2016) Trends and management implications of human-influenced life-history changes in marine ectotherms. Fish and Fisheries, **17**, 1005–1028.

Barson, N.J., Aykanat, T., Hindar, K., et al. (2015) Sex-dependent dominance at a single locus maintains variation in age at maturity in Atlantic salmon. Nature, **528**, 405–408.

Charnov, E.L. (1993) Life-history invariants. Oxford University Press, Oxford.

Chevin, L.-M., Lande, R., and Mace, G.M. (2010) Adaptation, plasticity, and extinction in a changing environment: towards a predictive theory. PLOS Biology, **8**, e1000357.

Dulvy, N.K., Ellis, J.R., Goodwin, N.B., Grant, A., Reynolds, J.D., and Jennings, S. (2004) Methods of assessing extinction risk in marine fishes. Fish and Fisheries, **5**, 255–276.

Dunlop, E.S., Heino, M., and Dieckmann, U. (2009) Eco-genetic modeling of contemporary life-history evolution. Ecological Applications, **19**, 1815–1834.

Enberg, K., Jørgensen, C., Dunlop, E.S., Heino, M., and Dieckmann, U. (2009) Implications of fisheries-induced evolution for stock rebuilding and recovery. Evolutionary Applications, **2**, 394–414.

Gienapp, P., Teplitsky, C., Alho, J., Mills, J.A., and Merilä, J. (2008) Climate change and evolution: disentangling environmental and genetic responses. Molecular Ecology, **17**, 167–178.

Hardie, D.C. (2007) The evolutionary ecology of Atlantic cod (*Gadus morhua*) in Canadian Arctic lakes. PhD thesis, Dalhousie University, Halifax.

Hendry A.P. (2016) Eco-evolutionary Dynamics, Princeton University Press, Princeton.

Jensen., A.L. (1997) Origin of the relation between K and Linf and synthesis of relations among life history parameters. Canadian Journal of Fisheries and Aquatic Sciences, **54**, 987–989.

Jørgensen, C., Enberg, K., Dunlop, E.S., et al. (2007) Managing evolving fish stocks. Science, **318**, 1247–1248.

Kuparinen, A. and Hutchings, J.A. (2012) Consequences of fisheries-induced evolution for population productivity and recovery potential. Proceedings of the Royal Society B, **279**, 2571–2579.

Kuparinen, A. and Hutchings, J.A. (2017) Genetic architecture of age at maturity can generate divergent and disruptive harvest-induced evolution. Philosophical Transactions of the Royal Society B, **372**, 20160035.

Kuparinen, A., Savolainen, O., and Schurr, F.M. (2010) Increased mortality can promote evolutionary adaptation of forest trees to climate change. Forest Ecology and Management, **259**, 1003–1008.

Kuparinen, A., Stenseth, N.C., and Hutchings, J.A. (2014) Fundamental population–productivity relationships can be modified through density-dependent feedbacks of life-history evolution. Evolutionary Applications, **7**, 1218–1225.

Kuparinen, A., Hutchings, J.A., and Waples, R. (2016) Harvest-induced evolution and effective population size. Evolutionary Applications, **9**, 658–672

Kuparinen, A., Perälä, T., Martinez, N.D.M., and Valdovinos, F.S. (2019) Environmentally-induced noise dampens and reddens with increasing trophic level in a complex food web. Oikos, **128**, 608–620.

Law, R. (2007) Fisheries-induced evolution: present status and future directions. Marine Ecology Progress Series, **335**, 271–277.

Mace, G.M., Collar, N.J., Gaston, K.J., et al. (2008) Quantification of extinction risk: IUCN's system for classifying threatened species. Conservation Biology, **22**, 1424–1442.

Morrissey, M.B., Kruuk, L.E.B., and Wilson, A.J. (2010) The danger of applying the Breeder's equation in observational studies of natural populations. Journal of Evolutionary Biology, **23**, 2277–2288.

Mousseau, T.A. and Roff, D.A. (1987) Natural selection and the heritability of fitness components. Heredity, **59**,181–197.

Notivol, E., Garcia-Gil, M.R., Alia, R., and Savolainen, O. (2007) Genetic variation of growth rhythm traits in the limits of a latitudinal cline in Scots pine. Canadian Journal of Forest Research, **37**, 540–551.

Olsen, E.M., Heino, M. Lilly, G.R., et al. (2004) Maturation trends indicative of rapid evolution preceded the collapse of northern cod. Nature, **428**, 932–935.

Oomen, R.A., Kuparinen, A., and Hutchings, J.A. (2020) Consequences of single-locus and tightly linked genomic architectures for evolutionary responses to environmental change. bioRxiv, BIORXIV/2020/928770.

Reznick, D.N., Shaw, F.H., Rodd, F.H., and Shaw, R.G. (1997) Evaluation of the rate of evolution in natural populations of guppies (*Poecilia reticulata*). Science, **275**, 1934–1937.

Roff, D.A. (1992) Life history evolution. Sinauer Associates, Sunderland, MA.

Savolainen, O., Bokma, F., Garcia-Gil, R., Komulainen, P., and Repo, T. (2004) Genetic variation in cessation of growth and frost hardiness and consequences for adaptation of *Pinus sylvestris* to climatic changes. Forest Ecology and Management, **197**, 79–89.

Savolainen, O., Pyhäjärvi, T., and Knürr, T. (2007) Gene flow and local adaptation in trees. Annals of Reviews in Ecolology, Evolution and Systematic, **38**, 595–619.

Von Bertalanffy, L. (1938) A quantitative theory of organic growth (inquiries on growth laws II). Human Biology, **10**, 181–213.

Wood, Z.T., Palkovacs E.P., and Kinnison, M.T. (2018) Eco-evolutionary feedbacks from non-target species influence harvest yield and sustainability. Scientific Reports, **8**, 6389.

World Wildlife Fund (2001) The status of wild Atlantic salmon: a river by river assessment. [http://d2ouvy59p0-dg6k.cloudfront.net/downloads/salmon2.pdf; accessed 19 August 2019]

CHAPTER 21

Demographic processes in socially structured populations

Maria Paniw, Gabriele Cozzi, Stefan Sommer, and Arpat Ozgul

21.1 Introduction

In animals, a staggering diversity of social organi-sations has evolved, ranging from solitary living (Kap-peler and van Schaik 2002) to group decision-making in eusocial species (Wilson and Hölldobler 2005). Among these life history strategies, sociality, where individ-uals form cohesive groups that are typically struc-tured by kinship, has evolved in numerous clades (Alexander 1974; Clutton-Brock 2002). Such group living confers direct and indirect fitness benefits to group members (Hamilton 1964; Clutton-Brock 2002; Silk 2007). Moreover, population ecologists increasing-ly recognise that accounting for social complexity in population models can improve population viability estimates (Walters et al. 2002) and inference from pop-ulation genetics (Parreira and Chikhi 2015) and there-fore lead to better conservation management decisions (Courchamp and Macdonald 2001; Berec et al. 2007).

In socially structured groups, vital rates of indi-viduals are affected by social rank, group size, and group structure (Courchamp and Macdonald 2001; Clutton-Brock 2016; Angulo et al. 2018). Depending on the degree of sociality, reproduction within a group may be monopolised by a dominant pair (Weldon McNutt and Silk 2008; Armitage 2014; Clutton-Brock and Manser 2016), and their offspring are then co-reared by nonreproducing group members (Cour-champ et al. 1999; Griesser et al. 2017), which gain indirect fitness benefits by raising close kin (Hamilton 1964). As a consequence of such cooperative breeding, reproductive success increases with group size (Cour-champ and Macdonald 2001; Clutton-Brock 2002). Group size also regulates the survival of individuals

depending on their social status, with dominants typi-cally benefitting from larger groups (Ozgul et al. 2014). In addition, group cohesion is highest at intermediate group sizes. At such sizes, the benefits for helpers of staying at home despite reproductive suppression outweigh the costs of dispersing and forming a new group, where the chances of reproducing are higher (Packer et al. 2005; Bonte et al. 2012; Hoogland 2013; Markham et al. 2015). These examples illustrate that social structure plays an important role in regulating vital rates. To accurately infer group- and population-level processes in social species, social structure therefore needs to be included in the respective demo-graphic models (Courchamp et al. 1999; Bateman et al. 2013); this is particularly true when addressing environmental change questions, as changes in group structure and traits of group members may precede population collapse (Paniw et al. 2019).

One straightforward way to account for direct demographic effects of social structure on population dynamics (e.g. differential responses in survival and reproduction depending on social state) is to parame-terise matrix population models (MPMs) (Chapter 9). MPMs have been used in numerous systems to study differences in demography among social stages (Figure 21.1). For instance, Brault and Caswell (1993) investigated population dynamics of killer whales (*Orcinus orca*) accounting for the structure of differ-ent pods. MPMs that include reproductive dominant and subordinate stages have been developed for the group-living Siberian jay (*Perisoreus infaustus*) (Layton-Matthews et al. 2018); and they have been used to derive optimal group size in the obligate cooperatively

Maria Paniw et al., *Demographic processes in socially structured populations*. In: *Demographic Methods Across the Tree of Life.*
Edited by Roberto Salguero-Gómez and Marlène Gamelon, Oxford University Press.
© Oxford University Press (2021). DOI: 10.1093/oso/9780198838609.003.0021

(a)   (b)   (c)

Figure 21.1 Three examples of social animals for which structured population models are available: (a) the yellow-bellied marmot (*Marmota flaviventris*), a facultative cooperative breeder; (b) the Kalahari meerkat (*Suricata suricatta*), an obligate cooperative breeder; and (c) the Siberian jay (*Perisoreus infaustus*), in which breeding within a group is monopolised by a dominant pair but helpers do not occur. (credits: Arpat Ozgul (a, b) and Michael Griesser (c).)

breeding meerkat (*Suricata suricatta*) (Bateman et al. 2018).

In addition to direct demographic effects of social structure, individual phenotypic traits can mediate these effects. For instance, body size can play an important role in acquiring and maintaining social status, where the largest individuals gain reproductive dominance (Hodge et al. 2008). Integral projection models (IPMs) (Chapter 10) can account for age- or stage-specific dynamics of phenotypic traits such as body size (Ellner et al. 2016). In social species, IPMs have been developed for marmots (Ozgul et al. 2010; Maldonado-Chaparro et al. 2018) and meerkats (Paniw et al. 2019). In both species, the IPMs have shown the critical role of body mass in mitigating effects of environmental factors on vital rates.

Despite their widespread use, both MPMs and IPMs may be a poor modelling choice for socially structured populations (Zeigler and Walters 2014). This is because these two modelling frameworks cannot easily account for demographic stochasticity inherent to small groups and complex behavioural mechanisms generating group structure. For instance, the acquisition of a high social status such as breeding dominance, is typically reserved for a specific individual within a group, such as the oldest or strongest female (Armitage 2014; Ozgul et al. 2014) and therefore constitutes a deterministic process. MPMs and IPMs, however, define transitions among social stages as probabilistic processes, assigning a likelihood to gain reproductive dominance based on, for example, age, mass, group size, or environmental covariates (e.g. Paniw et al. 2019). It is possible to incorporate demographic stochasticity into MPMs and IPMs, but this can jeopardise the analytical tractability of these models (Ovaskainen and Meerson 2010). Modelling deterministic transitions among social stages, in contrast,

requires individual-based models (IBMs) of group and population dynamics. Such models have been developed for meerkats (Ozgul et al. 2014) and red-cockaded woodpeckers (Zeigler and Walters 2014).

Despite increasing calls for an individual-based ecology (Stillman et al. 2015), the scarcity of mechanistic modelling frameworks for socially structured populations partly results from the type of information required to parameterise IBMs, that is, long-term, individual-based data spanning several family groups. Furthermore, these models tend to require increased computational capacity and lack analytical tractability because population-level metrics can be obtained through simulations only. Thus, in many cases, ecologists prefer simplified structured population models such as MPMs. The crucial question is then: To what extent does simplification of mechanistic processes governing social structure affect key inferences of group and population dynamics?

In this chapter, we try to provide an answer by testing population models of increasing complexity. To this end, we test population models that can be applied to socially structured, individual-based data sets and quantify consequences of model simplification.

In the following sections, we implement several approaches, increasing in complexity, to incorporate social structure into population models and assess how much model complexity is required to adequately capture the dynamics of a socially structured population. To parameterise all models, we use simulated individual-based data of a hypothetical population of a cooperative breeder, for example the African wild dog, *Lycaon pictus*, or Kalahari meerkat, *Suricata suricatta* (Figure 21.2). From these data, we first develop a simple stage-structured MPM, in which stages correspond to social status, whereby group dynamics, such as the processes determining the extinction and formation of

(a)

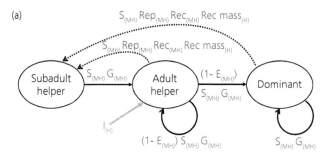

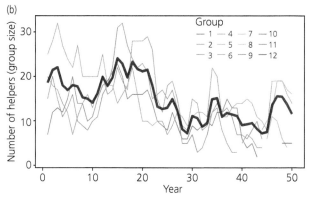

Figure 21.2 Three-stage life cycle (a) of a socially structured population simulated (b) and analysed in this chapter. Life cycle (social) stages are represented by circles. Transitions among stages and reproductive output are depicted by black solid and dashed arrows, respectively; they depend on stage-specific vital rates: that is, survival (S), emigration (E), reproduction (Rep), and recruitment (Rec) (blue); and trait transitions, that is, growth (G) and recruit mass (Rec mass) (red). Vital rates and trait transitions (i.e. demographic processes), in turn, are functions of mass (M) and number of helpers (H). Adult helpers can also immigrate (I) into the population when the number of helpers falls below a threshold. Using the life cycle, individual stage transitions and fates are simulated for 50 years, resulting in a data set (b) consisting of 12 groups of different sizes (average group size is plotted in black).

new groups, are ignored. We then expand this basic MPM into an IPM by including individual traits. In doing so, we first ignore and then incorporate group dynamics. Lastly, we attempt to reconstruct group dynamics with a flexible IBM approach. We compare the fit of each modelling framework to the observed data using the following metrics: instantaneous population growth rate, relative abundances of social stages, stage-specific mean masses, and average group sizes. These metrics are typically used to assess goodness-of-fit of individual-based or structured population models of social animals (e.g. Ozgul et al. 2014; Paniw et al. 2020).

We perform all analyses in *R* and provide the corresponding scripts (in the Appendix; (please go to www.oup.com/companion/SalgueroGamelonDM)). These scripts represent tools for researchers working with socially structured populations to both simulate data that best describe their study system and assess the level of model complexity needed to capture the dynamics of the system.

21.2 A simulated, socially structured population

The simulated, individual-based data set used to develop socially structured population models loosely follows the life cycle structure of meerkats presented

in Ozgul et al. (2014) (see also Paniw et al. 2019). Our adaptation (Figure 21.2a) is representative of life cycle dynamics in numerous other cooperative breeders such as African wild dogs (Courchamp and Macdonald 2001) and Arabian babblers (Ridley 2007). Our simulated population consists of females in different groups. In each group, one dominant monopolises most reproduction (although this reproductive dominance is relaxed when groups are small). The other group members then help co-rear the offspring. To gain reproductive opportunities, adult helpers can either 'queue' for dominance (i.e. wait for the death of the dominant to take

her position) or emigrate from their natal group and form new groups. Our simulation represents these social dynamics and is detailed in the *R* script gen-erate_social_data.R. The script is designed to simulate fates and body masses of individuals, while taking into account the effects of individual (life history stage) and social factors (group size) on vital rates (survival, reproduction, recruitment, and emigration).

The simulation follows individual females in discrete time-steps of 1 year over 50 years. It comprises three hierarchical levels: individual, social group, and population. All females are characterised by the following state variables: age; mass (i.e. log of body mass, in grams); group identity and associated group size (i.e. number of subadult and adult helpers); social stage (i.e. dominant, subadult helper, and adult helper); reproductive or pregnancy stage (pregnant or nonpregnant); and recruitment success (number of offspring that survive to become subadults). Only adults can reproduce, and adult subordinate helpers can emigrate from a given group (e.g. McNutt 1996; Maag et al. 2018). The vital rates of survival, emigration, reproduction, and recruitment determine the fate of individuals, while growth and recruit-mass functions determine changes in body mass (Figure 21.2a). We base the parameterisation of the vital rates and trait transitions (collectively referred to as *demographic processes* henceforth) on published models of meerkat demographic processes (Bateman et al. 2013; Ozgul et al. 2014; Paniw et al. 2019). In these models, all demographic processes increase with mass. Helper survival and dominant recruitment also increase with group size, but emigration is higher and adult helper reproduction and recruitment are lower when groups are larger. Therefore, an intermediate group size is favoured in long-term projections of group dynamics (Figure 21.2b).

The simulation determines group extinction at the beginning of each time-step. A group is considered extinct if it consists of maximally one dominant female and one adult helper. If a group goes extinct, the available pool of individuals that emigrate at a given time-step form a new group. More specifically, the largest number of emigrants from one of the four resident groups settles as a new group. All other emigrants disperse and leave the population at the end of the time-step. If no local emigrants are available at a given time-step, immigrants from outside the focal population form a new group. These immigrants consist of a dominant (i.e. founding) female and a random sample of six to eight adult helpers (Maag et al., pers. observation).

Individuals that survive from one time-step to the next change mass. Adult helpers can emigrate or stay in the group, and those that stay can become dominant or stay helpers. Surviving adult helpers and dominants can get pregnant, and those that do wean ≥ 0 subadult helpers with a certain body mass. Body mass and number of helpers in the group differentially affect each of these processes at each time-step. Due to the alloparental care of offspring, their body mass is not affected by the mass of their mother in the simulation. When all individuals have completed the demographic processes, their age is updated, and subadult helpers automatically become adult helpers; surviving dominants and adult helpers remain in their perspective stages unless a dominant dies, in which case the heaviest adult helper >2 years of age takes its place (Hodge et al. 2008).

We consider two levels of variation in demographic processes. First, within groups, the simulation implements demographic stochasticity by treating survival, emigration, and pregnancy as binomial probability events, and recruitment as a Poisson count event. Second, to account for environmental stochasticity, parameters that define the demographic-process models are sampled from 10-year-specific values, akin to random-year effects on demographic-process averages (Paniw et al. 2019). These year-specific values assume a spatially homogeneous environment and affect all groups simultaneously, that is, a specific random-effect structure is chosen at the beginning of each year in the simulation and is incorporated into all demographic-process models for all groups simultaneously. The simulated data result from one reproducible iteration of demographic and environmental stochasticity, that is, by setting a fixed random seed in *R* (see gen-erate_social_data.R). This data set follows 12 groups across the 50 years, with average group sizes of 14 helpers (Figure 21.2b). We later implement several stochastic iterations in the IBM in order to reconstruct group dynamics in the simulated data set (see 21.6).

We initialise the simulation of the individual-based data with four groups of different sizes; these groups constitute the starting population (Figure 21.2b). We assign masses to individuals within these initial groups based on their social stage. Demographic processes are then calculated for all groups simultaneously, upon which the simulation advances by one time-step. Demographic processes in existing and new groups are equivalent, with the exception that individuals do not emigrate from a new group in the year of founding.

21.3 A socially structured matrix population model

To assess how well the dynamics of a socially structured population can be captured by a simple structured population model, we parameterise an MPM using the simulated data set described in 21.2. Details on the parameterisation can be found in the *R* script social_MPM.R. The MPM consists of three stages—subadult helpers, adult helpers, and dominants—with transitions among stages, reproduction, and recruitment determined by stage-specific demographic processes (Figure 21.2a). We parameterise the demographic processes using generalised linear mixed models (GLMMs). We ignore group dynamics in the MPM but assume that population size can potentially affect demographic processes (Bateman et al. 2013; Paniw et al. 2019). We therefore model each demographic process as a function of population size, a fixed effect. We also include year as a random effect on the mean. We use Akaike's information criterion (AIC) (Burnham et al. 2011) to assess whether including population size provides a better model fit compared to the null model including only the random-year effect. We build an MPM for each of the 49 yearly transitions available from the data. Starting with the population vector in year 1, we project population dynamics until year 50 by multiplying an MPM, which describes demographic transitions from time t to $t+1$, by a population vector at time t to obtain a vector at $t+1$. We incorporate immigration by adding the number of observed immigrant adult helpers and dominants at time t to the modelled population vector at time t. From the projections, we obtain yearly instantaneous population growth rates and relative abundances of different social stages in the population.

21.4 A socially structured integral projection model

Next, we expand the MPM approach by explicitly considering the effect of a continuous phenotypic trait, that is, body mass, on demographic processes and population dynamics. For each of the three social stages that characterise our population, we test the effects of mass, population size, and their interaction on all demographic processes using GLMMs. We then parameterise a stage-mass classified IPM for each of the 49 annual demographic transitions. For each stage a, the IPM tracks the number of individuals (n_a) in 100 discretised mass classes at time t. Details on the parameterisation can be found in the R script `social_IPM_popSize.R`. The IPM approach allows to project trait and population dynamics simultaneously (Ellner et al. 2016). These projections follow the methods described in the MPM section (21.3). In the yearly IPMs, immigration is accounted for by adding the observed number of immigrant adult helpers and dominants in a given mass class at time t to the projected population vector at time t. In addition to yearly instantaneous population growth rates and relative abundances of different social stages in the population, we obtain the distribution of stage-specific masses.

21.5 Explicitly accounting for group sizes in a population model

As population size in our simulated data is correlated with average group size (see `generate_social_data.R`), the socially structured MPM and IPM

indirectly consider group dynamics by incorporating population size into demographic-process models and projecting population size through time. However, while in our simulated data, the interaction between mass and group size (i.e. number of subadult and adult helpers) determines adult helper survival, which is a critical demographic process (Paniw et al. 2019), modelling adult helper survival as a function of population size and mass does not capture this interactive effect (see the GLMMs in `social_IPM_popSize.R`). We therefore attempt to account for group dynamics more directly by refitting demographic-process GLMMs using group size instead of population size as the predictor variable. We then model average group size in the population as a linear function of population size and parameterise and project IPMs inferring average group size from population size (i.e. the sum of the population-vector elements) at each time-step of the projection. Details on the parameterisation can be found in the R script `social_IPM_groupSize.R`. As before, we obtain yearly instantaneous population growth rates, relative abundances of different social stages in the population, and the distribution of stage-specific masses.

Parameterising IPMs with average group sizes also allows for determining optimal group size analytically. At the optimal group size, average population fitness, that is, the long-term asymptotic population growth rate λ, is the highest. To obtain an estimate of optimal group size in our population, we calculate λ by performing an additional analysis (`optimal_groupSize.R`), in which we parameterise IPMs with group sizes ranging from 2 to 30 individuals, using averaged demographic processes across years.

21.6 An individual-based model as a mechanistic approach to modelling sociality

The most mechanistic approach to model the dynamics of a socially structured population is to parameterise an IBM, because IBMs account for various levels of stochasticity and can impose deterministic transitions among social stages (here, gaining dominance in a group). We implement an IBM for our simulated data, which essentially replicates the protocol of simulating life history data outlined in 21.2 (see `social_IBM.R`). However, unlike our simulated data set, which was generated using one random iteration of demographic and environmental stochasticity, we use 100 random iterations of the IBM to fully account for stochasticity. As in the structured models described in 21.3 and

21.4, we use parameters of the most parsimonious demographic-process models to parameterise the IBM.

21.7 Does increased model complexity improve inferences of population dynamics in a socially structured population?

For most metrics, the three modelling frameworks that we use to parameterise and project the dynamics of a socially structured population generate similar estimates (Figure 21.3). In particular, increasing model complexity does not reduce relative differences in projected and observed instantaneous population growth rates (Figure 21.3a). Model complexity also does little to reduce such differences in number of adult helpers and their mass. Here, the mechanistic model with the highest complexity (i.e. the IBM) only performs best (averaged across the 100 random iterations) for the first years of the projection (Figure 21.3b,c). In subsequent years, as projections diverge from observations, differences between the modelling frameworks decrease. With respect to average masses, the IPM based on population size slightly outperforms the IPM based on group size in projecting masses of adult helpers, but differences between IPM parameterisations do not translate into differences in projected average masses of subadult helpers and dominants (Figure 21.3c). The MPM performs worse than the IPMs in projecting the numbers of adult helpers (with the IPM parameterised with average group size outperforming the IPM parameterised with population size) (Figure 21.3b). However, for most years, the MPM actually does a better job than the IPMs in projecting the number of subadult helpers and dominants (Figure 21.3b). One metric in which the IBM clearly outperforms all other models is in its accuracy to project the number of dominants in the population, which, due to the deterministic nature of parameterising transitions to the dominant stage, is always equivalent to the number of groups (i.e. four).

It is important to emphasise that, although the average results of the IBM do not differ substantially from the results of the MPM and IPMs, the IBM captures the demographic and environmental stochasticity embedded in the demographic data (Figure 21.4). That is, observed dynamics are within the confidence interval of the 100 random iterations of the IBM (Figure 21.4; see also plots in `social_IBM.R`). Most variation among the iterations is seen in the number of adult helpers, which reflects the random sampling

of new group sizes when a group goes extinct in the population.

The structured modelling frameworks (here the IPMs), in contrast to the IBM, allow for determining optimal group size analytically (Figure 21.5). Average population fitness (i.e. the asymptotic population growth rate) is highest at a group size of nine helpers and declines sharply as group size increases (Figure 21.5). This sharp decline is in line with previous investigations of optimal group size in meerkats (Bateman et al. 2018) and other cooperative breeders (e.g. Markham et al. 2015). The analytically obtained optimal group size is lower than the average group size (13 individuals) in the observed data. One reason for this discrepancy in optimal group size is that the observed data are based on short-term simulations of 50 years and ignore asymptotic long-term dynamics. Another reason is that such asymptotic dynamics will likely never be reached due to the dynamic nature of changes in group size (Figure 21.2b). That is, calculating λ at different group sizes assumes that a given group size remains stable through time, which does not occur in the system.

21.8 Implications for life history models of socially structured populations

Perhaps due to perceived similarities to human societies, the behaviour and life histories of social animals have been studied extensively (Emery et al. 2007; Clutton-Brock 2016). Accounting for sociality, however, introduces additional levels of complexity into population models, questioning the relative usefulness of different modelling frameworks for investigating social species. Here we show that structured population models (MPMs and IPMs) can be a reasonable choice for projecting the dynamics of socially structured populations even though they don't consider deterministic changes in social stages and demographic stochasticity. Most importantly, the MPM did not perform worse than the IPMs in quantifying most metrics that inform on the goodness-of-fit of population models, suggesting that relatively simpler structured models do not necessarily compromise the accuracy of results.

We recommend using an IBM if assessing total group number in a population is crucial for an analysis. The most useful attribute of highly mechanistic models, such as IBMs, in projecting socially structured populations is their ability to quantify deterministic transitions among social stages in a population (Ozgul et al. 2014). In our simulations,

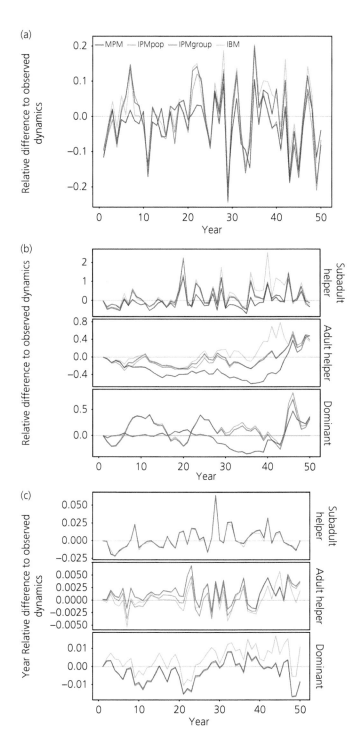

Figure 21.3 The three modelling frameworks used to describe the dynamics of a socially structured population: MPM; IPM accounting for population size (pop) or average group size (group); and IBM. The three models perform similarly in quantifying (a) instantaneous population growth rates, (b) relative abundance, and (c) average mass in each of the three social stages modelled. For the IBM, results averaged across 100 random iterations of the model are shown. Note that the relative difference in the number of dominants is 0 for the IBM.

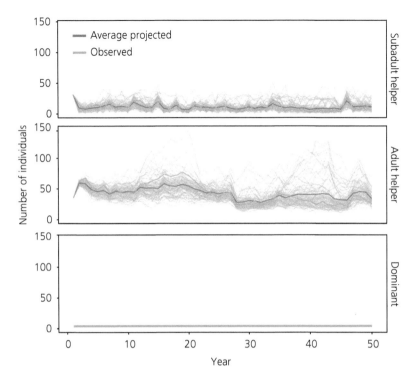

Figure 21.4 The IBM accurately simulates the stochastic dynamics of a socially structured population projected for 50 years. The grey lines show projections under each of the 100 random iterations of the IBM; the blue line indicates the average across the 100 iterations; numbers of projected and observed dominant individuals coincide.

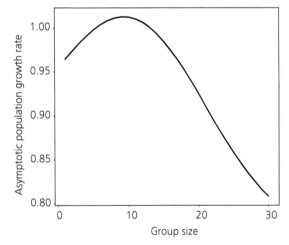

Figure 21.5 Asymptotic population growth rate as a function of group size. Population growth (i.e. fitness) peaks at a group size of nine individuals.

for example, the IBM is the only model that accurately projects the number of dominants and, as a consequence of model specifications, the number of social groups in the population. This accuracy is unsurprising as we parameterised the IBM to reflect the observed data in which maximally four social groups co-exist in the population at any given time. In addition to allowing deterministic transitions, an

IBM can account for various levels of stochasticity in a straightforward manner. This is important in systems in which demographic stochasticity may strongly affect modelling outcomes (Zeigler and Walters 2014). Despite such advantages, mechanistic models require detailed information on the processes that determine social interactions among individuals and group formation (e.g. when immigration occurs in our example). Due to these challenges, simpler models are often preferred (Ellner et al. 2002).

Among the structured population models that we use to parameterise and project the dynamics of a socially structured population, the MPM, which is the simplest in parameterisation, outperforms the IPM (and even the average results of the IBM) in the assessment of instantaneous population growth rates. This occurs because the MPM estimates the number of dominants and subadult helpers in the population more accurately in most years. Because simulated numbers of subadults and dominants are lower than simulated numbers of helpers, demographic stochasticity affects the former two life stages more strongly and, as a consequence, increases the uncertainty in predictions of demographic processes (Petchey et al. 2015). Moreover, models that project mass change in addition to stage change propagate additional uncertainty, resulting from assessments of mass change, to estimates of population dynamics. Thus, simple MPMs may prove useful for studying socially structured populations, if projecting mass change is not a critical component of the study.

If projecting mass change is a critical component of the study, IPMs can be a valuable alternative to IBMs. However, although we obtain similar results from IPMs parameterised with either population size or average group size, we caution against using these two metrics interchangeably. Even under a strong correlation of group and population size, demographic processes do not necessarily respond equally to these two metrics. In our example, while the probability to reproduce of heavy dominants is negatively affected by population size, this interaction effect is not significant when reproduction is modelled as a function of group size. In addition, when simulating the observed data, we consider the effect of group size on demographic processes but not the effects of population size. For instance, these two metrics have opposite effects on meerkat emigration (Bateman et al. 2013), since females emigrate more readily at large group sizes, when within-group competition is high, but not at high population densities, when between-group competition is high (Paniw et al. 2019).

The three modelling frameworks that we describe in this chapter represent commonly used approaches for socially structured populations. This is, however, no exhaustive list, and alternative models that can address some of the challenges associated with modelling sociality have recently been developed (Plard et al. 2019; Shizuka and Johnson 2019). For instance, integrated population models (IPM2), particularly if fitted in a Bayesian framework, can incorporate demographic stochasticity and account for error propagation due to parameter uncertainty in population projections (Kery and Schaub 2011; Plard et al. 2019). Furthermore, individual-based models do not need to be constructed 'from scratch' but can be implemented in various programmes (e.g. Vortex) (Zeigler and Walters 2014), which can facilitate their inference and perturbations. Therefore, we encourage readers to use the *R* code that accompanies this chapter as a possible foundation for analyses to be expanded and integrated into novel applications.

References

Alexander, R.D., 1974. The evolution of social behavior. *Annual Review of Ecology and Systematics*, **5**, 325–383.

Angulo, E. et al., 2018. Allee effects in social species. *Journal of Animal Ecology*, **87**, 47–58.

Armitage, K.B., 2014. *Marmot Biology: Sociality, Individual Fitness, and Population Dynamics*, Cambridge University Press.

Bateman, A.W. et al., 2018. Matrix models of hierarchical demography: linking group- and population-level dynamics in cooperative breeders. *The American Naturalist*, **192**, 188–203.

Bateman, A.W. et al., 2013. Social structure mediates environmental effects on group size in an obligate cooperative breeder, *Suricata suricatta*. *Ecology*, **94**, 587–597.

Berec, L., Angulo, E., and Courchamp, F., 2007. Multiple allee effects and population management. *Trends in Ecology and Evolution*, **22**, 185–191.

Bonte, D. et al., 2012. Costs of dispersal. *Biological Reviews of the Cambridge Philosophical Society*, **87**, 290–312.

Brault, S. and Caswell, H., 1993. Pod-specific demography of killer whales (*Orcinus orca*). *Ecology*, **74**, 1444–1454.

Burnham, K.P., Anderson, D.R., and Huyvaert, K.P., 2011. AIC model selection and multimodel inference in behavioral ecology: some background, observations, and comparisons. *Behavioral Ecology and Sociobiology*, **65**, 23–35.

Clutton-Brock, T., 2002. Breeding together: kin selection and mutualism in cooperative vertebrates. *Science*, **296**, 69–72.

Clutton-Brock, T., 2016. *Mammal Societies*, John Wiley & Sons.

Clutton-Brock, T.H. and Manser, M., 2016. Meerkats: cooperative breeding in the Kalahari. In: W.D. Koenig and J.L. Dickinson (eds) *Cooperative Breeding in Vertebrates: Studies of Ecology, Evolution, and Behavior*, 294–317, Cambridge University Press.

Courchamp, F. and Macdonald, D.W., 2001. Crucial importance of pack size in the African wild dog *Lycaon pictus*. *Animal Conservation*, **4**, 169–174.

Courchamp, F., Grenfell, B., and Clutton-Brock, T., 1999. Population dynamics of obligate cooperators. *Proceedings of the Royal Society of London B: Biological Sciences*, **266**, 557–563.

Ellner, S.P. et al., 2002. Precision of population viability analysis. *Conservation Biology*, **16**, 258–261.

Ellner, S.P., Childs, D.Z., and Rees, M., 2016. *Data-Driven Modelling of Structured Populations: A Practical Guide to the Integral Projection Model*, Springer.

Emery, N.J. et al., 2007. Cognitive adaptations of social bonding in birds. *Philosophical Transactions of the Royal Society of London. Series B, Biological Sciences*, **362**, 489–505.

Griesser, M. et al., 2017. Family living sets the stage for cooperative breeding and ecological resilience in birds. *PLOS Biology*, **15**, e2000483.

Hamilton, W.D., 1964. The genetical evolution of social behaviour. I. *Journal of Theoretical Biology*, **7**, 1–16.

Hodge, S.J. et al., 2008. Determinants of reproductive success in dominant female meerkats. *Journal of Animal Ecology*, **77**, 92–102.

Hoogland, J.L., 2013. Prairie dogs disperse when all close kin have disappeared. *Science*, **339**, 1205–1207.

Kappeler, P.M. and Van Schaik, C.P., 2002. Evolution of primate social systems. *International Journal of Primatology*, **23**, 707–740.

Kery, M. and Schaub, M., 2011. *Bayesian Population Analysis Using WinBUGS: A Hierarchical Perspective*, Academic Press.

Layton-Matthews, K., Ozgul, A., and Griesser, M., 2018. The interacting effects of forestry and climate change on the demography of a group-living bird population. *Oecologia*, **186**, 907–918.

Maag, N. et al., 2018. Density-dependent dispersal strategies in a cooperative breeder. *Ecology*, **99**, 1932–1941.

Maldonado-Chaparro, A.A. et al., 2018. Transient LTRE analysis reveals the demographic and trait-mediated processes that buffer population growth. *Ecology Letters*, **23**, 1353.

Markham, A.C. et al., 2015. Optimal group size in a highly social mammal. Proceedings of the National Academy of Sciences of the United States of America, **112**, 14882–14887.

McNutt, J.W. 1996. Sex-biased dispersal in African wild dogs, Lycaon pictus. *Animal Behaviour*, **52**, 1067–1077.

Ovaskainen, O. and Meerson, B., 2010. Stochastic models of population extinction. *Trends in Ecology and Evolution*, **25**, 643–652.

Ozgul, A. et al., 2010. Coupled dynamics of body mass and population growth in response to environmental change. *Nature*, **466**, 482–485.

Ozgul, A. et al., 2014. Linking body mass and group dynamics in an obligate cooperative breeder. *Journal of Animal Ecology*, **83**, 1357–1366.

Packer, C. et al., 2005. Ecological change, group territoriality, and population dynamics in Serengeti lions. *Science*, **307**, 390–393.

Paniw, M. et al., 2019. Life history responses of meerkats to seasonal changes in extreme environments. *Science*, **363**, 631–635.

Paniw, M. et al., 2020. Assessing seasonal demographic covariation to understand environmental-change impacts on a hibernating mammal, *Ecology Letters*, **23**, 588–597.

Parreira, B.R. and Chikhi, L., 2015. On some genetic consequences of social structure, mating systems, dispersal, and sampling. *Proceedings of the National Academy of Sciences of the United States of America*, **112**, E3318–E3326.

Petchey, O.L. et al., 2015. The ecological forecast horizon, and examples of its uses and determinants. *Ecology Letters*, **18**, 597–611.

Plard, F. et al., 2019. IPM 2: toward better understanding and forecasting of population dynamics. *Ecological Monographs*, **89**, e01364.

Ridley, A.R. 2007. Factors affecting offspring survival and development in a cooperative bird: social, maternal and environmental effects. *Journal of Animal Ecology*, **76**, 750–760.

Shizuka, D. and Johnson, A.E., 2019. How demographic processes shape animal social networks. *Behavioral Ecology*. http://dx.doi.org/10.1093/beheco/arz083.

Silk, J.B., 2007. The adaptive value of sociality in mammalian groups. *Philosophical Transactions of the Royal Society of London. Series B, Biological Sciences*, **362**, 539–559.

Stillman, R.A. et al., 2015. Making predictions in a changing world: the benefits of individual-based ecology. *Bioscience*, **65**, 140–150.

Walters, J.R., Crowder, L.B., and Priddy, J.A., 2002. Population viability analysis for red-cockaded woodpeckers using an individual-based model. *Ecological Applications*, **12**, 249–260.

Weldon Mcnutt, J. and Silk, J.B., 2008. Pup production, sex ratios, and survivorship in African wild dogs, *Lycaon pictus*. *Behavioral Ecology and Sociobiology*, **62**, 1061–1067.

Wilson, E.O. and Hölldobler, B., 2005. Eusociality: origin and consequences. Proceedings of the National Academy of Sciences of the United States of America, **102**, 13367–13371.

Zeigler, S.L. and Walters, J.R., 2014. Population models for social species: lessons learned from models of red-cockaded woodpeckers (*Picoides borealis*). *Ecological Applications*, **24**, 2144–2154.

CHAPTER 22

Demographic methods in epidemiology

Petra Klepac and C. Jessica E. Metcalf

22.1 Why demography is important to epidemic dynamics

Bernoulli's work on smallpox in the 1760s perhaps represents the earliest infectious disease model (Dietz and Heesterbeek 2002; Bernoulli and Blower 2004). Bernoulli estimated the gain in life expectancy that would emerge from smallpox elimination using life tables constructed by Edmond Halley (who also discovered Halley's comet). This early work reveals one half of the inextricable link between demography and infectious disease epidemiology: Infectious disease can shape host demography directly by changing host survival or fertility. For example, smallpox mortality was estimated to be as high as 30% (Kotar and Gessler 2013), plant pathogens like the *Citrus tristeza* virus may be invariably fatal (Dawson et al. 2013), pathogens like brucellosis cause spontaneous abortion in many ungulates (Moreno 2014), and fungal pathogens transmitted by pollinators can be sterilising for many species of plants (Antonovics 2005). However, just as host demography may be shaped by infectious disease dynamics, host demography may in turn shape the spread of infections.

Many features of infectious pathogen and host life history shape this reverse dependence. In vertebrates, infection can leave hosts protected from subsequent infection by the same pathogen as a result of adaptive immunity. The time span of this protection can range from a few years (for antigenically variable pathogens, like influenza) to lifelong (for childhood infections such as measles). Focusing on the latter case (lifelong immunity), following a large outbreak, most individuals will have been infected, and thus immunised, and are consequently no longer susceptible to infection. Because

infection can only spread via contact between a susceptible and infected individual, unless new susceptible individuals are introduced into the population, the pathogen will go extinct. For immunising infections, in many settings, the main source of new susceptible individuals will be via a core demographic process, that is, births (infants may be transiently protected by maternal immunity, but this generally wanes rapidly). Birth rates thus crucially determine pathogen dynamics. As an example, in England and Wales, during the baby boom that followed World War II, measles outbreaks occurred every year. As the birth rate fell and susceptibles were replenished more slowly, dynamics switched to outbreaks occurring every other year (Earn et al. 2000; Bjørnstad et al. 2002). The magnitude of such effects diminishes as the duration of immunity declines (Morris et al. 2015), but signatures of host demographic features such as sudden increases in the birth rate on pathogen dynamics are still expected across a broad array of pathogen life histories. In species like plants that lack adaptive immunity, pathogen-linked mortality and spatial heterogeneity might also result in susceptible depletion sufficient to limit pathogen spread.

Another important feature of pathogen and host life history that shapes dependence of pathogen dynamics on host demography is the fact that both transmission (or transfer of infection between infected and susceptible individuals) and recovery can differ by demographic features. For example, in human populations, younger individuals may have higher transmission rates than older individuals because they have greater rates of contact with other individuals (Mossong et al. 2008), and transmission rates depend on both rates of contact between a susceptible and infected individual

Petra Klepac and C. Jessica E. Metcalf, *Demographic methods in epidemiology*. In: *Demographic Methods Across the Tree of Life*.
Edited by Roberto Salguero-Gómez and Marlène Gamelon, Oxford University Press.
© Oxford University Press (2021). DOI: 10.1093/oso/9780198838609.003.0022

and probabilities of transfer of the infectious agent. In this situation, populations with higher birth rates, and thus a higher density of younger individuals, might experience faster pathogen spread. Susceptibility to infection, recovery, and/or duration of shedding can also be age dependent. For example, juvenile plants in a species of alpine carnations appear far more susceptible to fungal pathogens than adults (Bruns et al. 2017). In humans, duration of shedding of respiratory syncytial virus declines with age (Munywoki et al. 2015), which might directly reflect age effects or reflect reduced shedding following a previous exposure to the virus.

Age is not the only demographic feature that can contribute to variation in transmission and recovery (or indeed growth, survival, or fertility). An individual's ontogenetic trajectory can be measured by aspects of size, such as height or weight, or discrete life history phases, such as pupal stage. These aspects may not map perfectly to an individual's age (Caswell 2001 see Chapter 9), yet they may importantly affect both demography and pathogen-relevant processes. For example, risk of infection can be shaped by size of floral display in plants as a result of insect foraging behaviour (Ferrari et al. 2006b); and nematode-linked sheep mortality is higher in lower-weight individuals (Gulland 1992). Where such effects occur, the linked demographic and epidemiological dynamics may be better captured by tracking 'stage' rather than (or as well as) 'age' within a model. All of these scenarios represent the effect of demography on disease dynamics.

Another example of an important link between infectious disease dynamics and demography is captured by the distinction made between 'density-' and 'frequency-'dependent infectious disease transmission. In the former case (density-dependent transmission), as the population grows, so does the number of individuals infected by one infectious individual, resulting in a direct dependence of pathogen dynamics on host population dynamics. This might broadly capture the dynamics of directly transmitted infections like tuberculosis (Ramsey et al. 2002). In the latter case (frequency-dependent transmission), the number of individuals infected by one infectious individual is approximately fixed. This is often thought of in the context of sexually transmitted infections (where the number of partners an individual has is not dependent on population size) or the dynamics of vector-transmitted infections (where the number of hosts visited also does not scale with population size but saturates at a certain point). It is increasingly recognised that this dichotomy is a simplification, and dynamics may be somewhere between the two extremes (Wonham et al. 2006; Smith et al. 2009). However, whatever the scale of modulation, this additional dependency of pathogen dynamics on host demography needs to be considered when modelling infectious disease dynamics.

To conclude, host demography both depends on and shapes pathogen dynamics. Host fertility shapes the introduction of new susceptible individuals into the population (Earn et al. 2000); host demographic structure may be crucial to allowing pathogen spread, as a result of the form of transmission (density or frequency dependent; McCallum et al. 2001), age- (or stage-) specific patterns of contact (Mossong et al. 2008; Klepac et al. 2009), or differences in susceptibility (Bruns et al. 2017) or recovery (Munywoki et al. 2015) across age (or stage). Further, beyond direct effects of pathogens on host survival or fertility, the burden of infection may vary as a function of age (e.g. rubella infection is of greatest concern in women of child-bearing age (Metcalf et al. 2012a); SARS-CoV-2 shows striking increases in severity with age). These dependencies require modelling strategies that account for population structure in both host and pathogen and how they affect each other. Furthermore, model structures have to be able to encompass the fact that the relevant time-scales for host and pathogen may be very different. For example, human demography unfolds on a scale of years, while the generation time (or average interlude from a primary infection to a secondary infection) of a pathogen like measles is about 2 weeks. In this chapter, following an overview of core concepts in infectious disease modelling, including a brief overview of the parallel between the concepts of R_0 in demography and infectious disease dynamics, we introduce a methodological approach for formally addressing the combined interaction between demography and infectious disease dynamics and illustrate this approach using an example. We conclude by pointing to the broad spectrum of options available for modelling the intersection between host demography and infectious disease dynamics.

22.2 Modelling the intersection between epidemiology and demography

In a basic unstructured infectious disease model, initially ignoring demography, individuals are classified only according to their epidemiological state, for example, 'Susceptible' (S), 'Infected' (I), or 'Recovered' (R)—noting that further states are possible, dependent on the biology of the focal host–pathogen system (e.g. 'Maternally immune', 'Exposed', 'Vaccinated'). Individuals' movement between these epidemiological states is tracked using tools such as differential

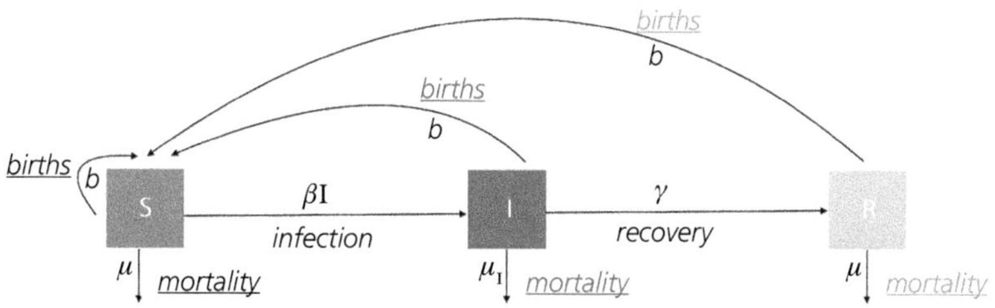

Figure 22.1 A susceptible–infected–recovered model showing transitions governed by host demography (text underlined) and infection processes (not underlined); parameters as in section 22.2. Colours reflect the fact that both infection and its aftermath (individuals in the R compartment) might also modify host demography, illustrated by the changed mortality for infected individuals μ_I.

equations. Susceptible individuals may become infected at a rate defined by the product of a transmission rate (often denoted β) and the density (or frequency; see above) of infected individuals; infected individuals recover (at a rate often denoted γ such that the average duration of the infectious period is equal to $1/\gamma$); and recovered individuals may lose immunity and return to the susceptible state. This may be represented as:

$$\frac{dS}{dt} = -\beta SI$$
$$\frac{dI}{dt} = \beta SI - \gamma I$$
$$\frac{dR}{dt} = \gamma I,$$

which is the classic Kermack and McKendrick (1927) compartmental model with constant rates. It is simple to extend this so-called closed population framing to encompass core demographic processes, such that individuals may enter or leave the population. All epidemiological classes (S, I, and R) are subject to mortality (e.g. at rate μ), which may be higher for infected individuals (e.g. μ_I). Births generally enter the susceptible class (or may briefly enter a maternally immune class before this maternal immunity wanes and they become susceptible). This could be dependent on the population size:[1] for example, at a rate $b \times$ the population size (shown below), or a constant rate, depending on assumptions about fertility:

$$\frac{dS}{dt} = b(S + I + R) - \beta SI - \mu S$$
$$\frac{dI}{dt} = \beta SI - \gamma I - \mu_I I$$
$$\frac{dR}{dt} = \gamma I - \mu R$$

[1] Discrete time analogues of capturing demography of births arguably date back to Leonardo Bonacci (or Leonardo of Pisa), now known as Fibonacci, in 1202, where number of births at time B_t is governed by the population size at two previous time-steps, $B_t = B_{t-1} + B_{t-2}$.

This framing indicates that how the infectious pathogen affects host demography will depend on both the pathogen transmission rate (β) and pathogen-related mortality (μ_I) because together these determine the excess mortality within the host population (Figure 22.1). However, the framing also reveals that host demography can modulate the pathogen's spread, which may complicate this relationship.

To describe pathogen spread, we can derive an expression for R_0, the expected number of new infections per infectious individual in a completely susceptible population, $R_0 = \beta / (\gamma + \mu_I)$, which reveals that pathogen population growth rate will decline as infected host mortality μ_I increases. The conditions for 'herd immunity', or the fraction of the population that must be immunised p_c such that one infected individual results in less than one new infection (i.e. the pathogen is doomed to extinction), require that $(1 - p_c) R_0$ is less than one (Cobey 2020), that is, $p_c = 1 - 1/R_0$. The underlying logic is as follows: the proportion of the population immunised by vaccination must be such that the effective reproduction number, R_E, or number of new infections per one infectious individual in one generation time of the disease in a population that is only partially susceptible, $R_E = R_0 S$, must be less than 1, where S is the proportion of the population that is susceptible (here assuming a frequency-dependent transmission formulation, so that the proportion, S, is the relevant quantity). From this definition of p_c, we can see that, all else being equal, higher pathogen-related mortality can reduce the threshold for pathogen elimination by inflating the denominator of R_0. Moving beyond pathogen-related mortality, how the threshold for elimination is approached will also depend on fertility and mortality more broadly. Assuming for simplicity that $\mu = \mu_I$ and the host population is at equilibrium, that is, $b = \mu$, the equilibrium pathogen incidence $I^* = \mu (R_0 - 1) / (R_0 (\gamma + \mu))$, so increasing the rate of birth or mortality can increase

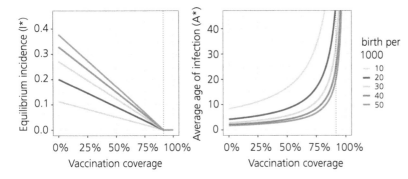

Figure 22.2 Impact of birth rate and vaccination coverage on equilibrium incidence and average age of infection. Dotted line shows the critical elimination threshold, or the proportion of the population we need to vaccinate in order to interrupt disease transmission in a population, $p_c = 1 - 1/R_0$. Parameters: $R_0 = 12$; infectious period is 14 days.

the equilibrium incidence of the pathogen (Figure 22.2, left).

Further relevant demographic information is embedded in this simple structure. For example, at equilibrium, the rate at which susceptible individuals are infected is (broadly) βI^*, and this will define the average age of infection, A^*, that is, $A^* = 1/\beta I^*$, so that infections with higher transmission rates β should have lower ages of infection, and, given the dependence of I^* on birth rates, birth rates can have a similar effect (Figure 22.2, right). This illustration is essentially a version of the Kermack–McKendrick model, with births and constant rates studied in 1932 (Kermack and McKendrick 1932), which notes the link between increasing births and increasing rate of disease and corresponding endemic level. Related expressions that explicitly encompass demography can also be obtained (McLean and Anderson 1988).

However, these basic models do not allow for different rates (infection, mortality, waning of immunity) for individuals of different demographic ages or stages. The clear next step is to further classify individuals according to their demographic stage or age, as well as their epidemiological state. In this scenario, patterns of ageing or ontogeny must be modelled alongside epidemiological processes.

One approach is to use partial differential equations (PDEs), where age is modelled as a continuous variable. This approach dates again to McKendrick (1925) (though it is commonly known as the McKendrick–von Foerster equation) and can be linked to Lotka's integral equation for the number of births in the populations by the method of characteristics to solve PDEs (Keyfitz and Keyfitz 1997). Technical challenges associated with such formulations (ranging from issues with convergence in simulations, to challenges in aligning models with data, and sheer computational intensiveness) mean that in practice PDE framings are often reduced to ordinary differential equations

(ODEs) (e.g. see Tudor (1985) for a measles example). Numerical solutions are generally obtained using finite difference methods, essentially discretising the system. Since most formulations of these models are equivalent to assuming exponential waiting times for transitions between states, it is important to verify that model outcomes do not reflect unrealistic patterns of ageing (i.e. some individuals may age very quickly into older age groups, and individuals may even be lost if model age groups are too large).

'Cohort ageing' is one possible alternative approach to modelling ageing as a continuous processes—under 'cohort ageing' all individuals are aged at a single time point in the year. This may be appropriate for hosts with very sharply seasonal life cycles (Peel et al. 2014) or human infections circulating in school settings (Schenzle 1984). Effectively, cohort ageing corresponds to using what is referred to as a 'birth pulse model', where, for example, all individuals are born on 1 January and age at the end of the year. Clear dynamical consequences (e.g. birth pulses shape pathogen persistence [Peel et al. 2014]) indicate that care must be taken in defining the form of ageing used by these models. Similar considerations around the framing of ageing apply to discrete-time matrix-based model frameworks (Klepac and Caswell 2011) which represent an alternative approach to capturing both demographic and epidemiological dynamics. The basic principles developed can be expanded to consider further structure within populations (e.g. households [Hilton and Keeling 2019]). Finally, as computational power increases, individual-based or agent-based models represent an increasingly tractable option to represent infectious disease transmission in demographically heterogeneous populations (e.g. Iozzi et al. 2010).

To conclude, many model frameworks can capture the intersection between demography and infectious disease dynamics. All have in common explicit expression of epidemiological and demographic rates (transmission, mortality, etc.) and may additionally include cross-classification of individuals by epidemiological and demographic state (age or stage); nuances in terms of how ageing or ontogeny is modelled must generally be considered. Before providing the mathematical details underpinning one particular approach (a discrete time-matrix-based framing [Klepac and Caswell 2011]), to provide further intuition on the links between demography and infectious disease dynamics, we expand further on how classic demography modelling relates to an essential parameter in infectious disease dynamics, R_0, briefly introduced above.

22.3 Demographic underpinnings of a core parameter, R_0

In infectious disease dynamics, the basic reproduction number, R_0, is defined as the expected number of secondary cases produced by a single (typical) infection in a completely susceptible population. As a result, R_0 acts as a threshold of whether or not disease will spread in a population and is a central quantity in infectious disease dynamics. If $R_0 > 1$ one person will on average infect more than one other person and the infection will spread within a population, while when $R_0 < 1$ the infection dies out. This has important implications for control, as $R_0 < 1$ indicates the conditions for successful containment and eventual elimination of an infection.

Even though the concept of a critical threshold arose from analysis of mathematical models for vector-borne diseases (Ross 1911, 1916; MacDonald 1952) and directly transmitted diseases (Kermack and McKendrick 1927, 1932) at the beginning of the last century, the potential use of R_0 in epidemiology to inform control of infectious diseases was only fully recognised in the 1970s (Dietz 1975; Hethcote 1975). Kermack and McKendrick noted in 1927 that disease won't spread if the population density is below the ratio of recovery rate and contact transmission rate (here designated γ/β or $1/R_0$). This is directly linked to the *critical elimination threshold* designating the proportion of the population we need to vaccinate in order to interrupt disease transmission in a population, $p_c = 1 - 1/R_0$ (Dietz 1975; Anderson and May 1991).

The analogous concept of 'net reproductive rate' (denoted by R_0 by Dublin and Lotka (1925)) had widespread applications in demography and ecology (Sharpe and Lotka 1911; Dublin and Lotka 1925)

long before epidemic R_0 took its prominent place in infectious disease control. In population biology, R_0 is defined as the expected number of offspring that an individual will produce during its lifetime, or the population growth rate from one generation to the next (Caswell 2001). The link with epidemic R_0 is evident even from the survival function. In demography,

$$R_0 = \int_0^\infty l(a)\, m(a)\, da$$

where $l(a)$ is survival probably from birth to age a, and $m(a)$ is reproduction rate at age a. However, if we let $l(a)$ instead be the probability that a newly infected individual remains infected at least for time a, and $m(a)$ be the rate of infectiousness of an individual infected a units of time, the same expression gives us the epidemic R_0 (Heesterbeek and Roberts 1995). Even in a model without explicit age structure, epidemic R_0 is influenced by demography and can be expressed by $R_0 \simeq L/A$ (derived in Dietz 1975), where L is the life expectancy ($1/\mu$ in our case) and A is the average age of infection. While framing R_0 is straightforward in a population structured only by epidemiological states (e.g. in the basic 'closed population' SIR defined in section 22.2 $R_0 = \beta/\gamma$), if individuals are also structured by age (or stage), the issue of defining a 'typical' infection is more complex, and infections at different ages (or stages) may have different consequences, depending on both their characteristics (e.g. how much they transmit) and the population context (e.g. concentrations of susceptibles into age or stage classes that a focal infected age or stage may have more or less contact with). The challenge of estimating R_0 in these settings was resolved via the development of next-generation-matrix approaches (Dietz 1975, 1993; Hethcote 1975; Diekmann et al. 1990; Van Den Driessche and Watmough 2002), which essentially characterises the duration of infectiousness for each age or stage class and the number of new infected individuals that will result from each.

22.4 A discrete-time multistate matrix approach

We can classify individuals according to their epidemiological state (which could include, for example, maternally immune, susceptible, infected, recovered, vaccinated); the epidemiological rates between those states determine the epidemiological dynamics. We focus here on discrete time formulations. It is straightforward to take the expressions for the SIR models and frame them in discrete time using a matrix formation (Klepac and Caswell 2011). Taking the first example illustrated (SIR with no demography) and setting the discrete time-step to the generation time of the infection, such that all individuals recover in one time-step, we discretise the probability of infection as $\phi_t = 1 - exp\,(-\beta I_t)$, additionally scaling β to the appropriate time-step. We can then express the matrix $\mathbf{T}$ that can be used to project the population vector $\mathbf{n}_t$ (of length 3, corresponding to numbers of individuals within the S, I, and R epidemiological states respectively) to $\mathbf{n}_{t+1}$ (likewise) according to:

$$\underset{\substack{n \\ \mathbf{n}_{t+1}}}{\begin{pmatrix} S \\ I \\ R \end{pmatrix}_{t+1}} = \underset{\substack{n \\ \mathbf{T}}}{\begin{pmatrix} (1-\phi_t) & 0 & 0 \\ \phi_t & 0 & 0 \\ 0 & 1 & 0 \end{pmatrix}} \underset{\substack{n \\ \mathbf{n}_t}}{\begin{pmatrix} S \\ I \\ R \end{pmatrix}_t}$$

Many operations used for matrix population models can be deployed (sensitivity analysis, etc. [Klepac et al. 2009]). Other important demographic processes could be added, for example, by multiplying each of the rates in $\mathbf{T}$ by the probability of survival over the focal time-step, and adding a birth rate b to each cell of the the first row of the matrix to capture the fact that all individuals produce offspring that are all born susceptible.

Going beyond this basic framework to add more detailed demography requires accounting for the fact that individuals also differ in age, maturity, and reproductive status, and might die at different rates as a function of age. Demographic models capture movement of individuals as they age, grow, and reproduce and the resulting population dynamics. One exemplar is the Leslie matrix approach [∧2], a discrete-time model focusing on the survival and fertility of each age class, which we take inspiration from below. The number of births here is determined by the fertility, F_s, of each age class (s) and their year-to-year survival probabilities, P_s. Then a number of individuals of each age s, $\mathbf{n}_s$, at time t (or the age distribution at time t) is easy to project to time $t + 1$ by multiplying the population vector $\mathbf{n}_s$ with the matrix of fertilities and survival probabilities known as the Leslie matrix, $\mathbf{L}$ (Leslie 1945, 1948) [∧2].

$$\underset{\substack{n \\ \mathbf{n}_{t+1}}}{\begin{pmatrix} n_0 \\ n_1 \\ n_2 \\ M \\ n_s \end{pmatrix}_{t+1}} = \underset{\substack{n \\ \mathbf{L}}}{\begin{pmatrix} F_0 & F_1 & L & F_s \\ P_0 & 0 & L & 0 \\ 0 & P_1 & L & 0 \\ M & & & \\ 0 & L & P_{s-1} & 0 \end{pmatrix}} \underset{\substack{n \\ \mathbf{n}_t}}{\begin{pmatrix} n_0 \\ n_1 \\ n_2 \\ M \\ n_s \end{pmatrix}_t}$$

$$\mathbf{n}_{t+1} = \mathbf{L}\,\mathbf{n}_t$$

Note that the Leslie matrix approach can be interpreted as an example of an approximation of the PDE approach. However, epidemiological and demographic transitions are not independent. Infection can affect survival and reproduction rates, while rates of transmission and recovery from infection can differ by age. Capturing this interaction between epidemiological and demographic processes requires a combined model structure. In a combined model, each individual is classified by both their infection status and demographic stage. If there are s demographic states and c epidemic states, the population at time t can be described by an $s \times c$ array.

$$N(t) = \begin{pmatrix} n_{11} & \cdots & n_{1c} \\ \vdots & & \vdots \\ n_{s1} & \cdots & n_{sc} \end{pmatrix}$$

The population vector can be organised either into epidemic categories by stacking columns on top of one another (grouping demographic stages within epidemic categories) or into demographic stages by stacking transposed rows of $N(t)$ using the vec operator (a function that stacks the columns of the matrix) (Hunter and Caswell 2005; Klepac and Caswell 2011). The vec-permutation matrix approach allows modular structure of the model, where demographic and epidemic processes are separately captured with block-diagonal matrices (see Klepac and Caswell 2011). In this way, additional processes could be conveniently captured with additional block-diagonal matrices. A challenge here is to ensure the right orientation of the population vector.

22.5 A specific example

To provide a concrete example, we illustrate the development of a model structure for completely immunising childhood infections like measles, mumps, and rubella. Focusing first on the epidemiological transitions, we start by setting the time-step to be equal to the average infectious period (i.e. $\gamma = 1$ discrete time-step), such that all infections recover at the end of each time-step. This duration is typically between a week and a month for many childhood infections (Metcalf et al. 2009). We also expand the set of epidemiological states considered (M), susceptible (S), infected (I), recovered (R), and vaccinated (V) state. For our focal pathogens, a key driver of dynamics is highly variable patterns of contact and thus opportunities for transmission between individuals of different ages (Mossong et al. 2008). The epidemic transitions in the model can be described by the following matrix:

$$\mathbf{A}(\mathbf{n}) = \begin{pmatrix} 1-d_m & 0 & 0 & 0 & 0 \\ d_m & (1-\Phi(\mathbf{n}))(1-v) & 0 & 0 & 0 \\ 0 & \Phi(\mathbf{n})(1-v) & 0 & 0 & 0 \\ 0 & 0 & 1 & 1 & 0 \\ 0 & v & 0 & 0 & 1 \end{pmatrix}$$

where d_m is loss of maternal immunity, v is the proportion of susceptible individuals that are vaccinated, the infection probability is given by

$$\Phi(n) = 1 - \exp(-\Lambda(\mathbf{n}(t)))$$

and the force of infection for stage i, Λ_i is given by

$$\Lambda_i(\mathbf{n}(t)) = \beta \frac{\sum_j C_{ij} I_j^\alpha(t)}{\sum^\mathbf{n}(t)}.$$

Here, β scales overall transmission which is set to be frequency dependent (the population size appears in the denominator of the expression), C_{ij} is the age-specific mixing matrix, capturing relative encounter and transmission probability between, and α captures heterogeneities in mixing not directly modelled and the effects of discretisation ($\alpha = 0.97$ for measles [Bjørnstad et al. 2002]). Since the time-step is taken to reflect the infectious period, and assuming that rate of mortality is likely to be negligible at this time-scale, the overall magnitude of transmission should approximate R_0. For this to be appropriately reflected in the expression for the force of infection, given that infection is completely immunising, the product of the contact matrix $\mathbf{C}$ and susceptible population is equivalent to the next-generation matrix because the duration of infection is set to one time-step in this scenario. We therefore scale $\mathbf{Cn}$ so that its dominant eigenvalue is equal to 1; the magnitude of β should then be close to that of R_0.

Ignoring births and epidemic transitions for the moment, the process of ageing (growth) and survival can be modelled with the following transition matrix:

$$\mathbf{D} = \begin{pmatrix} (1-u_1)s_1 & 0 & 0 & \cdots & 0 \\ u_1 s_1 & (1-u_2)s_2 & 0 & & 0 \\ & & & & \vdots \\ 0 & & & \cdots & s_s \end{pmatrix}$$

Here, s_i is the probability of surviving age i for the duration of the time-step, and u_i is the rate of ageing out of that class. If the age class i is equal to the time step, all of the individuals in that class move on to the next

age class ($i + 1$) at the end of the time-step, so $u_i = 1$ (e.g. for cohort ageing). If the age class i is longer than the duration of the time-step, u_i is set so that the average duration in that age class is equal to the duration of the age class. There is no growth out of the last age class s, so the last entry in the matrix only has mortality (or survival), $D_{ss} = s_s$. Here we use the vector that groups the infection categories within demographic stages (by stacking transposed rows of $N(t)$). The resulting population vector $\mathbf{n}(t)$ is:

$$\mathbf{n}(t) = \text{vec}(N^\top)$$
$$= (M_1, S_1, I_1, R_1, V_1, \ldots, M_s, S_s, I_s, R_s, V_s)(t)^\top.$$

The full transition matrix includes ageing, mortality, and infection dynamics:

$$A(\mathbf{n}(t))$$
$$= \begin{pmatrix} s_1(1-u_1)\mathbf{A}_1 & 0 & 0 & L & 0 \\ s_1 u_1 \mathbf{A}_1 & s_2(1-u_2)\mathbf{A}_2 & 0 & L & 0 \\ M & & & & \\ 0 & 0 & L & s_{s-1}u_{s-1}\mathbf{A}_{s-1} & s_s\mathbf{A}_s \end{pmatrix}$$

where A_i is the transmission matrix for demographic stage i.

We can define a corresponding fertility matrix, focusing first within an age group. Here we illustrate a scenario where fertility does not depend on epidemiological state but only on age. However, recovered or vaccinated mothers give birth to maternally immune offspring; all other mothers give birth to susceptible offspring:

$$\mathbf{F_i}(\mathbf{n}) = \begin{pmatrix} 0 & 0 & 0 & b_1 & b_1 \\ b_1 & b_1 & b_1 & 0 & 0 \\ 0 & 0 & 0 & 0 & 0 \\ 0 & 0 & 0 & 1 & 0 \\ 0 & 0 & 0 & 0 & 1 \end{pmatrix}$$

Here, b_i is the probability that individuals of age i produce one offspring over the duration of the time-step, and these are born susceptible (second row of matrix) unless mothers were recovered or vaccinated, in which cases they are born maternally immune (first row of matrix). The full transition matrix is:

$$F = \begin{pmatrix} \mathbf{F}_1 & \mathbf{F}_2 & \mathbf{F}_3 & L & \mathbf{F}_s \\ 0 & 0 & 0 & L & 0 \\ M & M & M & L & M \\ 0 & 0 & 0 & L & 0 \end{pmatrix}$$

The full dynamics can be captured according to:

$$\mathbf{n}_{t+1} = (A(\mathbf{n}(t)) + F)\mathbf{n}_t$$

which will project the population forwards, encompassing transitions between demographic stages (either stage or age) and epidemiological stages (maternally immune, susceptible, infected, etc.).

22.6 Case study

A key contribution that infectious disease models have brought to public policy is the concept of 'herd immunity' (also referred to as 'community immunity') and the related insight that not everyone need be immunised in order to eliminate an infection, as some individuals will be indirectly protected by the immunity of the 'herd' (Cobey 2020). In the simplest case, the critical threshold for population immunity achieved by vaccination is $p_c = 1 - 1/R_0$, but this expression ignores age (and other forms of) structure in the population. Heterogeneity in individual contacts can lead to herd immunity being achieved at a lower threshold than classical $1 - 1/R_0$ (Gomes et al. 2020), while reduced variation in contacts can lead to a herd immunity threshold higher than $1 - 1/R_0$ (Ferrari et al. 2006a).

Here we focus on the scenario where contacts and/ or magnitude of transmission between individuals vary across ages, and we work to identify how the vaccination coverage required to achieve pathogen elimination is altered. The structure of the proportion of the population that is susceptible, rather than the absolute proportion of the population that is susceptible, is what will shape the number of new infections following introduction of one (average) infected individual and, conversely, the pattern of immunity (and thus vaccination coverage) that is required for elimination.

To illustrate this, we take the example of vaccination coverage required for measles elimination using the

model structure. Measles is a directly transmitted, completely immunising infection, with a generation time of ~2 weeks (Bjørnstad et al 2002). The existence of a safe, inexpensive, and completely immunising vaccine, in conjunction with the high case fatality rate of this infection, means that every World Health Organization region currently has a measles elimination goal. Estimates of $R_0 \sim 15$ for measles indicate that $p_c = 0.93$ is the simplest (unstructured) analysis. How does this change if we add detail to the demographic context?

For directly transmitted infections like measles, patterns of contact over age have been suggested to play an important role in shaping spread (Edmunds et al. 2000). Patterns of contact over age have been inferred from diary studies, where individuals are asked to record the age and number of people they interact with over the course of some time period, for example a week (Mossong et al. 2008). We start by assuming that such measurements provide a reasonable reflection of relative patterns of contact over age of relevance for disease transmission (while noting that this does not necessarily follow) and use results from an existing study of age contact patterns (Mossong et al. 2008) as a starting point (Figure 22.3, left). We focus our investigation on a country whose demography is assumed to still be experiencing population growth, with a population pyramid that narrows as age increases, and one with a much flatter population profile (determined arbitrarily, for illustration; Figure 22.3, right).

Taking 2 weeks as our discrete time-step, as this reflects the life history of measles, we scale the pattern of contacts reported from the diary studies, C_{ij}, and

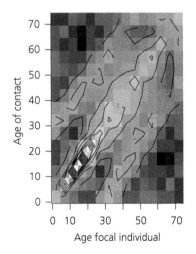
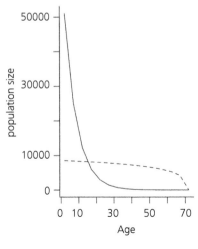

Figure 22.3 (Left) Age contact data, where colours indicate intensity of recorded contacts between different age classes (x and y axes), ranging from dark (low) to white (high) contacts, and showing that individuals tend to mostly have contacts with individuals of their own age (diagonal white coloured) or their own parents/offspring (off-diagonals light-coloured white). (Right) Population structures explored to evaluate the required vaccination profile: solid line—exponential population structure, typical for developing world; dashed line—rectangular population structure, commonly observed in developed countries.

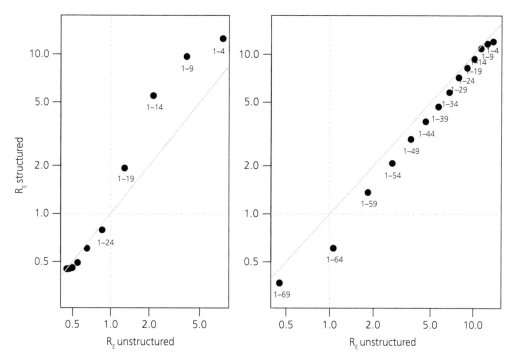

Figure 22.4 Comparison of estimated R_E under different age windows of vaccination (text on the figure indicates the age range of vaccination corresponding to each point) for unstructured (x axis) and structured (y axis) population models in either a growing (left, corresponding to the solid line in Figure 22.3) or static population (right, corresponding to the dashed line in Figure 22.3)

the transmission term β to obtain $R_0 = 15$, using the first population (Figure 22.3, right, solid line) as our reference. To illustrate the importance of age profiles of vaccination, we assume for simplicity that measles does not circulate in this population (i.e. that there is no natural immunity) and that there has been no vaccination to date (there is also no vaccine-derived immunity; both are unrealistic). We then define **v** as a vector reflecting individuals immunised by vaccination across age. We assume that every individual within a targeted age group is reached by vaccination, but, in line with estimates for the measles vaccine, we assume that vaccination efficacy is 97%, and we consider delivery of only a single dose. The public policy decision might then be what the upper age of the vaccination campaign should be to achieve measles elimination—are we targeting children up to age 5 years? Up to 10 years? Even higher?

In the unstructured model, where we include no differences in patterns of contact over age, the effective reproduction number R_E is defined by $R_E = R_0 s = R_0 \mathbf{vn}/\sum \mathbf{n}(t)$, where n is the number of individuals in each age class. For the structured population model, R_E is the dominant eigenvalue of the product of $R_0(1 - \mathbf{v})\,\mathbf{nC}$. We can plot the impact on R_E

across a range of upper ages targeted in the vaccination campaign, comparing results for structured and unstructured populations across the two population structures (Figure 22.4).

In the first example, the upper age class that must be targeted to achieve herd immunity is the same (up to age 24) for both the structured and the unstructured case; but one can also see very different patterns in each. In particular, the structured model shows much larger expected numbers of new cases for corresponding scales of vaccination of younger individuals (right-hand size of x = 1) and smaller expected numbers of new cases if older individuals are targeted (lefthand size of x = 1). In the second example, very wide range age campaigns are required, and the threshold for immunity differs between unstructured and structured models.

22.7 Conclusions

Developing models that tractably engage with the varying time-scales and dynamics of infectious diseases and host biology is a challenge. However, the intricate interdependence of demography and epidemiology makes engaging this challenge essential to appropriately characterising a range of biological scenarios, from comparing different management stragies of infected ungulates (Jennelle et al. 2014), to quantifying how changing demographic patterns such as declining birth rates might alter the burden of a human infection (Metcalf et al. 2012a), to identifying life stages critical to pathogen transmission in partially observed outbreaks in wildlife (Klepac et al. 2009), to anticipating perennial grassland community invasion by annual plants with and without a pathogenic virus (Borer et al. 2007). In this chapter, we detail one potential framework rooted in discrete time approaches (noting that there are many other possibilities, as briefly described). The balance of availability of analytic methods, such as sensitivity (see Klepac and Caswell 2011; Metcalf et al. 2012a) and tractable inference from data makes such discrete-time a powerful tool in addressing applied questions from a range of fields. Exciting directions for future research include extension of such tools and relevant statistical approaches for a range of data-streams.

References

Anderson, R. M. and R. M. May. 1991. Infectious *Diseases of Humans: Dynamics* and *Control*. Oxford: Oxford University Press.

Antonovics, J. 2005. Plant Venereal Diseases: Insights from a Messy Metaphor. *New Phytologist* **165** (1): 71–80.

Bernoulli, D. and S. Blower. 2004. An Attempt at a New Analysis of the Mortality Caused by Smallpox and of the Advantages of Inoculation to Prevent It. *Reviews in Medical Virology* **14** (5): 275–288.

Bjørnstad, O. N., B. F. Finkenstädt, and B. T. Grenfell. 2002. Dynamics of Measles Epidemics: Estimating Scaling of Transmission Rates Using a Time Series SIR Model. *Ecological Monographs* **72** (2): 169–184.

Borer, E. T., P. R Hosseini, E. W. Seabloom, and A. P. Dobson. 2007. Pathogen-Induced Reversal of Native Dominance in a Grassland Community. *Proceedings of the National Academy of Sciences* **104** (13): 5473–5478.

Bruns, E. L., J. Antonovics, V. Carasso, and M. Hood. 2017. Transmission and Temporal Dynamics of Anther-Smut Disease (Microbotryum) on Alpine Carnation (*Dianthus pavonius*). *Journal of Ecology* **105** (5): 1413–1424.

Caswell, H. 2001. Matrix *Population Models: Construction, Analysis*, and *Interpretation*. Sunderland, MA: Sinauer Associates.

Cobey, S. 2020. Modeling Infectious Disease Dynamics. *Science* **368** (6492): 713–714.

Dawson, W. O., S. M. Garnsey, S. Tatineni, S. Y. Folimonova, S. J. Harper, and S. Gowda. 2013. Citrus Tristeza Virus–Host Interactions. *Frontiers in Microbiology* **4**: 88.

Diekmann, O., J. A. P. Heesterbeek, and J. A. J. Metz. 1990. On the Definition and the Computation of the Basic Reproduction Ratio R0 in Models for Infectious Diseases in Heterogeneous Populations. *Journal of Mathematical Biology* **28** (4): 365–382.

Dietz, K. 1975. Transmission and Control of Arbovirus Diseases. SIMS 1974 Utah Conference Proceedings, January 1974: 104–121.

Dietz, K. 1993. The Estimation of the Basic Reproduction Number for Infectious Diseases. *Statistical Methods in Medical Research* **2** (1): 23–41.

Dietz, K. and J. A. P. Heesterbeek. 2002. Daniel Bernoulli's Epidemiological Model Revisited. *Mathematical Biosciences* **180** (1–2): 1–21.

Dublin, L. I. and A. J. Lotka. 1925. On the True Rate of Natural Increase: As Exemplified by the Population of the United States, 1920. *Journal of the American Statistical Association* **20** (151): 305–339.

Earn, D. J. D., P. Rohani, B. M. Bolker, and B. T. Grenfell. 2000. A Simple Model for Complex Dynamical Transitions in Epidemics. *Nature* **287**: 667–670.

Edmunds, W. J., N. J. Gay, M. Kretzschmar, R. G. Pebody, and H. Wachmann. 2000. The Pre-Vaccination Epidemiology of Measles, Mumps and Rubella in Europe: Implications for Modelling Studies. *Epidemiology & Infection* **125** (3): 635–650.

Ferrari, M. J., S. Bansal, L. A. Meyers, and O. N. Bjørnstad. 2006a. Network Frailty and the Geometry of Herd Immunity. *Proceedings of the Royal Society B: Biological Sciences* **273** (1602): 2743–2748.

Ferrari, M. J., O. N. Bjørnstad, J. L. Partain, and J. Antonovics. 2006b. A Gravity Model for the Spread of a Pollinator-Borne Plant Pathogen. *The American Naturalist* **168** (3): 294–303.

Gomes, M. G. M., R. Aguas, R. M. Corder, et al. 2020. Individual Variation in Susceptibility or Exposure to Sars-Cov-2 Lowers the Herd Immunity Threshold. *medRxiv*. https://www.medrxiv.org/content/10.1101/2020.04.27.20081893v3.

Gulland, F. M. D. 1992. The Role of Nematode Parasites in Soay Sheep (*Ovis aries* L.) Mortality During a Population Crash. *Parasitology* **105** (3): 493–503.

Heesterbeek, J. A. P. and M. G. Roberts. 1995. Mathematical Models for Microparasites of Wildlife. In *Ecology of Infectious Diseases in Natural Populations*, edited by B. T. Grenfell and A. P. Dobson, 90–122. Cambridge: Cambridge University Press.

Hethcote, H. W. 1975. Mathematical models for the spread of infectious diseases. In *Epidemiology*, edited by D. Ludwig & K. L. Cooke, 122–131. Philadelphia: Society for Industrial and Applied Mathematics.

Hilton, J. and M. J. Keeling. 2019. Incorporating Household Structure and Demography into Models of Endemic Disease. *Journal of the Royal Society Interface* **16** (157): 20190317.

Hunter, C. M. and H. Caswell. 2005. The Use of the Vec-Permutation Matrix in Spatial Matrix Population Models. *Ecological Modelling* **188** (1): 15–21.

Iozzi, F., F. Trusiano, M. Chinazzi, et al. 2010. Little Italy: An Agent-Based Approach to the Estimation of Contact

Patterns—Fitting Predicted Matrices to Serological Data. *PLOS Computational Biology* **6** (12): e1001021.

Jennelle, C. S., C. Henaux, G. Wasserberg, B. Thiagarajan, R. E. Rolley, and M. D. Samuel. 2014. Transmission of Chronic Wasting Disease in Wisconsin White-Tailed Deer: Implications for Disease Spread and Management. *PLOS One* **9** (3): e91043.

Kermack, W. O. and A. G. McKendrick. 1927. A contribution to the mathematical theory of epidemics. *Proceedings of the Royal Society of London. Series A* **115** (772): 700–721.

Kermack, W. O. and A. G. McKendrick. 1932. Mathematical Theory of Epidemics. II. The Problem of Endemicity. *Proceedings of the Royal Society of London. Series A* **138** (834): 55–73.

Keyfitz, B. L. and N. Keyfitz. 1997. The McKendrick Partial Differential Equation and its Uses in Epidemiology and Population Study. *Mathematical and Computer Modelling* **26** (6): 1–9.

Klepac, P. and H. Caswell. 2011. The Stage-Structured Epidemic: Linking Disease and Demography with a Multi-State Matrix Approach. *Theoretical Ecology* **4**:301–319.

Klepac, P., L. W. Pomeroy, O. N. Bjørnstad, T. Kuiken, A. Osterhaus, and J. M. Rijks. 2009. Stage Structured Transmission of Phocine Distemper Virus in the Dutch 2002 Outbreak. Proceedings of the Royal Society *of London*. Series B **276**: 2469–2476.

Kotar, S. L. and J. E. Gessler. 2013. *Smallpox: A History*. Jefferson, NC: McFarland.

Leslie, P. H. 1945. On the Use of Matrices in Certain Population Mathematics. *Biometrika* **33** (3): 183–212.

Leslie, P. H.. 1948. Some Further Notes on the Use of Matrices in Population Mathematics. *Biometrika* **35** (3/4): 213–245.

MacDonald, G. 1952. The Analysis of Equilibrium in Malaria. *Tropical Diseases Bulletin* **49** (9): 813–829.

McCallum, H., N. Barlow, and J. Hone. 2001. How Should Pathogen Transmission Be Modelled? *Trends in Ecology & Evolution* **16** (6): 295–300.

McKendrick, A. G. 1925. Applications of Mathematics to Medical Problems. Proceedings of the Edinburgh Mathematical Society **44**: 98–130.

McLean, A. R. and R. M. Anderson. 1988. Measles in Developing Countries. Part II. The Predicted Impact of Mass Vaccination. *Epidemiology & Infection* **100** (3): 419–442.

Metcalf, C. J. E., O. N. Bjørnstad, B. T. Grenfell, and V. Andreasen. 2009. Seasonality and Comparative Dynamics of Six Childhood Infections in Pre-Vaccination Copenhagen. *Proceedings of the Royal Society B: Biological Sciences* **276** (1676): 4111–4118.

Metcalf, C. J. E., J. Lessler, P. Klepac, F. T. Cutts, and B. T. Grenfell. 2012a. Impact of Birth Rate, Seasonality and Transmission Rate on Minimum Levels of Coverage Needed for Rubella Vaccination. *Epidemiology and Infection* **16**: 1–12.

Metcalf, C. J. E., J. Lessler, P. Klepac, A. Morice, B. T. Grenfell, and O. N. Bjornstad. 2012b. Structured Models of

Infectious Disease: Inference with Discrete Data. *Theoretical Population Biology* **82** (4): 275–282.

Moreno, E. 2014. Retrospective and Prospective Perspectives on Zoonotic Brucellosis. *Frontiers in Microbiology* **5**: 213.

Morris, S. E., V. E. Pitzer, C. Viboud, C. J. E. Metcalf, O. N. Bjørnstad, and B. T. Grenfell. 2015. Demographic Buffering: Titrating the Effects of Birth Rate and Imperfect Immunity on Epidemic Dynamics. *Journal of the Royal Society Interface* **12** (104): 20141245.

Mossong, J., N. Hens, M. Jit, et al. 2008. Social Contacts and Mixing Patterns Relevant to the Spread of Infectious Diseases. *PLOS Medicine* **5** (3): e74.

Munywoki, P. K., D. C. Koech, C. N. Agoti, et al. 2015. Influence of Age, Severity of Infection, and Co-Infection on the Duration of Respiratory Syncytial Virus (RSV) Shedding. *Epidemiology & Infection* **143** (4): 804–812.

Peel, A. J., J. R. C. Pulliam, A. D. Luis, et al. 2014. The Effect of Seasonal Birth Pulses on Pathogen Persistence in Wild Mammal Populations. *Proceedings of the Royal Society B: Biological Sciences* **281** (1786): 20132962.

Ramsey, D., N. Spencer, P. Caley, et al. 2002. The Effects of Reducing Population Density on Contact Rates Between Brushtail Possums: Implications for Transmission of Bovine Tuberculosis. *Journal of Applied Ecology* **39** (5): 806–818.

Ross, R. 1911. Some Quantitative Studies in Epidemiology. *Nature* **87** (2188): 466–467.

Ross, R. 1916. An Application of the Theory of Probabilities to the Study of a priori Pathometry. Part I. *Proceedings of the Royal Society of London: Series A* **92** (638): 204–230.

Schenzle, D. 1984. An Age-Structured Model of Pre- and Post-Vaccination Measles Transmission. *Mathematical Medicine and Biology: A Journal of the IMA* **1** (2): 169–191.

Sharpe, F. R. and A. J. Lotka. 1911. A Problem in Age-distribution. The London, Edinburgh, and Dublin Philosophical Magazine and Journal of Science 21: 435–438.

Smith, M J., S. Telfer, E. R. Kallio, et al. 2009. Host–Pathogen Time Series Data in Wildlife Support a Transmission Function Between Density and Frequency Dependence. *Proceedings of the National Academy of Sciences* **106** (19): 7905–7909.

Tudor, D. W. 1985. An Age-Dependent Epidemic Model with Application to Measles. *Mathematical Biosciences* **73** (1): 131–147.

Van Den Driessche, P. and J. Watmough. 2002. Reproduction Numbers and Sub-Threshold Endemic Equilibria for Compartmental Models of Disease Transmission. *Mathematical Biosciences* 180 (1–2): 29–48.

Wonham, M. J., M. A. Lewis, J. Rencławowicz, and P. Van Den Driessche. 2006. Transmission Assumptions Generate Conflicting Predictions in Host–Vector Disease Models: A Case Study in West Nile Virus. *Ecology Letters* **9** (6): 706–725.

Subject Index

Tables, figures, and boxes are indicated by an italic *t*, *f*, and *b* following the page number.